Lecture Notes in Computer

T0238136

Commenced Publication in 1973
Founding and Former Series Editors:
Gerhard Goos, Juris Hartmanis, and Jan van Leeuwen

Josef Pieprzyk (Ed.)

Topics in Cryptology – CT-RSA 2010

The Cryptographers' Track at the RSA Conference 2010
San Francisco, CA, USA, March 1-5, 2010
Proceedings

 Springer

Volume Editor

Josef Pieprzyk
Macquarie University
Department of Computing
Sydney, NSW 2109, Australia
E-mail: josef@science.mq.edu.au

Library of Congress Control Number: Applied for

CR Subject Classification (1998): E.3, D.4.6, K.6.5, C.2, K.4.4

LNCS Sublibrary: SL 4 – Security and Cryptology

ISSN	0302-9743
ISBN-10	3-642-11924-7 Springer Berlin Heidelberg New York
ISBN-13	978-3-642-11924-8 Springer Berlin Heidelberg New York

springer.com

© Springer-Verlag Berlin Heidelberg 2010
Printed in Germany

Typesetting: Camera-ready by author, data conversion by Scientific Publishing Services, Chennai, India
Printed on acid-free paper   06/3180   5 4 3 2 1 0

Preface

The RSA Conference is an annual event that attracts hundreds of vendors and thousands of participants from industry and academia. Since 2001, the conference has included an academic Cryptographers' Track (CT-RSA). This year was the 10th anniversary of CT-RSA. Since its conception, the CT-RSA conference has become a major avenue for publishing high-quality research papers. The RSA conference was held in San Francisco, California, during March 1–5, 2010.

This year we received 94 submissions. Each paper got assigned to three referees. Papers submitted by the members of the Program Committee got assigned to five referees. In the first stage of the review process, the submitted papers were read and evaluated by the Program Committee members and then in the second stage, the papers were scrutinized during an extensive discussion. Finally, the Program Committee chose 25 papers to be included in the conference program. The authors of the accepted papers had two weeks for revision and preparation of final versions. The revised papers were not subject to editorial review and the authors bear full responsibility for their contents. The submission and review process was supported by the iChair conference submission server. We thank Matthiew Finiasz and Thomas Baignères for letting us use iChair. The conference proceedings were published by Springer in this volume of *Lecture Notes in Computer Science*.

The Program Committee invited two distinguished researchers to deliver their keynote talks. The first speaker was Bart Preneel from Katholieke Universiteit Leuven, Belgium. His talk was entitled "The First 30 Years of Cryptographic Hash Functions and the NIST SHA-3 Competition." The second speaker was Craig Gentry from IBM Research, USA who gave a talk on "Computing on Encrypted Data."

There are many people who contributed to the success of the 10th edition of CT-RSA. First we would like to thank the authors of all papers (both accepted and rejected) for submitting their papers to the conference. A special thanks to the members of the Program Committee and the external referees who gave their time, expertise and enthusiasm in order to ensure that each paper received a thorough and fair review. We are thankful to Vijayakrishnan Pasupathinathan for taking care of the iChair server. I thank the CT-RSA Steering Committee for giving me an opportunity to serve as the Program Chair. Last but not least, I would like to thank the RSA conference team, especially Bree LaBollita, for their help.

March 2010   Josef Pieprzyk

CT-RSA 2010

The 10th Cryptographers' Track at the RSA Conference

The Mascone Center, San Francisco, California, USA
March 1–5, 2010

Program Chair

Josef Pieprzyk   Macquarie University, Australia

Program Committee

Joonsang Baek   Institute for Infocomm Research, Singapore
Josh Benaloh   Microsoft Research, USA
Alex Biryukov   University of Luxembourg, Luxembourg
Colin Boyd   QUT, Australia
Xavier Boyen   Stanford University, USA
Alex Dent   RHUL, UK
Christophe Doche   Macquarie University, Australia
Orr Dunkelman   Weizmann Institute, Israel, ENS, France
Serge Fehr   CWI, The Netherlands
Marc Fischlin   Darmstadt University of Technology,
                         Germany
Goichiro Hanaoka   AIST, Japan
Stanisław Jarecki   UC, Irvine, USA
Jonathan Katz   University of Maryland, USA
Aggelos Kiayias   University of Connecticut, USA
Kwangjo Kim   KAIST, Korea
Mirosław Kutyłowski   Wroclaw University of Technology, Poland
Helger Lipmaa   Cybernetica AS, Estonia
Stefan Lucks   University of Weimar, Germany
Tal Malkin   Columbia University, USA
Ilya Mironov   Microsoft Research, USA
David Naccache   ENS, France
Giuseppe Persiano   University of Salerno, Italy
Vincent Rijmen   K.U. Leuven, Belgium, TU Graz, Austria
Matt Robshaw   France Telecom, France
Kazue Sako   NEC, Japan
Berry Schoenmakers   Eindhoven University, The Netherlands
Ron Steinfeld   Macquarie University, Australia
Huaxiong Wang   NTU, Singapore

Steering Committee

Masayuki Abe	NTT, Japan
Marc Fischlin	Darmstadt University of Technology, Germany
Tal Malkin	Columbia University, USA
Ron Rivest	MIT, USA
Moti Yung	Google Inc. and Columbia University, USA

External Reviewers

Abdalla, Michel
Alford, Amy
Avanzi, Roberto
Blömer, Johannes
Brzuska, Christina
Choi, Seung Geol
Chow, Sherman
Chu, Cheng-Kang
Crampton, Jason
Dachman-Soled, Dana
Dagdelen, Özgür
D'Arco, Paolo
Etrog, Jonathan
Farashahi,
   Reza Rezaeian
Farshim, Pooya
Fleischmann, Ewan
Forler, Christian
Furukawa, Jun
Galdi, Clemente
Gallais, Jean-Francois
Gierlichs, Benedikt
Gogolewski, Marcin
Golicz, Mateusz
Gonzalez, Juan
Gorantla, Choudary
Gordon, Dov
Gorski, Michael
Groschädl, Johann

Guo, Jian
Hinek, Jason
Isshiki, Toshiyuki
Izu, Tetsuya
Jiang, Shaoquan
Kiltz, Eike
Kizhvatov, Ilya
Klonowski, Marek
Konidala, Divyan M.
Koshiba, Takeshi
Krzywiecki, Łukasz
Kubiak, Przemysław
Lehmann, Anja
Liu, Joseph K.
Liu, Xiaomin
Marcello, Sandra
Marchwicki, Karol
Matsuda, Takahiro
Masucci, Barbara
Müller, Volker
Nikolić, Ivica
Nogami, Yasuyuki
Olsen, Josh
Onete, Maria Cristina
Pehlivanoglu, Serdar
Poschmann, Axel
Qiu, Ying
Raykova, Mariana
Sakai, Yasuyuki

Schmitd, Jörn-Marc
Schröder, Dominique
Seurin, Yannick
Shao, Jun
Shin, Sungmok
Standaert,
   Francois-Xavier
Stehlé, Damien
Stevens, Marc
Struminski, Tomasz
Sugita, Makoto
Szymanski, Piotr
Tadaki, Kohtaro
Takagi, Tsuyoshi
Tartary, Christophe
Teranishi, Isamu
Tillich, Stefan
Vercauteren, Frederik
Wan, Andrew
Wee, Hoeteck
Wu, Hongjun
Wu, Mu-En
Yang, Guomin
Yen, Sung-Ming
Yoneyama, Kazuki
Yoo, Myunghan
Zagórski, Filip
Zhou, Hong-Sheng

Table of Contents

Invited Talk

Public-Key Cryptography

Side-Channel Attacks

Cryptographic Protocols

Cryptanalysis

Symmetric Cryptography

The First 30 Years of Cryptographic Hash Functions and the NIST SHA-3 Competition

Bart Preneel

Katholieke Universiteit Leuven and IBBT
Dept. Electrical Engineering-ESAT/COSIC,
Kasteelpark Arenberg 10 Bus 2446, B-3001 Leuven, Belgium
bart.preneel@esat.kuleuven.be

Abstract. The first designs of cryptographic hash functions date back to the late 1970s; more proposals emerged in the 1980s. During the 1990s, the number of hash function designs grew very quickly, but for many of these proposals security flaws were identified. MD5 and SHA-1 were deployed in an ever increasing number of applications, resulting in the name "Swiss army knifes" of cryptography. In spite of the importance of hash functions, only limited effort was spent on studying their formal definitions and foundations. In 2004 Wang et al. perfected differential cryptanalysis to a point that finding collisions for MD5 became very easy; for SHA-1 a substantial reduction of the security margin was obtained. This breakthrough has resulted in a flurry of research, resulting in new constructions and a growing body of foundational research. NIST announced in November 2007 that it would organize the SHA-3 competition, with as goal to select a new hash function family by 2012. From the 64 candidates submitted by October 2008, 14 have made it to the second round. This paper presents a brief overview of the state of hash functions 30 years after their introduction; it also discusses the progress of the SHA-3 competition.

1 Early History and Definitions

Cryptographic hash functions map input strings of arbitrary (or very large) length to short fixed length output strings. In their 1976 seminal paper on public-key cryptography [31], Diffie and Hellman identified the need for a one-way hash function as a building block of a digital signature scheme. The first definitions, analysis and constructions for cryptographic hash functions can be found in the work of Rabin [74], Yuval [99], and Merkle [60] of the late 1970s. Rabin proposed a design with a 64-bit result based on the block cipher DES [37], Yuval showed how to find collisions for an n-bit hash function in time $2^{n/2}$ with the birthday paradox, and Merkle's work introduced the requirements of collision resistance, second preimage resistance, and preimage resistance. In 1987, Damgård [26] formalized the definition of collision resistance, and two years later Naor and Yung defined a variant of seoncd preimage resistant functions called Universal One Way Hash Functions (UOWHFs) [66] (also known as functions

J. Pieprzyk (Ed.): CT-RSA 2010, LNCS 5985, pp. 1–14, 2010.

offering eSEC [79]). In 2004 Rogaway and Shrimpton [79] formally studied the relations between collision resistance and several flavors of preimage resistance and second preimage resistance. Hash functions should also destroy the algebraic structure of the signature scheme; typical examples are the Fiat-Shamir heuristic [36] and Coppersmith's attack on the hash function in X.509 Annex D [24] that was intended for use with RSA [77] (this attack breaks the signature scheme by constructing message pairs (x, x') for which $h(x) = 256 \cdot h(x')$). This development resulted in the requirement that hash functions need an 'ideal' behavior which would allow them to instantiate the theoretical concept of random oracles (see e.g. Bellare and Rogaway [10]). Constructions of MAC algorithms based on hash functions (such as HMAC) have resulted in the requirement that the hash function can be used to construct pseudo-random functions, which has a.o. been studied by Bellare et al. [8,6].

This paper is organized as follows. Section 2 describes brute force attacks and generic constructions for iterated hash functions, while Sect. 3 gives an overview of three types of hash function constructions. Section 4 presents the status of NIST's SHA-3 competition 1 year after the submission deadline and presents the planning for the future. Our concluding remarks are presented in Sect. 5. As cryptographic hash functions have become a rich subject, we don't attempt to be complete in this short contribution. We mostly provide some pointers to the literature, with an emphasis on very early work and on the most recent results.

2   Generic Analysis and Design

2.1   Brute Force Attacks

For an ideal hash function with a hash result of bitlength n, finding a (second) preimage takes $\Theta(2^n)$ evaluations of the hash function. However, if one considers multiple targets, then the expected cost to find a (second) preimage for one of these 2^t targets is reduced to $\Theta(2^{n-t})$ (note that for $t = n/2$ this corresponds to $\Theta(2^{n/2})$). If one intends to find a (second) preimage for all 2^t targets, one can apply Hellman's time-memory tradeoff [42]: after a precomputation of $\Theta(2^n)$, additional (second) preimages can be found at a cost of $\Theta(2^{2n/3})$; this method requires a storage of $\Theta(2^{2n/3})$. Wiener provides a detailed analysis in the full cost model [96]. The answer to this degradation in security is to parameterize the hash function with a salt (also known as spice, tweak or key) [60], so that each application can get a different function.

For an n-bit hash function, collisions can be found in time $\Theta(2^{n/2})$; there exist algorithms with low memory that are highly parallellizable [92]. This shows that for long term collision resistance (10 years or more), a hash result of 192 or 256 bits is required.

In practice, collision resistance is much harder to achieve than (second) preimage resistance. Simon [84] also proved that there is no black box reduction from preimage resistance to collision resistance. Fortunately, only few applications need collision resistance: the most notable ones are digital signatures (where either the signer or the verifier can cheat) and binding commitments. It is important however to understand that circumventing the requirement of collision resistance is harder than expected (see for example the attack on the RMX mode in [40]).

2.2   Iterated Hash Functions

From the first designs (including the Rabin function [74]), it was understood that a hash function h should be constructed by iterating a compression function f with fixed size inputs. The input is first padded such that the length of the input is multiple of the block length. Next it is divided into t blocks x_1 through x_t. The hash result is then computed as follows:

$$H_0 = IV \tag{1}$$
$$H_i = f(x_i, H_{i-1}) \qquad i = 1, 2, \ldots t \tag{2}$$
$$h(x) = g(H_t) \,. \tag{3}$$

Here IV is the abbreviation of *Initial Value*, H_i is called the chaining variable, and the function g is called the output transformation. While many MAC algorithms have an output transformation, this is a relatively new feature for hash functions. However, it is easy to see that the absence of an output transformation leads to an extension attack, that is, one can compute $h(x\|y)$ from $h(x)$ and y (without knowing x), which is undesirable for some applications.

In two articles presented at Crypto'89, Damgård [27] and Merkle [61] show under which conditions collision resistance of the compression function f is sufficient to obtain collision resistance of the function h. The standard way to satisfy these conditions is to fix the IV and to append the message length at the end; Lai and Massey [54] coined the name Merkle-Damgård strengthening for this construction.[1] Naor and Yung [66] obtained similar results for Universal One-Way Hash Functions, which is the eSEC variant of a second preimage resistant hash function. Lai and Massey [54] present a necessary and sufficient condition for ideal second preimage resistance of an iterated hash function (that is, finding a second preimage takes about 2^n evaluations of the compression function f); unfortunately later on their result turned out to be incorrect.

During the last five years, a number of limitations have been identified for these iterated constructions, for example the work on long-message second preimages by Dean [28] and Kelsey and Schneier [51], the multicollisions by Joux [47] and the herding attack by Kelsey and Kohno [50]. The (surprising) implication of the multicollision attack is that the concatenation of two iterated hash functions $(h(x) = h_1(x)\|h_2(x))$ is as most as strong as the strongest of the two; more precisely, if the result of h_i has bitlength n_i, the cost of a collision attack on h is at most $n_1 \cdot 2^{n_2/2} + 2^{n_1/2}$ (here we assume w.l.o.g. that $n_1 \leq n_2$). This

---

[1] Some authors refer to any linear iterated hash function as described above as "the Merkle-Damgård construction," which is clearly not appropriate since this approach dates back to the earlier work by Rabin in 1978 [74].

complexity is much lower than one would expect intuitively, that is $2^{(n_1+n_2)/2}$. On the other hand, a large number of improvements have been proposed to these constructions including work by Andreeva et al. (ROX [2]), Bellare and Ristenpart (EMD [9]), Biham and Dunkelman (HAIFA [15], see also [19]), and Yasuda [98]. Maurer et al. [59] generalize the concept of indistinguishability to indifferentiability from random oracles. Coron et al. [25] have studied how the Merkle-Damgård construction can be modified to satisfy indifferentiability from random oracles. Other work in this direction can be found in [13,65].

Merkle has introduced the so-called Merkle trees [60] for constructing digital signature schemes. Damgård has shown that the domain of a collision resistant compression function can also be extended by a tree construction [27]; an optimized version was proposed by Pal and Sarkar [68]. While the tree construction offers increased parallelism, it has the unfortunate property that for every size of the tree one obtains a different hash function, which is undesirable from an interoperability point of view.

3 Hash Function Constructions

During the 1980s, the need for an efficient and secure hash function was well understood (see for example the note presented at Eurocrypt'86 [70]). In the late 1980s and early 1990s a large number of designs was created; about 50 proposals were known in 1993, but and at least two thirds of them were broken (see the PhD thesis of the author for the status at that time [71]). After fifteen years of cryptanalysis, very few of those early schemes remain secure. Since then, about hundred new hash function designs have been proposed; 64 of these have been submitted to the SHA-3 competition (cf. Sect. 4). Many of them have not survived for long either.

Next we describe the status of the three main classes of hash functions: hash functions based on block ciphers, hash functions based on modular arithmetic and dedicated hash functions.

3.1 Hash Functions Based on Block Ciphers

The first constructions for hash functions were all based on block ciphers, more in particular based on DES [37]. This approach has several advantages: the design and evaluation effort of a block cipher can be reused, and one may obtain very compact implementations. However, it may well be that a block cipher has weaknesses in the key schedule which have only very limited impact on its use for encryption, but which may be undesirable when it is used in a hash function construction. Examples are the weak keys of DES [64] and the key schedule weaknesses of AES-192 and AES-256 [16,17].

After cryptanalysis of several proposals, a more systematic approach for cryptanalysis has been used by Preneel et al. [72] and for security proofs in the ideal cipher model by Winternitz [97], Black et al. [18] and Stam [87]. The more difficult problem is how to construct hash functions with a result that is larger than the block length, since most block ciphers have a block length of 64 or 128 bits, which is clearly not sufficient to obtain collision resistance. This area turned out to be very difficult; substantial progress has been made from the point of view of cryptanalysis (e.g. Knudsen et al. [52]) and design (e.g. MDC-2 [20,88], Merkle [61] and Hirose [43]). Recent work by Rogaway and Steinberger [80] and Stam [86] has studied constructions based on permutations. It is fair to state that we are improving our understanding of the problem on how to construct hash functions from small building blocks; on the other hand, it is not clear that the most efficient hash functions can be designed by starting from a block cipher.

3.2   Hash Functions Based on Arithmetic Primitives

Public key cryptology, and in particular modular arithmetic, has also been a source of inspiration for hash function constructions. This has resulted in hash functions with a security proof based on number theoretic assumptions such as factoring and discrete logarithm. One example is the construction by Bellare et al. [8] based on the discrete logarithm problem in a group of large prime order. An interesting construction is VSH [23], for which finding collisions is provably related to factoring; however, due to structural properties identified a.o. by Saarinen [81], VSH does not have the properties expected from a general purpose hash function. In the area of 'ad hoc' constructions, a large number of proposals was broken; eventually MASH-1 and MASH-2 were standardized in ISO/IEC 10118-4 [46]; they use squaring and raising to the power $2^8 + 1$ respectively. Schemes based on additive or multiplicative knapsacks offer attractive performance results. However, in spite of theoretical support (e.g. Ajtai's work [1]), practical constructions have not fared well until now: see for example the attack by Patarin [69] on an additive knapsack scheme, the attack by Tillich and Zémor [90] on the LPS hash function [22] and the cryptanalysis by Grass et al. [41] of the 1994 scheme of Tillich and Zémor [89].

3.3   Dedicated Hash Functions

The limitations of block cipher based hash functions resulted in a series of designs from scratch. These hash functions were among the first algorithms to be designed to be efficient in software on microprocessors rather than in hardware implementations. The Binary Condensing Algorithm [91] and MD2 of Rivest [49] use 8-bit to 8-bit S-boxes, while N-Hash [63] is based on 8-bit additions. The first 32-bit proposals date back to the beginning of the 1990s and include MD4 [75], MD5 [76] and Snefru [62]. Around the same time, differential cryptanalysis of block ciphers was developed by Biham and Shamir [14]; they applied these techniques to cryptanalyze N-hash and Snefru.

MD5 was proposed by Rivest in 1991 as a strengthened version of MD4. As it was optimized for software implementations, MD5 was about 10 times faster than DES in software. Moreover, MD5 was available without any licenses and it was easier to export than an encryption algorithm. As a consequence, MD5 was adopted very quickly in many applications.[2] Unfortunately, weaknesses were identified early on: in 1992, den Boer and Bosselaers [30] found collisions for the compression function and in 1996, Dobbertin found collisions for MD5 but with a random *IV* rather than the fixed *IV* from the specifications [32]; his attack combined differential attacks with techniques such as continuous approximations and genetic programming. In 2004, Wang et al. [93,94,95] made a breakthrough with enhanced differential attacks that combine improved differential paths with clever message modification techniques. Optimized versions of their attacks can find collisions for MD5 in milliseconds [85] and collisions for MD4 by hand. It is important to point out that MD4 and MD5 have a 128-bit result: this implies that a brute force collision search with a budget of US$ 100,000 would find a collision in a few days [92]. In spite of these weaknesses, it was still unexpected to some that Sotirov et al. [85] announced on December 31, 2008 that they managed to create a rogue CA certificate using MD5; such a certificate makes it possible to impersonate any website on the Internet. While their attack required some cryptanalytic improvements (as CAs insert a serial number into the message before signing), the main surprise seems that more than four years after the announcements by Wang et al., the most popular CAs had not yet removed MD5 from their offerings.

NIST (National Institute for Standards and Technology, USA) was apparently not confident in the security of MD5 and proposed in 1993 a strengthened version of it called SHA (Secure Hash Algorithm) with a 160-bit result; it is now frequently called SHA-0. In 1995, NIST discovered a certificational weakness in SHA-0 (no details were published), which resulted in a new release of the standard published under the name SHA-1 [38]. In 2002, NIST published three new hash functions with longer hash results that are commonly called SHA-2: SHA-256, SHA-384 and SHA-512 [39]. In December 2003, SHA-224 has been added in a change notice to [39]. In 1998 Chabaud and Joux [21] showed how collisions for SHA-0 can be found in 2^{61} steps compared to 2^{80} for a brute force attack. Wang et al. [93,95] present a major improvement in 2005 by showing that finding a collision for SHA-0/SHA-1 takes only $2^{39}/2^{69}$ steps. The best collision attack for SHA-0 by Manuel and Peyrin [57] takes only 2^{33} steps. For SHA-1 the situation is more complex: at least four teams have announced improved collision attacks with complexity between 2^{52} and 2^{63}; however, at this stage no one has found a collision and there is some doubt about the complexities of these attacks. On the other hand, Joux and Peyrin have found collisions for 70 (out of 80) steps of SHA-1 in time 2^{39} (4 days on a PC) [48].

There are still some older proposals that have withstood cryptanalysis, such as RIPEMD-160 [33] and Whirlpool [5] (both designs have been included in ISO 10118 [46], together with SHA-1 and SHA-2); for the most recent status of attacks on Whirlpool, see [55]. Moreover, early cryptanalysis of the SHA-2 family suggests that this second generation functions has a substantial security margin against collision attacks (the results by Indesteege et al. [45] and Sanadhya and Sarkar [82] can only break 24 out of 64 steps of SHA-256).

---

[2] In 2005, there were about 800 uses of MD5 in Microsoft Windows.

However, the breakthrough collision attacks on MD5 and SHA-1 have resulted in a serious concern about the robustness of our current hash functions. With the exception of the recent rogue CA attack of [85], the practical impact of these attacks has so far been rather limited, as most applications rely on (second) preimage resistance rather than collision resistance. For MD2, Knudsen et al. [53] find preimages in time 2^{73}. Leurent [56] has shown that preimages for MD4 can be found in 2^{102} steps, and Sasaki and Aoki have developed a shortcut preimage attack for MD5 [83] with complexity 2^{123}. Preimage attacks for SHA-1 seem to be completely beyond reach today: the best attack by Aoki et al. [4] works for 48 out of 80 steps). Somewhat surprisingly, preimages for SHA-256 can be found faster than brute force for 43 out of 64 steps [3].

In view of these developments, the cryptographic community agrees that we need new hash functions that offer an adequate security margin for the next 30 years or more; in view of this it would be prudent to develop alternatives for SHA-2. This has motivated NIST to call for an open competition; this is a procedure commonly used in cryptography, a.o. for the block ciphers DES and AES; there were also the European competitions NESSIE [73] and eSTREAM [78] as well as the Japanese Cryptrec initiative [44]. While the industry is currently migrating to SHA-256 as a replacement for MD5 and SHA-1, some players seem to be waiting for SHA-3.

4   The NIST SHA-3 Competition

After two open workshops and a public consultation period, NIST has published on November 2, 2007 an open call for contributions for SHA-3, a new cryptographic hash family [67]. The deadline for the call for contributions was October 31, 2008. A SHA-3 submission needs to support hash results of 224, 256, 384 and 512 bits to allow substitution for the SHA-2 family. It should work with legacy applications such as DSA and HMAC. Designers should present detailed design documentation, including a reference implementation, optimized implementations for 32-bit and 64-bit machines; they should also evaluate hardware performance. If an algorithm is selected, it needs to be available worldwide without royalties or other intellectual property restrictions.

Even if preparing a submission required a substantial effort, NIST received 64 submissions. Early December 2008, NIST has announced that 51 designs have been selected for the first round. Five of the 13 rejected designs have been published by their designers (see [35]); it is perhaps not surprising that four of these five designs have been broken very quickly. From the 51 Round 1 candidates, about half were broken in early July 2009. This illustrates that designing a secure and efficient hash function is a challenging task.

On July 24, 2009, NIST announced that 14 algorithms have been selected for Round 2, namely Blake, Blue Midnight Wish, CubeHash, ECHO, Fugue, Grøstl, Hamsi, JH, Keccak, Luffa, Shabal, SHAvite-3, SIMD and Skein. By mid September 2009, several of these algorithms have been tweaked, which means that small modifications have been made that should not invalidate earlier analysis. The majority of these designs use an iterated approach as described in Sect. 2.2 or a variant thereof: four Round 2 candidates (Blue Midnight Wish, Grøstl, Shabal, and SIMD) use a modification of the Merkle-Damgård construction with a larger internal memory, also known as a wide-pipe construction, and three use the HAIFA approach [15] (Blake, ECHO, and SHAvite-3). Five candidates (CubeHash, Fugue, Hamsi, Keccak, and Luffa) use a (variant of a) sponge construction [13]. Several designs (ECHO, SHAvite-3, Fugue, and Grøstl) employ AES-based building blocks; the first two benefit substantially from the AES instructions that will be offered in the 2010 Intel Westmere processor (see [12] for details). The hash functions Blue Midnight Wish, CubeHash, Blake and Skein are of the ARX (Addition, Rotate, XOR) type; they derive their non-linearity from the carries in the modular addition.

About half the Round 1 candidates originate from Europe, one third from North America, and one in six from Asia; two designs are from the Southern Hemisphere. Note that this is only an approximation as some algorithms have designers from multiple components and some designers have moved. A very large part of the Round 1 cryptanalysis was performed by researchers in Europe. In Round 2, 9 out of 16 (64%) of the designs are European, while 3 are from North America and 2 from Asia.

Two designs were expected for Round 2 but did not make it. MD6 by Rivest was probably not selected because of the slower performance; moreover, an error was found by the designer in the proof of security against differential attacks. Lane was probably removed because of the rebound attack on its compression function in [58]; it should be pointed out that this attack has a very high memory complexity, which makes it questionable whether it is more efficient than a brute force attack.

Two designs in Round 1 had remarkable security results: SWIFFT admits an asymptotic proof of security against collision and preimage attacks under worst-case assumptions about the complexity of certain lattice problems; the collision and preimage security of FSE can be reduced to hard problems in coding theory. However, both designs are rather slow; moreover, they require additional building blocks to achieve other security properties.

It is notably difficult to make reliable performance comparisons; all the Round 2 candidates have a speed that varies between 5 and 35 cycles per byte. It should be pointed out that due to additional implementation efforts, the best current SHA-2 implementations have a speed of about 15 cycles/byte; it will thus become more difficult for SHA-3 to be faster than SHA-2. The reader is referred to the SHA-3 Zoo and eBASH for security and performance updates; these sites are maintained by the ECRYPT II project [35].

The following tentative time line has been announced for the remainder of the competition: NIST intends to select approximately 5 finalists in Q4 of 2010. The third and final conference will take place in early 2012; it will be followed by an announcement of the decision in Q2 of 2012. Overall, it seems that there are many interesting candidates and the review and selection process will be extremely challenging. As a consequence of this competition, both the theory and practice of hash functions will make a significant step forward.

5 Concluding Remarks

During the last five years, we have seen a cryptographic meltdown in the security of widely used hash functions. Fortunately the practical implications have been limited, as most applications rely on (second) preimage resistance rather than on collision resistance. However, we have learned that upgrading cryptographic algorithms is always more difficult than anticipated. This is surprising, since in software implementations cryptographic algorithms are typically negotiated during the first phase of the protocol; Bellovin and Rescorla [11] explain the shortcomings of TLS in this context.

We can only regret that SHA-1 was not designed with 128 or 160 steps instead of 80; this would have avoided many of the problems we face today. While RIPEMD-160 seems a more secure alternative, its adoption is still limited: most users are upgrading to SHA-256, because of the longer hash result.

During the last five years, the theory and practice of cryptographic hash functions has advanced substantially. In view of this, one can expect that the SHA-3 competition will result in a robust hash function with a good performance. It is essential that the selection is not driven too much by performance; sufficient attention should be paid to the assurance in the security evaluation (that is, how easy or hard is the analysis of the design). Finally, note that (except for some tweaks), the design of SHA-3 will reflect the state of the art in 2008, rather than the state of the art in 2012.

For the long term, we face the challenging problem to design an efficient hash function for which the security can be reduced to a mathematical problem that is elegant and/or better understood.

References

1. Ajtai, M.: Generating hard instances of lattice problems. In: Proceedings 28th ACM Symposium on the Theory of Computing, pp. 99–108 (1996)
2. Andreeva, E., Neven, G., Preneel, B., Shrimpton, T.: Seven-property-preserving iterated hashing: ROX. In: Kurosawa, K. (ed.) ASIACRYPT 2007. LNCS, vol. 4833, pp. 130–146. Springer, Heidelberg (2007)
3. Aoki, K., Guo, J., Matusiewicz, K., Sasaki, Y., Wang, L.: Preimages for step-reduced SHA-2. In: Matsui, M. (ed.) ASIACRYPT 2009. LNCS, vol. 5912, pp. 578–597. Springer, Heidelberg (2009)
4. Aoki, K., Sasaki, Y.: Meet-in-the-middle preimage attacks against reduced SHA-0 and SHA-1. In: Halevi, S. (ed.) CRYPTO 2009. LNCS, vol. 5677, pp. 70–89. Springer, Heidelberg (2009)
5. Barreto, P.S.L.M., Rijmen, V.: The Whirlpool hashing function. NESSIE submission (September 2000)
6. Bellare, M.: New proofs for NMAC and HMAC: security without collision-resistance. In: Dwork, C. (ed.) CRYPTO 2006. LNCS, vol. 4117, pp. 602–619. Springer, Heidelberg (2006)

7. Bellare, M., Canetti, R., Krawczyk, H.: Keying hash functions for message authentication. In: Koblitz, N. (ed.) CRYPTO 1996. LNCS, vol. 1109, pp. 1–15. Springer, Heidelberg (1996)
8. Bellare, M., Goldreich, O., Goldwasser, S.: Incremental cryptography: the case of hashing and signing. In: Desmedt, Y.G. (ed.) CRYPTO 1994. LNCS, vol. 839, pp. 216–233. Springer, Heidelberg (1994)
9. Bellare, M., Ristenpart, T.: Multi-property-preserving hash domain extension and the EMD transform. In: Lai, X., Chen, K. (eds.) ASIACRYPT 2006. LNCS, vol. 4284, pp. 299–314. Springer, Heidelberg (2006)
10. Bellare, M., Rogaway, P.: Random oracles are practical: a paradigm for designing efficient protocols. In: ACM Conference on Computer and Communications Security, pp. 62–73. ACM, New York (1993)
11. Bellovin, S.M., Rescorla, E.K.: Deploying a new hash algorithm. In: Proceedings of the Network and Distributed System Security Symposium, NDSS 2006, The Internet Society (2006)
12. Benadjila, R., Billet, O., Gueron, S., Robshaw, M.J.B.: The Intel AES instructions set and the SHA-3 candidates. In: Matsui, M. (ed.) ASIACRYPT 2009. LNCS, vol. 5912, pp. 162–178. Springer, Heidelberg (2009)
13. Bertoni, G., Daemen, J., Peeters, M., Van Assche, G.: On the indifferentiability of the sponge construction. In: Smart, N.P. (ed.) EUROCRYPT 2008. LNCS, vol. 4965, pp. 181–197. Springer, Heidelberg (2008)
14. Biham, E., Shamir, A.: Differential Cryptanalysis of the Data Encryption Standard. Springer, Heidelberg (1993)
15. Biham, E., Dunkelman, O.: A framework for iterative hash functions – HAIFA. In: Proceedings Second NIST Hash Functions Workshop 2006, Santa Barbara, CA, USA (August 2006)
16. Biryukov, A., Dunkelman, O., Keller, N., Khovratovich, D., Shamir, A.: Key recovery attacks of practical complexity on AES variants with up to 10 rounds. IACR Eprint 2009/374, August 19 (2009)
17. Biryukov, A., Khovratovich, D.: Related-key cryptanalysis of the full AES-192 and AES-256. In: Matsui, M. (ed.) ASIACRYPT 2009. LNCS, vol. 5912, pp. 1–18. Springer, Heidelberg (2009)
18. Black, J., Rogaway, P., Shrimpton, T.: Black-box analysis of the block-cipher-based hash-function constructions from PGV. In: Yung, M. (ed.) CRYPTO 2002. LNCS, vol. 2442, pp. 320–355. Springer, Heidelberg (2002)
19. Bouillaguet, C., Dunkelman, O., Fouque, P.-A., Joux, A.: On the security of iterated hashing based on forgery-resistant compression functions. IACR Eprint 2009/077, February 6 (2009)
20. Brachtl, B.O., Coppersmith, D., Hyden, M.M., Matyas, S.M., Meyer, C.H., Oseas, J., Pilpel, S., Schilling, M.: Data Authentication Using Modification Detection Codes Based on a Public One Way Encryption Function, U.S. Patent Number 4,908,861, March 13 (1990)
21. Chabaud, F., Joux, A.: Differential collisions: an explanation for SHA-1. In: Krawczyk, H. (ed.) CRYPTO 1998. LNCS, vol. 1462, pp. 56–71. Springer, Heidelberg (1998)
22. Charles, D.X., Goren, E.Z., Lauter, K.E.: Cryptographic hash functions from expander graphs. In: Proceedings Second NIST Hash Functions Workshop 2006, Santa Barbara, CA, USA (August 2006)
23. Contini, S., Lenstra, A.K., Steinfeld, R.: VSH, an efficient and provable collision-resistant hash function. In: Vaudenay, S. (ed.) EUROCRYPT 2006. LNCS, vol. 4004, pp. 165–182. Springer, Heidelberg (2006)

24. Coppersmith, D.: Analysis of ISO/CCITT Document X.509 Annex D. IBM T.J. Watson Center, Yorktown Heights, N.Y., 10598, Internal Memo, June 11 (1989) (also ISO/IEC JTC1/SC20/WG2/N160)
25. Coron, J.-S., Dodis, Y., Malinaud, C., Puniya, P.: Merkle-Damgård revisited: how to construct a hash function. In: Shoup, V. (ed.) CRYPTO 2005. LNCS, vol. 3621, pp. 430–448. Springer, Heidelberg (2005)
26. Damgård, I.B.: Collision free hash functions and public key signature schemes. In: Price, W.L., Chaum, D. (eds.) EUROCRYPT 1987. LNCS, vol. 304, pp. 203–216. Springer, Heidelberg (1988)
27. Damgård, I.B.: A design principle for hash functions. In: Brassard, G. (ed.) CRYPTO 1989. LNCS, vol. 435, pp. 416–427. Springer, Heidelberg (1990)
28. Dean, R.D.: Formal aspects of mobile code security, PhD thesis, Princeton University (January 1999)
29. De Cannière, C., Rechberger, C.: Preimages for reduced SHA-0 and SHA-1. In: Wagner, D. (ed.) CRYPTO 2008. LNCS, vol. 5157, pp. 179–202. Springer, Heidelberg (2008)
30. den Boer, B., Bosselaers, A.: Collisions for the compression function of MD5. In: Helleseth, T. (ed.) EUROCRYPT 1993. LNCS, vol. 765, pp. 293–304. Springer, Heidelberg (1994)
31. Diffie, W., Hellman, M.E.: New directions in cryptography. IEEE Trans. on Information Theory IT-22(6), 644–654 (1976)
32. Dobbertin, H.: The status of MD5 after a recent attack. CryptoBytes 2(2), 1–6 (Summer 1996)
33. Dobbertin, H., Bosselaers, A., Preneel, B.: RIPEMD-160: a strengthened version of RIPEMD. In: Gollmann, D. (ed.) FSE 1996. LNCS, vol. 1039, pp. 71–82. Springer, Heidelberg (1996), http://www.esat.kuleuven.ac.be/~bosselae/ripemd160
34. Dodis, Y., Ristenpart, T., Shrimpton, T.: Salvaging Merkle-Damgård for practical applications. In: Joux, A. (ed.) Eurocrypt 2008. LNCS, vol. 5479, pp. 371–388. Springer, Heidelberg (2009)
35. ECRYPT II, The SHA-3 Zoo, http://ehash.iaik.tugraz.at/wiki/The_SHA-3_Zoo
36. Fiat, A., Shamir, A.: How to prove yourself: practical solutions to identification and signature problems. In: Odlyzko, A.M. (ed.) CRYPTO 1986. LNCS, vol. 263, pp. 186–194. Springer, Heidelberg (1987)
37. FIPS 46, Data Encryption Standard, Federal Information Processing Standard, NBS, U.S. Department of Commerce (January 1977) (revised as FIPS 46-1(1988); FIPS 46-2(1993), FIPS 46-3(1999))
38. FIPS 180-1, Secure Hash Standard, Federal Information Processing Standard (FIPS), Publication 180-1, National Institute of Standards and Technology, US Department of Commerce, Washington D.C., April 17 (1995)
39. FIPS 180-2, Secure Hash Standard, Federal Information Processing Standard (FIPS), Publication 180-2, National Institute of Standards and Technology, US Department of Commerce, Washington D.C., August 26 (2002) (Change notice 1 published on December 1, 2003)
40. Gauravaram, P., Knudsen, L.R.: On randomizing hash functions to strengthen the security of digital signatures. In: Joux, A. (ed.) EUROCRYPT 2008. LNCS, vol. 5479, pp. 88–105. Springer, Heidelberg (2009)
41. Grassl, M., Ilic, I., Magliveras, S., Steinwandt, R.: Cryptanalysis of the Tillich-Zémor hash function, IACR Eprint 2009/376, July 30 (2009)
42. Hellman, M.E.: A cryptanalytic time-memory trade-off. IEEE Trans. on Information Theory IT-26(4), 401–406 (1980)

43. Hirose, S.: Some plausible constructions of double-block-length hash functions. In: Robshaw, M.J.B. (ed.) FSE 2006. LNCS, vol. 4047, pp. 210–225. Springer, Heidelberg (2006)
44. Imai, H., Yamagishi, A.: "Cryptrec". In: van Tilborg, H.C.A. (ed.) Encyclopedia of Cryptography and Security, pp. 119–123 (2005)
45. Indesteege, S., Mendel, F., Preneel, B., Rechberger, C.: Collisions and other non-random properties for step-reduced SHA-256. In: Avanzi, R.M., Keliher, L., Sica, F. (eds.) SAC 2009. LNCS, vol. 5381, pp. 276–293. Springer, Heidelberg (2009)
46. ISO/IEC 10118, Information technology – Security techniques – Hash-functions, Part 1: General (2000); Part 2: Hash-functions using an n-bit block cipher algorithm (2000); Part 3: Dedicated hash-functions (2003); Part 4: Hash-functions using modular arithmetic (1998)
47. Joux, A.: Multicollisions in iterated hash functions. Application to cascaded constructions. In: Franklin, M. (ed.) CRYPTO 2004. LNCS, vol. 3152, pp. 306–316. Springer, Heidelberg (2004)
48. Joux, A., Peyrin, T.: Hash functions and the (amplified) boomerang attack. In: Menezes, A. (ed.) CRYPTO 2007. LNCS, vol. 4622, pp. 244–263. Springer, Heidelberg (2007)
49. Kaliski Jr., B.S.: The MD2 Message-Digest algorithm, Request for Comments (RFC) 1319, Internet Activities Board, Internet Privacy Task Force (April 1992)
50. Kelsey, J., Kohno, T.: Herding hash functions and the Nostradamus attack. In: Vaudenay, S. (ed.) EUROCRYPT 2006. LNCS, vol. 4004, pp. 183–200. Springer, Heidelberg (2006)
51. Kelsey, J., Schneier, B.: Second preimages on n-bit hash functions for much less than 2^n work. In: Cramer, R. (ed.) EUROCRYPT 2005. LNCS, vol. 3494, pp. 474–490. Springer, Heidelberg (2005)
52. Knudsen, L.R., Lai, X., Preneel, B.: Attacks on fast double block length hash functions. Journal of Cryptology 11(1), 59–72 (Winter 1998)
53. Knudsen, L.R., Mathiassen, J.E., Muller, F., Thomsen, S.S.: Cryptanalysis of MD2. Journal of Cryptology, 19 p. (in print, 2010)
54. Lai, X., Massey, J.L.: Hash functions based on block ciphers. In: Rueppel, R.A. (ed.) EUROCRYPT 1992. LNCS, vol. 658, pp. 55–70. Springer, Heidelberg (1993)
55. Lamberger, M., Mendel, F., Rechberger, C., Rijmen, V., Schläffer, M.: Rebound distinguishers: results on the full Whirlpool compression function. In: Matsui, M. (ed.) ASIACRYPT 2009. LNCS, vol. 5912, pp. 126–143. Springer, Heidelberg (2009)
56. Leurent, G.: MD4 is not one-way. In: Nyberg, K. (ed.) FSE 2008. LNCS, vol. 5086, pp. 412–428. Springer, Heidelberg (2008)
57. Manuel, S., Peyrin, T.: Collisions on SHA-0 in one hour. In: Nyberg, K. (ed.) FSE 2008. LNCS, vol. 5086, pp. 16–35. Springer, Heidelberg (2008)
58. Matusiewicz, K., Naya-Plasencia, M., Nikolic, I., Sasaki, Y., Schläffer, M.: Rebound attack on the full Lane compression function. In: Matsui, M. (ed.) ASIACRYPT 2009. LNCS, vol. 5912, pp. 106–125. Springer, Heidelberg (2009)
59. Maurer, U.M., Renner, R., Holenstein, C.: Indifferentiability, impossibility results on reductions, and applications to the random oracle methodology. In: Naor, M. (ed.) TCC 2004. LNCS, vol. 2951, pp. 21–39. Springer, Heidelberg (2004)
60. Merkle, R.: Secrecy, Authentication, and Public Key Systems. UMI Research Press (1979)
61. Merkle, R.: One way hash functions and DES. In: Brassard, G. (ed.) CRYPTO 1989. LNCS, vol. 435, pp. 428–446. Springer, Heidelberg (1990)
62. Merkle, R.: A fast software one-way hash function. Journal of Cryptology 3(1), 43–58 (1990)

63. Miyaguchi, S., Iwata, M., Ohta, K.: New 128-bit hash function. In: Proceedings 4th International Joint Workshop on Computer Communications, Tokyo, Japan, July 13–15, pp. 279–288 (1989)

64. Moore, J.H., Simmons, G.J.: Cycle structure of the DES for keys having palindromic (or antipalindromic) sequences of round keys. IEEE Transactions on Software Engineering 13, 262–273 (1987)

65. Naito, Y., Yoneyama, K., Wang, L., Ohta, K.: How to confirm cryptosystems security: the original Merkle-Damgård is still alive! In: Matsui, M. (ed.) ASIACRYPT 2009. LNCS, vol. 5912, pp. 382–398. Springer, Heidelberg (2009)

66. Naor, M., Yung, M.: Universal one-way hash functions and their cryptographic applications. In: Proceedings 21st ACM Symposium on the Theory of Computing, pp. 387–394 (1990)

67. NIST SHA-3 Competition, http://csrc.nist.gov/groups/ST/hash/

68. Pal, P., Sarkar, P.: PARSHA-256 – A new parallelizable hash function and a multi-threaded implementation. In: Johansson, T. (ed.) FSE 2003. LNCS, vol. 2887, pp. 347–361. Springer, Heidelberg (2003)

69. Patarin, J.: Collisions and inversions for Damgård's whole hash function. In: Safavi-Naini, R., Pieprzyk, J.P. (eds.) ASIACRYPT 1994. LNCS, vol. 917, pp. 307–321. Springer, Heidelberg (1995)

70. Pinkas, D.: The need for a standardized compression algorithm for digital signatures. In: Ingemarsson, I. (ed.) Abstracts of Papers: Eurocrypt 1986, A Workshop on the Theory and Application of Cryptographic Techniques, May 20-22, 1986, p. 7 (1986)

71. Preneel, B.: Analysis and design of cryptographic hash functions. Doctoral Dissertation, Katholieke Universiteit Leuven (1993)

72. Preneel, B., Govaerts, R., Vandewalle, J.: Hash functions based on block ciphers: a synthetic approach. In: Stinson, D.R. (ed.) CRYPTO 1993. LNCS, vol. 773, pp. 368–378. Springer, Heidelberg (1994)

73. Preneel, B.: NESSIE project. In: van Tilborg, H.C.A. (ed.) Encyclopedia of Cryptography and Security, pp. 408–413 (2005)

74. Rabin, M.O.: Digitalized signatures. In: Lipton, R., DeMillo, R. (eds.) Foundations of Secure Computation, pp. 155–166. Academic Press, New York (1978)

75. Rivest, R.L.: The MD4 message digest algorithm. In: Menezes, A., Vanstone, S.A. (eds.) CRYPTO 1990. LNCS, vol. 537, pp. 303–311. Springer, Heidelberg (1991)

76. Rivest, R.L.: The MD5 message-digest algorithm. Request for Comments (RFC) 1321, Internet Activities Board, Internet Privacy Task Force (April 1992)

77. Rivest, R.L., Shamir, A., Adleman, L.: A method for obtaining digital signatures and public-key cryptosystems. Communications ACM 21, 120–126 (1978)

78. Robshaw, M.J.B., Billet, O. (eds.): New Stream Cipher Designs. LNCS, vol. 4986. Springer, Heidelberg (2008)

79. Rogaway, P., Shrimpton, T.: Cryptographic hash function basics: definitions, implications, and separations for preimage resistance, second-preimage resistance, and collision resistance. In: Roy, B., Meier, W. (eds.) FSE 2004. LNCS, vol. 3017, pp. 371–388. Springer, Heidelberg (2004)

80. Rogaway, P., Steinberger, J.P.: Constructing cryptographic hash functions from fixed-key blockciphers. In: Wagner, D. (ed.) CRYPTO 2008. LNCS, vol. 5157, pp. 433–450. Springer, Heidelberg (2008)

81. Saarinen, M.-J.O.: Security of VSH in the real world. In: Barua, R., Lange, T. (eds.) INDOCRYPT 2006. LNCS, vol. 4329, pp. 95–103. Springer, Heidelberg (2006)

82. Sanadhya, S.K., Sarkar, P.: New collision attacks against up to 24-step SHA-2. In: Chowdhury, D.R., Rijmen, V., Das, A. (eds.) INDOCRYPT 2008. LNCS, vol. 5365, pp. 91–103. Springer, Heidelberg (2008)

83. Sasaki, Y., Aoki, K.: Finding preimages in full MD5 faster than exhaustive search. In: Joux, A. (ed.) EUROCRYPT 2008. LNCS, vol. 5479, pp. 134–152. Springer, Heidelberg (2008)

84. Simon, D.: Finding collisions on a one-way street: can secure hash functions be based on general assumptions? In: Nyberg, K. (ed.) EUROCRYPT 1998. LNCS, vol. 1403, pp. 334–345. Springer, Heidelberg (1998)

85. Stevens, M., Sotirov, A., Appelbaum, J., Lenstra, A., Molnar, D., Osvik, D.A., de Weger, B.: Short chosen-prefix collisions for MD5 and the creation of a rogue CA certificate. In: Halevi, S. (ed.) CRYPTO 2009. LNCS, vol. 5677, pp. 55–69. Springer, Heidelberg (2009)

86. Stam, M.: Beyond uniformity: better security/efficiency tradeoffs for compression functions. In: Wagner, D. (ed.) CRYPTO 2008. LNCS, vol. 5157, pp. 397–412. Springer, Heidelberg (2008)

87. Stam, M.: Blockcipher based hashing revisited. In: Dunkelman, O. (ed.) Fast Software Encryption. LNCS, vol. 5665, pp. 67–83. Springer, Heidelberg (2009)

88. Steinberger, J.P.: The collision intractability of MDC-2 in the ideal-cipher model. In: Naor, M. (ed.) EUROCRYPT 2007. LNCS, vol. 4515, pp. 34–51. Springer, Heidelberg (2007)

89. Tillich, J.-P., Zémor, G.: Hashing with SL_2. In: Desmedt, Y.G. (ed.) CRYPTO 1994. LNCS, vol. 839, pp. 40–49. Springer, Heidelberg (1994)

90. Tillich, J.-P., Zémor, G.: Collisions for the LPS expander graph hash function. In: Smart, N.P. (ed.) EUROCRYPT 2008. LNCS, vol. 4965, pp. 254–269. Springer, Heidelberg (2008)

91. Van Heurck, P.: Trasec: Belgian security system for electronic funds transfers. Computers & Security 6, 261–268 (1987)

92. van Oorschot, P.C., Wiener, M.J.: Parallel collision search with cryptanalytic applications. Journal of Cryptology 12(1), 1–28 (1999)

93. Wang, X., Yin, Y.L., Yu, H.: Finding collisions in the full SHA-1. In: Shoup, V. (ed.) CRYPTO 2005. LNCS, vol. 3621, pp. 1–16. Springer, Heidelberg (2005)

94. Wang, X., Yu, H.: How to break MD5 and other hash functions. In: Cramer, R. (ed.) EUROCRYPT 2005. LNCS, vol. 3494, pp. 19–35. Springer, Heidelberg (2005)

95. Wang, X., Yu, H., Yin, Y.L.: Efficient collision search attacks on SHA-0. In: Shoup, V. (ed.) CRYPTO 2005. LNCS, vol. 3621, pp. 1–16. Springer, Heidelberg (2005)

96. Wiener, M.J.: The full cost of cryptanalytic attacks. Journal of Cryptology 17(2), 105–124 (2004)

97. Winternitz, R.: A secure one-way hash function built from DES. In: Proceedings IEEE Symposium on Information Security and Privacy, pp. 88–90. IEEE Press, Los Alamitos (1984)

98. Yasuda, K.: How to fill up Merkle-Damgård hash functions. In: Pieprzyk, J. (ed.) ASIACRYPT 2008. LNCS, vol. 5350, pp. 272–289. Springer, Heidelberg (2008)

99. Yuval, G.: How to swindle Rabin. Cryptologia 3, 187–189 (1979)

Errors Matter: Breaking RSA-Based PIN Encryption with Thirty Ciphertext Validity Queries

Nigel P. Smart

Dept. Computer Science,
University of Bristol,
Merchant Venturers Building,
Woodland Road,
Bristol, BS8 1UB,
United Kingdom
nigel@cs.bris.ac.uk

Abstract. We show that one can recover the PIN from a standardized RSA-based PIN encryption algorithm from a small number of queries to a ciphertext validity checking oracle. The validity checking oracle required is rather special and we discuss whether such oracles could be obtained in the real world. Our method works using a minor extension to the ideas of Bleichenbacher and Manger, in particular we obtain information from negative, as well as positive, responses from the validity checking oracle.

1 Introduction

Despite advances in provably secure cryptographic systems over the last decade or so, there are still a large number of systems deployed which do not use provably secure algorithms. This is mainly due to legacy reasons, and the problems of replacing or updating already deployed systems. In this paper we focus on a particular example of a non-provably secure encryption method based on RSA, namely the PIN encryption method in the EMV card payment system. EMV (Europay, Mastercard, Visa) is an industrial consortium which develops card payment standards for the banking industry. The chip-and-pin system used in most credit card payment systems is the main example deployed instantiation of their standards.

The RSA algorithm, being the earliest public key algorithm, was deployed in a number of systems before the advent of the provable security methodology. Probably the most famous example was the early adoption of the PKCS-v1.0 encryption method employed in SSL. This was famously attacked by Bleichenbacher [2], who used a ciphertext validity checking oracle, for the PKCS-v1.0 padding scheme, to recover the underlying RSA plaintext for a given challenge. The number of queries required by Bleichenbacher was very large; [2] claims 2^{20} queries needed to break an RSA key size as used in SSL.

J. Pieprzyk (Ed.): CT-RSA 2010, LNCS 5985, pp. 15–25, 2010.

With the advent of provably secure systems such as OAEP [1] it appeared that RSA based systems could be deployed in a provably secure manner. However, Manger [7] showed how an extension to Bleichenbacher's attack could be deployed against RSA-OAEP when the ciphertext validity checking oracle returned different error messages when it failed for different events (similar data can be obtained via timing attacks on poor implementations). In particular the OAEP padding scheme when applied to RSA can result in two possible errors occurring on decrypting an invalid ciphertext. The first error is that upon inverting the RSA function one obtains a number which is too large for the OAEP padding mechanism. The second error is that the OAEP padding mechanism returns an invalid ciphertext result. The security proof for OAEP assumes that the first error does not occur. But Manger, using an oracle which tells him whether the first error occurs as opposed to the second, managed to break the RSA-OAEP scheme, again with a number of ciphertext queries which was linear in (but strictly larger than) the bit-size of the public key. For example [7] claims 1100 queries were needed to break a 1024-bit RSA message, and 2200 queries for a 2048 bit message.

In our paper we take a similar approach to Manger, in that we look at one out of a possible set of failure events. We examine the PIN encryption method in the EMV standard [4,5]. This method is used to RSA encrypt four digit PIN numbers from a trusted terminal to a smart card. The smart card then decrypts the RSA ciphertext, recover the PIN and checks it against it's internally held PIN value. On decrypting the RSA ciphertext the card performs three validity checks. We assume, much as Manger does, that the failure or success of one of these checks, leaks to the attacker.

We then adapt the method of Bleichenbacher and Manger to this situation. In particular a number of simplifications can be performed. Firstly we are not interested in recovering the full RSA plaintext, we only care about the underlying PIN. Luckily, for the attacker, each PIN defines a distinct interval of possible RSA plaintexts. So our goal is to determine which of 10000 possible intervals the plaintext lies in. Secondly, and much for the same reason, we are able to use the failure of an oracle query to reduce the possibilities for the underlying PIN. In the Bleichenbacher and Manger attacks only positive oracle responses are used to reduce the space of underlying plaintexts.

We end the paper by discussing the practical impact of our attacks, and point out that it is highly unlikely that our method will provide a means for an attacker to obtain PINs for the EMV system. However, our work does once again point out the need for old legacy systems to be replaced with new ones (EMVCo is already engaged in a process to update their standards due to the need to increase key sizes). In addition our method of using negative responses from the validity checking oracle may be useful in other applications.

2   PIN Encryption Method

We assume an RSA modulus N of t bits in length, where t is a precise multiple of eight. We write $k = t/8$ for the number of bytes in N. In this section we outline how PIN encryption and verification is performed in the EMV standard [4,5] (the same method is used in the equivalent ISO standard [6]) First we cover the creation of the PIN block, then we discuss how an RSA message is formatted, and finally we discuss verification. Full details can be found in [4,5]. For later use we write

$$f(\mathfrak{a}, \mathfrak{b}, \mathfrak{u}, \mathfrak{r}) = \mathfrak{a} \cdot 2^{8(k-1)} + \mathfrak{b} \cdot 2^{8(k-9)} + \mathfrak{u} \cdot 2^{8(k-17)} + \mathfrak{r}$$

where $\mathfrak{a}$ is a one-byte value, $\mathfrak{b}$ and $\mathfrak{u}$ are eight bytes values and $\mathfrak{r}$ is a $k - 17$ byte value.

PIN Block Format: A PIN block is defined to be a sequence of eight bytes, or equivalently sixteen nibbles. A PIN is assumed to be a sequence of between four and twelve integers in the range '0' to '9'. The PIN block is then defined to be (in nibbles)

C	N	P	P	P	P	P/F	P/F	P/F	P/F	P/F	P/F	P/F	P/F	F	F

where

- C is the number 2.
- N encodes the length of the PIN, i.e. it is a number between 0 and 12.
- P is a PIN digit between 0 and 9.
- F is a filler digit of 15 (i.e. the nibble with all ones set).

In the common case of a four digit PIN, $P_1\|P_2\|P_3\|P_4$, then the value of the PIN block (as an integer) is given by

$$\begin{aligned}\mathfrak{b} &= 2 \cdot 2^{60} + 4 \cdot 2^{56} + (P_1 \cdot 2^{12} + P_2 \cdot 2^8 + P_3 \cdot 2^4 + P_4) \cdot 2^{40} + (2^{40} - 1) \\ &= (2^{61} + 2^{58} + 2^{40} - 1) + (P_1 \cdot 2^{12} + P_2 \cdot 2^8 + P_3 \cdot 2^4 + P_4) \cdot 2^{40}.\end{aligned}$$

RSA Message Format: Before encrypting a PIN the encryptor obtains an 8-byte nonce $\mathfrak{u}$ from the decryptor. This is used to avoid replay attacks. The encryptor then generates the pin block $\mathfrak{b}$ and creates a $k - 17$ byte random number $\mathfrak{r}$. The message $\mathfrak{m}$ is then created as

$$\mathfrak{m} = f(127, \mathfrak{b}, \mathfrak{u}, \mathfrak{r}).$$

The encryptor then generates

$$\mathfrak{c} = \mathfrak{m}^e \pmod{N}$$

and sends $\mathfrak{c}$ to the verifier.

PIN Verification: The verifier, who knows the RSA secret key d, on obtaining a ciphertext $\mathfrak{c}'$ first obtains the underlying RSA plaintext $\mathfrak{m}'$, via

$$\mathfrak{m}' = \mathfrak{c}^d \pmod{N}.$$

Then the verifier recovers the values of $\mathfrak{a}', \mathfrak{b}', \mathfrak{u}', \mathfrak{r}'$ from the equation

$$\mathfrak{m}' = f(\mathfrak{a}', \mathfrak{b}', \mathfrak{u}', \mathfrak{r}').$$

Then he performs the following tests:

1. Return fail if $u \neq u'$, i.e. the nonce recovered is not equal to the nonce sent.
2. Return fail if $a' \neq 127$.
3. Return fail if the PIN in the PIN block b' is not equal to the expected PIN.

We call the above tests Test 1, Test 2, and Test 3 in what follows. Note, that no mention is made in [4] of checking whether the PIN block is formatted correctly, only that the recovered PIN is correct.

In this paper we consider the situation when Test 2 is always executed for every ciphertext passed to the verifier and the attacker can determine whether the Test 2 returns fail or not. Later in Section 4 we shall discuss how realistic this is.

We note that there is a trivial attack on the PIN system, given a card one could run through all 10000 possible PIN blocks, obtain 10000 nonces, and then form 10000 ciphertexts. Each ciphertext is then passed to the card for checking, until one is returned as valid, in which case the attacker has determined the correct PIN. However, as we shall discuss later, the card has mechanisms built into it to protect against checking too many invalid PIN numbers. Hence, the question is how; few challenge ciphertexts are needed to recover a PIN?

3   The "Attack"

Our method is a simple extension of the methods of Bleichenbacher [2] as applied to the PKCS-v1.0 encryption scheme, and the method of Manger [7] as applied to the PKCS-v1.2 OAEP encryption scheme. In the original methods of Bleichenbacher and Manger the attacker is given a validity oracle which given a ciphertext returns whether the underlying plaintext lies in a given range (the range depending on the padding scheme being used). The attacker uses the positive responses from the oracle to essentially half the interval in which the plaintext behind a target ciphertext lies.

The main difference in our attack is that since we are only trying to determine one of a small number of PINs, rather than recovering the whole text, we are able to make use of negative results from our validity oracle. This comes at the expense of increasing the number of possible intervals for our target message. However, this does not result in an exponential blow-up since we have a small number of ranges in which a valid ciphertext could lie.

We describe our method in a bottom-up fashion by first describing Algorithm 1. This takes as input an interval $[a, b]$, and an integer s such that we know bounds L and U such that

$$L \leq s \cdot m \pmod{N} \leq U,$$

for a message $m \in [a, b]$. The algorithm works by evaluating all integers r such that

$$L \leq s \cdot m - r \cdot N \leq U,$$

which leads to

$$s \cdot a - U \leq s \cdot m - U \leq r \cdot N \leq s \cdot m - L \leq s \cdot b - L,$$

since $m \in [a, b]$. Then we deduce new bounds on m, for each of these values of r, from

$$L + r \cdot N \leq s \cdot m \leq U + r \cdot N.$$

Algorithm 1. UpdateInterval(a, b, L, U, s, N)

Input: a, b, L, U, s, N
Output: A list of intervals, *List*
$List \leftarrow \{\}$
for r **from** $\lceil (s \cdot a - U)/N \rceil$ **to** $\lfloor (s \cdot b - L)/N \rfloor$ **do**
$\quad a' \leftarrow \max(a, \lceil (L + r \cdot N)/s \rceil)$
$\quad b' \leftarrow \min(b, \lfloor (U + r \cdot N)/s \rfloor)$
$\quad$ **if** $a' \leq b'$ **then**
$\quad\quad List \leftarrow List \cup \{[a', b']\}$
return *List*

To describe the our next Algorithm 2, we assume we have an oracle $\mathcal{O}$ which on input of an RSA ciphertext $\mathfrak{c}'$ will return whether the underlying RSA message $\mathfrak{m}'$ lies in the interval $[\mathcal{L}, \mathcal{U}]$, for specific fixed integers $\mathcal{L}$ and $\mathcal{U}$. Algorithm 2 takes as input a ciphertext $\mathfrak{c}$ a list of intervals *List*, for which we know that the underlying message $\mathfrak{m}$ lies in one of the intervals contained in *List*, and a "test" integer s.

Algorithm 2. UpdateList$(List, s, \mathfrak{c}, N, e)$

Input: $List, s, \mathfrak{c}, N, e$
Output: A new list of intervals, *List'*
$List' \leftarrow \{\}$
$\mathfrak{c}' = s^e \cdot \mathfrak{c} \pmod{N}$
$flag \leftarrow \mathcal{O}(\mathfrak{c}')$
forall $[a, b] \in List$ **do**
$\quad$ **if** $flag$ **then**
$\quad\quad List' = List' \cup \text{UpdateInterval}(a, b, \mathcal{L}, \mathcal{U}, s, N)$
$\quad$ **else**
$\quad\quad List' = List' \cup \text{UpdateInterval}(a, b, 0, \mathcal{L}, s, N)$
$\quad\quad List' = List' \cup \text{UpdateInterval}(a, b, \mathcal{U}, N, s, N)$
return *List'*

Notice that UpdateList will create at most two intervals for every one in *List* if the oracle returns *false*. This is the main difference between our method and that of Bleichenbacher and Manger. However, this only works due to the nature of the underlying message we are trying to recover. It may appear that the size of *List'* could become very large if UpdateList is repeatedly called, but we control the size of $List'$ using another list of intervals $TList$, which contains one interval for each PIN number. The $TList$ is created by Algorithm 3 which takes as input the value of the nonce $\mathfrak{u}$ which underlies in the target ciphertext $\mathfrak{c}$. In Algorithm 3 we assume that PINs are four digits in length. Notice, that all the intervals created are distinct.

Algorithm 3. CreateTList$(\mathfrak{u}, k)$

Input: $\mathfrak{u}, k$
Output: $TList$
$TList \leftarrow \{\}$
forall $p_1, p_2, p_3, p_4 \in [0, 9]$ **do**
    $\mathfrak{b} \leftarrow$ the PIN block for this PIN.
    $a \leftarrow f(127, \mathfrak{b}, \mathfrak{u}, 0)$
    $b \leftarrow f(127, \mathfrak{b}, \mathfrak{u}, 2^{8(k-17)})$
    $TList \leftarrow TList \cup \{[[a, b], [p_1, p_2, p_3, p_4]]\}$
return $TList$

To filter the intervals we then use Algorithm 4. This takes a set of intervals $List$ for which one contains the target message, and a list $TList$ of intervals which contain the PINs. It then forms two new lists $List'$ and $TList'$. Clearly if Algorithm 4 ever finds returns $TList'$ containing only one element, then we have found the PIN.

Algorithm 4. Filter$(List, TList)$

Input: $List, TList$
Output: $List', TList'$
$List' \leftarrow \{\}, \qquad TList' \leftarrow \{\}$
forall $[a, b] \in List$ **do**
    $flag \leftarrow false$
    forall $[[a', b'], PIN] \in TList$ **do**
        if $a' \leq b$ and $a \leq b'$ **then**
            $TList' \leftarrow TList' \cup \{[[a', b'], PIN]\}$
            $flag \leftarrow true$
    if $flag$ **then**
        $List' \leftarrow List' \cup \{[a, b]\}$
return $List', TList'$

In Algorithm 5 we describe the main method, which takes as input a ciphertext $\mathfrak{c}$ corresponding to a PIN encryption (and the corresponding nonce $\mathfrak{u}$) and outputs the corresponding PIN. In this algorithm we choose the value of s to use in the call to the oracle so as to hopefully half the number of possible messages on each call to the oracle. For a particular value of r we require that

$$\mathcal{L} \leq s \cdot m - rN \leq \mathcal{U}$$

Algorithm 5. Main(c, u, N, e)

   Input: c
   Output: PIN
   $TList \leftarrow$ CreateTList(u, k)
   $b_0 \leftarrow$ the pin block corresponding to the PIN: $0, 0, 0, 0$
   $b_9 \leftarrow$ the pin block corresponding to the PIN: $9, 9, 9, 9$
   $a \leftarrow f(127, b_0, u, 0)$
   $b \leftarrow f(127, b_9, u, 2^{8(k-17)})$
   $List \leftarrow \{[a, b]\}$
   $r \leftarrow 0$
   while $\#TList > 1$ **do**
      $a \leftarrow \min\{a : [a, b] \in List\}$
      $b \leftarrow \max\{b : [a, b] \in List\}$
      $r \leftarrow \max(r + 1, \lceil (2 \cdot b \cdot (\mathcal{U} - \mathcal{L})/(b - a) - \mathcal{L})/N \rceil)$
      $s \leftarrow \lceil (\mathcal{L} + r \cdot N)/b \rceil$
      if $s \leq \lfloor (\mathcal{U} + r \cdot N)/a \rfloor$ **then**
         $List \leftarrow$ UpdateList($List, s, c, N, e$)
         $List, TList \leftarrow$ Filter($List, TList$)
   $PIN \leftarrow TList[1][1]$
   return PIN

---

and so we must have

$$\lceil (\mathcal{L} + r \cdot N)/b \rceil \leq s \leq \lfloor (\mathcal{U} + r \cdot N)/a \rfloor,$$

which explains the choice of s in Algorithm 5. However, the key is the choice of r.

If we assume an oracle call is successful then the range of the new maximum possible interval is given by $(\mathcal{U} - \mathcal{L})/s$. Which given our above bound on s is upper bounded by

$$b \left(\frac{\mathcal{U} - \mathcal{L}}{\mathcal{L} + r \cdot N} \right).$$

But we really want this last value to be less than $(b - a)/2$. Thus we require

$$b \left(\frac{\mathcal{U} - \mathcal{L}}{\mathcal{L} + r \cdot N} \right) \leq \frac{b - a}{2},$$

which implies

$$r \geq \frac{2 \cdot b \cdot (\mathcal{U} - \mathcal{L})/(b - a) - \mathcal{L}}{N}.$$

To run the algorithm, which works for any oracle $\mathcal{O}$, although possibly not very efficiently, we need to specify the values of $\mathcal{L}$ and $\mathcal{U}$, which depend on precise validity checking oracle $\mathcal{O}$ that we use.

3.1 Experimental Results

We assume the validity checking oracle simply returns true if the leading byte of the plaintext is equal to 127, and it returns false otherwise. This means that our values for $\mathcal{L}$ and $\mathcal{U}$ are given by

$$\mathcal{L} = 127 \cdot 2^{8(k-1)},$$
$$\mathcal{U} = 127 \cdot 2^{8(k-1)} + 2^{8(k-1)} - 1.$$

If we use the parameters for RSA keys as defined in the EMVCo specification, then we find that the RSA key size is only 896 bits, i.e. $k = 112$. This is the size of the RSA modulus for the chip-and-pin cards public key.

   We ran a series of 1000 experiments to emulate the above attack and recorded how many oracle queries were needed to recover the PIN. The percentages can be found in Table 1. Notice, that the number of oracle queries is incredibly low by modern cryptographic standards (even for a modulus of 896 bits). The reason is that although the message appears to be padded with a large amount of randomness, this randomness is not mixed in with the plaintext (like it would be with OAEP). In particular the combination of this with the small number of possible plaintexts (i.e. 10000) leads to a highly reduced number of oracle queries.

Table 1. Percentage of attacks with a given number of oracle queries

Range	%		Range	%		Range	%
$0 - 4$	0.0		$5 - 9$	0.3		$10 - 14$	6.9
$15 - 19$	25.9		$20 - 24$	28.2		$25 - 29$	13.7
$30 - 34$	7.4		$35 - 39$	4.6		$40 - 44$	3.0
$45 - 49$	2.5		$50 - 54$	0.3		$55 - 59$	1.3
$60 - 64$	1.2		$65 - 69$	0.6		$70 - 74$	0.4
$75 - 79$	0.5		$80 - 84$	0.1		$85 - 89$	0.0
$90 - 94$	0.1		$95 - 99$	0.1		$100 - 104$	0.2
$105 - 109$	0.1		$110 - 114$	0.1		$115 - 119$	0.4
$120 - 124$	0.1		$125 - 129$	0.1		$130 - 134$	0.2
$135 - 139$	0.1		$140 - 144$	0.1		≥ 145	1.5

   So we see that 75% of all PINs can be recovered with less than 30 calls to the oracle.

4   Practical Attack Considerations

We now consider what are the practical consequences of the above analysis. Firstly we consider how the information which our oracle provides may (or may not) be obtained in real life. Then we consider, assuming the oracle is available, how one can obtain the required number of oracle queries.

4.1   Is the Oracle Practical?

Upon performing the RSA decryption function the card needs to execute three validity checking steps as explained earlier. The EMV standard [4] mentions these in the order 1, 2, 3 as specified earlier, however no warnings are given that executing them in a different order will make a difference to the security. A PIN encryption passes only if *all* tests pass.

In [8] the tests are given in a different order, namely 2, 1, 3. Indeed it might be tempting to implement the padding checking algorithm in this order with Test 2 before the other operations since then one is dealing with the leading byte first.

In performing three tests there are two basic ways this can be done. The first method tests the three conditions individually in sequence and aborts on the first occurrence of a failure, we call this the "individual" method. In the second method, one tests all three conditions and then returns failure at the end, we call this the "aggregate" method. The aggregate method is usually considered "best practice" since it avoids any timing analysis which could result from the individual method. However, any padding test is susceptible to a possible simple power analysis style attack in which the attacker observes whether the test passes or fails.

In the following table we give for each method of testing, and each order of the tests whether the timing analysis or simple power analysis *could* result in an attack against PIN encryption. We make no claim as to whether such an attack could be carried out in practice, only that it is possible if such an implementation choice is made. In all cases we ignore the trivial attack which requires 10000 ciphertext queries.

Order	Method	Timing	SPA
1, 2, 3	Individual	-	-
1, 3, 2	Individual	-	-
2, 1, 3	Individual	✔	✔
2, 3, 1	Individual	✔	✔
3, 2, 1	Individual	-	-
3, 1, 2	Individual	-	-
1, 2, 3	Aggregate	-	✔
1, 3, 2	Aggregate	-	✔
2, 1, 3	Aggregate	-	✔
2, 3, 1	Aggregate	-	✔
3, 2, 1	Aggregate	-	✔
3, 1, 2	Aggregate	-	✔

We note that if Test 2 is not performed first in the individual method then it is highly unlikely for the attacker to be able to apply the Bleichenbacher/Manger style attack above, since the card is highly likely to abort after performing Test 1. This means that is would be better practice to implement the individual test method in the order 1, 2, 3, rather than the aggregate test. However, it is clear from our analysis that any card which executes Test 2, for every ciphertext passed to it, is susceptible assuming a suitable side-channel.

4.2   Can One Obtain This Many Queries?

So from now on we assume that an attacker can obtain access to an oracle which tells him whether Test 2 passes or fails for a particular ciphertext. The card usually has two counters within it; a PIN try counter n and a PIN decipherment error counter m. The card which will lock as soon as n invalid PINs have been tested (or m decryption errors have occured) in a row. According to sources within the industry, the value of m is usually quite large and is independent of n, i.e. decrementing m does not decrement n, hence the PIN try counter only applies upon a valid decryption. In particular the PIN try counter is decremented only on the failure of what we called Test 3, whilst the PIN decipherment counter is decremented on failure of Test 1 or Test 2. The value of n is not stated in the standard but in the field it is generally set to be equal to three, whilst the use of a value for m is not alluded to at all within the standard. Thus according to our sources it is plausible to be able to query the card with many invalid ciphertexts before it locks out.

In the following we however assume the most pessimistic situation for the attacker, in that we assume that a PIN decipherment error is treated as a PIN try error and that the value of the PIN try counter is set to three. This means an attacker can execute at most two invalid ciphertext queries for every valid PIN number entered by the user. It is not unreasonable, given how most people use their cards, to assume that this can be increased to four, since users are used to things "going wrong" (for example by entering their PIN incorrectly). Hence at a electronic point of sale terminal they would not treat as suspicious the request for a second PIN entry request.

Hence, obtaining 30 such ciphertext queries for attacking a particular user would only be possible if one was able to find a user who repeatedly used his card at the attackers terminal. Such an attack might be economic for a high net-worth individual, but would still require around 7-8 such visits to obtain the PIN with a 75% chance of success. Alternatively, a high through-put supermarket could easily blame a software glitch for requiring all customers in one day to enter their PINs three times. This would give a total of 6 such queries per card. With enough customers passing through the door one would expect at least one PIN to be recovered within a few hours of trading.

5   Conclusion

We have presented a variant of Bleichenbacher and Manger's method to attack PIN encryption via RSA. The attack can be considered highly theoretical, since implementing it in practice would require a lot of work and would require the card to provide a validity checking oracle for Test 2. What is interesting however is the small number of oracle queries which result in a full PIN recovery.

In some sense theoretically our method could be improved. Information theoretically we are only trying to recover 13.28 bits of information, since we are trying to recover a 10000 bit PIN number. Each oracle query returns one bit of information, i.e. does a certain multiple of the message lie in a given interval?

It would be interesting whether the attack could be improved, for example by a better choice of parameters or by using a different oracle, so as to reduce the number of oracle queries even further.

However, finally we note that PIN encryption from the keypad to the card is not implemented in many geographic locations. For example it was decided in the UK chip-and-pin system to not implement PIN encryption so as to enable cheaper cards, but we note that this comes at an expense in security as the attacks on unencrypted PINs in [3] point out.

Acknowledgments

The author was partially supported by the eCrypt-2 Network of Excellence, and a Royal Society Wolfson Merit Award.

References

1. Bellare, M., Rogaway, P.: Optimal asymmetric encryption. In: De Santis, A. (ed.) EUROCRYPT 1994. LNCS, vol. 950, pp. 92–111. Springer, Heidelberg (1995)
2. Bleichenbacher, D.: Chosen ciphertext attacks against protocols based on the RSA encryption standard PKCS #1. In: Krawczyk, H. (ed.) CRYPTO 1998. LNCS, vol. 1462, pp. 1–12. Springer, Heidelberg (1998)
3. Drimer, S., Murdoch, S.J., Anderson, R.: Thinking inside the box: system-level failures of tamper proofing. In: IEEE Symposium on Security and Privacy, pp. 281–295 (2008)
4. EMV. Integrated circuit card specifications for payment systems, Book 2. Security and Key Management. Version 4.2 (June 2008), www.emvco.com
5. EMV. Integrated circuit card specifications for payment systems, Book 3. Application Specification. Version 4.2 (June 2008), www.emvco.com
6. ISO 9564-2. Banking – Personal Identification Number management and security – Part 2: Approved algorithm(s) for PIN encipherment (2005), www.iso.org
7. Manger, J.: A chosen ciphertext attack on RSA Optimal Asymmetric Encryption Padding (OAEP) as standardized in PKCS # 1 v2.0. In: Kilian, J. (ed.) CRYPTO 2001. LNCS, vol. 2139, pp. 230–238. Springer, Heidelberg (2001)
8. Radu, C.: Implementing electronic card payment systems. Artech House Publishers (2002)

Efficient CRT-RSA Decryption for Small Encryption Exponents

Subhamoy Maitra and Santanu Sarkar

Applied Statistics Unit, Indian Statistical Institute,
203 B T Road, Kolkata 700 108, India
{subho,santanu_r}@isical.ac.in

Abstract. Consider CRT-RSA with the parameters p, q, e, d_p, d_q, where p, q are secret primes, e is the public encryption exponent and d_p, d_q are the private decryption exponents. We present an efficient method to select CRT-RSA parameters in such a manner so that the decryption becomes faster for small encryption exponents. This is the most frequently used situation for application of RSA in commercial domain. Our idea is to choose e and the factors (with low Hamming weight) of d_p, d_q first and then applying the extended Euclidean algorithm, we obtain p, q of same bit size. For small e, we get an asymptotic reduction of the order of $\frac{1}{3}$ in the decryption time compared to standard CRT-RSA parameters for large $N = pq$. In case of practical parameters, with 1024 bits N and $e = 2^{16} + 1$, we achieve a reduction of more than 27%. Extensive security analysis is presented for our selected parameters and benchmark examples are also provided.

Keywords: RSA, CRT-RSA, Key Generation, Efficient Decryption, Primes, Exponents.

1 Introduction

Till date, RSA [17] is the most important public key cryptosystem in academics as well as commercial domain. Given the wide application of RSA, an important area of research is to explore how one can implement the encryption and decryption operations of RSA efficiently. The encryption and decryption operations (modular exponentiations) are based on modular square and multiplication of large integers. The overall cost of exponentiation depends on the bit pattern of the encryption and decryption exponents. In this paper we present certain strategies to choose the RSA exponents in such a manner so that the cost of decryption gets reduced significantly compared to the existing methods.

To explain our contribution more clearly, let us first present the RSA public key cryptosystem. By l_i, we denote the number of bits in an integer i, i.e., $l_i = \lceil \log_2 i \rceil$ when i is not a power of 2 and $l_i = \log_2 i + 1$, when i is a power of 2. In RSA, we need a large integer N such that $N = pq$, where p, q are primes. For better security practice, the primes are so chosen that $l_p = l_q = \frac{l_N}{2}$. The encryption and decryption exponents are denoted by e, d and they are chosen in

J. Pieprzyk (Ed.): CT-RSA 2010, LNCS 5985, pp. 26–40, 2010.

such a manner that $ed \equiv 1 \bmod \phi(N)$, where $\phi(N) = \phi(pq) = (p-1)(q-1)$. This $\phi(.)$ is the well known Euler's totient function. In general, e, N are distributed as the public key and d is kept secret.

For encryption, the plaintext is managed in such a manner so that it can be expressed as $M_1, M_2, \ldots \in \mathbb{Z}_N$ and the encryption operation is $C_i = M_i^e \bmod N$, for $i = 1, 2, \ldots$. The ciphertext C_i can be communicated through the public channel and it can be decrypted by the valid receiver as $M_i = C_i^d \bmod N$.

The cost of modular exponentiation can be reduced if one can reduce e, d. However, as $ed > \phi(N)$, we have $l_e + l_d \geq l_N$ and one cannot make both e, d small. Consider that one likes to make the decryption process faster. Then the secret decryption exponent d has to be made small. However, Wiener [25] showed that when $d < \frac{N^{\frac{1}{4}}}{3}$ one can factor N efficiently making RSA insecure. This result has been improved by Boneh and Durfee [4] till the upper bound $N^{0.292}$.

To reduce the cost of encryption, one can take a small e. It has been pointed out by Coppersmith [6], that RSA with very small e, e.g., $e = 3$, is not secure. On the other hand, it is believed that little larger encryption exponents are quite secure, as example it is a common practice to use $e = 2^{16} + 1$. In such a case, d becomes quite large, i.e., of the order of N, and the decryption process will be much less efficient than the encryption. As an approach to make RSA decryption faster in such a scenario, Wiener [25] proposed the application of Chinese Remainder Theorem (CRT) as described earlier by Quisquater and Couvreur [16]. This is known as CRT-RSA. Our final result will explain how the designed parameters in this paper will provide substantial improvement in the decryption phase of CRT-RSA while the encryption exponent is small (this is the most popular commercial scenario). However, at this point let us motivate how we can improve the basic RSA with our idea. This will help in building the background of this work.

In the basic RSA, after choosing e, if one determines p, q first, then in most of the cases d will be $O(N)$. However, without deciding the primes first, one may choose a large factor of d, say d_1 and then try to find out the primes. In such a case $d = d_1 d_2$, where the designer has no control over the choice of d_2. For some ciphertext C, the decryption can be done as $\Psi = C^{d_1} \bmod N$ followed by $M = \Psi^{d_2} \bmod N$. Consider that d_1 is chosen in such a manner so that there are very few ones in its bit pattern. Then the number of multiplications will be reduced significantly while calculating $\Psi = C^{d_1} \bmod N$. Our strategy to find these parameters properly using Extended Euclidean Algorithm (EEA), to analyse the security of these parameters and the advantages related to the speed-up of our proposal are presented in Section 2.

For a quick reference, we present the famous square and multiply algorithm for modular exponentiation in Algorithm 1. If one looks at the step 4, it is clear to note that the number of multiplications required is equal to the number of ones in the binary pattern of y. If the number of multiplications can be reduced, then the process becomes faster and also consumption of power is less. Thus our idea above in choosing d_1 with few ones in the binary pattern provides significant improvement. Note that there may be some special purpose algorithms for

Algorithm 1. The fast square and multiply algorithm for modular exponentiation

---

    Input: x, y, N
    Output: $x^y \bmod N$
1   $z = y, u = 1, v = x$;
2   **while** $z > 0$ **do**
3   **if** $z \equiv 1 \bmod 2$ **then**
4         $u = uv \bmod N$;
      end
5      $v = v^2 \bmod N; z = \lfloor \frac{z}{2} \rfloor$;
    end
6 **return** u.

---

exponentiation that may work efficiently in some specific cases, but in general, the fast square and multiply strategy presented in Algorithm 1 is considered to be the most efficient and popular for modular exponentiation. Thus, while analysing the advantage of our proposal, we will consider this algorithm as the benchmark.

For any integer i, let the number of ones in its binary form be w_i. Calculation of $x^y \bmod N$ requires l_y many squares and w_y many multiplications. Considering the square and multiplication operation of large integers are of same time complexity [15, Section 14.18], the calculation of $x^y \bmod N$ needs $l_y + w_y$ many multiplications following Algorithm 1. In average case, one may consider $w_y = \frac{l_y}{2}$ and hence the total cost of calculating $x^y \bmod N$ is around $\frac{3}{2} l_y$ many modulo multiplications.

An alternative approach of RSA decryption is by Chinese Remainder Theorem, which is popular as CRT-RSA [16,25]. The encryption technique is similar to the standard RSA, but the decryption process is little different. Instead of one decryption exponent as in standard RSA, here one needs two decryption exponents (d_p, d_q) where $d_p \equiv d \bmod (p-1)$ and $d_q \equiv d \bmod (q-1)$. To decrypt the ciphertext C, one needs to calculate both $C_p \equiv C^{d_p} \bmod p$ and $C_q \equiv C^{d_q} \bmod q$. From C_p, C_q one can get the plaintext M by the application of CRT. The CRT-RSA decryption is 4 times faster than standard RSA on an average using the schoolbook multiplication. However, with the large size integers, it may be useful to try the Karatsuba multiplication and in such a case CRT-RSA decryption is around 3 times faster than standard RSA on an average.

Later, in our discussion, it will analyse the decryption process of RSA as well as CRT-RSA for small encryption exponent. The average number of modular multiplications for both the situations are same which is $\frac{3 l_N}{2}$. In the case of RSA, the operations are with l_N bit integers, but for CRT-RSA the operations are with $\frac{l_N}{2}$ bit integers. This provides the advantage of CRT-RSA over RSA in terms of execution efficiency. However, one should note that the secret primes p, q need to be available at the decryption side.

Boneh described the meet-in-the middle attack [3] on CRT-RSA, where one can factor N in time and space $O(\min\{\sqrt{d_p}, \sqrt{d_q}\})$. Jochemsz and May [11] pointed out that it is not secure to use the CRT decryption exponents smaller than $N^{0.073}$. So far, there is no serious threat to CRT-RSA when e is small (but not very small). In such a case, d_p, d_q will be $O(\sqrt{N})$. Similar to our approach in the standard RSA case, we will try to construct d_p, d_q in such a manner so that we will have some controlled factors of both d_p, d_q with less number of ones that will help to speed-up CRT-RSA. This is presented in Section 3 with associated efficiency measure and security analysis. We also present examples for explaining our idea. Further, in Appendix A, benchmark examples are presented as (e, N) pairs for possible attempts to analyse the security.

Since the Extended Euclidean Algorithm (EEA) is used frequently in our paper, let us briefly discuss how do we exploit it. The readers are referred to [19, Chapter 5] for a detailed discussion on EEA. Given two relatively prime positive integers a, b, one can find a unique pair of integers x_i, y_i such that $ax_i - by_i = 1$, where $(i-1)b < x_i < ib$ and $(i-1)a < y_i < ia$ for $i \geq 1$. For maintaining the bit-sizes of several parameters, we will consider the case $i = 2$. In fact, one can get x_1, y_1 such that $0 < x_1 < b$ and $0 < y_1 < a$ using EEA and then get $x_i = x_1 + (i-1)b$, $y_i = y_1 + (i-1)a$.

There are related works on optimizing RSA or CRT-RSA parameters using EEA for efficient encryption and decryption [9,20,21] and some of these proposals have been revisited [10,22] due to cryptanalytic results [7,1]. However, none of these proposals [9,10,20,21,22] consider small e around the value of $2^{16} + 1$, which is the most popular in commercial RSA applications. In [20, Section 7], it has been commented that "our variants can not provide better performance than RSA-CRT". In our strategy, we can take care of such small encryption exponents and still provide significant improvement during the decryption process. We follow similar kind of strategy has also been used in [21, Section 4.1, Scheme-A]. However, the strategy of [21] did not concentrate on small e and the low weight factors d_{p_1}, d_{q_1} of d_p, d_q. This is the idea that helped us to accommodate fast encryption with small e, as well as efficient decryption with small weight factors of the CRT-RSA decryption exponents d_p, d_q.

Remark 1. With respect to RSA/CRT-RSA public key cryptosystem, there are two aspects of efficiency, one is the efficiency in the key generation phase and the other is the efficiency during the encryption or decryption process.

Experimental results in support of our claim shows that one can write simple programs to generate the RSA keys efficiently using our Algorithms 2, 3. We have implemented the programs in SAGE 3.1.1 over Linux Ubuntu 8.04 on a Compaq laptop with Dual CORE Intel(R) Pentium(R) D CPU 1.83 GHz, 2 GB RAM and 2 MB Cache. Our strategies require some more steps of EEA than the standard RSA/CRT-RSA key generation algorithms and our proposals are in a similar line to the proposals of [9,10,21] which are efficient and unlike [20] where costly operations like factorization are required in the key generation process. However, we have not tried to present optimized implementation for the key generation part and only outlined proof of concept implementation to justify the efficiency of the key generation process.

Our main contribution in this paper is to show (by exactly counting the number of arithmetic operations) that the keys generated by our methods provide significant improvement in the decryption process without compromising security.

ROAD MAP. After the introductory discussion in this section, we get into our contributions in Sections 2, 3. Section 2 presents our proposal for RSA, while Section 3 identifies the improvements in case of CRT-RSA. Each of these sections are organized as follows. We start with our algorithm to generate the RSA/CRT-RSA keys, followed by the comparison of efficiency in the decryption process considering the keys generated from our strategies with the standard RSA/CRT-RSA decryption for small encryption exponents; then we explain a comprehensive security analysis for our proposals and conclude each section with examples and experimental details. The conclusion of this paper is presented in Section 4.

2   Our RSA Key Generation Algorithm

Let us present our proposal towards the key generation algorithm for standard RSA in Algorithm 2.

Algorithm 2. Our RSA key generation algorithm

Input: e, the encryption exponent and b, the bit size of the prime p.
Output: primes p, q and decryption exponent d such that $d_1 | d$ and w_{d_1} is low.
1 Choose a random prime p with $l_p = b$ such that $\gcd(e, p-1) = 1$;
2 Choose an odd d_1 with low weight (we will later discuss in exact detail how this weight should be determined) such that $l_{ed_1} = l_p$;
3 Choose a random integer k such that $k < e$;
4 if $\gcd(ed_1, k(p-1)) \neq 1$ go to step 2;
5 Using EEA find d_2, y with $k(p-1) < d_2 < 2k(p-1)$ and $ed_1 < y < 2ed_1$ such that $ed_1 d_2 - k(p-1)y = 1$;
6 if $y + 1$ is not prime go to step 2;
7 Report $p, q = y + 1$ and d_1, d_2 where $d = d_1 d_2$;

It is clear that Algorithm 2 is a probabilistic polynomial time algorithm in $\log_2 N$. Note that ed_1 has l_p many bits. Since $ed_1 < y < 2ed_1$, y can be of l_p or $l_p +$ 1 many bits. One can assume that y is distributed uniformly at random in $ed_1 < y < 2ed_1$ for uniformly random choices of e, d_1, k, p (we have also checked it with detailed experiments). Thus, following Prime Number Theorem [19, Chapter 5], $y + 1$ will be prime on an average of order of l_p many iterations of steps 2 to 6. If one requires some more properties on the primes, such that $\frac{p-1}{2}, \frac{q-1}{2}$ are also primes, then one may incorporate checking for the additional properties in Algorithm 2. Given that it is efficient to obtain such primes from random choices, we also assume that those kinds of primes will be captured using our algorithm. We have also confirmed that from actual implementations. It is also not guaranteed that Algorithm 2 will provide p, q of same bit size; in fact $l_q = l_p$ or $l_p + 1$. However, with a few attempts of Algorithm 2, it is possible to get p, q of same bit size.

2.1 Efficiency of Decryption Process

Now let us analyse the efficiency of our strategy in the decryption process. For the analysis, let us consider that each of p, q are of $\frac{l_N}{2}$ many bits where $N = pq$ (the analysis will change very little if $l_q = l_p + 1$).

Since, $l_{ed_1} = \frac{l_N}{2}$, we have $l_{d_1} = \frac{l_N}{2} - l_e$ or $\frac{l_N}{2} - l_e + 1$. For our analysis, let us assume $l_{d_1} = \frac{l_N}{2} - l_e$. Since $k(p-1) < d_2 < 2k(p-1)$, we have $l_{d_2} = l_k + \frac{l_N}{2}$ or $l_{d_2} = l_k + \frac{l_N}{2} \pm 1$. Let us consider $l_{d_2} = l_k + \frac{l_N}{2}$ for the analysis.

The decryption process can be written as $C^d \bmod N = (C^{d_1})^{d_2} \bmod N$. To calculate $C^{d_1} \bmod N$, we need $\frac{l_N}{2} - l_e$ many square and w_{d_1} many multiplication operations. Let $C^{d_1} \bmod N = \Psi$. To calculate $\Psi^{d_2} \bmod N$, we require $l_k + \frac{l_N}{2}$ many square and w_{d_2} many multiplications. If one assumes that d_2 is selected uniformly at random among all the $l_k + \frac{l_N}{2}$ bit integers, then on an average, $w_{d_2} = \frac{l_k}{2} + \frac{l_N}{4}$. Thus, the total number of multiplications (considering modulo square and multiplication operations are of same time complexity for large integers) under this scenario is $\left(\frac{l_N}{2} - l_e + w_{d_1}\right) + \left(l_k + \frac{l_N}{2} + \frac{l_k}{2} + \frac{l_N}{4}\right) = \frac{5}{4}l_N - l_e + w_{d_1} + \frac{3}{2}l_k \leq \frac{5}{4}l_N - l_e + w_{d_1} + \frac{3}{2}l_e$ (as $k < e$) $= \frac{5}{4}l_N + \frac{1}{2}l_e + w_{d_1}$.

As we have discussed in the introduction, on an average, RSA decryption requires $\frac{3}{2}l_N$ many modular multiplications. Thus, in asymptotic sense, i.e., when w_{d_1} and l_e are negligible with respect to l_N, we get a reduction of $1 - \frac{\frac{5}{4}l_N}{\frac{3}{2}l_N} = \frac{1}{6}$, which is more than 16%.

In practical scenario, one can take $l_N = 1024$, for $e = 2^{16} + 1$, $l_e = 17$ and $w_{d_1} = 40$. Later, in Section 2.2, we will discuss that taking a low weight d_1 will not pose any security problem given the state-of-the-art literature in RSA cryptanalysis. Thus, in such a case the advantage is $1 - \frac{\frac{5}{4} \cdot 1024 + \frac{1}{2} \cdot 17 + 40}{\frac{3}{2} \cdot 1024} > 0.13$.

Since all the exponentiation operation are modular, one needs to study how the $\bmod N$ part in the calculation of $v^2 \bmod N$ or $uv \bmod N$ can be done efficiently, where $u, v \in \mathbb{Z}_N$. It has been pointed out by A. K. Lenstra [12] that the operation becomes efficient when N is of the form $N = 2^{l_N-1} + t$ for some positive integer t which is significantly smaller than N. It has been pointed out that one may get 30% improvement for encryption and decryption with 1024-bit RSA moduli [12].

One should note, that the idea of putting large number of zeros in N [12] works well for application in standard RSA. In the proposal of [12], N of the form $2^{1024} \pm t$ has been considered where t is not much smaller than 2^{500}. Thus, in such a case, the upper half of the bits in N are zero (except the MSB). The CRT-RSA decryption can be made faster in this manner when one can put a large number of zeros in the secret primes. However, it is well known that if half of the bits of p is known from any side, then N can be factorized easily [6]. Putting large number of zeros in p, q is not recommendable and hence the idea of [12] cannot be exploited for CRT-RSA. Our work provides improvement from a different direction that that of [12], where we do not use any constraint on N. Thus, our strategy can be exploited to achieve efficiency in the CRT-RSA case as described in Section 3.

2.2 Security Analysis

Below we discuss the existing attacks to show that the RSA parameters obtained from our scheme are quite secure.

Wiener's Attack and its extensions: Here, $|\frac{e}{N} - \frac{k}{d}| = \frac{k}{d} \times \frac{p+q-1-\frac{1}{k}}{N} > \frac{k}{d} \cdot \frac{q}{N}$. Wiener's [25] attack will be successful if $|\frac{e}{N} - \frac{k}{d}| < \frac{1}{2d^2}$. Thus, Wiener's attack fails if $\frac{k}{d} \cdot \frac{q}{N} >> \frac{1}{2d^2}$ i.e., when $2kd >> \frac{N}{q}$, which is true in our case as we consider d as $O(N)$. Similar kind of attacks by Verheul and Tilborg [23] and Weger [24] do not work too.

Boneh-Durfee attack: Since for our method $e << p + q$ so Boneh-Durfee attack [4] does not work.

We have $ed = 1+k(N+1-p-q)$. In our case $d = d_1 d_2$. So $1+k(N+1-p-q) \equiv 0 \bmod (ed_1)$. Hence $p+q \equiv (k^{-1}+N+1) \bmod ed_1$. As both $p+q$ and ed_1 are of $O(\sqrt{N})$, one can find out $p+q$ easily when both d_1 and k are known. Since $k < e$ and e is small integer one can try every integer in $[1, \ldots, e-1]$ as k. However, the attack won't be successful unless d_1 is known. By proper choice of d_1, we will guarantee that it cannot be exhaustively searched in a complexity less than factoring N with the state of the art knowledge [13].

Number Field Sieve (NFS) [13] is the fastest known factorization algorithm that requires around 2^{86} time complexity to factor a 1024-bit RSA modulus. When $l_N = 1024$ and $e = 2^{16} + 1$ then $l_{d_1} = \frac{1024}{2} - l_e = 512 - 17 = 495$. Now consider $w_{d_1} = 40$. Each of the MSB and LSB of d_1 is 1. Since, $\binom{493}{38} > 2^{190}$, searching d_1 is impractical.

There are certain partial key exposure attacks [2,8] based on knowledge of some bits of the secret decryption exponent d. In a similar line, consider that either the lower half or the upper half in the bit pattern of d_1 may be available to the attacker. Even in such a scenario, we like that the other half of d_1 cannot be exhaustively searched with a complexity less than factoring the modulus N. Since, $\binom{246}{19} > 2^{94}$, this is taken care of. Further, the lattice based attack presented in [8] won't work as long as d_2 is not known to the adversary.

2.3 Examples and Particulars of Implementation for Key Generation

For experimental purposes, we use the platform as described in Remark 1.

Example 1. The inputs to Algorithm 2 are $l_p = b = 512$, $e = 2^{16} + 1$.

We randomly choose the prime p of 512 bits with the constraint that $\gcd(e, p-1) = 1$ as follows:

84720546042485958248788975205816609909142885193053162949602876424216014501173987574462025540409529224589197090368025968335187937773811404742000874986542199.

Then we have randomly chosen d_1 such that $l_{ed_1} = 512$ and $w_{d_1} = 40$ with the constraint that the number of ones in the most and least significant halves in the binary pattern of d_1 are same, i.e., 20. The chosen d_1 is:

1022934635466485045802894351683149486306996424256878080946212893712733116183457234861106118690881613705738188223497430582527961276691228722010826997799.

Next select a random integer $k = 21779$, less than e, such that $\gcd(ed_1, k(p - 1)) = 1$. Following Algorithm 2, we get a 512-bit prime $q = y + 1$ as

8133479254716248646809517452068728838775076046279893270182604728283118990484688165461622349951654469352370222634999637399535924902445388498746028926875923 and d_2 as

2238559300016443763252531334422479959665311950574647993856180096046662459357873326043966651922700841484372085625172819478099250703566362251596507475150851443375.   □

We have also experimented with 100 many runs of Algorithm 2, with $e = 2^{16} + 1$ and $l_p = b = 512$. We found that in each run, on an average,

the loop upto step 3, i.e., random choice of k is executed 441 times and among them

the loop upto step 5, i.e., when $\gcd(ed_1, k(p - 1)) = 1$, and y is selected by EEA is executed 287 times,

before finding a prime $q = y+1$. The average time for execution of Algorithm 2 is 3.01 seconds for generation of RSA parameters when $l_p = 512$, i.e., with RSA moduli of 1023 or 1024 bits.

3   Our CRT-RSA Key Generation Algorithm

Like the discussion after Algorithm 2 in the previous section, one can note that Algorithm 3 is a probabilistic polynomial time algorithm in $\log_2 N$ and one may put further constraints on the choice of the primes p, q. This will be demonstrated in Example 3.3, Section 3.3 later.

3.1   Efficiency of Decryption Process

Now, we compare the decryption process with the keys available from Algorithm 3 with that of standard CRT-RSA. For the analysis, consider that p, q are of $\frac{l_N}{2}$ many bits where $N = pq$. Let the encryption exponent e has l_e many bits. Then $\frac{l_N}{2} - l_e \leq l_{d_p}, l_{d_q} \leq \frac{l_N}{2}$ and in most of the cases, the corresponding decryption exponents d_p, d_q will be of $O(\sqrt{N})$ and thus $\frac{l_N}{2}$ bit long. To

Algorithm 3. Our CRT-RSA key generation algorithm

Input: e, the encryption exponent and b, such that the bit size of primes p, q are b or $b + 1$.

Output: primes p, q and decryption exponents d_p, d_q such that $d_{p_1}|d_p$, $d_{q_1}|d_q$ and $w_{d_{p_1}}, w_{d_{q_1}}$ are low.

1 Choose an odd d_{p_1} with low weight such that $l_{ed_{p_1}} = b$;
2 Choose a random integer k_p such that $k_p < e$;
3 if $\gcd(ed_{p_1}, k_p) \neq 1$ go to step 2;
4 Using EEA find d_{p_2}, y with $k_p < d_{p_2} < 2k_p$ and $ed_{p_1} < y < 2ed_{p_1}$ such that $ed_{p_1}d_{p_2} - k_p y = 1$;
5 if $y + 1$ is not prime go to step 2;
6 Report $d_p = d_{p_1}d_{p_2}, p = y + 1$;
7 In a similar manner from step 1 to step 5, generate $q, d_q = d_{q_1}d_{q_2}$;

---

calculate $C^{d_p} \bmod p$ one needs $\frac{l_N}{2}$ many squares and w_{d_p} many multiplications. In average case $w_{d_p} = \frac{1}{2} \times \frac{l_N}{2} = \frac{l_N}{4}$. Hence total cost of calculating $C^{d_p} \bmod p$ is around $\frac{l_N}{2} + \frac{l_N}{4} = \frac{3l_N}{4}$ many modular multiplications. Similarly to calculate $C^{d_q} \bmod q$ one needs $\frac{3l_N}{4}$ many modular multiplications. So total cost in decryption will be $2 \times \frac{3l_N}{4} = \frac{3l_N}{2}$ many modular multiplications.

In Algorithm 3 we have $l_{d_{p_1}} = \frac{l_N}{2} - l_e$ or $l_{d_{p_1}} = \frac{l_N}{2} - l_e + 1$. For our analysis we take $l_{d_{p_1}} = \frac{l_N}{2} - l_e$. Since $k_p < d_{p_2} < 2k_p$, $l_{d_{p_2}}$ is l_{k_p} or $l_{k_p} + 1$. For our analysis we take $l_{d_{p_2}} = l_{k_p}$. Now $C^{d_p} \bmod p = (C^{d_{p_1}})^{d_{p_2}} \bmod p$. To calculate $C^{d_{p_1}} \bmod p$, we need $\frac{l_N}{2} - l_e$ many square and $w_{d_{p_1}}$ many multiplication operations. Let $C^{d_{p_1}} \bmod p = \Psi$. To calculate $\Psi^{d_{p_2}} \bmod p$, one requires l_{k_p} many square and $w_{d_{p_2}}$ many multiplication operations. If we assume d_{p_2} is selected uniformly at random among all the l_{k_p} bit integers, then on an average $w_{d_{p_2}} = \frac{l_{k_p}}{2}$. Considering modular square and multiplication operations of large integers take equal amount of time, to calculate $C^{d_p} \bmod p$ in our proposed method requires $(\frac{l_N}{2} - l_e + w_{d_{p_1}}) + (l_{k_p} + \frac{l_{k_p}}{2}) = \frac{l_N}{2} - l_e + w_{d_{p_1}} + \frac{3l_{k_p}}{2} \leq \frac{l_N}{2} - l_e + w_{d_{p_1}} + \frac{3}{2}l_e$ (as $k_p < e$) $= \frac{l_N}{2} + \frac{l_e}{2} + w_{d_{p_1}}$ many multiplication operations. Similarly total number of multiplications in our proposed method to calculate $C^{d_q} \bmod q$ will be $\leq \frac{l_N}{2} + \frac{l_e}{2} + w_{d_{q_1}}$. Hence total number of multiplications in our decryption algorithm will be $\leq l_N + l_e + w_{d_{p_1}} + w_{d_{q_1}}$.

Thus, in asymptotic sense, i.e., when $l_e, w_{d_{p_1}}, w_{d_{q_1}}$ are negligible with respect to l_N, we get a reduction of $1 - \frac{l_N}{\frac{3}{2}l_N} = \frac{1}{3}$, which is 33%.

In practical scenario, one can take $l_N = 1024$, for $e = 2^{16} + 1, l_e = 17$ and $w_{d_{p_1}} = w_{d_{q_1}} = 40$. Later, in Section 3.2, we will discuss that taking a low weight d_{p_1}, d_{q_1} will not pose any security problem given the state-of-the-art literature in CRT-RSA cryptanalysis. Thus in such a case the advantage is $1 - \frac{1024 + 17 + 40 + 40}{\frac{3}{2} \cdot 1024} > 0.27$.

Application of CRT in the final step is considered negligible [5] compared to the calculations $C^{d_p} \bmod p$ and $C^{d_q} \bmod q$. Thus we also ignore this in our comparison.

3.2   Security Analysis

Below we discuss the existing attacks to show that the CRT-RSA parameters obtained from our scheme are quite secure.

Let $N = pq$ be an l_N-bit RSA modulus with $l_p = l_q = \frac{l_N}{2}$. Further let $ed_p \equiv 1 \bmod (p-1)$ and $ed_q \equiv 1 \bmod (q-1)$ with $l_e = \alpha \cdot l_N, l_{d_p} = l_{d_q} = \beta \cdot l_N$, where $0 \leq \alpha \leq 1, 0 < \beta \leq \frac{1}{2}$. Then N can be factored in polynomial time if any of the following holds: (i) $\beta < \frac{3}{8} - \frac{\alpha}{2}$ [14], (ii) $\beta < \frac{2}{5} - \frac{2}{5}\alpha$ [1]. For our case, e is small, e.g., $e = 2^{16} + 1$. Hence d_p, d_q are $O(\sqrt{N})$, i.e., $\beta = \frac{1}{2}$. Thus the attacks presented in [14,1] will not work in this case.

The following results have been noted in [2] towards CRT-RSA cryptanalysis. (i) Suppose an attacker knows the bits of d_p except $\delta \cdot l_N$ many LSBs. Then N can be factored in polynomial time of l_N if $\delta < \frac{1}{4} - \alpha$. (ii) Suppose an attacker knows the bits of d_p, except $\delta \cdot l_N$ many MSBs. Moreover e is small such that $k_p, k_q < e$ can be searched easily. Then N can be factored in polynomial time of l_N if $\delta < \frac{1}{4}$.

Now l_{d_p} is $\frac{l_N}{2} + l_{k_p} - l_e$ or $\frac{l_N}{2} + l_{k_p} - l_e \pm 1$. For the analysis, let us take $l_{d_p} = \frac{l_N}{2}$ considering e to be small. To resist the attack of [2], we need that attacker can not guess the bits either in the most significant half or in the least significant half of d_p as well as d_q. It is evident from Algorithm 3 that d_{p_2} or d_{q_2} are very small for small e and thus the security depends on proper choices of d_{p_1} and d_{q_1}. When $e = 2^{16} + 1$ and $l_N = 1024$, then $l_{d_{p_1}} \geq 495$; also both the MSB and LSB of d_{p_1} are 1. Since $\binom{246}{19} > 2^{94}$, one can choose $w_{d_{p_1}} = w_{d_{q_1}} = 38 + 2 = 40$ (the term 2 is added as the MSB and LSB of d_p, d_q will be 1) to resist the exhaustive search in a complexity lesser than factorization of 1024 RSA moduli. Thus, we take equal weights in the most significant as well as least significant halves of d_p, d_q.

Additionally, this kind of choice resists the attack on the CRT exponents with low Hamming weight as explained in [10, Section 6] following the idea of [18].

Since we are concentrating on small e, the attacks of [11,9] will not be applicable in our case.

In our strategy, when e is small then d_p (respectively d_q) has a small factor d_{p_2} (respectively d_{q_2}). In any method, when e is chosen first, then the choice of d_p, d_q cannot be controlled and thus they may well have small factors. We do not know about any result that small factors of d_p, d_q may pose any security problem.

One may be tempted to choose $d_{p_1} - d_{q_1}$ small for storage advantage. However, we like to point out that this will not be secure. We have $ed_p = 1 + k_p(p-1), ed_q = 1 + k_q(q-1)$. So $ed_{p_1}d_{p_2} - 1 + k_p = k_p p, ed_{q_1}d_{q_2} - 1 + k_q = k_q q$, as $d_p = d_{p_1}d_{p_2}$ and $d_q = d_{q_1}d_{q_2}$. Multiplying the above two equations and letting $c = d_{p_1} - d_{q_1}$ we get $e^2 d_{p_2}d_{q_2}d_{p_1}^2 + (ek_p d_{q_2} + ed_{p_2}k_q - e^2 d_{p_2}cd_{q_2} - ed_{p_2} - ed_{q_2})d_{p_1} - ek_p cd_{q_2} + ecd_{q_2} - k_p k_q N + k_p k_q - k_p - k_q + 1 = 0$. When the attacker knows $d_{p_2}, k_p, d_{q_2}, k_q$ and c, he can easily find out d_{p_1} by solving the equation $f(x) = e^2 d_{p_2}d_{q_2}x^2 + (ek_p d_{q_2} + ed_{p_2}k_q - e^2 d_{p_2}cd_{q_2} - ed_{p_2} - ed_{q_2})x - ek_p cd_{q_2} + ecd_{q_2} - k_p k_q N + k_p k_q - k_p - k_q + 1 = 0$. For small e, $d_{p_2}, k_p, d_{q_2}, k_q$ will be small. So the attacker may find these values by exhaustive search. Thus, in the actual design $c = d_{p_1} - d_{q_1}$ should be large enough so that finding c becomes impractical by exhaustive search.

3.3 Examples and Particulars of Implementation for Key Generation

For experimental purposes, we use the platform as described in Remark 1.

Example 2. We present a CRT-RSA instance using Algorithm 3 with $e = 2^{16} + 1$.

We have randomly chosen d_{p_1} such that $l_{ed_{p_1}} = 512$ and $w_{d_{p_1}} = 40$ with the constraint that the number of ones in the most and least significant halves in the binary pattern of d_{p_1} are same, i.e., 20. The chosen d_{p_1} is:

10229345649675461834170907801542126815396904650789266492292003495592584214143116198136429047108369066477735535674622465278585082570500337422339735533.

Next we have selected a positive random number $< e$ as $k_p = 54515$ such that $\gcd(ed_{p_1}, k_p) = 1$. Following Algorithm 3, we get a 512-bit prime $p = y + 1$ as

925033577483624877758165052473985025208424615264973702445966391820817633431628968129548056952716209571439174251182484580505109292574254840671505548670 3856 and $d_{p_2} = 75221$.

Similarly we take d_{q_1}, k_q as follows:

10229345653359679953397964558314197774154672247859695390447369735707009535744449233668871233464814577890983659906341326067390621178733052093151838412 9 and 11196.

Note that $l_{ed_{q_1}} = 512$ and $w_{d_{q_1}} = 40$ with the constraint that the number of ones in the most and least significant halves in the binary pattern of d_{q_1} are same, i.e., 20.

Following Algorithm 3, we get a 512-bit prime $q = y + 1$ as

76890089670843518975563793321177463010696968831203842284895894791074650554894446689992429876100396605578159767485698183082095701493282016772115064393 03703 and $d_{q_2} = 12841$.   □

We have experimented with 100 many runs of Algorithm 3, with $e = 2^{16} + 1$ and $b = 512$. We found that in each run, on an average,

the loop upto step 2, i.e., random choice of k_p is executed 404 times and among them

the loop upto step 4, i.e., when $\gcd(ed_{p_1}, k_p) = 1$, and y is selected by EEA is executed 328 times,

before finding the prime $p = y + 1$. The average time for execution of Algorithm 3 is 2.87 seconds for the generation of one prime p when $l_p = 512$, i.e., with RSA moduli of 1023 or 1024 bits. Similar effort is required to generate q.

As we have discussed earlier that one may incorporate some other properties while generating the primes. Below we show how Algorithm 3 runs when we put additional constraints on the primes p, q such that $\frac{p-1}{2}, \frac{q-1}{2}$ are primes too. These primes $\frac{p-1}{2}, \frac{q-1}{2}$ are well known as Sophie-Germain primes.

In these cases the algorithm has to try more to get such primes. With $e = 2^{16} + 1$ and $b = 512$, we find that for Example 3.3,

the loop upto step 2, i.e., random choice of k_p is executed 77548 (respectively 99405) times and among them,

the loop upto step 4, i.e., when $\gcd(ed_{p_1}, k_p) = 1$, and y is selected by EEA is executed 62907 (respectively 80752) times,

before finding the prime $p = y + 1$ (respectively $q = y + 1$). The time for execution of Algorithm 3 in this case took 501.59 and 673.29 seconds for the generation of p, q respectively.

Example 3. We present a CRT-RSA instance using Algorithm 3 with $e = 2^{16} + 1$ with the constraint on the primes p, q such that $\frac{p-1}{2}, \frac{q-1}{2}$ are also primes.

We have randomly chosen d_{p_1} such that $l_{ed_{p_1}} = 512$ and $w_{d_{p_1}} = 40$ with the constraint that the number of ones in the most and least significant halves in the binary pattern of d_{p_1} are same, i.e., 20. The chosen d_{p_1} is:

102293651756383964719249763767825082839022557563711539885076457266765922065218356942848525482843805498444447334594562620436894433026411062425379688577.

Next we have selected a positive random number $< e$ as $k_p = 33590$ such that $\gcd(ed_{p_1}, k_p) = 1$. Following Algorithm 3, we get a 512-bit prime $p = y + 1$ as

72786207479178939560622134536792692481599472401734597962485394473617964385205069834158547738470308242281099590715490777173691057946576945718295285214847502 and $d_{p_2} = 36469$. One may check that $\frac{q-1}{2}$ is also prime.

Similarly we take d_{q_1}, k_q as follows:

102295029660628890489918025067861937988247526555436922234444277690560935916262878224630012547147030418168005660727122770496341597038718898720334685185 and 13004.

We have $l_{ed_{q_1}} = 512$ and $w_{d_{q_1}} = 40$ with the constraint that the number of ones in the most and least significant halves in the binary pattern of d_{q_1} are same, i.e., 20.

Following Algorithm 3, we get a 512-bit prime $q = y + 1$ as

5316785590434653868189587374677157853747130116803908359041341335613341278477953871146627419673515278301454171147161522484491176394570117314674466654414246 and $d_{q_2} = 10313$. One can verify that $\frac{q-1}{2}$ is a prime too. □

Some benchmark examples of (e, N) pairs, generated from Algorithm 3, are presented in Appendix A so that one may use them for cryptanalytic attempt on our proposal.

4   Conclusion

Use of small encryption exponent involving CRT in the decryption phase is the most popular scenario in commercial RSA applications. In this paper we have presented a proposal for making the CRT-RSA decryption process efficient for small encryption exponents. The decryption exponents designed by our strategy make the modular exponentiations faster as well as less power consuming. We first introduce our method for standard RSA and then based on that idea we explain how the CRT-RSA parameters can be chosen for efficient decryption. We obtain an asymptotic reduction of the order of one-third in the decryption time compared to standard CRT-RSA parameters for large N; further we provide examples for a reduction of more than 27% for practical parameters with 1024 bits N and $e = 2^{16} + 1$.

Extensive security analysis is presented to study our algorithms. It is evident from the research history in cryptology that even after a careful security analysis, the security claims are mostly conjectures and the weaknesses are often identified at a later stage. This is the reason, we present some benchmark challenges that can be studied to cryptanalyse our proposals.

Acknowledgments. The second author likes to acknowledge the Council of Scientific and Industrial Research (CSIR), India for supporting his research fellowship. The authors also like to thank Mr. Sourav Sen Gupta of University of Washington, USA for comments and suggestions on the working draft of this paper.

References

1. Bleichenbacher, D., May, A.: New Attacks on RSA with Small Secret CRT-Exponents. In: Yung, M., Dodis, Y., Kiayias, A., Malkin, T.G. (eds.) PKC 2006. LNCS, vol. 3958, pp. 1–13. Springer, Heidelberg (2006)
2. Blömer, J., May, A.: New Partial Key Exposure Attacks on RSA. In: Boneh, D. (ed.) CRYPTO 2003. LNCS, vol. 2729, pp. 27–43. Springer, Heidelberg (2003)
3. Boneh, D.: Twenty Years of Attacks on the RSA Cryptosystem. Notices of the AMS 46(2), 203–213 (1999)
4. Boneh, D., Durfee, G.: Cryptanalysis of RSA with Private Key d Less Than $N^{0.292}$. IEEE Transactions on Information Theory 46(4), 1339–1349 (2000)
5. Boneh, D., Shacham, H.: Fast variants of RSA. CryptoBytes 5(1), 1–9 (2002)
6. Coppersmith, D.: Small Solutions to Polynomial Equations and Low Exponent Vulnerabilities. Journal of Cryptology 10(4), 223–260 (1997)
7. Durfee, G., Nguyen, P.: Cryptanalysis of the RSA schemes with short secret exponents from Asiacrypt 1999. In: Okamoto, T. (ed.) ASIACRYPT 2000. LNCS, vol. 1976, pp. 14–29. Springer, Heidelberg (2000)

8. Ernst, M., Jochemsz, E., May, A., de Weger, B.: Partial Key Exposure Attacks on RSA up to Full Size Exponents. In: Cramer, R. (ed.) EUROCRYPT 2005. LNCS, vol. 3494, pp. 371–386. Springer, Heidelberg (2005)
9. Galbraith, S., Heneghan, C., McKee, J.: Tunable Balancing RSA. In: Boyd, C., González Nieto, J.M. (eds.) ACISP 2005. LNCS, vol. 3574, pp. 280–292. Springer, Heidelberg (2005)
10. Galbraith, S., Heneghan, C., McKee, J.: Tunable Balancing RSA, http://www.isg.rhul.ac.uk/~sdg/full-tunable-rsa.pdf
11. Jochemsz, E., May, A.: A Polynomial Time Attack on RSA with Private CRT-Exponents Smaller Than $N^{0.073}$. In: Menezes, A. (ed.) CRYPTO 2007. LNCS, vol. 4622, pp. 395–411. Springer, Heidelberg (2007)
12. Lenstra, A.: Generating RSA moduli with a predetermined portion. In: Ohta, K., Pei, D. (eds.) ASIACRYPT 1998. LNCS, vol. 1514, pp. 1–10. Springer, Heidelberg (1998)
13. Lenstra, A.K., Lenstra Jr., H.W.: The Development of the Number Field Sieve. Springer, Heidelberg (1993)
14. May, A.: Cryptanalysis of unbalanced RSA with small CRT-exponent. In: Yung, M. (ed.) CRYPTO 2002. LNCS, vol. 2442, pp. 242–256. Springer, Heidelberg (2002)
15. Menezes, A., Van Oorschot, P., Vanstone, S.: Handbook of Applied Cryptography. CRC Press, Boca Raton (1997)
16. Quisquater, J.-J., Couvreur, C.: Fast decipherment algorithm for RSA public-key cryptosystem. Electronic Letters 18, 905–907 (1982)
17. Rivest, R.L., Shamir, A., Adleman, L.: A Method for Obtaining Digital Signatures and Public Key Cryptosystems. Communications of ACM 21(2), 158–164 (1978)
18. Stinson, D.R.: Some baby-step-giant-step algorithms for the low Hamming weight discrete logarithm problem. Math. Comp. 71(237), 379–391 (2001)
19. Stinson, D.R.: Cryptography - Theory and Practice, 2nd edn. Chapman & Hall/CRC, Boca Raton (2002)
20. Sun, H.M., Yang, C.T.: RSA with Balanced Short Exponents and Its Application to Entity Authentication. In: Vaudenay, S. (ed.) PKC 2005. LNCS, vol. 3386, pp. 199–215. Springer, Heidelberg (2005)
21. Sun, H.-M., Wu, M.-E.: Design of Rebalanced RSA-CRT for Fast Encryption. In: Proceedings of Information Security Conference, pp. 16–27 (2005), http://eprint.iacr.org/2005/053
22. Sun, H.-M., Hinek, M.J., Wu, M.-E.: On the Design of Rebalanced RSA-CRT, http://www.cacr.math.uwaterloo.ca/techreports/2005/cacr2005-35.pdf
23. Verheul, E., van Tilborg, H.: Cryptanalysis of less short RSA secret exponents. Applicable Algebra in Engineering, Communication and Computing 18, 425–435 (1997)
24. de Weger, B.: Cryptanalysis of RSA with small prime difference. Applicable Algebra in Engineering, Communication and Computing 13, 17–28 (2002)
25. Wiener, M.: Cryptanalysis of Short RSA Secret Exponents. IEEE Transactions on Information Theory 36(3), 553–558 (1990)

Appendix A: Benchmark Examples for Cryptanalytic Attempts

Below we present five different instances of N, given $e = 2^{16} + 1$. In all the cases, $w_{d_{p_1}} = w_{d_{p_2}} = 40$. These are presented as benchmarks so that one can try breaking CRT-RSA with the parameters chosen using our method in Algorithm 3.

85539835223707341271621812163519410252729240171560768323703987992231749685629965040514306583348280769654296802388032299435231999565296767233509979621723342497847662758992960420066173561859951862187542930497870718832189377583897278257801045584630307063612520069497549910437863841930016986937230562360427858251,

93041224225689554448199217082651898681406610874528582227326464135395627947017918249012610535836460090732466762862310712682359588133950479766435968716987106097585679455116271461963736569314139691003186344452789816334794620233599012578457687663361995136200118797857618932225959797689464988606510585860094234199,

74315398043752754412733586761112629224413448565513873207304590111223052751429469219052501838845799724046701440989702817272498309144105828100390634561990942328166147292792875134564896932764532925096618617384028067767404358445689193218573902617312942864472583723202660031594832645765609847087222538519252259113,

95289938549834934221894274094422728525358726357543016192496246393595277330182410413362373270424158090647491451049590221264490306809480183070528314336011063897581452472933276307508441023185399242013822173246890794014845224947192109540515410576187716302639672044663571312528455407908594733169616486132276307403,

61271526025326656965932530126354126698475570689048653733934483622548513582611826324067251533323158434557431470052384859563350594358849967511810056212226508827263396422089447048813293204449214341135563834953269811587856684431737954178728734582340374048123166288438604736925222636169070692181760562485925774758

1.

Resettable Public-Key Encryption:
How to Encrypt on a Virtual Machine

Scott Yilek

Department of Computer Science and Engineering,
University of California at San Diego,
9500 Gilman Drive, La Jolla, CA 92093, USA
syilek@cs.ucsd.edu

Abstract. Typical security models used for proving security of deployed cryptographic primitives do not allow adversaries to rewind or reset honest parties to an earlier state. Thus, it is common to see cryptographic protocols rely on the assumption that *fresh* random numbers can be continually generated. In this paper, we argue that because of the growing popularity of virtual machines and, specifically, their state snapshot and revert features, the security of cryptographic protocols proven under these assumptions is called into question. We focus on public-key encryption security in a setting where resetting is possible and random numbers might be reused. We show that existing schemes and security models are insufficient in this setting. We then provide new formal security models and show that making a simple and efficient modification to any existing PKE scheme gives us security under our new models.

1 Introduction

In the past few decades, cryptographers have modeled numerous cryptographic primitives and protocols in order to argue about their security. Because of this, we have strong tools to securely execute just about any desirable task. These tools include symmetric and public-key encryption, message authentication codes, digital signatures, key exchange protocols, and more. Moreover, the security guarantees are provable by reductions from problems conjectured to be difficult.

A typical security model has an adversary playing a game against an environment which may contain multiple honest parties. As the adversary interacts with the environment and time progresses, the states of the honest parties continually change to reflect events that take place. For example, if an adversary is executing an interactive protocol with an honest party (call her Alice) and the adversary sends a message to Alice, then Alice's next message will be a function of her current state, which will itself be a function of the past messages she has received from the adversary. This essentially means that an adversary cannot 'take back' messages and try others, effectively rewinding the protocol and resetting Alice's state.

J. Pieprzyk (Ed.): CT-RSA 2010, LNCS 5985, pp. 41–56, 2010.
© Springer-Verlag Berlin Heidelberg 2010

Modeling security of protocols in this way is natural because it fits our understanding of how the world works. The complex states of our computers are constantly changing every time we click a mouse or receive a packet from the network. Thus, it seems perfectly reasonable that protocols proven secure in this model will continue to be secure in the foreseeable future. However, as we argue in this paper, this may not be the case because of the increasing popularity of *virtual machines*.

VIRTUAL MACHINES. In the past few years, it has become common for systems to run on virtual machines. In short, a virtual machine (VM) is software that emulates a real machine. A VM consists of a virtual machine monitor (or hypervisor) which can emulate multiple virtual computers that can have varying instruction sets and run multiple operating systems. The VM monitor will then share the physical machine's resources among the virtual machines, translating machine instructions and acting as a simulator for the underlying operating systems.

It is especially common for servers to run on virtual machines. This will likely become even more typical in the future given the rising popularity of cloud computing services like Amazon's Elastic Compute Cloud (EC2) [1]. In this service, a user buys some compute time and receives access to a virtual machine on one of Amazon's servers. Within that VM, the user can run a fully functional OS and, in particular, run a web server for his or her business.

Thus far, virtual machines have often been seen as being *beneficial* to security. Because VMs provide a type of sandbox, they have been used to test potentially malicious code [13] and isolate web browsers from the rest of the system to mitigate the effects of browser vulnerabilities [14]. VMs have also been used to more easily create large honeypots [22]. Despite this, the focus of this paper is on how VMs can be *detrimental* to security. The reason, which also happens to be one of the most useful features of VMs, is the ability to take state snapshots.

STATE SNAPSHOTS. Virtual machines allow a user to take a snapshot of the current system state. This snapshot contains the contents of all the virtual machine's disks and the contents in memory at the time of the snapshot. At a later point in time, the VM can be reverted back to this previous state and restarted. To see why this may be useful, consider the following scenario. Alice, a system administrator, is running an important web server on a virtual machine, and at some point in time there is a crash or some other major problem. Instead of spending time diagnosing the problem and getting the system working again, Alice can instead revert the VM back to a 'good state' for which she has a snapshot. In other words, Alice takes a snapshot of the system when things are running smoothly, and then reverts back to this state whenever things go wrong. In this scenario, the server has effectively traveled back in time; program variables and other state that may have been in memory are now active again. Thus, if an adversary is attacking Alice's server and can make it crash (using, for example, a DoS attack), he essentially has the ability to rewind the server.

Virtual machine state snapshots can also help protect against malware when web browsing. A user who is concerned about their machine being compromised from visiting a malicious website can run a web browser inside of a virtual machine and take a snapshot of the fresh machine state with a browser window open and ready for a URL. Then, if the user visits a malicious site, he can simply "blow away" the current state of the machine and revert back to the fresh, uncompromised state captured in the snapshot to visit another website. Thus, every time the user wants to visit an important or potentially malicious website, he can do so starting from a fresh state.

Though seemingly useful, state snapshots on VMs raise some important issues. What happens to our supposedly secure cryptographic tools in a setting with resets? Are they still secure? Researchers have examined these questions before for zero-knowledge [12,21,3] and identification protocols [8], where the motivation was smart cards that cannot keep internal state. However, the growing popularity of virtual machines means we need to ask these questions for a wider range of cryptographic primitives.

To see why reset attacks can have negative effects on cryptographic protocols, consider a common assumption in cryptography: it is possible to continually generate fresh and unbiased random numbers. This is an assumption made in nearly every cryptographic protocol. It is, however, considered reasonable since pseudorandom number generators (PRNGs) are well-studied both in theory and practice (c.f., [20,16]). In deployed systems, PRNGs are often implemented in software and consist of numerous state variables and arrays that are occasionally seeded with entropy and used to generate pseudorandom numbers. For example, in OpenSSL [2], the software PRNG has a 1023-byte array (entropy pool) that is supposed to contain high entropy data from a variety of sources, as well as some variables with counters and other important state. At a high level, when random bytes are requested, data from the entropy pool and information in the state variables is continually mixed together using a cryptographic hash function and the result is the output of the PRNG. However, these arrays and variables will be captured by a state snapshot since they reside in memory. If the machine is later reset, the PRNG could output a string of "random" bytes that it already outputted sometime in the past before the machine was reset. These un-fresh coins might then be used in a cryptographic operation with potentially disastrous consequences.

Garfinkel and Rosenblum point out that this threat exists in theory [17]. Then, in recent work [24], Ristenpart and Yilek show the threat is in fact a problem in practice. They demonstrate attacks on both servers and clients run inside of virtual machines utilizing snapshots and resets. Attacking servers, they show that resetting a web server running Apache with mod_ssl leads to randomness reuse in DSA signing and thus secret key compromise. They then show that if a client runs a web browser inside of a virtual machine (as described above) to protect against malware, in particular resetting the virtual machine between browsing sessions, then the browser will send the same secret random keying material to two different websites. So, if from a saved state a user (Alice) visits a malicious site inside of the virtual machine, then resets the machine back to the saved state and visits her bank, the same secret key material will be sent by the browser to both the malicious site and the bank! An adversary in control of the malicious site can then compromise Alice's banking session.

OUR RESULTS. Due to the danger posed by virtual machines, we propose building cryptographic primitives that are more resilient in the face of randomness reuse. In this paper, we focus on one particular primitive, public-key encryption, and make the following contributions. First, we provide formal security definitions to model public-key encryption security in the face of resetting attacks. Second, we show that existing PKE schemes and their common security notions IND-CPA [18] and IND-CCA [23] are insufficient when such resetting attacks are possible. Third, we show that, perhaps somewhat surprisingly, a small and efficient modification can be made to *any* existing PKE scheme secure under the typical notions (e.g., IND-CCA) in order to ensure security against resetting attacks. Our modification does not rely on random oracles [10], requires no extra assumptions, and is very efficient.

A CLOSER LOOK. The generally accepted "right" notion of security for public-key encryption is indistinguishability under chosen-ciphertext attack (IND-CCA). Though this is a strong notion of security, it fails to suffice in a setting where randomness may be reused. At a high level, the reason is that for many schemes, given a ciphertext and the corresponding plaintext it is often possible to learn some of the coins (or some useful function of the coins) used to encrypt the message. If another ciphertext is generated using those same coins, it may be possible for an adversary to learn parts of the underlying plaintext. More specifically, consider an encryption scheme that applies a trapdoor one-way function to a random value r and then concatenates $H(r) \oplus m$ and $G(r \| m)$. This scheme is known to be IND-CCA secure if H and G are modeled as random oracles [10]. Now, if another message m' is encrypted using the same coins r, then m' will be xor'd *with the same pad* $H(r)$, and anyone who knows m will also know the pad and be able to learn m'.

Since IND-CCA is insufficient for our setting, we develop a new notion of security for PKE which we call IND-RA, for indistinguishability under resetting attack. Our security notion is similar to IND-CCA except that we allow the adversary to continually see encryptions under the same coins, as if the adversary is repeatedly resetting a server and observing new encryptions. An important aspect of our security definition is that we allow the adversary to see encryptions *under public keys of its choice* and using coins that are not fresh. In particular, the adversary could see a message encrypted under a public key for which it knows the secret key, allowing it to decrypt the ciphertext; because of this, it is important that in the process of decryption not too much information is leaked about the coins used to create the ciphertext, meaning that randomness-recovering encryption cannot meet our security definition. Allowing this power in the definition is important because it models the possibility that a machine sends an encrypted message to some user Bob, is reset by the adversary, and is then forced to encrypt a message to the adversary using the same coins. We want to ensure that even if this happens, the adversary does not learn any information about Bob's message. This is a strong security requirement, but nonetheless, we are able to meet it.

We note that though our security notion provides seemingly the best possible security guarantees for PKE under reset attacks, it may still be insufficient for some applications. This is due to an inherent limitation in a model that allows repeated randomness: if the same message is encrypted twice to the same public key using the same randomness, the resulting ciphertexts will be identical. Thus, plaintext equality may be leaked to an adversary, which could be problematic in some applications. Therefore, we are not proposing IND-RA secure encryption as a complete solution to virtual machine reset attacks. Instead, we believe that IND-RA secure encryption should be used in conjunction with systems solutions, some of which are discussed in [24]. In other words, similar to [6], our constructions are a way to hedge against system failures; in our case, if the randomness happens to be reused, then our schemes do not fail immediately, but instead still provide some meaningful, provable security guarantees.

PREVIOUS WORK. Resettability has been considered in cryptography in the setting of zero-knowledge proof systems [12,21,3], the related area of identification protocols [8], and multiparty computation [19]. Zero-knowledge proofs allow a prover to prove an assertion to a verifier without revealing any information other than whether or not the assertion is true. Proving the soundness[1] and zero-knowledge properties in a setting where provers and verifiers can rewind each other is a difficult and interesting theoretical question. To see why, consider the notion of resettable-soundness in the standard model, considered by [3]. Nearly all known zero-knowledge proofs are designed specifically so that the ability to rewind the verifier allows one to easily convince it of any statement; this is useful for proving the zero-knowledge property. Yet, if we then give the prover that same ability to rewind the verifier, it becomes problematic to prove soundness. This problem has also been studied extensively in other models (c.f., [21]). However, to the best of our knowledge, no one has previously looked at practical and deployed cryptographic primitives like public-key encryption in such a setting.

In the symmetric setting, Rogaway and Shrimpton [25] investigate secure key-wrap and discuss how their techniques can apply to handle IV misuse, where IVs, which should always be fresh, are reused (possibly because of a faulty implementation). Since IVs are typically counter variables or fresh random numbers, investigating their reuse is similar to investigating the effect of a state reset.

Our work is also loosely related to public-key encryption with randomness re-use [4] and stateful public-key encryption [9]. However, both are concerned with making PKE schemes more efficient by reusing *some, but not all* random coins and still require encrypting parties to have access to fresh and unbiased randomness.

---

[1] Informally, an interactive protocol is sound if it is difficult for a malicious prover to convince the verifier that a false statement is true.

Bellare et al. recently introduced *hedged public-key encryption* [6]. At a high level, they present encryption schemes that are IND-CPA secure when the randomness used to encrypt is good, while meeting a weaker notion they call IND-CDA when the randomness is bad but the message/randomness pairs still have high entropy. Interestingly, while their goal is similar to ours and reused randomness could be considered "bad" randomness, their definitions do not appear to apply to the resettability setting.

PAPER ORGANIZATION. In Section 2, we discuss important definitions and notation that will be needed in the rest of the paper. In Section 3, we define our new notion of security for public-key encryption that models resetting attacks. In Section 4, we give constructions for schemes that meet our new notion of security. For corrections and updates to this paper, please see [26].

2   Preliminaries

NOTATION. For an integer $n \in \mathbb{N}$, we let $[n]$ denote the set $\{1, \ldots, n\}$. For the rest of the paper, let $k \in \mathbb{N}$ denote the security parameter and 1^k its unary encoding. Unless stated otherwise, all algorithms in this paper are randomized. We use "PT" for polynomial-time.

Our security definitions use the code-based games from [11]. Security definitions are formulated by considering a game played with an adversary. Such a game consists of procedures **Initialize** and **Finalize** as well as procedures for handling oracle calls the adversary can make. At the start of the game, **Initialize** is run and its output is given to the adversary. The adversary then runs and may make oracle calls that are answered by the corresponding game procedures. When the adversary halts with output w, that becomes the input to the **Finalize** procedure and the resulting output of **Finalize** is called the output of the game. We denote by $G^A \Rightarrow y$ the event that game G, when run with adversary A, outputs y. Sometimes we let G^A denote the event $G^A \Rightarrow \text{true}$.

PUBLIC-KEY ENCRYPTION. A public-key encryption scheme $\mathcal{AE} = (\mathcal{K}, \mathcal{E}, \mathcal{D})$ is a triple of PT algorithms. The randomized key generation algorithm $\mathcal{K}$, on input the security parameter 1^k in unary, outputs a pair of keys (pk, sk). The randomized encryption algorithm $\mathcal{E}$, on input public key pk and message $m \in \{0,1\}^{\eta(k)}$, outputs a ciphertext c. We let $\rho(k)$ denote the number of coins $\mathcal{E}$ uses on messages of length $\eta(k)$. Finally, the deterministic decryption algorithm $\mathcal{D}$, on input a secret key sk and ciphertext c, outputs either $\perp$ in the case of failure, or $m \in \{0,1\}^{\eta(k)}$. We require that for all $k \in \mathbb{N}$, all (pk, sk) outputted by $\mathcal{K}(1^k)$ and for all $\{0,1\}^{\eta(k)}$, it is true that $\mathcal{D}(sk, \mathcal{E}(pk, m)) = m$.

We say the IND-advantage of an adversary A is

$$\mathbf{Adv}^{\text{ind}}_{\mathcal{AE}, A}(k) = 2 \cdot \Pr\left[\, \text{IND}^A_{\mathcal{AE}}(k) \Rightarrow \text{true} \,\right] - 1 \,,$$

where the security game is found in Figure 1. To differentiate between chosen-plaintext and chosen-ciphertext attacks, we consider adversary classes. Let $\mathcal{A}^{\text{CPA}}_{\text{ind}}$

proc. Initialize(k):	proc. LR(m_0, m_1):
$b \leftarrow_\$ \{0,1\}$; $(pk, sk) \leftarrow_\$ \mathcal{K}(1^k)$	$c \leftarrow_\$ \mathcal{E}(pk, m_b)$
$S \leftarrow \emptyset$	$S \leftarrow S \cup \{c\}$
Ret pk	Return c
proc. Dec(c):	proc. Finalize(b'):
If $c \in S$ then return $\perp$	Ret $(b = b')$
Else return $\mathcal{D}(sk, c)$	

Fig. 1. Security game $\text{IND}_{\mathcal{AE}}(k)$

proc. Initialize(k):	proc. Fun(x):	Game $\text{REAL}_{\mathcal{F}}(k)$
$K \leftarrow_\$ \text{Keys}_k$	Return $\text{Fun}(K, x)$	
Ret 1^k		proc. Finalize(a):
		Ret a
proc. Initialize(k):	proc. Fun(x):	Game $\text{RAND}_{\mathcal{F}}(k)$
$\text{FunTab} \leftarrow \emptyset$	If $\text{FunTab}[x] = \perp$ then	
Ret 1^k	$\quad \text{FunTab}[x] \leftarrow_\$ \text{Rng}_k$	proc. Finalize(a):
	Return $\text{FunTab}[x]$	Ret a

Fig. 2. Security games for pseudorandom function security

be the class of all PT ind-adversaries making 1 **LR** query and 0 **Dec** queries.[2] Let $\mathcal{A}_{\text{ind}}^{\text{CCA}}$ be the class of all PT ind-adversaries making 1 **LR** query and any number of **Dec** queries.

Finally, we let IND-XXX be the set of all PKE schemes $\mathcal{AE}$ such that $\text{Adv}_{\mathcal{AE}, A}^{\text{ind}}(k)$ is a negligible function in k for all $A \in \mathcal{A}_{\text{ind}}^{\text{XXX}}$, for XXX $\in$ {CPA, CCA}.

PSEUDORANDOM FUNCTIONS. Let $\text{Fun} : \text{Keys}_k \times \text{Dom}_k \rightarrow \text{Rng}_k$ be a family of functions indexed by a security parameter k. We say the PRF-advantage of a prf-adversary D is

$$\text{Adv}_{\text{Fun}, D}^{\text{prf}}(k) = \Pr\left[\, \text{REAL}_{\text{Fun}}^{D}(k) \Rightarrow 1 \,\right]$$
$$- \Pr\left[\, \text{RAND}_{\text{Fun}}^{D}(k) \Rightarrow 1 \,\right] ,$$

where the security games can be found in Figure 2. While Keys_k, Dom_k, and Rng_k can be arbitrary finite sets, in this paper we will always consider families of functions with $\text{Keys}_k = \{0,1\}^{\ell(k)}$, $\text{Dom}_k = \{0,1\}^{n(k)}$, and $\text{Rng}_k = \{0,1\}^{t(k)}$ for some polynomials $\ell(\cdot)$, $n(\cdot)$, and $t(\cdot)$.

---

[2] Recall that it is well known that allowing multiple **LR** queries is equivalent by a standard hybrid argument.

proc. Initialize(k):	**proc. LR(m_0, m_1):**
$b \leftarrow_\$ \{0,1\}$; $(pk^*, sk^*) \leftarrow_\$ \mathcal{K}(1^k)$	$c \leftarrow \mathcal{E}(pk^*, m_b; r^*)$
$r^* \leftarrow_\$ \{0,1\}^{\rho(k)}$; $S \leftarrow \emptyset$	$S \leftarrow S \cup \{c\}$
Ret pk^*	Return c
proc. Enc(pk, m):	**proc. Dec(c):**
$c \leftarrow \mathcal{E}(pk, m; r^*)$	If $c \in S$ then return $\perp$
Return c	Else return $\mathcal{D}(sk^*, c)$
	proc. Finalize(b'):
	Ret $(b = b')$

Fig. 3. Game $\mathrm{RA}_{\mathcal{AE}}(k)$

3   Security Definition

Let $\mathcal{AE} = (\mathcal{K}, \mathcal{E}, \mathcal{D})$ be a PKE scheme. Consider game RA in Figure 3. We say an RA-adversary is one who plays game RA and makes one query to its **LR** oracle, zero or more queries to the **Enc** oracle, and zero or more queries to the **Dec** oracle. We then say the RA-advantage of an RA-adversary A is

$$\mathbf{Adv}^{\mathrm{ra}}_{\mathcal{AE}, A}(k) = 2 \cdot \Pr\left[\, \mathrm{RA}^A_{\mathcal{AE}}(k) \Rightarrow \mathsf{true} \,\right] - 1 \,.$$

In game RA, the adversary is given a target public key pk^* and can make queries to three oracles. It can query the **LR** oracle with messages m_0 and m_1. In response, the adversary receives the encryption of m_b under the target public key pk^* using the coins r^* chosen by **Initialize**. The adversary is also given an **Enc** oracle which takes as input a public key pk and message m. The oracle returns the encryption of m under public key pk, again using the coins r^* chosen in **Initialize**. It is important that the adversary can choose the public key pk. In particular, the adversary can query **Enc** with a public key for which it knows the corresponding secret key. Notice the game is similar to IND, but with the addition of the **Enc** oracle so that the adversary can continually see messages encrypted under the same coins used by **LR**. This is how we model resetting attacks. The adversary can also query a **Dec** oracle with a ciphertext (not returned by **LR**) and receive its decryption. Finally, the adversary outputs a guess bit.

EQUALITY PATTERNS. As we mentioned above, if there are no restrictions on the **LR** queries that an ra-adversary A can make, then A can trivially win the game. To see this, consider an ra-adversary that first queries **Enc**(pk^*, m) and then queries **LR**(m, m'), where $m, m' \in \{0,1\}^\eta$ and $m \neq m'$. The **Enc** query will give the adversary the encryption of m under coins r^*, and the **LR** query will give the adversary either encryption of the same message m under the same coins r^*, or it will give the adversary the encryption of m' under coins r^*. Clearly the adversary only needs to compare the two oracle answers and guess 0 if they are the same and guess 1 otherwise.

This attack is an inherent limitation of the resettable PKE setting, since for fixed coins encryption becomes a deterministic function. (It is also similar to limitations in the setting of deterministic PKE [5].) Nevertheless, as we said earlier, we are interested in achieving the best security possible in this situation. Therefore, we consider security against all adversaries "that don't trivially win". This informal notion is captured formally by the following definition:

Let A be an RA-adversary making **LR** query (m, m') and q queries (pk_1, m_1) to (pk_q, m_q) to **Enc**. Then we say that A is equality-pattern respecting if for all $i \in [q]$, $pk_i = pk^*$ if and only if $m_i \notin \{m, m'\}$. In other words, an equality-pattern respecting adversary never queries **Enc** on the target public key pk^* and a message that appears in its **LR** query.

ADVERSARY CLASSES. To differentiate between various kinds of attacks, we use classes of adversaries. Let $\mathcal{A}_{\mathrm{ra}}^{\mathrm{XXX}}$ be the class of all PT equality-pattern respecting adversaries that make one **LR** query and make 0 **Dec** queries if XXX = CPA and 0 or more **Dec** queries if XXX = CCA.

Finally, let RA-XXX be the set of all PKE schemes $\mathcal{AE}$ such that $\mathbf{Adv}_{\mathcal{AE},A}^{\mathrm{ra}}(k)$ is a negligible function in k for all $A \in \mathcal{A}_{\mathrm{ra}}^{\mathrm{XXX}}$, for XXX $\in \{\mathrm{CPA}, \mathrm{CCA}\}$.

ALTERNATIVE DEFINITIONS. We could instead consider a more complex definition in which there is more than one randomness r^* under which the adversary gets to see encryptions. Additionally, we could also allow the adversary more than one **LR** query. We present this more complex definition in Appendix 4 and show that security under it is implied by security under the simpler definition given in this section.

RELATION TO IND. Now that we have formally defined resettable security for public-key encryption, it is useful to compare it to indistinguishability under chosen plaintext and chosen-ciphertext attacks, the typical notions of security for PKE. First, it is easy to see that any scheme that is RA-XXX is also IND-XXX, since RA is identical to IND except for the additional **Enc** oracle. Thus, any IND-adversary can easily be turned into an RA-adversary making zero **Enc** oracle queries. Second, we prove the following:

Proposition 1. *For $XXX \in \{\mathrm{CPA}, \mathrm{CCA}\}$, if there exists a scheme $\overline{\mathcal{AE}}$ that is IND-XXX secure, then there exists scheme $\mathcal{AE}$ that is IND-XXX secure but is not RA-XXX secure.* $\square$

Proof. We will prove for XXX=CCA, but the proof easily extends to the CPA setting. Let $\overline{\mathcal{AE}} = (\overline{\mathcal{K}}, \overline{\mathcal{E}}, \overline{\mathcal{D}})$ be an arbitrary IND-CCA scheme. We construct a new PKE scheme $\mathcal{AE} = (\mathcal{K}, \mathcal{E}, \mathcal{D})$ such that $\mathcal{AE}$ is still in IND-CCA, but $\mathcal{AE}$ is not in RA-XXX. The scheme $\mathcal{AE}$ has encryption algorithm $\mathcal{E}(pk, m; r \parallel K \parallel K')$ that outputs $c_1 \parallel c_2 \parallel c_3$, where $c_1 = \overline{\mathcal{E}}(pk, K \parallel K'; r)$, $c_2 = K \oplus m$ and $c_3 = \mathsf{MAC}_{K'}(c_2)$. The IND-CCA security of $\mathcal{AE}$ follows from the well-known KEM/DEM composition theorem of [15]. We can construct an ra-adversary A with advantage 1 against $\mathcal{AE}$. Adversary A, upon receiving target public key pk^*, queries the **Enc** oracle with $(pk^*, 0^{\eta(k)})$ and immediately learns K from the response, since K is xor'd with all 0s. Then, A queries $\mathbf{LR}(m_0, m_1)$ for unique messages m_0 and m_1

(which do not equal the string of all zeroes). A can then use K to decrypt the response and win the game.   □

DISCUSSION. There are a few important aspects of our security definition that require more discussion.

First, as shown in Proposition 1, our definition is stronger than previous notions of security. Since we are concerned about random coins being reused, one might ask why we even need a new definition and do not just use deterministic public-key encryption [5], eliminating the coins altogether. The reason is that we still want our schemes to meet the previous definitions (i.e., IND-CCA) to ensure they have as much security as possible, and it is well-known that no deterministic scheme can ever be IND-CCA (or IND-CPA) secure.

Second, we allow the adversary to give arbitrary public keys to the **Enc** oracle and see the resulting ciphertexts under those keys and the reused coins. As mentioned in the introduction, this is important to model the situation in which a machine is reset and then an encryption is sent to the adversary; we want to make sure other encryptions using the same coins still maintain their privacy. This aspect of our definition resembles a similar ability allowed in the definition of stateful PKE [9].

Third, one might wonder what our equality pattern restriction means in practice. It simply reflects the fact that if a message is encrypted twice using the same public key and the same coins, then the resulting ciphertexts will be the same. An adversary observing the two ciphertexts will know that the underlying plaintexts are the same. This attack is unavoidable in the resettability setting, and whether or not it is a problem will depend on the application.

4   Achieving IND-RA Security

In this section we show that we can make a simple and efficient modification to any IND-XXX PKE scheme and immediately get an RA-XXX secure scheme. Our transformation relies only on the existence of pseudorandom functions and thus we do not require the random oracle model [10]. This means that if we take a PKE scheme that is IND-XXX secure in the standard model, our modified scheme will be RA-XXX secure in the standard model.

Let $\overline{\mathcal{AE}} = (\overline{\mathcal{K}}, \overline{\mathcal{E}}, \overline{\mathcal{D}})$ be a PKE scheme and let $\mathsf{Fun} : \mathsf{Keys}_k \times \mathsf{Dom}_k \to \mathsf{Rng}_k$ be a family of functions with $\mathsf{Keys}_k = \{0,1\}^{\rho(k)}$, $\mathsf{Dom}_k = \{0,1\}^{n(k)}$, and $\mathsf{Rng}_k = \{0,1\}^{\rho(k)}$. The domain size $\{0,1\}^{n(k)}$ should be large enough to encode any public key generated from $\overline{\mathcal{K}}(1^k)$ and a message in $\{0,1\}^{n(k)}$. We build a PKE scheme $\mathcal{AE} = (\mathcal{K}, \mathcal{E}, \mathcal{D})$ from $\overline{\mathcal{AE}}$ and $\mathcal{F}$ as follows. Key generation and decryption are the same as in $\overline{\mathcal{AE}}$, and $\mathcal{E}(pk, m; r)$ computes $\bar{r} \leftarrow \mathsf{Fun}(r, (pk \parallel m))$ and returns $\overline{\mathcal{E}}(pk, m; \bar{r})$.

Theorem 1. *If $\overline{\mathcal{AE}}$ is IND-XXX and Fun is a secure PRF, then $\mathcal{AE}$ is RA-XXX.*   □

proc. Enc(pk, m):	**proc. LR**(m_0, m_1): Game G_0
$\bar{r} \leftarrow \mathsf{Fun}(r^*, (pk \parallel m))$	$\bar{r} \leftarrow \mathsf{Fun}(r^*, (pk^* \parallel m_b))$
$c \leftarrow \overline{\mathcal{E}}(pk, m; \bar{r})$	$c \leftarrow \overline{\mathcal{E}}(pk^*, m_b; \bar{r})$
Return c	$S \leftarrow S \cup \{c\}$
	Return c

proc. Enc(pk, m):	**proc. LR**(j, m_0, m_1): Game G_1
$\bar{r} \leftarrow\!\!{}^{\$} \{0,1\}^{\rho(k)}$	$\bar{r} \leftarrow\!\!{}^{\$} \{0,1\}^{\rho(k)}$
$c \leftarrow \overline{\mathcal{E}}(pk, m; \bar{r})$	$c \leftarrow \overline{\mathcal{E}}(pk^*, m_b; \bar{r})$
Return c	$S \leftarrow S \cup \{c\}$
	Return c

Fig. 4. Games for the proof of Theorem 1. The procedures **Initialize**, **Finalize**, and **Dec** are omitted for brevity.

Proof. Let $\mathcal{AE}$ be constructed from Fun and $\overline{\mathcal{AE}}$ as above. Let $A \in \mathcal{A}_{\mathrm{ra}}^{\mathrm{XXX}}$ be an efficient RA-adversary attacking $\mathcal{AE}$. We assume that A never makes duplicate queries to the **Enc** oracle; this is without loss because all such queries will return the same response. It is easy to see that this fact combined with the fact that A is equality-pattern respecting means that every query A makes to **Enc** and **LR** results in a unique combination of pk and m; the game will never encrypt the same message twice under the same public key. Now, denote by G_0 the game RA defined in Section 3. Thus by definition,

$$\mathbf{Adv}_{\mathcal{AE},A}^{\mathrm{ra}}(k) = 2 \cdot \Pr\left[\, G_0^A \,\right] - 1 \,.$$

Now consider game G_1. The relevant procedures from games G_0 and G_1 are shown in Figure 4. In G_0, oracles **LR** and **Enc** use Fun to derive the randomness used to encrypt since this is what $\mathcal{AE}$ does. However, in G_1, those oracles choose fresh random coins and use those to encrypt the messages. We claim that these games appear close to adversary A by showing there exists an efficient prf-adversary B such that

$$\Pr\left[\, G_0^A \,\right] - \Pr\left[\, G_1^A \,\right] \le \mathbf{Adv}_{\mathsf{Fun},B}^{\mathrm{prf}}(k) \,.$$

The adversary B, attempting to decide if it is in the real or random world, flips a bit b and chooses a target public key pk^* by running the key generation algorithm. B then runs A just as in G_0 and G_1. On **Enc** and **LR** queries, B uses its **Fun** oracle to derive the randomness for encryption; in the case of the **LR** query, B encrypts the message corresponding to the bit b that it chose. In the CCA case, B answers **Dec** queries simply by using the secret key sk^* (which it knows because it chooses pk^* and sk^*). When A eventually outputs a guess bit b', B outputs 1 if $b = b'$ and 0 otherwise. We can see that when B is in the 'real' world (i.e., its **Fun** queries are answered using Fun), it perfectly simulates G_0 for A, while if B is in the 'random' world (i.e., its **Fun** queries are answered with random range points) then it perfectly simulates G_1 for A. The claim follows.

We then claim that there exists an efficient ind-adversary C such that

$$\mathbf{Adv}^{\mathrm{ind}}_{\mathcal{AE},C}(k) = 2 \cdot \Pr\left[\, \mathrm{G}^A_1 \,\right] - 1 \, .$$

The ind-adversary C is given a target public key pk^* and access to an **LR** oracle to which it can make a single query. In the CCA setting it also has access to a **Dec** oracle. Adversary C runs A as in G_1, answering its oracle queries as follows. On A's single **LR** query, C simply answers with its own **LR** oracle. In the CCA setting C answers A's **Dec** queries using its own **Dec** oracle. On **Enc** queries from A, C encrypts the messages itself using fresh randomness and returns the resulting ciphertexts to A. At the end of execution, C outputs the same bit that A guesses. It is easy to see that C perfectly simulates the G_1 game for A and the claim follows.

Combining the above equations we can see that

$$\begin{aligned}
\mathbf{Adv}^{\mathrm{ra}}_{\mathcal{AE},A}(k) &= 2 \cdot \Pr\left[\, \mathrm{G}^A_0 \,\right] - 1 \\
&\leq 2 \cdot \left(\mathbf{Adv}^{\mathrm{prf}}_{\mathsf{Fun},B}(k) + \Pr\left[\, \mathrm{G}^A_1 \,\right]\right) - 1 \\
&\leq 2 \cdot \mathbf{Adv}^{\mathrm{prf}}_{\mathsf{Fun},B}(k) + \mathbf{Adv}^{\mathrm{ind}}_{\mathcal{AE},C}(k) \, . \qquad \square
\end{aligned}$$

The existence of secure PRFs is implied by the existence of one-way functions (which are necessary for PKE to exist), so we do not need any additional assumptions. In practice, one would want to instantiate the PRF using HMAC [7] or a block-cipher such as AES. Notice that in the random oracle model we can replace $\mathsf{Fun}(r, (pk \parallel m))$ with $H(r, pk, m)$, where H is a random oracle, since a random oracle gives us a simple way to construct a PRF. Of course, if we did so, we would lose the standard model guarantees.

EXTENSIONS. A natural question to ask is what happens to security if the same randomness is used across multiple different primitives. For example, what if some randomness r is used for public-key encryption, but then after a virtual machine reset r is used for DSA signing? Formally modeling this situation is an interesting open problem. However, we conjecture that our PRF approach in this section will generalize and provide security in such a setting. Specifically, to protect a primitive P against reused randomness, one would want to apply the PRF not only to P's inputs, but also to some unique value identifying P, e.g., an algorithm ID. This should guarantee that different primitives use distinct randomness, even after a reset.

Acknowledgements

We would like to thank Barath Raghavan for originally expressing to us his concern about the effects that resetting VMs might have on cryptographic primitives. We would also like to thank Mihir Bellare, Thomas Ristenpart, and Daniele Micciancio for providing useful feedback on an earlier version of this paper. Additionally, we thank the CT-RSA 2010 anonymous referees for many useful comments. Scott Yilek is supported by Daniele Micciancio's NSF grant CNS–0831536 and Mihir Bellare's NSF grant CNS–0627779.

References

1. http://aws.amazon.com/ec2/
2. http://www.openssl.org/
3. Barak, B., Goldreich, O., Goldwasser, S., Lindell, Y.: Resettably-sound zero-knowledge and its applications. In: 42nd Annual Symposium on Foundations of Computer Science – FOCS 2001, pp. 116–125. IEEE, Los Alamitos (2001)
4. Bellare, M., Boldyreva, A., Kurosawa, K., Staddon, J.: Multi-recipient encryption schemes: Efficient constructions and their security. IEEE Transactions on Information Theory 53(11) (2007)
5. Bellare, M., Boldyreva, A., O'Neill, A.: Deterministic and efficiently searchable encryption. In: Menezes, A. (ed.) CRYPTO 2007. LNCS, vol. 4622, pp. 535–552. Springer, Heidelberg (2007)
6. Bellare, M., Brakerski, Z., Naor, M., Ristenpart, T., Segev, G., Shacham, H., Yilek, S.: Hedged public-key encryption: How to protect against bad randomness. In: ASIACRYPT 2009. LNCS, pp. 232–249. Springer, Heidelberg (2009)
7. Bellare, M., Canetti, R., Krawczyk, H.: Keying hash functions for message authentication. In: Koblitz, N. (ed.) CRYPTO 1996. LNCS, vol. 1109, pp. 1–15. Springer, Heidelberg (1996)
8. Bellare, M., Fischlin, M., Goldwasser, S., Micali, S.: Identification protocols secure against reset attacks. In: Pfitzmann, B. (ed.) EUROCRYPT 2001. LNCS, vol. 2045, pp. 495–511. Springer, Heidelberg (2001)
9. Bellare, M., Kohno, T., Shoup, V.: Stateful public-key cryptosystems: How to encrypt with one 160-bit exponentiation. In: Proceedings of the 13th ACM Conference on Computer and Communications Security – CCS 2006, pp. 380–389. ACM, New York (2006)
10. Bellare, M., Rogaway, P.: Random oracles are practical: A paradigm for designing efficient protocols. In: Proceedings of 1st ACM Conference on Computer and Communications Security – CCS 1993, pp. 62–73. ACM, New York (1993)
11. Bellare, M., Rogaway, P.: Code-based game-playing proofs and the security of triple encryption. In: Vaudenay, S. (ed.) EUROCRYPT 2006. LNCS, vol. 4004, pp. 409–426. Springer, Heidelberg (2006)
12. Canetti, R., Goldreich, O., Goldwasser, S., Micali, S.: Resettable zero-knowledge. In: Proceedings of the Thirty-Second Annual ACM Symposium on Theory of Computing – STOC 2000, pp. 235–244. ACM, New York (2000)
13. Chen, P.M., Noble, B.D.: When virtual is better than real. In: Proceedings of the 2001 Workshop on Hot Topics in Operating Systems, pp. 133–138 (2001)
14. Cox, R.S., Gribble, S.D., Levy, H.M., Hansen, J.G.: A safety-oriented platform for web applications. In: Proceedings of the 2006 IEEE Symposium on Security and Privacy, pp. 350–364. IEEE, Los Alamitos (2006)
15. Cramer, R., Shoup, V.: Design and analysis of practical public-key encryption schemes secure against adaptive chosen ciphertext attack. SIAM Journal on Computing 33(1), 167–226 (2003)
16. Desai, A., Hevia, A., Yin, Y.L.: A practice-oriented treatment of pseudorandom number generators. In: Knudsen, L.R. (ed.) EUROCRYPT 2002. LNCS, vol. 2332, pp. 368–383. Springer, Heidelberg (2002)
17. Garfinkel, T., Rosenblum, M.: When virtual is harder than real: Security challenges in virtual machine based computing environments. In: Proceedings of the 10th Workshop on Hot Topics in Operating Systems – HotOS-X (May 2005)

18. Goldwasser, S., Micali, S.: Probabilistic encryption. Journal of Computer and System Sciences 28(2), 270–299 (1984)
19. Goyal, V., Sahai, A.: Resettably secure computation. In: EUROCRYPT 2009. Springer, Heidelberg (2009)
20. Håstad, J., Impagliazzo, R., Levin, L.A., Luby, M.: A pseudorandom generator from any one-way function. SIAM Journal on Computing 28(4), 1364–1396 (1999)
21. Micali, S., Reyzin, L.: Soundness in the public-key model. In: Kilian, J. (ed.) CRYPTO 2001. LNCS, vol. 2139, p. 542. Springer, Heidelberg (2001)
22. Provos, N.: A virtual honeypot framework. In: Proceedings of the 13th USENIX Security Symposium, pp. 1–14 (2004)
23. Rackoff, C., Simon, D.R.: Non-interactive zero-knowledge proof of knowledge and chosen ciphertext attack. In: Feigenbaum, J. (ed.) CRYPTO 1991. LNCS, vol. 576, pp. 433–444. Springer, Heidelberg (1992)
24. Ristenpart, T., Yilek, S.: When good randomness goes bad: Virtual machine reset vulnerabilities and hedging deployed cryptography. In: Proceedings of the Network and Distributed System Security Symposium – NDSS 2010. Internet Society (to appear, 2010)
25. Rogaway, P., Shrimpton, T.: Deterministic authenticated-encryption: A provable-security treatment of the key-wrap problem. In: Vaudenay, S. (ed.) EUROCRYPT 2006. LNCS, vol. 4004, pp. 373–390. Springer, Heidelberg (2006)
26. Yilek, S.: Resettable public-key encryption: How to encrypt on a virtual machine. Cryptology ePrint Archive, Report 2009/474 (2009), http://eprint.iacr.org/2009/474

Appendix

A   An Equivalent Security Definition

As we mentioned in Section 3, we can consider a more complicated security game to capture IND-RA security. We give a more detailed discussion in [26], but briefly compare the definitions here.

Let $\mathcal{AE} = (\mathcal{K}, \mathcal{E}, \mathcal{D})$ be a PKE scheme. We say the RA-advantage of an adversary A is

$$\mathbf{Adv}^{\mathrm{ra}}_{\mathcal{AE},A}(k) = 2 \cdot \Pr\left[\, \mathrm{RA2}^{A}_{\mathcal{AE}}(k) \Rightarrow \mathsf{true} \,\right] - 1 \ .$$

The security game RA2 can be found in Figure 5. In the game, the adversary is given a target public key pk^* and can make queries to three oracles. It can query the **LR** oracle with index j and messages m_0 and m_1. In response, the adversary receives the encryption of m_b under the target public key pk^* using the coins indexed by j. The adversary is also given an **Enc** oracle which takes as input a public key pk, index j, and message m. The oracle returns the encryption of m under public key pk using the coins indexed by j. It is important that the adversary can choose the public key pk. In particular, the adversary can query **Enc** with a public key for which it knows the corresponding secret key. With both the **LR** and **Enc** oracles, an adversary can continually see messages encrypted under the same coins by repeatedly querying the same index. This is how we model resetting attacks. Of course, the adversary can also see messages

---

proc. Initialize(k):

$b \leftarrow\!\! \$ \; \{0,1\}$; $(pk^*, sk^*) \leftarrow\!\! \$ \; \mathcal{K}(1^k)$
CoinTab $\leftarrow \emptyset$; $S \leftarrow \emptyset$
Ret pk^*

proc. Enc(pk, j, m):

If CoinTab[j] $= \bot$ then
   CoinTab[j] $\leftarrow\!\! \$ \; \{0,1\}^{\rho(k)}$
$r_j \leftarrow$ CoinTab[j]
$c \leftarrow \mathcal{E}(pk, m; r_j)$
Return c

proc. LR(j, m_0, m_1):

If CoinTab[j] $= \bot$ then
   CoinTab[j] $\leftarrow\!\! \$ \; \{0,1\}^{\rho(k)}$
$r_j \leftarrow$ CoinTab[j]
$c \leftarrow \mathcal{E}(pk^*, m_b; r_j)$
$S \leftarrow S \cup \{c\}$
Return c

proc. Dec(c):

If $c \in S$ then return $\bot$
Else return $\mathcal{D}(sk^*, c)$

proc. Finalize(b'):

Ret $(b = b')$

Fig. 5. Game RA2$_{\mathcal{AE}}(k)$

encrypted under other coins by querying other indices. Finally, the adversary can also query a **Dec** oracle with a ciphertext and receive its decryption.

EQUALITY PATTERNS. For our alternate definition, we need a much more complicated notion of equality patterns. Let A be any adversary that queries I different indices to its **LR** and **Enc** oracles and makes q_i queries to the **LR** oracle with index i. Let E_i be the set of all messages m such that A makes query **Enc**(pk^*, i, m). Let $(m_0^{i,1}, m_1^{i,1})$ to $(m_0^{i,q_i}, m_1^{i,q_i})$ be A's **LR** queries for index $i \in [I]$. Then, if for all $i \in [I]$ and for all $j \neq k \in [q_i]$,

$$m_0^{i,j} = m_0^{i,k} \text{ iff } m_1^{i,j} = m_1^{i,k} ,$$

and for all $i \in [I]$ and all $j \in [q_i]$

$$m_0^{i,j} \notin E_i \wedge m_1^{i,j} \notin E_i,$$

then we say that A is equality-pattern respecting.

ADVERSARY CLASSES. To differentiate between various kinds of attacks, we use classes of adversaries. Let $\mathcal{A}_{ra}^{XXX}$ be the class of all PT equality-pattern respecting adversaries that make 0 **Dec** queries if XXX = CPA and 0 or more **Dec** queries if XXX = CCA.

It is easy to see that if we only consider adversaries that query **LR** once and query **LR** and **Enc** on only a single randomness index, then the definition becomes equivalent to the definition in Section 3. (CoinTab has only one entry, which we call r^* in the simpler definition.) To justify our use of the simpler security game, we use hybrid arguments to prove the following two lemmas:

Lemma 1. *Let $\mathcal{AE}$ be a PKE scheme and $A_{q,i} \in \mathcal{A}_{ra}^{XXX}$ be an adversary querying **LR** q times and querying **LR** and **Enc** on combined at most i different indices. Then there exists an adversary $A_{q,1} \in \mathcal{A}_{ra}^{XXX}$ making at most q **LR** queries and querying **LR** and **Enc** on only a single index, such that*

$$\mathbf{Adv}^{\mathrm{ra}}_{\mathcal{AE},A_{q,i}}(k) \le i \cdot \mathbf{Adv}^{\mathrm{ra}}_{\mathcal{AE},A_{q,1}}(k) \, ,$$

where the running time of $A_{q,1}$ is about the same as that of $A_{q,i}$.   □

Lemma 2. Let $\mathcal{AE}$ be a PKE scheme and $A_{q,1} \in \mathcal{A}^{\mathrm{XXX}}_{ra}$ be an adversary querying $\mathbf{LR}$ q times and querying $\mathbf{LR}$ and $\mathbf{Enc}$ on combined at most 1 different index. Then there exists an adversary $A_{1,1} \in \mathcal{A}^{\mathrm{XXX}}_{ra}$ making 1 $\mathbf{LR}$ query and querying $\mathbf{LR}$ and $\mathbf{Enc}$ on only a single index, such that

$$\mathbf{Adv}^{\mathrm{ra}}_{\mathcal{AE},A_{q,1}}(k) \le q \cdot \mathbf{Adv}^{\mathrm{ra}}_{\mathcal{AE},A_{1,1}}(k) \, ,$$

where the running time of $A_{1,1}$ is about the same as that of $A_{q,1}$.   □

Proof of Lemma 1 (Sketch). Let $A_{q,i}$ be any equality pattern respecting adversary making q queries to $\mathbf{LR}$ and querying $\mathbf{LR}$ and $\mathbf{Enc}$ on at most i different randomness indices. We will build an adversary $A_{q,1}$ making at most q queries to the $\mathbf{LR}$ oracle and querying $\mathbf{LR}$ and $\mathbf{Enc}$ on at most 1 randomness index. The adversary $A_{q,1}$ runs $A_{q,i}$ and guesses an index $j \in \{1, \ldots, i\}$. It then uniformly chooses coins r_ℓ for $\ell \ne j$ and answers $\mathbf{Enc}$ and $\mathbf{LR}$ queries from $A_{q,i}$ as follows. On query $\mathbf{Enc}(pk, k, m)$, if $k = j$ then $A_{q,1}$ replies with its own $\mathbf{Enc}$ oracle and otherwise uses coins r_k to answer the query. On query $\mathbf{LR}(k, m_0, m_1)$, if $k < j$ (resp. $k > j$) then $A_{q,1}$ replies with the encryption of m_0 (resp. m_1) under coins r_k; if $k = j$ then $A_{q,1}$ replies using its own $\mathbf{LR}$ oracle. At the end of the simulation, $A_{q,1}$ outputs the same guess bit as $A_{q,i}$.   □

Proof of Lemma 2 (Sketch). Let $A_{q,1}$ be any equality pattern respecting adversary making q queries to $\mathbf{LR}$ and querying $\mathbf{LR}$ and $\mathbf{Enc}$ on only a single randomness index. We can build an adversary $A_{1,1}$ making only a single $\mathbf{LR}$ query. Adversary $A_{1,1}$ guesses a query $t \in \{1, \ldots, q\}$ and answers the first $t - 1$ $\mathbf{LR}$ queries using its $\mathbf{Enc}$ oracle applied to m_0, the tth query using its own $\mathbf{LR}$ oracle, and the rest of the queries again using $\mathbf{Enc}$, this time applied to m_1. As above, $A_{1,1}$ outputs the same answer as $A_{q,1}$.   □

Plaintext-Awareness of Hybrid Encryption

Shaoquan Jiang[1] and Huaxiong Wang[2]

[1] School of Computer Science and Engineering
University of Electronic Science and Technology of China
shaoquan.jiang@gmail.com
[2] School of Physical and Mathematical Sciences
Nanyang Technological University, Singapore
HXWang@ntu.edu.sg

Abstract. We study plaintext awareness for hybrid encryptions. Based on a binary relation R, we define a new notion of PA2 (or R-PA2 for short) and a notion of IND-CCA2 (or R-IND-CCA2 for short) for key encapsulation mechanism (KEM). We define a relation R_{DEM} from the description of data encryption mechanism (DEM). We prove two composition results, which holds with or without (public) random oracles.

a. When KEM, with R_{DEM}-PA2 and R_{DEM}-IND-CCA2 security, composes with a one-time pseudorandom and unforgeable (OT-PUE) DEM, the resulting hybrid encryption is PA2 secure. OT-PUE is weak and even unnecessarily passively secure and can be realized by a one-time pad encryption followed by a pseudorandom function.
b. If KEM is R_{DEM}-IND-CCA and DEM is passively secure and unforgeable, the hybrid encryption (KEM, DEM) is IND-CCA2 secure.

As an application, we show that DHIES, a public key encryption scheme by Abdalla et al. [1] and now in IEEE P1361a and ANSI X.963, is PA2 secure. As another application, we prove that a hash proof system based hybrid encryption is PA2. Consequently, this especially implies that the concrete Kurosawa-Desmedt hybrid encryption (CRYPTO04) is PA2.

1 Introduction

Plaintext-awareness (PA) for an encryption system intuitively means that the only way for one to generate a valid ciphertext is to apply the encryption algorithm to a message. In other words, when one produces a ciphertext, he must know the plaintext. ElGamal encryption is certainly not plaintext-aware since for any public key (g, h), one can generate a valid ciphertext (A, B) by simply taking $A, B \leftarrow \langle g \rangle$ (if $d = \log_g h$, then (A, B) is an encryption of $m = A^{-d}B$ but DDH assumption asserts that m is unknown to its encrypter). PA has important applications in some security systems. For instance, Di Raimondo et al. [18] uses the plaintext-awareness [17] of Cramer-Shoup [15] to prove the deniability of SKEME key exchange protocol. Hybrid encryption [15] is a recently proposed framework for constructing efficient public key encryption schemes. It consists of a key encapsulation mechanism (KEM) and a data encryption mechanism (DEM). The former, based on the public key, encapsulates a temporary

J. Pieprzyk (Ed.): CT-RSA 2010, LNCS 5985, pp. 57–72, 2010.

secret key and can be decapsulated only with the private key. The latter uses the temporary key to encrypt the real data. The final ciphertext consists of the ciphertexts generated by both mechanisms. The decryption works in the obvious way. Note there are hybrid schemes in which KEM mixes with DEM or KEM is based on a tag (e.g., [4,23,2,3]). In this paper, we will not consider this type of hybrid encryption. Since a hybrid encryption is a new encryption paradigm, it is interesting to study its plaintext-awareness, which is the task of this work.

1.1   Related Works

The notion of PA was first formally proposed by Bellare and Rogaway [9], while the intuition can be dated back to [11,12]. The formalization [9] is in the random oracle. Under their definition, PA plus IND-CPA does not imply IND-CCA2. This is not very natural since if CCA2 attacker knows the plaintext of his ciphertext, then decryption oracle should be useless and hence IND-CCA2 is equivalent to IND-CPA. Bellare et al. [6] filled this gap by allowing ciphertext eavesdropping. PA without random oracle was considered by Herzog et al. [21] but in the key-registration setting, where any user owns a public key. Bellare et al. [7] formalized plaintext-awareness in the classical setting (i.e., without a key registration). They formalized three notions PA2, PA1 and PA0 and the adversary goal is to forge a ciphertext for which he does not know the plaintext. PA2 admits adaptive chosen ciphertext attacks and eavesdropping attacks while PA1 and PA0 only admit adaptive chosen ciphertext attacks. PA1 differs from PA0 in that PA0 only allows one decryption query. Dent [17] showed that Cramer-Shoup hybrid encryption is PA2 secure and this is the first proof of PA2 for a practical encryption. His paper established some techniques for proving PA2. Beyond this, PA of hybrid encryption is not well studied in the literature. Even the result of Dent [17] is "not a practical tool" (quoted from [17]) for studying PA2 of many practical schemes since it implicitly assumes that KEM and DEM both are at least IND-CCA2. So to study PA2 for many *practical* schemes in which KEM and DEM are not strongly secure, we have to look for new tools. One of such practical schems is DHIES by Bellare et al. [1,10], which appeared in standard drafts [5,13,19] and PA of DHIES was considered but unproved by its authors. Regarding this, [1] states:

"In [10], a claim is made that DHIES should achieve plaintext awareness if this hash function is modeled as a public random oracle and one assumes the computational DH assumption. In fact, technical problems would seem to thwart any possibility of pushing through such a result...."

1.2   Our Contribution

We define a new notion of PA2 with respect to a relation R (or R-PA2 for short) for KEM, which is weaker than PA2. We also define a notion of IND-CCA2 with respect to a relation R (or R-IND-CCA2 for short) for KEM, which turns out to be an alternative of LCCA [4]. R-IND-CCA2 is weaker than IND-CCA2. We associate relation R_{DEM} with DEM and prove two composition results below.

a. When KEM, with R_{DEM}-PA2 and R_{DEM}-IND-CCA2 security, composes with a one-time pseudorandom and unforgeable (OT-PUE) DEM, the resulting hybrid encryption is PA2 secure. OT-PUE is a weak notion and even does not guarantee the passive security and can be realized by a one-time pad encryption followed by a pseudorandom function.

b. If KEM is R_{DEM}-IND-CCA and DEM is passively secure and unforgeable, the hybrid encryption (KEM, DEM) is IND-CCA2 secure.

These compositions hold with or without a (public) random oracle. To show the usefulness of these compositions, we consider two applications. As an application, we prove that DHIES is PA2 secure in the public random oracle model, under CDH and DHK assumptions and when DEM is OT-PUE. As another application, we prove that a hash proof system based hybrid encryption is PA2 secure, if it uses a computational universal$_2$ projective hash family for an *extractable* hard subset membership problem and DEM is OT-PUE and passively secure. An important implication of this application is that the concrete Kurosawa-Desmedt hybrid encryption [25] is PA2 secure. Two applications seem unlikely to be proven PA2 under the results in Dent [17] since KEM of the former does not appear to be IND-CCA2 and KEM of the latter is not IND-CCA2 [14].

2 Preliminaries

Notations. $x \leftarrow S$ samples x uniformly random from a set S. For two random variables X, Y over a finite set V, the probability distance between them is $\mathsf{Dist}[X, Y] = \frac{1}{2} \sum_{v \in V} |\Pr[X = v] - \Pr[Y = v]|$. Function $\epsilon : \mathbb{N} \to \mathbb{R}$ is negligible if $\lim_{n \to \infty} p(n)\epsilon(n) = 0$ for any polynomial $p(n)$. We usually use $negl(\kappa)$ to denote a *negligible function*. PPT means probabilistic polynomial time. Random variables $X \approx Y$ means that they are computationally indistinguishable.

2.1 Diffie-Hellman Knowledge Assumption

$p = 2q + 1$ and q are large primes. $\mathbb{G}$ is the subgroup of $\mathbb{Z}_p^*$ of order q. g is a generator of $\mathbb{G}$. For any PPT adversary $\mathcal{H}$, there exists a PPT extractor $\mathcal{H}^*$ such that the experiment below terminates with 0 for probability $1 - negl(\kappa)$.

Let $a \leftarrow \mathbb{Z}_q$, $A = g^a$. Take $r, r^* \leftarrow \{0,1\}^*$ as a random tape respectively for $\mathcal{H}$, $\mathcal{H}^*$. Input (p, g, A, r) to $\mathcal{H}$ and (p, g, A, r, r^*) to $\mathcal{H}^*$. $\mathcal{H}$ can query $\mathcal{H}^*$ as follows.

- $\mathcal{H}$ *issues query* (B, C) *to* $\mathcal{H}^*$. $\mathcal{H}^*$ *responds with some* $b \in \mathbb{Z}_q$ *to* $\mathcal{H}$. *If* $B^a = C$ *but* $B \neq g^b$, *then the experiment terminates with output 1; otherwise, continue.*

If the experiment does not terminate until $\mathcal{H}$ halts, the experiment outputs 0.

2.2 Simulatable Random Variable

Let V be a finite set and Z is a random variable over V. $\ell \in \mathbb{N}$. $\Phi : \{0,1\}^\ell \to V$ is a deterministic function and $\Phi^* : V \to \{0,1\}^\ell$ is a probabilistic function. Then, Z is said to be simulatable by (Φ, Φ^*), if $\mathsf{Dist}[Z, \Phi(U_\ell)] = negl(\kappa)$ for

$U_\ell \leftarrow \{0,1\}^\ell$ and $\mathsf{Dist}[\Phi^*(z), U_\ell(z)] = negl(\kappa)$ for any $z \in V$, where $U_\ell(z)$ is uniformly distributed over $\left\{ u_\ell \mid \Phi(u_\ell) = z, u_\ell \in \{0,1\}^\ell \right\}$. From the definition, we can sample Z using U_ℓ by $Z = \Phi(U_\ell)$, and recover the randomness used by Φ to sample $Z = z$ by computing $\Phi^*(z)$. The following fact is immediate.

Fact 1.   Keep the notions above. Then $\Phi^*(Z)$ and U_ℓ are statistically close.

2.3   Hybrid Encryption and Key Encapsulation Mechanism

A hybrid encryption system [15] is a public key encryption system that consists of two components: Key Encapsulation Mechanism (KEM) and Data Encryption Mechanism (DEM). KEM generates a ciphertext c that encodes a secret key K. DEM encrypts the data into a ciphertext e using K. The final ciphertext for the hybrid encryption is (c, e). The decryption works in an obvious way. Formally, a hybrid encryption $\mathcal{PKE} = (\mathsf{KEM}, \mathsf{DEM})$ is defined as follows. KEM=(KEM.Gen, KEM.Key, KEM.Enc, KEM.Dec) and DEM=(DEM.Enc, DEM.Dec). Initially, take $\mathsf{sp} \leftarrow \mathsf{KEM.Gen}(1^\kappa)$ to generate system parameter sp.

KEM.Key(sp). Take $(pk, sk) \leftarrow \mathsf{KEM.Key}(\mathsf{sp})$ to generate public key pk and private key sk.
KEM.Enc(pk). Take $(K, c) \leftarrow \mathsf{KEM.Enc}(pk)$ to generate session key K and ciphertext c that encapsulates K.
KEM.Dec(c). Given c, use sk to decapsulate $K = \mathsf{KEM.Dec}(sk, c)$.

DEM=(DEM.Enc, DEM.Dec) is a pair of a symmetric encryption/decryption algorithms.

$\mathcal{PKE}$ works as follows. Run $\mathsf{sp} \leftarrow \mathsf{KEM.Gen}(1^\kappa), (pk, sk) \leftarrow \mathsf{KEM.Key}(\mathsf{sp})$ to generate public key pk and private key sk. To encrypt m, compute $(K, c) \leftarrow \mathsf{KEM.Enc}(pk)$ and $e \leftarrow \mathsf{DEM.Enc}(K, m)$. The ciphertext is (c, e). To decrypt (c, e) with sk, compute $K = \mathsf{KEM.Dec}(sk, c)$ and $m = \mathsf{DEM.Dec}(K, e)$.

2.4   Security of KEM

In this section, we introduce two security notions of KEM used in this work.

Chosen plaintext security (IND-CPA). Let $\mathsf{sp} \leftarrow \mathsf{KEM.Gen}(1^\kappa), (pk, sk) \leftarrow$ KEM.Key(sp), $(K_0, c) \leftarrow \mathsf{KEM.Enc}(pk), b \leftarrow \{0,1\}, K_1 \leftarrow \mathcal{K}$, where $\mathcal{K}$ is the key space of encapsulation by KEM. KEM is IND-CPA if given (K_b, c), no PPT adversary can guess b non-negligibly better than $1/2$.

IND-CCA2 with Respect to a Relation. We now introduce a security notion for KEM, called chosen ciphertext security with respect to a binary relation R (or R-IND-CCA2 for short). This notion turns out to be an alternative formation of LCCA by [4] and weaker than Constrained CCA in [22]. We keep our relation based formulation for consistency with our relation based plaintext-awareness.

Let $R \subseteq \mathcal{K} \times \{0,1\}^*$ be a binary relation. R-IND-CCA2 is defined through a game between an attacker $\mathcal{A}$ and a challenger. Challenger samples $\mathsf{sp} \leftarrow$ KEM.Gen(1^κ), $(pk, sk) \leftarrow \mathsf{KEM.Key}(\mathsf{sp})$, gives pk to $\mathcal{A}$ and answers his queries.

– $\mathcal{A}$ can issue a challenge query at any time but just once. In turn, Challenger takes $(K_0, c^*) \leftarrow \mathsf{KEM.Enc}(pk)$, $b \leftarrow \{0,1\}$, $K_1 \leftarrow \mathcal{K}$ and provides (K_b, c^*) to $\mathcal{A}$.

– $\mathcal{A}$ can issue a decryption query (c, α) at any time. Upon this, if $c = c^*$, outputs $\bot$; otherwise, he first computes $K = \mathsf{KEM.Dec}(sk, c)$. If $(K, \alpha) \in R$, return K to $\mathcal{A}$; otherwise, return $\bot$.

At the end of the game, $\mathcal{A}$ outputs guess b' for b. He succeeds if $b' = b$. Denote the above game by Γ_R. R-IND-CCA2 security is defined as follows.

Definition 1. *R is a binary relation. A* key encapsulation mechanism KEM *is said to be* adaptive chosen ciphertext secure with respect to R *(or R-IND-CCA2) if* $\Pr[\mathsf{Succ}(\mathcal{A}, \Gamma_R)] = 1/2 + negl(\kappa)$ *for any PPT adversary $\mathcal{A}$.*

2.5   Public Random Oracle

Public random oracle is an idealized object for hash function $H : \{0,1\}^* \rightarrow \{0,1\}^\ell$. Specifically, for any input x, $H(x)$ is uniformly random in $\{0,1\}^\ell$, except that the same input gets the same output. Algorithmically, it can be described in a query model below. Keep a set H-list Ω, which is initially empty. Upon any query x, if x was not recorded in Ω, take $y \leftarrow \{0,1\}^\ell$ randomly and put (x, y) into Ω. In any case, if $(x, y) \in \Omega$ for some $y \in \{0,1\}^*$, return y as $H(x)$. This idealized object was first proposed in Bellare and Rogaway [8]. It was then popularly used in the literature to prove the security for many practical systems. In our work, we will sometimes adopt this model for plaintext-awareness. But we should be careful since PA is defined in terms of a plaintext extractor, which, upon a decryption query, plays as a decryption oracle to extract the plaintext encrypted in a ciphertext while using an adversary's knowledge only. That is, the extractor's code should be executable by the adversary himself. Especially, the extractor can not choose the value of $H(x)$ (since an adversary can not). Thus, the extractor can not maintain H-oracle by himself. In other words, H is non-programmable. In our work, H-oracle is maintained by a trusted third party $\mathcal{H}$. This is called a **public random oracle model.** Under this, when any participant (e.g., extractor, adversary) wishes to compute $H(x)$, he has to query $\mathcal{H}$. This model was previously adopted in [28], where they allow the simulator to see the oracle inputs of the adversary. In our paper, we remove this condition as our simulator sees the random tape of the adversary and all of his messages received (thus his entire view) and so he can generate these H-queries himself.

2.6   Plaintext-Awareness

Plaintext-awareness essentially means that when one generates a ciphertext he should know the plaintext. Bellare and Palacio [7] formalized three levels of plaintext-awareness, denoted by PA0, PA1 and PA2. In the following, we will introduce them, first in the standard model and then in the public random oracle model. We will introduce a new notion of R-PA2 for KEM with relation R.

Plaintext-awareness for public key encryption in the standard model
PA2 essentially states that an adversary can not create a *new* ciphertext without

knowing its plaintext, even if he has eavesdropped some other ciphertexts. The formal definition is described using two games. In Game one, the adversary can access to a real decryption oracle and an encryption oracle. The former captures the CCA2 attack and the latter captures the eavesdropping attack. In our model, eavesdropped ciphertexts are modeled as outputs of normal encryptions, which is different from Bellare et al. [7] where they are generated by any PPT algorithm. Our formulation is reasonable since a ciphertext without following the specification does not have a security guarantee and so a normal encrypter is unlikely to do so. In Game one, the adversary finally generates an arbitrary output (e.g. his entire view). Game two is similar to Game one, except that the decryption oracle is answered by a plaintext extractor, who is given the public key, the adversary's random tape and the ciphertext history generated by the encryption oracle. Especially, he is NOT given the decryption key. Finally, the encryption scheme is said PA2 if no efficient algorithm can distinguish the adversary outputs in these games. When a scheme is PA2, the extractor conceivably always extracts the plaintext. Since the extractor only uses the adversary's knowledge, the latter should 'know' the plaintext since he can run the extractor's code himself. In both games, the adversary is not allowed to issue decryption queries with the eavesdropped ciphertexts; otherwise, the scheme is not PA2 unless it is insecure. Let $S = (S.\text{Gen}, S.\text{Key}, S.\text{Enc}, S.\text{Dec})$ be a public key encryption and κ be the security parameter. The two games proceed as follows.

Game G_0 :
$\text{sp} \leftarrow S.\text{Gen}(1^\kappa), (pk, sk) \leftarrow S.\text{Key}(\text{sp}); \Omega = \{\}.$
Let r_A, r_P be the random tapes for PPT algorithms $\mathcal{A}$ and P, respectively.
Run $\mathcal{A}$ with input pk and coins r_A and answer his queries until it halts.

- If $\mathcal{A}$ issues a decryption query with c for $c \notin \Omega$, computes $m = S.\text{Dec}(sk, c)$ and returns m to $\mathcal{A}$. If $c \in \Omega$, ignore it.
- If $\mathcal{A}$ issues an encryption query with a message distribution $\mathcal{M}$, P takes $m \leftarrow \mathcal{M}$, computes $c = S.\text{Enc}(pk, m)$ and returns c to $\mathcal{A}$. Update $\Omega = \Omega \cup \{c\}$.

Finally, $\mathcal{A}$ outputs a string x.

Game G_1 :
$\text{sp} \leftarrow S.\text{Gen}(1^\kappa), (pk, sk) \leftarrow S.\text{Key}(\text{sp}); \Omega = \{\}.$
Let r_A, r_P, r_{A^*} be random tapes for PPT algorithms $\mathcal{A}$, P and $\mathcal{A}^*$, respectively.
Run $\mathcal{A}$ with input pk and coins r_A and answer his queries below until it halts.

- If $\mathcal{A}$ issues a decryption query c for $c \notin \Omega$, compute $m = \mathcal{A}^*(pk, c, \Omega, r_A, r_{A^*})$ and return m to $\mathcal{A}$. If $c \in \Omega$, ignore it.
- If $\mathcal{A}$ issues an encryption query with a message distribution $\mathcal{M}$, P takes $m \leftarrow \mathcal{M}$, computes and returns $c = S.\text{Enc}(pk, m)$ to $\mathcal{A}$. Update $\Omega = \Omega \cup \{c\}$.

Finally, $\mathcal{A}$ outputs a string x.

Use $out(\mathbf{G}_i)$ to denote the output of $\mathcal{A}$ in $\mathbf{G}_i, i = 0, 1$.

Definition 2. *A public-key encryption* S *is* computationally PA2 secure *if for any PPT* $\mathcal{A}$*, there exists a PPT* $\mathcal{A}^*$ *such that* $out(\mathbf{G}_0) \approx out(\mathbf{G}_1)$.

Plaintext-awareness in the public random oracle model. PA2 in the public random oracle model is similar to PA2 in the standard model above, *except a public random oracle* $\mathcal{H}$ *is added into the games*, where when any participant (P, $\mathcal{A}$, $\mathcal{A}^*$, distinguisher, or challenger) wants to compute $H(x)$, he sends x to $\mathcal{H}$ and receives $H(x)$. P could issue a H-query in order to compute a ciphertext; the challenger could issue a H-query in order to answer the decryption query; a *out* distinguisher may issue a H-query to maximize his advantage. $\mathcal{H}$ oracle answers a H-query by maintaining a H-list as mentioned before. Denote $\mathbf{G}_0, \mathbf{G}_1$ in the public random oracle model by $\mathbf{G}_0^H, \mathbf{G}_1^H$, respectively. Then, PA2 in the public random oracle model is stated as follows.

Definition 3. $H : \{0,1\}^* \rightarrow \{0,1\}^\ell$ *is a public random oracle. A public-key encryption* S *is* computationally PA2 secure in the public random oracle model *if for any PPT* $\mathcal{A}$*, there exists a PPT* $\mathcal{A}^*$ *such that* $out(\mathbf{G}_0^H) \approx out(\mathbf{G}_1^H)$.

Plaintext-awareness for KEM with respect to a relation. The objective of KEM is to encapsulate a secret key in the ciphertext. Hence, its PA2 definition should naturally capture the following intuition: when one generates a KEM ciphertext c, he should know the key encapsulated in it. However, we find that this intuition is too strong to be useful since many practical KEMs do not satisfy this. We thus relax it to the following: if one generates a ciphertext c and knows *partial* information about the encapsulated key, he must know the whole key. In our specification, partial information is interpreted as satisfying a pre-defined binary relation R. Specifically, when the adversary submits a ciphertext for decryption, he also submits a string α as a proof that he knows partial information about the encapsulated key K. The decryption oracle decrypts K (if any) and verifies if $(K, \alpha) \in R$. If yes, the adversary is said to know the partial information and is given K; otherwise, he is not given K. We call this relaxed plaintext-awareness for KEM, R-PA2. As for public-key encryption, we formalize R-PA2 in terms of two games. Let KEM be a key encapsulation mechanism and $\mathcal{K}$ be the space of the encapsulated key. Let $R \subseteq \mathcal{K} \times \{0,1\}^*$ be a binary relation. We define two games $\mathbf{G}_{i,R}$ ($i = 0, 1$), parameterized by R.

Game $\mathbf{G}_{0,R}$:

$\mathsf{sp} \leftarrow \mathsf{KEM.Gen}(1^\kappa), (pk, sk) \leftarrow \mathsf{KEM.Key}(\mathsf{sp})$.

Let r_A, r_P be the random tapes for PPT algorithms $\mathcal{A}$ and P, respectively.

Run $\mathcal{A}$ with input pk and coins r_A and answer his queries until it halts.

- If $\mathcal{A}$ issues a decryption query with (α, c), compute $K = \mathsf{KEM.Dec}(sk, c)$. If $K = \bot$ or $(K, \alpha) \notin R$, then return $\bot$; otherwise, return K to $\mathcal{A}$.
- If $\mathcal{A}$ issues an encryption query, P computes $(K, c) = \mathsf{KEM.Enc}(pk)$ and returns c.

Finally, $\mathcal{A}$ outputs a string x.

Game $\mathbf{G}_{1,R}$:
sp $\leftarrow$ KEM.Gen(1^κ), $(pk, sk) \leftarrow$ KEM.Key(sp); $\Omega = \{\}$.
Let r_A, r_P, r_{A^*} be random tapes for PPT algorithms $\mathcal{A}$, P and A^*, respectively.
Run $\mathcal{A}$ with input pk and coins r_A and answer his queries below until it halts.

- If $\mathcal{A}$ issues a decryption query with (α, c), $\mathcal{A}^*$ computes and returns K to $\mathcal{A}$, where $K = \mathcal{A}^*(R, pk, \alpha, c, \Omega, r_A, r_{A^*})$.
- If $\mathcal{A}$ issues an encryption query, P computes $(K, c) = $ KEM.Enc(pk) and returns c to $\mathcal{A}$. Update $\Omega = \Omega \cup \{c\}$.

Finally, $\mathcal{A}$ outputs a string x.

Now we are ready to formally state R-PA2.

Definition 4. *A key encapsulation mechanism* KEM *is* PA2 *secure with respect to a binary relation R (or R-PA2, for short) if for any PPT $\mathcal{A}$, there exists a PPT $\mathcal{A}^*$ such that $out(\mathbf{G}_{0,R}) \approx out(\mathbf{G}_{1,R})$.*

R-PA2 for KEM in the public random oracle model.   Similar to the public key encryption case, we can define R-PA2 for KEM in the public random oracle model, by adding public random oracle into games $\mathbf{G}_{0,R}$ and $\mathbf{G}_{1,R}$. We summarize the revised definition as follows.

Definition 5. *Let $H : \{0,1\}^* \to \{0,1\}^\ell$ be a public random oracle. A key encapsulation mechanism* KEM *is* PA2 *secure with respect to a binary relation R (or R-PA2, for short) in the public random oracle model if for any PPT $\mathcal{A}$, there exists a PPT $\mathcal{A}^*$ such that $out(\mathbf{G}_{0,R}^H) \approx out(\mathbf{G}_{1,R}^H)$.*

PA1/PA0. PA1/PA0 for these systems are simply defined by removing the encryption oracle in the respective setting. That is, $\mathcal{A}$ looses the ability of eavesdropping ciphertexts. PA0 is a special case of PA1, where $\mathcal{A}$ is only allowed to issue one decryption query.

Remark. In all of the PA definitions above, $\mathcal{A}$ outputs an arbitrary string and PA is defined as indistinguishability of $\mathcal{A}$'s outputs in two games. As stated in Bellare [7], separating the attacker $\mathcal{A}$ and the distinguisher is important since the extractor can obtain the coins of $\mathcal{A}$ but not that of the distinguisher.

2.7   One-Time Pseudorandom Unforgeable Encryption (OT-PUE)

One-time pseudorandom unforgeable encryption essentially states that the ciphertext is pseudorandom and unforageable, even if the adversary can issue a single decryption query. Formally,

Let $\mathcal{PUE} = $ (PUE.Key, PUE.Enc, PUE.Dec) be a *symmetric* encryption. Consider a game between an adversary $\mathcal{A}$ and a challenger.

- $\mathcal{A}$ can issue a challenge query *once* with a message m. Challenger takes $K \leftarrow$ PUE.Key(1^κ) and $b \leftarrow \{0,1\}$. If $b = 0$, let $c^* =$ PUE.Enc(K, m); otherwise, $c^* \leftarrow \{0,1\}^\ell$, for $\ell = |$PUE.Enc$(K, m)|$. Finally, return c^* to $\mathcal{A}$.
- Receiving c^*, $\mathcal{A}$ can issue a *single* query $c \neq c^*$ to the challenger. If $b = 0$, the latter returns $m =$ PUE.Dec(K, c) (note: by default $m = \perp$ if c is invalid); if $b = 1$, he simply returns $\perp$.

At the end of the game, $\mathcal{A}$ outputs a guess bit b' for b. Let Γ_b be the above game when the challenge bit is b. The security definition is as follows.

Definition 6. *A symmetric encryption scheme $\mathcal{PUE}$ is a* one-time pseudorandom unforgeable encryption (OT-PUE) *if for any PPT adversary $\mathcal{A}$, $\Pr[\mathcal{A}(\Gamma_0) = 1] = \Pr[\mathcal{A}(\Gamma_1) = 1] + negl(\kappa)$. If the challenge query is removed, then a scheme satisfying this is called* one-time unforgeable encryption *(or* OT-UE *for short).*

Note that this notion is rather weak: it does not imply passive security. But on the other hand, it is not hard to find an IND-CCA2 security scheme which is not OT-PUE. Hence, it is not comparable with IND-CCA2.

OT-PUE vs One-Time Authenticated Encryption. One-time authenticated encryption (OT-AE) was introduced in Hofheinz and Kiltz [22]. Essentially, an encryption (E, D) is OT-AE if it is passively secure and unforgeable. In terms of a game, adversary $\mathcal{A}$ first submits messages m_0, m_1 of equal length. Challenger takes $b \leftarrow \{0,1\}$ and returns $c^* = E_K(m_b)$. Then, $\mathcal{A}$ can make a single decryption query $c \neq c^*$. Challenger returns $m = D_K(c)$ if $b = 0$; $\perp$ otherwise. Finally, $\mathcal{A}$ outputs a guess bit b' and succeeds if $b' = b$. The following is simple.

Lemma 1. *If symmetric encryption S is passively secure and OT-PUE, then it is OT-AE.*

3 Composition for Secrecy

Bellare et al. [7] showed that when a public encryption is IND-CPA and PA2, it must also be IND-CCA2. This provides an alterative (i.e., via PA2) to prove IND-CCA2 for a public encryption, especially for a hybrid encryption. Dent [17] presented results for proving PA2 of a hybrid encryption, for which KEM is IND-CCA2. In many practical hybrid encryptions, KEMs are not IND-CCA2. Hence, we look for results suitable to prove IND-CCA2 of a hybrid encryption (via PA), in which KEM is not necessarily IND-CCA2. Toward, we will prove that R-IND-CCA2 KEM plus a proper DEM is an IND-CCA2 hybrid encryption. In the next section, we will show KEM, with R-PA2 and R-IND-CCA2, plus a proper DEM gives a PA2 hybrid encryption. Hence, R-PA2 KEM and R-IND-CCA2 are essential for a hybrid encryption to be both IND-CCA2 and PA2.

Let K be the key space of DEM, $R_{DEM} = \Big\{ (K, \alpha) | $DEM.Dec$(K, \alpha) \neq \perp, K \in \mathcal{K}, \alpha \in \{0,1\}^* \Big\}$. We show that if KEM is R_{DEM}-IND-CCA2 and DEM is OT-AE, then (KEM, DEM) hybrid encryption is IND-CCA2. This result is extended from

Hofheinz and Kiltz [22, Theorem 3.1], where they require KEM is constrained IND-CCA2 which is not hard to see stronger than R_{DEM}-IND-CCA2. Our proof strategy is as follows. If the theorem is wrong, we can build a R_{DEM}-IND-CCA2 attacker $\mathcal{B}$ that uses an IND-CCA2 attacker $\mathcal{A}$ as a subroutine. $\mathcal{B}$ mainly needs to answer the decryption query (c, e) from $\mathcal{A}$. The idea is to let $\mathcal{B}$ ask his decapsulation oracle to decapsulate c and then uses the returned K to decrypt e, except when c is in his challenge (K_b, c) he can not ask. However, in this case, he can decrypt e using K_b. The formal proof can be found in the full paper [24].

Theorem 1. *Let* KEM *is* OT-AE. *If* KEM *is* R_{DEM}-IND-CCA2 *in the random oracle model (resp. standard model), then* (KEM, DEM) *hybrid encryption is* IND-CCA2 *in the random oracle model (resp. standard model).*

4   Composition for Plaintext-Awareness

In this section, we study the question: which type of plaintext-awareness for KEM plus a reasonable DEM can guarantee PA2 of the hybrid encryption? Dent [17] provided an answer, where KEM does not seem weaker than PA2. In many practical schemes such as DHIES, the PA2 conidition for KEM is too strong. We hence look for a suitable composition that works with a weak KEM.

We present a composition theorem, where KEM is only R_{DEM}-PA2. We show that, if KEM is R_{DEM}-PA2 and R_{DEM}-IND-CCA2 and DEM is pseudorandom and unforgeable, then the hybrid encryption is PA2. The impact of this result can be stated as follows: if we want to study PA2 of hybrid encryption (KEM, DEM), we only need to study KEM's R-PA2 and R-IND-CCA2 properties.

To prove the result, we need to construct a PA2 extractor $\mathcal{A}^*$ that answers an adversary $\mathcal{A}$'s decryption query (c, e). Our idea is, $\mathcal{A}^*$ can internally simulate R_{DEM}-PA2 game of KEM and use its KEM extractor $\mathcal{B}^*$ to extract the key K in c and then use K to decrypt e. In doing so, we must be careful about two subtle issues. Firstly, the simulated R_{DEM}-PA2 must be self-contained; otherwise, we cannot guarantee the R_{DEM}-PA2 attacker (say, $\mathcal{B}$) outputs $\mathcal{A}$'s decryption query (c, e). To avoid this, we simulate the R_{DEM}-PA2 game such that the view of $\mathcal{A}$ is deterministic in the view of $\mathcal{B}$. Secondly, for a decryption query (c, e) by $\mathcal{A}$ such that c is output by encryption oracle of $\mathcal{B}$, $\mathcal{B}^*$ can not output K. So how can $\mathcal{A}^*$ decrypt e? This is not a problem since in this case K must be computationally random in view of $\mathcal{A}$ and hence unforgeability of DEM implies $(K, e) \notin R_{DEM}$. So $\mathcal{A}^*$ can simply reject. The formal proof is in the full paper [24].

Theorem 2. (KEM, DEM) *is a hybrid encryption, where* KEM *is* R_{DEM}-*PA2 in the public random oracle model (resp. the standard model) and* R_{DEM}-IND-CCA2 *in the random oracle model (resp. the standard model) and* DEM *is* OT-PUE. *Then* (KEM, DEM) *is PA2 in the public random oracle model (resp. the standard model).*

5   Applications

5.1   DHIES

DHIES public key encryption was proposed by Abdalla et al. [1]. The earlier version appeared in [10]. It is now in the draft standards of IEEE P1361a and ANSI X.963 [5,19]. In this section, we will prove its PA2 via composition results obtained in previous sections. Our result also implies a new proof for IND-CCA2 under DHK and CDH assumptions in the random oracle model although IND-CCA2 for DHIES is not new [1,15]. We first review DHIES. Let data encryption mechanism of DHIES be DEM=(DEM.Enc, DEM.Dec). Its KEM, KEM_{hE}, is described as follows.

Let $p = 2q + 1$ and q be two large primes. g is a generator of order q in $\mathbb{Z}_p^*$. $H : \{0,1\}^* \leftarrow \{0,1\}^\kappa$ is a hash function, where κ is the security parameter. So its system parameter is sp=(p, g).

$\mathsf{KEM}_{hE}.\mathsf{Key}(\mathsf{sp})$. Let $d \leftarrow \mathbb{Z}_q$ and $h = g^d$. The public key is (p, g, h) and the secret key is d.

$\mathsf{KEM}_{hE}.\mathsf{Enc}(pk)$. Take $r \leftarrow \mathbb{Z}_q$ and compute $u = g^r$. The ciphertext is u and encapsulated key $K = H(h^r)$.

$\mathsf{KEM}_{hE}.\mathsf{Dec}(u)$. To decrypt u, compute $K = H(u^d)$.

Plaintext-Awareness. Using Theorem 2, we show that DHIES is PA2. Toward this, we first show that KEM_{hE} is R_{DEM}-PA2. That is, we construct a KEM_{hE} key extractor without using d. Our idea is to deploy a DHK extractor as a subroutine. More specifically, for decryption query (g^t, e) by adversary $\mathcal{A}$, if h^t was not queried to random oracle by $\mathcal{A}$, then e is unlikely to be valid and hence reject; if h^t was queried by $\mathcal{A}$, we can find it by issuing a DHK (g^t, x) query for each random oracle query x from $\mathcal{A}$:   since DHK extractor never errs, if $x = h^t$ is Diffie-Hellman, then t can be extracted; otherwise, it can not output t (since it does not exist). When t will be extracted, the decryption key $H(h^t)$ can be computed by the extractor $\mathcal{A}^*$ and hence the decryption will be correct. The formal proof is available in the full paper [24].

Lemma 2. *Let* DEM *be* OT-UE. *Then, under* DHK *assumption,* KEM_{hE} *is* R_{DEM}-PA2 *secure in the public random oracle model.*

Next, we show that KEM_{hE} is R_{DEM}-IND-CCA2 in the random oracle model. The proof can be seen in the full paper.

Lemma 3. *If* DEM *is* OT-UE, *then, under* DHK *and* CDH *assumptions,* KEM_{hE} *is* R_{DEM}-IND-CCA2 *in the random oracle model.*

From Theorem 2 and Lemmas 2, 3, we conclude the following theorem.

Theorem 3. *Let* DEM *be* OT-PUE. *Then, under* DHK *and* CDH *assumptions,* DHIES *is PA2 secure in the public random oracle model.*

IND-CCA2 (*revisited*). IND-CCA2 security of DHIES is not new. Abdalla et al. [1] proved it under oracle Diffie-Hellman assumption and Cramer-Shoup [15] implied a proof under a strong Diffie-Hellman assumption in the random oracle model. From Lemma 3 and Theorem 1, when DEM is OT-PUE and passively secure, we get a new proof under CDH and DHK assumptions in the random oracle model.

5.2   Hash Proof System Based Hybrid Encryption

5.2.1   Hash Proof System

Now we introduce the hash proof system, which was initially introduced by Cramer and Shoup [15]. To cater our use, we slightly modify the definition. We also add a notion of *extractability* introduced in our separate paper.

(a) **Hard Subset Membership Problem.** A hard subset membership problem essentially is a problem, in which one can efficiently sample a hard instance. More formally, a subset membership problem $\mathcal{I}$ is a collection $\{\mathcal{I}_n\}_{n\in\mathbb{N}}$, where $\mathcal{I}_n$ is a probability distribution for a random variable Λ_n that is efficiently sampled by a polynomial time algorithm as follows.

- Generate a finite non-empty set $X_n, L_n \subseteq \{0,1\}^{poly(n)}$ s.t. $L_n \subset X_n$, and distribution $D(L_n)$ over L_n and distribution $D(X_n\backslash L_n)$ over $X_n\backslash L_n$.
- Generate a witness set $W_n \subseteq \{0,1\}^{poly(n)}$ and a NP-relation $R_n \subseteq X_n \times W_n$ such that $x \in L_n$ if and only if there exists $w \in W_n$ s.t. $(x,w) \in R_n$. There exists a polynomial time algorithm that samples x according to $D(L_n)$ and outputs a witness $w \in W_n$ s.t. $(x,w) \in W_n$. Further, there exists a polynomial time algorithm that samples x according to $D(X_n\backslash L_n)$.

Denote $\Lambda_n = (X_n, L_n, W_n, R_n, D(L_n), D(X_n\backslash L_n))$. $\mathcal{I} = \{\mathcal{I}_n\}_{n\in\mathbb{N}}$ is a **hard subset membership problem** if for $\Lambda_n \leftarrow \mathcal{I}_n$, we have that $x \leftarrow D(L_n)$ and $y \leftarrow D(X_n\backslash L_n)$ are indistinguishable.

(b) **Extractable Hard Subset Membership Problem.** Now we introduce a notion of extractability for $\mathcal{I}$. A hard subset membership problem $\mathcal{I} = \{\mathcal{I}_n\}_n$ is extractable if for any PPT adversary $\mathcal{A}$, there exists a PPT extractor $\mathcal{A}^*$ such that the following experiment terminates with 0 for probability $1 - negl(n)$.

Let $\Lambda = (X, L, W, R, D(L), D(X\backslash L)) \leftarrow \mathcal{I}_n$. Let $desc(\Lambda)$ be the description of Λ. Sample $r, r^* \leftarrow \{0,1\}^*$ as a random tape for $\mathcal{A}$ and $\mathcal{A}^*$, respectively. Input $(desc(\Lambda), r)$ to $\mathcal{A}$ and $(desc(\Lambda), r, r^*)$ to $\mathcal{A}^*$. Then $\mathcal{A}$ can query $\mathcal{A}^*$ as follows.

- $\mathcal{A}$ queries $x \in X$ to $\mathcal{A}^*$. $\mathcal{A}^*$ responds with some $w \in W$ to $\mathcal{A}$. If $x \in L$ but $(x,w) \notin W$, the experiment terminates with output 1; otherwise, it continues.

If the experiment does not terminate until $\mathcal{A}$ halts, the experiment outputs 0.

(c) **Projective Hash Functions.** Let $\Lambda = (X, L, W, R, D(L), D(X\backslash L))$ be sampled from a subset membership problem $\mathcal{I}_n$. Consider a function family

$\langle \mathcal{H}, \mathcal{K}, X, L, G, S, \alpha \rangle$, which is described by $desc(\Lambda)$ and $\lambda \leftarrow \{0,1\}^n$ [1], where $G, S, \mathcal{K}$ are finite, non-empty sets, $\mathcal{H} = \{H_k \mid k \in \mathcal{K}\}$ is a set of hash functions from X to G and $\alpha : \mathcal{K} \rightarrow S$ is a deterministic function. $\mathcal{K}$ is called a *key space*, $k \in \mathcal{K}$ is called the *projection key*, S is called the *projection space* for α. The family $\langle \mathcal{H}, \mathcal{K}, X, L, G, S, \alpha \rangle$ is called a projective hash family (PHF) for Λ, if a random instance of it is determined by $desc(\Lambda)$ and a uniformly random string λ and if $H_k(x)$ for $x \in L$, is uniquely determined by $\alpha(k)$ and x. It is called an efficient PHF, if $\alpha(k)$ and $H_k(x)$ are both polynomially computable for any (k, x) and if $H_k(x)$ can be polynomially computable from $x, w, \alpha(k)$ for $(x, w) \in R$. Now we define the following.

Definition 7. $\{I_n\}_n$ *is a hard subset membership problem. Sample an instance* $\Lambda = \langle X, L, W, R, D(L), D(X \backslash L) \rangle \leftarrow \mathcal{I}_n$. *PHF* $= \langle \mathcal{H}, \mathcal{K}, X, L, G, S, \alpha \rangle$ *is a projective hash family for* Λ. *PHF is* computational universal$_2$ *if any PPT* $\mathcal{A}$ *has a negligible advantage in the following game. Sample an instance of PHF by* $desc(\Lambda)$ *and* $\lambda \leftarrow \{0,1\}^n$. *Take* $k \leftarrow \mathcal{K}$. *Provide* $(\lambda, desc(\Lambda), \alpha(k))$ *to* $\mathcal{A}$.

- $\mathcal{A}$ *is given* $x_1 \leftarrow D(X \backslash L)$ *and* $H_k(x_1)$.
- $\mathcal{A}$ *can adaptively issue an* **Evalu** *query with* $x \in X$, *where oracle* **Evalu** *does the following. It first checks if* $x \in L$ *(maybe in exponential time). If yes, return* $H_k(x)$; *otherwise* $\perp$.
- *Throughout the game,* $\mathcal{A}$ *can come up with a challenge* $x_2 \in X \backslash L$ *for* $x_2 \neq x_1$. *He receives* K_b, *where* $b \leftarrow \{0,1\}, K_0 = H_k(x_2)$ *and* $K_1 \leftarrow \mathcal{K}$. *After query* x_2, $\mathcal{A}$ *can still query any* x *to* **Evalu**.

At the end of game, $\mathcal{A}$ *outputs a guess bit* b' *for* b. *He succeeds if* $b' = b$.

If we only require $(x_2, \alpha(k), H_k(x_2))$ *to be indistinguishable from* $(x_2, \alpha(k), g)$ *for* $g \leftarrow G$ *and any* $x_2 \in D(X \backslash L)$ *(i.e., with access to* **Evalu** *oracle and without obtaining* $(x_1, H_k(x_1))$*), then HPF is called* smooth.

5.2.2   Key Encapsulation Mechanism [25,22]

Now we describe KEM from hash proof system [25]. Use KEM_{hps} to denote it. Initially, take $\Lambda = \langle X, L, W, R, D(L), D(X \backslash L) \rangle \leftarrow \mathcal{I}_\kappa$. Let $\mathcal{PHF} = \langle \mathcal{H}, \mathcal{K}, X, L, G, S, \alpha \rangle$ be the projective hash family for Λ. Sample an instance $(\lambda, desc(\Lambda))$ from $\mathcal{PHF}$. Then the system parameter $\mathsf{sp} = (\lambda, desc(\Lambda))$.

$\mathsf{KEM}_{hps}.\mathsf{Key}(\mathsf{sp})$. Take $k \leftarrow \mathcal{K}$ and compute $pk = (\alpha(k), \lambda, desc(\Lambda))$. Then pk is the public key and k is the secret key.

$\mathsf{KEM}_{hps}.\mathsf{Enc}(pk)$. Take $x \leftarrow D(L)$ with witness w such that $(x, w) \in W$. The ciphertext is x and the encapsulated key is $H_k(x)$. Note that a sender can compute $H_k(x)$ using x, w, pk.

$\mathsf{KEM}_{hps}.\mathsf{Dec}(k, x)$. To decrypt x, compute $K = H_k(x)$ using (k, x).

---

[1] Note here we require that in addition to $desc(\Lambda)$, PHF can be described by a parameter $\lambda \leftarrow \{0,1\}^\kappa$. The requirement $\lambda \leftarrow \{0,1\}^\kappa$ is not essential. It can be relaxed as any simulatable variable (for results in this paper to hold).

Plaintext-Awareness. The following lemma essentially states that KEM_{hps} is R_{DEM}-PA2 if $\mathcal{PHF}$ is smooth and $\mathcal{I}$ is an extractable hard subset membership problem. The formal proof is in the full paper.

Lemma 4. *Let* $\mathcal{I} = \{\mathcal{I}_\kappa\}_{k \in \mathbb{N}}$ *be an extracble hard subset membership problem.* DEM *is* OT-UE. *Let* $\Lambda = \langle X, L, W, R, D(L), D(X \backslash L)\rangle \leftarrow \mathcal{I}_\kappa$. $\mathcal{PHF} = \langle \mathcal{H}, \mathcal{K}, X, L, G, S, \alpha\rangle$ *is a smooth projective hash family for* Λ. $\alpha(k)$ *for* $k \leftarrow \mathcal{K}$ *is simulatable by* (Φ_1, Φ_1^*) *and* $x \leftarrow D(X \backslash L)$ *is simulatable by* (Φ_2, Φ_2^*). *Then,* KEM_{hps} *is* R_{DEM}*-PA2.*

The following lemma shows that the computational universal$_2$ of HPF implies R_{DEM}-IND-CCA2 for KEM_{hps}. Since R_{DEM}-IND-CCA2 is weaker than constrained-CCAs in [22], where it was shown that KEM_{hps} is constrained CCA2 [22, Theorem 6.2], the following is implied by this.

Lemma 5. $\mathcal{I} = \{\mathcal{I}_k\}$ *is a hard subset membership problem.* HPF *is computational universal$_2$ HPF for* $\mathcal{I}$. DEM *is* OT-AE. *Then,* KEM_{hps} *is* R_{DEM}*-IND-CCA2 secure.*

From Lemmas 4, 5 and Theorem 2, we immediately have

Theorem 4. $\{\mathcal{I}_\kappa\}$ *is an extractable hard subset membership problem.* DEM *is* OT-PUE. $\Lambda = \langle X, L, W, R, D(L), D(X \backslash L)\rangle \leftarrow \mathcal{I}_\kappa$. $\mathcal{PHF} = \langle \mathcal{H}, \mathcal{K}, X, L, G, S, \alpha\rangle$ *is computational universal$_2$ for* Λ. $K \leftarrow G$, $x \leftarrow D(X \backslash L)$, $\alpha(k)$ *for* $k \leftarrow \mathcal{K}$, *are all simulatable. Then* $(\text{KEM}_{hps}, \text{DEM})$ *is PA2.*

5.2.3   Concrete Kurosawa-Desmedt Scheme [25]

Kurosawa and Desmedt [25] used the following KEM as an example for their HPS based hybrid encryption. It is important since this hybrid encryption is more efficient than Cramer-Shoup scheme. Denote its KEM by KEM_{kd}.

- **Description of** $\mathcal{I}_k$. Sample a prime $p = 2q + 1$ where q is also a large prime. Let $\mathbb{G}$ be the prime group of $\mathbb{Z}_p^*$ of order q. Take $g_1, g_2 \leftarrow \mathbb{G}$. The set $X = \{(g_1^{r_1}, g_2^{r_2}) \mid r_1, r_2 \in \mathbb{Z}_q\}$. Language L is defined as $L = \{(g_1^r, g_2^r) \mid r \in \mathbb{Z}_q\}$. $D(L)$ is defined as taking $r \leftarrow \mathbb{Z}_q$ and outputting (g_1^r, g_2^r). Similarly define $D(X \backslash L)$. $\mathcal{I}$ is a hard subset membership problem from DDH assumption by $\mathbb{G}$. Also, based on DHK assumption, $\mathcal{I}$ is an extractable hard subset membership problem.
- **Description of** $\mathcal{PHF}$. Let $G = S = \mathbb{G}$ and $\mathcal{K} = \{(x_1, x_2, y_1, y_2) \mid x_1, x_2, y_1, y_2 \in \mathbb{Z}_q\}$. $\alpha(k) = (c, d) = (g_1^{x_1} g_2^{x_2}, g_1^{y_1} g_2^{y_2})$. Let h_λ be a target collision resistent hash function, indexed by $\lambda \leftarrow \{0, 1\}^\kappa$. For $(u_1, u_2) \in X$, define $H_k(u_1, u_2) = u_1^{x_1 + y_1 \tau} u_2^{x_2 + y_2 \tau}$, where $\tau = h_\lambda(u_1, u_2)$. If $(u_1, u_2) = (g_1^r, g_2^r)$, then $H_k(u_1, u_2) = u_1^{x_1 + y_1 \tau} u_2^{x_2 + y_2 \tau} = (g_1^{x_1 + y_1 \tau} g_2^{x_2 + y_2 \tau})^r = (cd^\tau)^r = \alpha(k)^r$. Hence, this is a projective hash family for $\mathcal{I}$. Further, it is known that this hash family is computational universal$_2$ [22, Lemma 6.3]. $desc(\Lambda) = (p, g_1, g_2)$. $desc(\mathcal{PHF}) = (\lambda, c, d, desc(\Lambda))$. Besides, c, d are easily shown to be simulatable (also see Dent [17]). Hence, $\mathcal{PHF}$ is an extractable and computational universal$_2$ projective hash family.

Theorems 4 and the discussions above, we have

Theorem 5. *Let* DEM *be* OT-PUE *and passively secure.* h_λ *is target collision-resistant. Then* DHK *and* DDH *assumptions, hybrid encryption* (KEM$_{kd}$, DEM) *is* IND-CCA2 *secure and PA2 secure.*

Acknowledgements. The authors are grateful to anonymous referees for invaluable comments and for pointing out that an independent work by James Birkett at RHUL also achieves PA2 for Korusawa-Desmedt scheme. S. Jiang is supported by NSFCs (No. 60673075, 60973161) and UESTC Young Faculty Plans. H. Wang is supported in part by the Singapore National Research Foundation under Research Grant NRF-CRP2-2007-03 and the Singapore Ministry of Education under Research Grant T206B2204.

References

1. Abdalla, M., Bellare, M., Rogaway, P.: The Oracle Diffie-Hellman Assumptions and an Analysis of DHIES. In: Naccache, D. (ed.) CT-RSA 2001. LNCS, vol. 2020, pp. 143–158. Springer, Heidelberg (2001)
2. Abe, M., Kiltz, E., Okamoto, T.: Compact CCA-Secure Encryption for Messages of Arbitrary Length. In: Public Key Cryptography 2009. LNCS, vol. 5443, pp. 377–392. Springer, Heidelberg (2009)
3. Abe, M., Kiltz, E., Okamoto, T.: Chosen Ciphertext Security with Optimal Ciphertext Overhead. In: Pieprzyk, J. (ed.) ASIACRYPT 2008. LNCS, vol. 5350, pp. 355–371. Springer, Heidelberg (2008)
4. Abe, M., Gennaro, R., Kurosawa, K.: Tag-KEM/DEM: A New Framework for Hybrid Encryption. J. Cryptology 21(1), 97–130 (2008)
5. American National Standards Institute (ANSI) X9.F1 subcommittee, ANSI X9.63 Public key cryptography for the Financial Services Industry: Elliptic curve key agreement and key transport schemes, Working draft, January 8 (1999)
6. Bellare, M., Desai, A., Pointcheval, D., Rogaway, P.: Relations among notions of security for public-key encryption schemes. In: Krawczyk, H. (ed.) CRYPTO 1998. LNCS, vol. 1462, pp. 26–45. Springer, Heidelberg (1998)
7. Bellare, M., Palacio, A.: Towards Plaintext-Aware Public-key Encryption without Random Oracles. In: Lee, P.J. (ed.) ASIACRYPT 2004. LNCS, vol. 3329, pp. 48–62. Springer, Heidelberg (2004)
8. Bellare, M., Rogaway, P.: Random Oracle is Practical: A Paradigm for Designing Efficient Protocols. In: Proceedings of the 1st ACM Symposium on Computer and Communication Security, CCS 1993, pp. 62–73 (1993)
9. Bellare, M., Rogaway, P.: Optimal asymmetric encryption. In: De Santis, A. (ed.) EUROCRYPT 1994. LNCS, vol. 950, pp. 92–111. Springer, Heidelberg (1995)
10. Bellare, M., Rogaway, P.: Minimizing the use of random oracles in authen- ticated encryption schemes. In: Han, Y., Qing, S. (eds.) ICICS 1997. LNCS, vol. 1334, pp. 1–16. Springer, Heidelberg (1997)
11. Blum, M., Feldman, P., Micali, S.: Non-interactive zero knowledge and its applications. In: Proceedings of the 20th Annual ACM Symposium on Theory of Computing, STOC 1988, pp. 103–112 (1988)
12. Blum, M., Feldman, P., Micali, S.: Proving security against chosen ciphertext attacks. In: Goldwasser, S. (ed.) CRYPTO 1988. LNCS, vol. 403, pp. 256–268. Springer, Heidelberg (1990)

13. Certicom Research, Standards for Efficient Crpytography Group (SECG) - SEC 1: Elliptic Curve Cryptography. Version 1.0, September 20 (2000)
14. Choi, S., Herranz, J., Hofheinz, D., Hwang, J.Y., Kiltz, E., Lee, D.H., Yung, M.: The Kurosawa-Desmedt Key Encapsulation is not Chosen-Ciphertext Secure. Information Processing Letters 109(16), 897–901 (2009)
15. Cramer, R., Shoup, V.: Design and Analysis of Practical Public-Key Encryption Schemes Secure Against Adaptive Chosen Ciphertext Attack. SIAM Journal on Computing 33, 167–226 (2003)
16. Desai, A.: New Paradigms for Constructing Symmetric Encryption Schemes Secure against Chosen-Ciphertext Attack. In: Bellare, M. (ed.) CRYPTO 2000. LNCS, vol. 1880, pp. 394–412. Springer, Heidelberg (2000)
17. Dent, A.: The Cramer-Shoup Encryption Scheme is Plaintext Aware in the Standard Model. In: Vaudenay, S. (ed.) EUROCRYPT 2006. LNCS, vol. 4004, pp. 289–307. Springer, Heidelberg (2006)
18. Di Raimondo, M., Gennaro, R., Krawczyk, H.: Deniable Authentication and Key Exchange. In: Proceedings of the 13th ACM Computer and Communication Security, CCS 2006, pp. 400–409 (2006)
19. IEEE P1363a Committee, IEEE P1363a, Version D6, November 9, 2000. Standard specifications for public-key cryptography
20. Goldwasser, S., Micali, S.: Probabilitic encryption. J. Comput. Syst. Sci. 28(2), 270–299 (1984)
21. Herzog, J., Lizkov, M., Micali, S.: Plaintext Awareness via Key Registration. In: Boneh, D. (ed.) CRYPTO 2003. LNCS, vol. 2729, pp. 548–564. Springer, Heidelberg (2003)
22. Hofheinz, D., Kiltz, E.: Secure Hybrid Encryption from Weakened Key Encapsulation. In: Menezes, A. (ed.) CRYPTO 2007. LNCS, vol. 4622, pp. 553–571. Springer, Heidelberg (2007)
23. Hofheinz, D., Kiltz, E.: Practical Chosen Ciphertext Secure Encryption from Factoring. In: EUROCRYPT 2009. LNCS, vol. 5479, pp. 313–332. Springer, Heidelberg (2009)
24. Jiang, S., Wang, H.: Plaintext-Awareness of Hybrid Encryption. Full version of this work, http://sites.google.com/site/shaoquan0825
25. Kurosawa, K., Desmedt, Y.: A New Paradigm of Hybrid Encryption Scheme. In: Franklin, M. (ed.) CRYPTO 2004. LNCS, vol. 3152, pp. 426–442. Springer, Heidelberg (2004)
26. Kurosawa, K., Matsuo, T.: How to Remove MAC from DHIES. In: Wang, H., Pieprzyk, J., Varadharajan, V. (eds.) ACISP 2004. LNCS, vol. 3108, pp. 236–247. Springer, Heidelberg (2004)
27. Möller, B.: A Public-Key Encryption Scheme with Pseudo-random Ciphertexts. In: Samarati, P., Ryan, P.Y.A., Gollmann, D., Molva, R. (eds.) ESORICS 2004. LNCS, vol. 3193, pp. 335–351. Springer, Heidelberg (2004)
28. Pass, R.: On the deniability in the common reference string and random oracle model. In: Boneh, D. (ed.) CRYPTO 2003. LNCS, vol. 2729, pp. 316–337. Springer, Heidelberg (2003)
29. Phan, D.H., Pointcheval, D.: About the Security of Ciphers (Semantic Security and Pseudo-Random Permutations). In: Handschuh, H., Hasan, M.A. (eds.) SAC 2004. LNCS, vol. 3357, pp. 182–197. Springer, Heidelberg (2004)

Speed Records for NTRU*

Jens Hermans**, Frederik Vercauteren***, and Bart Preneel

Katholieke Universiteit Leuven, ESAT/SCD-COSIC and IBBT
Kasteelpark Arenberg 10
B-3001 Leuven-Heverlee, Belgium
{jens.hermans,frederik.vercauteren,bart.preneel}@esat.kuleuven.be

Abstract. In this paper NTRUEncrypt is implemented for the first time on a GPU using the CUDA platform. As is shown, this operation lends itself perfectly for parallelization and performs extremely well compared to similar security levels for ECC and RSA giving speedups of around three to five orders of magnitude. The focus is on achieving a high throughput, in this case performing a large number of encryptions/decryptions in parallel. Using a modern GTX280 GPU a throughput of up to 200 000 encryptions per second can be reached at a security level of 256 bits. This gives a theoretical data throughput of 47.8 MB/s. Comparing this to a symmetric cipher (not a very common comparison), this is only around 20 times slower than a recent AES implementation on a GPU.

Keywords: NTRU encryption, Graphical Processing Unit, Parallelization, CUDA.

1   Introduction

Graphical Processing Units (GPUs) have long been used only for the rendering of games and other graphical applications. More recent GPUs are also used for general purpose parallel programming, using new programming models. A General Purpose GPU (GPGPU) contains a large number of processor cores (240 for the GTX280 [23]) that run at frequencies that are mostly lower than CPUs (1.2 GHz for the GTX280 GPU compared to 3.8 GHz for a recent Intel Pentium 4 [18]). Compared to a CPU a GPU provides a much larger computing power (several GFlops, or even TFlops for multiple GPUs) for specific parallel applications, because of the large number of cores. The recent change towards general scalar processor cores, that support 32- or 64-bit integer and bitwise operations, offers a new opportunity to implement cryptographic applications on GPUs.

---

* This work was supported in part by the IAP Programme P6/26 BCRYPT of the Belgian State (Belgian Science Policy).
** Research assistant, sponsored by the Fund for Scientific Research - Flanders (FWO).
*** Postdoctoral Fellow of the Fund for Scientific Research - Flanders (FWO).

J. Pieprzyk (Ed.): CT-RSA 2010, LNCS 5985, pp. 73–88, 2010.

There are several cryptographic ciphers that have a high level of parallelism, making them suitable for implementation on GPU. For performing a single encryption/decryption GPUs might not be very well suited: there is a latency compared to a CPU, because of the transfer of the data between main memory and GPU memory. In many applications the focus is not on the latency of a single cryptographic application, but on the throughput: one wants to perform a large number of encryptions/decryptions as fast as possible. In the case of a symmetric block cipher this will be the case when operating on a large block of data (using a suitable block cipher mode). Asymmetric ciphers are not often used in such a mode of operation, but more likely on servers that need to process many different secured connections where a large number of asymmetric cryptographic operations need to be performed. Currently cryptographic co-processors are used to speed up these operations, but a GPU might provide an alternative for these co-processors. An advantage is the fact that GPUs are almost by default present in modern computers and are also much underused. Another advantage is the flexibility: GPUs are easy to reprogram, making it an interesting co-processor to add to large computing farms. The large power consumption of the fastest GPUs is however a disadvantage, especially with a growing focus on the energy performance of data centers. One of the most likely uses of GPUs will be performing attacks on ciphers. GPUs have a very good computing power / price ratio, making them very economic for bulk computations. One can have around 200 GFlops (around 1 TFlops theoretically) for less than €500. In many attacks multiple cryptographic operations need to be performed, or at least part of these operations, so implementing and optimizing the original cryptographic operation on GPU is also of great use for attacks.

The choice for NTRUEncrypt (in short: NTRU) as the cryptographic cipher is less obvious: RSA [24] and ECC [7] are currently the respectively dominant and rising ciphers. NTRU has a large potential as a future cipher, given the very simple nature of it's core operation: the convolution (compared to a modular exponentiation for RSA and repeated squaring/doubling for ECC). This simple operation makes it very suitable for embedded devices with limited computing power, but also for parallelization, since a convolution can be split up over several processors. NTRU also has a good (asymptotic) performance of $O(N^2)$ (or even $O(N \log N)$ using FFT), compared to, for example, $O(N^3)$ for RSA. So, NTRU is expected to outperform RSA and ECC at similar security levels, and NTRU will also provide a good scalability for the future.

Because of these properties of NTRU, it was chosen as the cipher to be implemented on a GPU in this paper. For this paper the *ees1171ep1* parameter set is used, a high security ($k = 256$, the symmetric key size in bits) parameter set as claimed in [27]. Besides this parameter set a special version using product-form polynomials is also implemented. Product-form polynomials improve performance even further. Taking this high security level into account, NTRU performs very well when comparing with RSA (which would require a 15360 bit modulus) and ECC. For high throughput applications a speedup of three to five orders of magnitude is reached compared to RSA and ECC. The

GPU implementation reaches a throughput of up to 200 000 encryptions per second which is equivalent to a theoretical data throughput of 47.8 MB/s.

Organization. In Section 2 previous work on cryptography on GPUs and NTRU implementations is discussed. Next, a brief introduction is given to the NTRU cryptosystem in Section 3, especially on the parameter sets that have been proposed in the literature. In Section 4 the basics of GPGPU programming are explained, with a focus on the CUDA platform. This knowledge of NTRU and CUDA is combined to make an optimized GPU implementation of NTRU in Section 5. Finally the performance of the implementation is evaluated and compared to other implementations and other ciphers in Section 6.

2   Related Work

There is already much software available for GPUs, ranging from simple linear algebra packages to complex physical simulations. There has not been much development of cryptographic applications for the GPU, until recently when GPUs started supporting integer and bitwise operations. For example, AES was implemented on GPU [20] [14] [8], offering a maximum throughput of 831 MBytes/s (128 bit key, [20]).

RSA [24] has been implemented before the introduction of recent GPGPU platforms, using the OpenGL API [21] and more recently using modern platforms [26] [11], reaching up to 813 modular exponentiations (1024-bit integers) per second [26]. GPUs are also used to launch attacks, for example elliptic curve integer factoring [5] and brute force attacks, like for wireless networks [25].

There are no GPU implementations for NTRU. NTRU has however been implemented on a variety of platforms, like embedded devices, FPGAs [3] and custom hardware [1]. NTRU turns out to perform very well on devices with limited computing capabilities, given the simple nature of the convolution that is the central encryption/decryption operation. Compared to other modern cryptosystems like ECC, NTRU turns out to be very fast [19].

3   NTRUEncrypt

In this section the basics of NTRU are briefly introduced, based upon [2], to which we refer for further, more complete, information.

Let $\mathbb{Z}$ denote the ring of integers and $\mathbb{Z}_q$ the integers modulo q. NTRUEncrypt is a public-key cryptosystem that works with the polynomial ring $P(N) = \mathbb{Z}[X]/(X^N - 1)$ (and $P_q(N) = \mathbb{Z}_q[X]/(X^N - 1)$), where N is a positive prime. A vector from $\mathbb{Z}^N$ (resp. $\mathbb{Z}_q^N$) can be associated with a polynomial by $f = (f_0, f_1, \ldots, f_{N-1}) = \sum_{i=0}^{N-1} f_i X^i$.

The multiplication of two polynomials $h = f \star g$ is defined as the cyclic convolution of their coefficients:

$$h_k = (f \star g)_k = \sum_{i+j \equiv k \mod N} f_i \cdot g_j \ (0 \leq k < N) \tag{1}$$

which is the ordinary polynomial multiplication modulo $X^N - 1$.

The polynomials used in NTRU are selected from several polynomial sets $\mathcal{L}_f, \mathcal{L}_g, \mathcal{L}_r$ and $\mathcal{L}_m$. First the basic operations (key creation, encryption and decryption) of NTRU are introduced and afterwards, in Section 3.1, the structure of the polynomials and the parameter sets are discussed.

Key creation. The private key is a polynomial f, chosen at random from the set $\mathcal{L}_f$. Another polynomial $g \in \mathcal{L}_g$ is also chosen at random, but is not needed anymore after key generation. From these polynomials the public key h can be computed as

$$h = p \star f_q^{-1} \star g \quad \mathrm{mod}\ q \qquad (2)$$

where f_q^{-1} is the inverse of f in $P_q(N)$ and p is a polynomial (usually 3 or $X+2$).
   The polynomials f and g generally have small coefficients, while h has large coefficients.

Encryption. The message $m \in \mathcal{L}_m$ can be encrypted by choosing a random polynomial $r \in \mathcal{L}_r$ as a blinding factor and computing the ciphertext as

$$e = r \star h + m \quad \mathrm{mod}\ q. \qquad (3)$$

In practical schemes the message is padded with random bits and masked. For this paper, these steps are ignored, and only the computation of $r \star h + m \ \mathrm{mod}\ q$ is considered.

Decryption. Decryption can be done by convolving the ciphertext e with the private key f

$$a \equiv e \star f \equiv p \star r \star g + m \star f \quad \mathrm{mod}\ q \qquad (4)$$

and next convolving by $f_p^{-1} \ \mathrm{mod}\ p$. By a careful choice of f it can be assured that $f_p^{-1} = 1$, so only a reduction mod p is needed.
   One of the problems NTRU faces are decryption failures: the first step of the decryption only computes $a \ \mathrm{mod}\ q$ and not a. The problem is that knowing $a \ \mathrm{mod}\ q$ is not enough to know $a \ \mathrm{mod}\ p$. The problem of decryption failures has been studied extensively in [17]. In this paper it suffices to pick the coefficients of a from $(-q/2, q/2]$ and assume the probability of decryption failures is negligibly low.

3.1   Parameter Sets

The parameter N must always be chosen to be prime, since composites allows the problem to be decomposed [13]. The parameter q is mostly chosen as a power of 2, to ease the computations modulo q. The parameter p must be relatively prime to q, but it is not necessary that p is an integer, it can be a polynomial. Popular choices for p are 3 and $X + 2$.

Besides the parameters N, p, q there are the sets of polynomials $\mathcal{L}_f, \mathcal{L}_g, \mathcal{L}_m, \mathcal{L}_r$ that have to be defined. The message space $\mathcal{L}_m$ is defined as $P_p(N)$, since the message is obtained during the decryption after reducing modulo p.

The other sets of polynomials are chosen as ternary (for $p = 3$) or binary (for $p = X + 2$) polynomials.

Ternary polynomials. Define $\mathcal{L}(d_x, d_y)$ as the set of all ternary polynomials that have d_x coefficients set to 1 and d_y coefficients set to -1 (all other coefficients are 0).

One of the most natural choices for the polynomial sets is

$$\mathcal{L}_f = \{1 + p \star F : F \in \mathcal{L}(d_f, d_f)\} , \ \mathcal{L}_r = \mathcal{L}(d_r, d_r) , \ \mathcal{L}_g = \mathcal{L}(d_g, d_g)$$

which is also used in the most recent standards draft [27]. The choice of $\mathcal{L}_f$ as $1 + p \star F$ guarantees that $f_p^{-1} = 1$.

For ternary polynomials p is set to 3.

Binary polynomials. Binary polynomials offer an alternative for ternary polynomials and are much easier to implement in hardware and software. A disadvantage is that binary polynomials are by definition unbalanced, so $f(1) \neq 0$. As a consequence information on m, namely $m(1)$, leaks.

In [12] the following parameters are used:

$$\mathcal{L}_f = \{1 + p \star F : F \in \mathcal{L}(d_f, d_f)\} , \ \mathcal{L}_r = \mathcal{L}(d_r, 0) , \ \mathcal{L}_g = \mathcal{L}(d_g, 0)$$

Product-form polynomials. The central operation when encrypting is a convolution with a binary/ternary polynomial. The number of non-zero elements in $r \in \mathcal{L}_r$ is crucial for the performance of the encryption operation. A smaller number of non-zero elements will make the convolution faster (and lower memory usage, depending on the storage strategy) but will also degrade the security. By taking

$$\mathcal{L}_r = \{r_1 \star r_2 + r_3 : r_1 \in \mathcal{L}_{r_1}, r_2 \in \mathcal{L}_{r_2}, r_3 \in \mathcal{L}_{r_3}\}$$

with $d_{r_1}, d_{r_2}, d_{r_3} \ll d_r$ the convolution is still secure, since $r_1 \star r_2 + r_3$ still contains roughly the same amount of randomness as a single random r [15]. For our implementation $d_{r_1} = d_{r_2} = d_{r_3} = 5$, so each polynomial r_i has 10 non-zero coordinates. The performance is however increased drastically. The convolution $t = r \star h \mod q$ can be computed in several steps as in [3]:

$$t_1 \leftarrow r_2 \star h \ ; \quad t_2 \leftarrow r_1 \star t_1 \ ; \quad t_3 \leftarrow r_3 \star h \ ; \quad t \leftarrow t_2 + t_3 \mod q \quad (5)$$

Since each of r_1, r_2, r_3 have a low number of non-zero elements, the convolutions in (5) are much faster, requiring less additions than $r \star h$. Another advantage is the lower storage requirement.

4   GPU Programming

4.1   The CUDA Platform

The CUDA programming guide [22] explains in detail all aspects of the platform and programming model and was used as a basis for the following sections. The GTX280 that was used for this paper is a GPU that belongs to the range of Tesla Architecture GPUs from Nvidia. The Tesla architecture is based upon an array of multiprocessors (30 for the GTX280) that each contain 8 scalar processors. A multiprocessor is operated as a SIMT-processor (Single-Instruction, Multiple-Thread): a single instruction uploaded to the GPU causes multiple threads to perform the same scalar operation (on different data). The CUDA programming model from Nvidia, that is used to program their GPUs, provides a C-like language to program the GPU.

Programming model. As stated above, all programming is done using scalar operations: one needs to program a single thread which will then be executed in multitude on the GPU. Threads are grouped into blocks. All blocks together form a 'grid' of blocks. Threads within the same block can use shared memory. Both threads and blocks can be addressed in a multi-dimensional way. All scheduling of instructions (threads) on the multiprocessors is hidden from the programmer and is done on-chip. Threads are scheduled in *warps* of 32 threads. For optimal performance divergent branching inside the same half-warp (16 threads) must be avoided: each thread in a half-warp must execute the same instruction, otherwise the execution will be serialized. If divergent branching occurs, one possible strategy is to ensure that the thread ID for which divergence occurs coincides with a change of half-warp.

Memory. A multiprocessor contains fast on-chip memory in the form of registers, shared memory and caches. Off-chip memory is also available in the form of global memory and specialized texture and constant memory. The global memory is not cached. The GTX280 provides 1GB of off-chip memory.

Each of the memory types has specific features and caveats[1]:

- Global memory: as the global memory is off-chip there is a large performance hit (hundreds of cycles). Another issue is that multiple threads might access different global memory addresses at the same time, which creates a bottle-neck and forces the scheduler to stop the execution of the block until all memory is loaded. It is recommended to run a large number of blocks, to ensure the scheduler can keep the multiprocessors busy, while memory loading takes place. One way to avoid such large performance penalties are coalesced memory reads, in which all threads from a half-warp access either the same address or a block of consecutive addresses. In the case of loading a single address the total cost is only one memory load.

---

[1] Texture memory is not used in this paper, so details have been omitted.

- Registers: care has to be taken to limit the number of registers per thread as the registers are shared among all threads and blocks running on the same multiprocessor.
- Shared memory: shared memory is stored in banks, such that consecutive 32 bits are stored in consecutive banks. When accessing shared memory one needs to ensure that threads within the same warp access different banks, to avoid 'bank conflicts'. Bank conflicts result in serialization of the execution.
- Constant memory: the advantage of using constant memory is the presence of a special read-only cache, which allows for fast access times.

Instructions. Almost all operations that are available in C can be used in CUDA. CUDA only uses 32-bit (`int`, `float`) and 64-bit variables (`long`, `double`) for arithmetic, other types are converted first. In this paper, we will refer to 32-bit integers as 'int' (or just 'integer') and to 64-bit integers as a 'long'. Integer addition and bitwise operations take 4 clock cycles. 32-bit integer multiplication takes 16 cycles. Integer division and modulo are expensive and should be avoided.

5   The Implementation

For the implementation the *ees1171ep1* parameter set from [27] is used. This parameter set (with ternary polynomials and $N = 1171, p = 3, q = 2048 = 2^{11}, d_r = 106$) is one of the three strongest from the draft standard. Considering the relatively young age of NTRU and recent attacks (e.g. [16]), it is better to be rather conservative in the parameter choices and take one of the strong parameter sets.

Two implementations were made: one using the default ternary polynomials, the other using product-form ternary polynomials. In the last case $d_{r_1} = d_{r_2} = d_{r_3} = 5$.

The generation of random data (needed for encryption) is performed by the CPU, although parallel implementations exist for CUDA. There are several reasons for this choice: first of all it is the goal of this paper to compare the central NTRU operation, the convolution, and not to compare choices of random number generators. By computing the random numbers beforehand on CPU, any influence of the choice of the random generator is excluded. Second, one might consider an attack strategy in which the opponent would explicitly choose r, instead of using random numbers. Another advantage of performing the generation of r on CPU is exploiting the parallel computation by using both CPU and GPU.

5.1   Operations

Both parallel encryption (two variants) and parallel decryption are implemented on CUDA. The superscript i in m^i denotes the i-th message that is used in the parallel computation. The operations are defined as follows:

- **Encryption:** given $r^i \in \mathcal{L}_r$, h^i and $m^i \in \mathcal{L}_m$ (for $i \in [0, P)$, with P the number of parallel encryption operations) the kernel computes $e^i = r^i \star h^i + m^i \mod q$. Two strategies for the public key are considered: one which uses the same public key for all encryptions ($\forall i : h_i = h$) and one with different public keys for every operation.
- **Decryption:** given e^i and f, compute m^i. The private key is the same for all decryptions.

Key generation was not implemented, although situations exist where one would like to generate multiple keys in parallel.

For encryption both ordinary and product-form ternary polynomials are used as r.

The decryption operation can be written as

$$e \star f \equiv e \star (1 + p \star F) \equiv e + (e \star F) + (e \star F) \ll 1 \mod q \qquad (6)$$

where "$\ll$" is a left bit shift. Besides some extra scalar operations for each coefficient, one can reuse the encryption algorithm. In the next sections only encryption is discussed. The results section only includes results for the case that F is an ordinary ternary polynomial. Because there was no performance difference compared to encryption, decryption was not implemented for product-form polynomials.

5.2   Memory Usage - Bit Packing

Since all data must be transferred from the main computer memory to the GPU (device) memory, it is in the best interest to limit the amount of memory used.

One standard technique is bit packing. The ternary coefficients of r can be encoded as 2 bits, of which 32 can be packed into a 64-bit `long`. The coefficients of h are each 11 bit long, allowing for up to 5 coefficients to be stored in a `long`. We however pack only 4 elements of h in a long. The extra unused bits come in handy when performing an addition on the entire long, so that the overflow does not corrupt one of the packed values stored higher in the bit array. Note that although the polynomial m also has ternary coefficients we choose to store it using 11 bits per element. This way, the result of the encryption e (which is mod q) can be written in the same space as m, which results in a smaller memory usage. In total 623 `long`'s are required to store h, m and r.

For the implementation with product-form polynomials the values of r_1, r_2 and r_3 can be stored in a different way. Instead of encoding each ternary coefficient as two bits, the indices of the non-zero coefficients are stored, as in [3]. Since each index is in $[0, N-1]$, $\lceil \log_2(N) \rceil = 11$ bits are needed to store each index. These indices are again packed, but not aligned to 16 bit multiples, since the access is sequential (see further). The memory consumption is only lowered moderately to 592 `longs`, but the new structure of the convolution has a large impact on the construction of the loops and thus the performance.

Since multiple encryption/decryption operations are performed, multiple messages m and blinds r need to be uploaded to the device. All variables are stored in one large array of long's, e.g. a single m^b is packed to 293 longs, with the total array being $293 \times P$ long's. Note that the time for bit packing the data on CPU is not included in the timing results and that all host-memory is page-locked.

In the next sections and the algorithms in Appendix 7, we use the notation $x_{\text{packed},i}$ to refer to the long containing the i-th element of the x polynomial (which is denoted as x_i). $P(i)$ is used to denote the index of the long that contains x_i. When there is a reference to x_i in the pseudo-code, the index calculation and decoding are implicit.

5.3   Encoding

The coefficients of h are encoded as 11 bit integers, in the range $[0, q-1]$. The blind r, consisting of ternary coefficients, is encoded by mapping $\{0, 1, -1\}$ to 2-bit values (which can be chosen arbitrarily). The message m also consists of ternary coefficients, but for efficient computation, these are loaded in the memory space that will contain the result e. Because of this, the ternary coefficients are stored as 11-bit values in two's complement (e.g. $(-1)_3 = 2^{11} - 1$).

5.4   Blocks, Threads and Loop Nesting

Parallelism is present at two levels: at the level of a single encryption, which is split over multiple threads, and at the level of the parallel encryptions, which are split over the blocks. When performing a single encryption, one needs to access all elements r_i^b, h_j^b and e_k^b. Each block (block index denoted with the superscript b) is responsible for doing a single encryption/decryption operation. To make storing e_k as efficient as possible, each thread is responsible of computing 4 coefficients of e, which implies that each thread writes only one long.

For the normal ternary polynomials, the algorithm executed by each thread is presented in Algorithm 3. There is an implicit 'outer loop' that iterates over k (the parallel threads). The middle loop (over i) selects the element from r^b and then uses simple branching and addition (no multiplications).

Algorithm 1 shows the algorithm for the product-form ternary polynomials. The implicit outer loop is the same, but the computation inside is completely different. The computation of $r_2 \star h$ is split over all threads and the results are stored (packed) in shared memory. Unlike the other convolutions in Algorithm 1, all threads need all indices of $r_2 \star h$ and not just the $k \ldots k+3$-th coefficients.

Since r_1, r_2 and r_3 are stored using indices, the convolution algorithm is different from that used for ordinary polynomials. Algorithm 2 describes part of such a convolution. Again, only 4 coefficients of the result are computed, which matches the division of the convolution among the threads.

5.5   Memory Access

Since the convolutions are very simple operations, using only addition and some index-keeping operations, the memory access will be the bottleneck. One of the solutions is to explicitly cache the elements of r and h in registers (the GPU does not have a cache for global memory). Especially for r this turns out to be a good strategy, since each `long` contains 32 coefficients, thereby reducing the number of accesses to global memory with a factor 32. For h no significant benefits were observed, so the caching was omitted. The main reason is that the packed coefficients of h are less often accessed (many of the r_i are zero) and they are accessed in a more or less random pattern, so caching them for more than one iteration (over i in Algorithm 3) makes no sense. There is however a benefit from executing multiple threads in parallel: when thread t accesses h_j, thread $t+1$ will at the same time access h_{j+4}, which is always stored in the next `long`. This means that memory access is coalesced, although bad alignment of the memory blocks will prevent the full speedup.

For product-form polynomials the number of memory accesses is much lower: the space used to store r is smaller. As r_1, r_2 and r_3 are accessed only once, this means a drop in memory access from 296 to 48 bytes per block. The number of accesses to h also goes down: only the convolutions $r_2 \star h$ and $r_3 \star h$ need access to h. r_2 and r_3 each have 10 non-zero coefficients, giving a total of 20 accesses to h for each element in the result, so $20 \times 1171 = 23420$ `longs` per block, compared to $2Nd_r = 248252$ `longs` per block for ordinary polynomials.

Note that the access to e is coalesced, since each thread accesses a consecutive `long`.

5.6   Branching

Almost no divergent branching occurs during the execution of the algorithms. In the case of normal polynomials branching on r_i is not divergent, as each thread has the same value for i. The only divergent branches are for the modulo computation. There is one aspect when using product-form polynomials in Algorithm 1 that might cause a performance hit: the thread synchronization. Since the intermediate result t_{shared} is shared among all threads, all threads should wait for the completion of that computation.

6   Results

In this section the results of the GPU implementations are compared to a simple CPU implementation in C and other implementations found in the literature. The CPU tests were performed on an Intel Core2 Extreme CPU, running at 3.00GHz. This processor has four cores, but only one of these cores is used as the CPU implementation is not parallel. The GPU simulations were performed on a GTX280. To verify that all implementations were correct, the output was verified (with success) against a reference implementation in Magma [6].

Table 1 shows the results expressed as milliseconds per operation (or operations per second). Results for different h_i are, obviously, only available for the GPU when doing multiple (20000) operations in parallel. The times in Table 1 are the minimal times over 10 identical experiments. All results are expressed as wall clock time, since this is the only way to be able to compare CPU and

Table 1. Performance comparison of NTRU on an Intel Core2 CPU and a Nvidia GTX280 GPU using ordinary and product-form ternary polynomials ($N = 1171, q = 2048, p = 3$)

	Encryption (different h^i)		Encryption (same h^i)		Decryption	
	μs/op	op/s	μs/op	op/s	μs/op	op/s
Ordinary						
CPU	-	-	$10.5 \cdot 10^3$	(95)	$10.5 \cdot 10^3$	(95)
GPU, 1 op.	-	-	$1.75 \cdot 10^3$	-	$1.87 \cdot 10^3$	-
GPU, 20000 ops	41.3	24 213	40.0	25 025	41.1	24 331
Product-form						
CPU	-	-	$0.31 \cdot 10^3$	(3225.8)	-	-
GPU, 1 op.	-	-	$0.16 \cdot 10^3$	-	-	-
GPU, $\sim 2^{16}$ ops	4.58	218 204	4.51	221 845	-	-

GPU. Taking the minimum time ensures that clearing of the cache or context switches do not bias the results. Clearing of the cache and context switches depend heavily on the environment in which the program is used, so it would not be fair to include these in the measurements. Overall, the GPU times had a small variance, so the difference between average time and minimal time was negligible. The time for copying data from main to GPU memory is included in the GPU performance figures.

The CPU implementation does not use any optimizations like bit packing and just consists of a few nested loops. The CPU implementation only performs one single encryption/decryption. Despite the fact that the CPU implementation is not optimized, we use it as a rough basis for comparison for the GPU version. The available performance results for previous implementations are for different (less secure) parameter sets, which makes it very hard to compare.

From Table 1 it is clear that encryption and decryption have roughly the same performance: the extra element-wise operations for decryption do not take much time. This is also the reason that decryption was not implemented separately for product-form ternary polynomials, since it would show the same performance. Encryption with the same h is slightly faster than using different h^i, although an explanation for this has not been found[2].

Figure 1 shows the subsequent gain in performance when increasing the number of parallel encryptions (for ordinary polynomials). Around 2^{11} encryptions the GPU approaches its maximum performance, larger numbers of parallel encryptions yield only a slight improvement in the number of operations per second.

Table 1 shows that for all implementations, product-form polynomials are much faster, as expected by the lower number of memory accesses in Section 5.5. The performance increases by almost a factor 10 compared to ordinary polynomials. Again a small difference is observed between encryption with the same and different h^i.

---

[2] The opposite result was expected. As h was not stored in constant memory, there should be no benefit from caching.

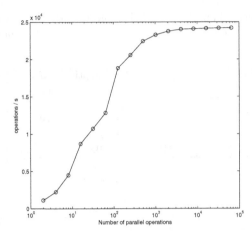

Fig. 1. NTRU encryption operations per second using ordinary polynomials and the same h ($N = 1171, q = 2048, p = 3$)

Table 2. Comparison of several NTRU, RSA and ECC implementations. The chosen parameter set and claimed security level (k) is listed for all ciphers. The number of operations per second is listed, together with the amount of data encrypted/decrypted per operation (excluding all padding, headers...).

Platform		(N, q, p)	Enc/s	Dec/s	bit/op
FPGA [3]	Xilinx Virtex 1000EFG860 @ 50 MHz	$(251, 128, X + 2)$	$193 \cdot 10^3$	-	251
Palm [3]	Dragonball @ 20 MHz (C)	Product form	21	11	
Palm [3]	Dragonball @ 20 MHz (ASM)	($k < 80$)	30	16	
ARM C [3]	ARM7TDMI @ 37 MHz		307	148	
FPGA [1]	Xilinx Virtex 1000EFG860 @ 500kHz	$(167, 128, 3)$	18	8.4	250
		($k \ll 80$)			
C	Intel Core2 Extreme @ 3.00GHz	$(1171, 2048, 3)$	95	95	1756
CUDA	GTX280 (1 op)	($k = 256$ [27])	571	546	
CUDA	GTX280 (20000 ops)		$24 \cdot 10^3$	$24 \cdot 10^3$	
C	Intel Core2 Extreme @ 3.00GHz	$(1171, 2048, 3)$	$3.22 \cdot 10^3$	-	1756
CUDA	GTX280 (1 op)	Product form	$6.25 \cdot 10^3$	-	
CUDA	GTX280 ($\sim 2^{16}$ ops)	($k = 256$ [27])	$218 \cdot 10^3$	-	
RSA comparison					
CUDA [26] Nvidia 8800GTS		1024 bit		813	1024
C++ [9]	Intel Core2 @ 1.83GHz	($k = 80$ [4])	($14 \cdot 10^3$)	657	1024
CUDA [26] Nvidia 8800GTS		2048 bit		104	2048
C++ [9]	Intel Core2 @ 1.83GHz	($k = 112$ [4])	($6.66 \cdot 10^3$)	168	2048
ECC comparison					
CUDA [26] Nvidia 8800GTS (PointMul)		ECC NIST-224		$1.41 \cdot 10^3$	
C [10]	Intel Core2 @ 1.83 GHz (ECDSA)	($k = 112$ [4])		$1.86 \cdot 10^3$	

Table 2 compares the CPU and GPU implementations with previous work on NTRU and to some RSA and ECC implementations. A note of caution is due, since the previous NTRU implementations use much lower security parameters and because the platforms that are used are totally different. Also note that the amount of data encrypted per operation is different. As a very rough extrapolation to convert the results for the other NTRU implementations to the security level of our implementation one can use the $O(N^2)$ asymptotic performance of NTRU. This drastically lowers the performance measures for the other NTRU implementations, ignoring even the increase of q and d_r. For applications with a focus on high throughput (many op/s), the CUDA implementation for product-form polynomials outperforms all other NTRU implementations (taking the higher security parameters and amount of data into account). The implementation with product-form polynomials gives a speed of more than 200 000 encryptions per second or 41.8 MByte/s. For applications that need to perform a small number of encryptions with low latency, the parallelization of CUDA does not give much speedup compared to the CPU. However, when comparing NTRU with RSA and ECC, the speedup is large: up to 1300 times faster than 2048-bit RSA and 117 times faster than ECC NIST-224 when comparing the number of encryptions per second (or up to 1113 times faster than 2048-bit RSA when comparing the data throughput). In addition, the security level of NTRU is much higher: when extrapolating to RSA and ECC with $k = 256$ bit security, this would add an extra factor of around 10 for ECC and around 400 for RSA (assuming $O(N^3)$ operations for RSA and ECC, where N is the length of a message block). So, in this extrapolation, NTRU has a speedup of five orders of magnitude compared to RSA and three orders of magnitude compared to ECC. The results listed for RSA encryption on CPU are operations with a small public key ($e = 17$), which allows for further optimization that has not been done for the RSA GPU implementation.

7   Conclusion

In this paper NTRU encryption/decryption was implemented for the first time on GPU. Several design choices, such as the NTRU parameters sets, are compared. The exact implementation is analysed in detail against the CUDA platform, explaining the impact of every choice by looking at the underlying effects on branching, memory access, blocks & threads... Although the programming is done in C, the CUDA model has its own specific ins and outs that take some time to learn, making a good implementation not very straightforward.

Many external factors, like power consumption, cost, reprogrammability, context (latency vs throughput), space... besides the speed of the cipher influence the choice of platform. In areas in which reprogrammability, cost and throughput are important and power consumption is of lesser importance, a GPU implementation is a very good option.

For 2^{16} encryptions a peak performance of around 218 000 encryptions/s (or 4.58×10^{-6} s/encryption) is reached, using product-form polynomials. This corresponds to a theoretical data throughput of 47.8 MB/s. The GPU performs at its best when performing a large number of parallel NTRU operations. Parallel NTRU implementations could serve well on servers processing many secured connections or in various attack strategies in which many (partial) encryption operations are needed. A single NTRU operation on GPU is still faster than a (simple) CPU implementation, but the speedup is limited. Even then a GPU might be interesting to simply move load off the CPU.

Comparing NTRU to other cryptosystems like RSA and ECC shows that NTRU, at a high security level, is much faster than RSA (around five orders of magnitude) and ECC (around three orders of magnitude). Even when only performing a single operation NTRU is still faster by around a factor of 35 for 2048 bit RSA and 3 for ECC NIST-244. Because of the ways NTRU can be parallelized, NTRU also clearly outperforms RSA and ECC for high-throughput applications. So, both for low-latency (single operation) and high-throughput (multiple operations) applications NTRU on GPU outperforms RSA and ECC.

References

1. Atıcı, A.C., Batina, L., Fan, J., Verbauwhede, I., Yalçın, S.B.O.: Low-cost implementations of NTRU for pervasive security. In: ASAP 2008, pp. 79–84. IEEE Computer Society, Los Alamitos (2008)
2. ECRYPT AZTEC. Lightweight Asymmetric Cryptography and Alternatives to RSA (2005)
3. Bailey, D.V., Coffin, D., Elbirt, A.J., Silverman, J.H., Woodbury, A.D.: NTRU in Constrained Devices. In: Koç, Ç.K., Naccache, D., Paar, C. (eds.) CHES 2001. LNCS, vol. 2162, pp. 262–272. Springer, Heidelberg (2001)
4. Barker, E., Barker, W., Burr, W., Polk, W., Smid, M.: Recommendation for Key Management. NIST special publication 800, 57 (2007)
5. Bernstein, D.J., Chen, H.C., Chen, M.S., Cheng, C.M., Hsiao, C.H., Lange, T., Lin, Z.C., Yang, B.Y.: The Billion-Mulmod-Per-Second PC. In: SHARCS 2009, pp. 131–144 (2009)
6. Bosma, W., Cannon, J., Playoust, C.: The Magma Algebra System I: The User Language. Journal of Symbolic Computation 24(3-4), 235–265 (1997)
7. Cohen, H., Frey, G., Avanzi, R.: Handbook of Elliptic and Hyperelliptic Curve Cryptography. CRC Press, Boca Raton (2006)
8. Cook, D., Ioannidis, J., Keromytis, A.D., Luck, J.: Cryptographics: Secret key cryptography using graphics cards. In: Menezes, A. (ed.) CT-RSA 2005. LNCS, vol. 3376, pp. 334–350. Springer, Heidelberg (2005)
9. Dai, W.: Crypto++: benchmarks, http://www.cryptopp.com/benchmarks.html
10. Ecrypt Ebats. ECRYPT benchmarking of asymmetric systems (2007), http://www.ecrypt.eu.org/ebats/
11. Fleissner, S.: GPU-Accelerated Montgomery Exponentiation. In: Shi, Y., van Albada, G.D., Dongarra, J., Sloot, P.M.A. (eds.) ICCS 2007. LNCS, vol. 4487, pp. 213–220. Springer, Heidelberg (2007)
12. Consortium for Efficient Embedded Security. Efficient embedded security standards #1: Implementation aspects of NTRU and NSS, Version 1 (2002)
13. Gentry, C.: Key Recovery and Message Attacks on NTRU-Composite. In: Pfitzmann, B. (ed.) EUROCRYPT 2001. LNCS, vol. 2045, pp. 182–194. Springer, Heidelberg (2001)
14. Harrison, O., Waldron, J.: AES Encryption Implementation and Analysis on Commodity Graphics Processing Units. In: Paillier, P., Verbauwhede, I. (eds.) CHES 2007. LNCS, vol. 4727, pp. 209–226. Springer, Heidelberg (2007)
15. Hoffstein, J., Silverman, J.H.: Random small Hamming weight products with applications to cryptography. Discrete Applied Mathematics 130(1), 37–49 (2003)

16. Howgrave-Graham, N.: A Hybrid Lattice-Reduction and Meet-in-the-Middle Attack Against NTRU. In: Menezes, A. (ed.) CRYPTO 2007. LNCS, vol. 4622, pp. 150–169. Springer, Heidelberg (2007)
17. Howgrave-Graham, N., Nguyen, P.Q., Pointcheval, D., Proos, J., Silverman, J.H., Singer, A., Whyte, W.: The impact of decryption failures on the security of ntru encryption. In: Boneh, D. (ed.) CRYPTO 2003. LNCS, vol. 2729, pp. 226–246. Springer, Heidelberg (2003)
18. Intel. Intel Pentium 4 - SL8Q9 Datasheet (2008)
19. Karu, P., Loikkanen, J.: Practical Comparison of Fast Public-key Cryptosystems (2001), http://www.tml.tkk.fi/Opinnot/Tik-110.501/2000/papers/
20. Manavski, S.A.: CUDA Compatible GPU as an Efficient Hardware Accelerator for AES Cryptography. In: ICSPC 2007, November 2007, pp. 65–68. IEEE, Los Alamitos (2007)
21. Moss, A., Page, D., Smart, N.P.: Toward Acceleration of RSA Using 3D Graphics Hardware. In: Galbraith, S.D. (ed.) Cryptography and Coding 2007. LNCS, vol. 4887, pp. 364–383. Springer, Heidelberg (2007)
22. Nvidia. Compute Unified Device Architecture Programming Guide (2007)
23. Nvidia. GeForce GTX280 - GeForce GTX 200 GPU Datasheet (2008)
24. Rivest, R.L., Shamir, A., Adleman, L.M.: A Method for Obtaining Digital Signatures and Public-Key Cryptosystems. Commun. ACM 21(2), 120–126 (1978)
25. Settings, M.: Password crackers see bigger picture. Network Security 2007(12), 20 (2007)
26. Szerwinski, R., Güneysu, T.: Exploiting the Power of GPUs for Asymmetric Cryptography. In: Oswald, E., Rohatgi, P. (eds.) CHES 2008. LNCS, vol. 5154, pp. 79–99. Springer, Heidelberg (2008)
27. Whyte, W., Howgrave-Graham, N., Hoffstein, J., Pipher, J., Silverman, J.H., Hirschhorn, P.: IEEE P1363.1: Public Key Cryptographic Techniques Based on Hard Problems over Lattices

Appendix

A   Code Listings

Algorithm 1. Pseudo-code for a single NTRU encryption (product-form polynomials)

1: $b \leftarrow$ blockID
2: $k \leftarrow 4 * $ threadID
3: Allocate $e_{\text{temp}}[0 \dots 3] \leftarrow 0$
4: Allocate $t_{\text{shared}}[0 \dots N-1]$
5: $t_{\text{shared}}[k \dots k+3] \leftarrow Convolve(h^b, r^b_{2,+}, r^b_{2,-}, k, t_{\text{shared}}[k \dots k+3])$
6: Synchronize threads
7: $e_{\text{temp}}[0 \dots 3] \leftarrow Convolve(t_{\text{shared}}, r^b_{1,+}, r^b_{1,-}, k, e_{\text{temp}}[0 \dots 3])$
8: $e_{\text{temp}}[0 \dots 3] \leftarrow Convolve(h^b, r^b_{3,+}, r^b_{3,-}, k, e_{\text{temp}}[0 \dots 3])$
9: **for** $l = 0$ to 3 **do**
10:    $e^b_{k+l} \leftarrow m^b_{k+l} + e_{\text{temp}}[l] \mod q$
11: **end for**

Algorithm 2. Pseudo-code for a single product-form convolution. $Convolve(h, r^+, r^-, k, t)$

Require: h: an ordinary polynomial,
   r^+, r^-: the positions of the $+1$ and -1 elements in the polynomial r,
   t: result of the convolution,
   k: offset of the results that need to be calculated.
Ensure: $t[k \ldots k + 3] = \{h \star r\}_{k \ldots k+3}$
 1: $k \leftarrow 4 * \text{threadID}$
 2: **for** $l = 0$ to $d_{r-1} - 1$ **do**
 3:    $i \leftarrow r_l^+$
 4:    **for** $\delta_k = 0$ to 3 **do**
 5:       $t[k + \delta_k] \leftarrow t[k + \delta_k] + h_{(k+\delta_k - i \mod N)}$
 6:    **end for**
 7: **end for**
 8: **for** $l = 0$ to $d_{r-1} - 1$ **do**
 9:    $i \leftarrow r_l^-$
10:    **for** $\delta_k = 0$ to 3 **do**
11:       $t[k + \delta_k] \leftarrow t[k + \delta_k] - h_{(k+\delta_k - i \mod N)}$
12:    **end for**
13: **end for**
14: **return** $t[k \ldots k + 3] \mod q$

Algorithm 3. Pseudo-code for a single NTRU encryption (ordinary polynomials)

 1: $b \leftarrow \text{blockID}$
 2: $k \leftarrow 4 * \text{threadID}$
 3: Allocate $e_{\text{temp}}[0 \ldots 3] \leftarrow 0$
 4: **for** $i = 0$ to 10 **do**
 5:    **for** $l = 0$ to 3 **do**
 6:       **if** $P(i) \neq P(i-1)$ **then**
 7:          $r_{\text{cache}} \leftarrow r^b_{\text{packed},i}$
 8:       **end if**
 9:       $r_{\text{elem}} \leftarrow r_i$ (from r_{cache})
10:       $j \leftarrow k + l - i \mod N$
11:       **if** $r_{\text{elem}} = 1$ **then**
12:          $e_{\text{temp}}[l] \leftarrow e_{\text{temp}}[l] + h^b_j$
13:       **end if**
14:       **if** $r_{\text{elem}} = -1$ **then**
15:          $e_{\text{temp}}[l] \leftarrow e_{\text{temp}}[l] - h^b_j$
16:       **end if**
17:    **end for**
18: **end for**
19: **for** $l = 0$ to 3 **do**
20:    $e^b_{k+l} \leftarrow m^b_{k+l} + e_{\text{temp}}[l] \mod q$
21: **end for**

Algorithm 4. Pseudo-code for a single NTRU Decryption

Require: F: the private key ($f = 1 + p \star F$)
   e: the encrypted message
 1: $k \leftarrow 4 * \text{threadID}$
 2: Execute Algorithm 3, taking $m = 0$, $r = F$ and $h = e$ and obtaining $t[0 \ldots 3]$.
 3: **for** $l = 0$ to 3 **do**
 4:    $t[l] \leftarrow t[l] + (t[l] \ll 1) + e_{k+l}$
 5:    $tmp \leftarrow t[l] - p * ((p^{-1} \mod q) * t[l] \gg \log_2 q)$
 6:    $(t[l] > q) \Rightarrow (tmp \leftarrow tmp + 1)$
 7:    $t[l] \leftarrow tmp$
 8: **end for**

High-Speed Parallel Software Implementation of the η_T Pairing

Diego F. Aranha[1,*], Julio López[1], and Darrel Hankerson[2]

[1] University of Campinas
{dfaranha,jlopez}@ic.unicamp.br
[2] Auburn University
hankedr@auburn.edu

Abstract. We describe a high-speed software implementation of the η_T pairing over binary supersingular curves at the 128-bit security level. This implementation explores two types of parallelism found in modern multi-core platforms: vector instructions and multiprocessing. We first introduce novel techniques for implementing arithmetic in binary fields with vector instructions. We then devise a new parallelization of Miller's Algorithm to compute pairings. This parallelization provides an algorithm for pairing computation without increasing storage costs significantly. The combination of these acceleration techniques produce serial timings at least 24% faster and parallel timings 66% faster than the best previous result in an Intel Core platform, establishing a new state-of-the-art implementation of this pairing instantiation in this platform.

Keywords: Efficient software implementation, vector instructions, multi-core architectures, bilinear pairings, parallelization.

1 Introduction

The computation of bilinear pairings is the most expensive operation in Pairing-based Cryptography, especially for high levels of security. For this reason, implementations must employ all the resources found in the target platform to obtain maximum efficiency. A resource being increasingly introduced in computing platforms is parallelism, in the form of vector instructions (data parallelism) and multiprocessing (task parallelism). This trend is observed even in the embedded space, with proposals of resource-constrained multi-core architectures and vector instruction sets for multimedia processing in portable devices.

This work describes a high-performance implementation of the η_T pairing [1] over binary supersingular curves at the 128-bit security level which employs these two forms of parallelism in a very efficient way. The target platform is the Intel Core architecture [2], the most popular 64-bit computing platform. Our main contributions are:

---

* Supported by FAPESP under grant no. 2007/06950-0.

J. Pieprzyk (Ed.): CT-RSA 2010, LNCS 5985, pp. 89–105, 2010.

- *Novel techniques for implementing arithmetic in binary fields:* we explore powerful SIMD instructions to accelerate arithmetic in binary fields. We focus on the SSE family of vector instructions, but the same techniques can be employed with other SIMD instruction sets such as Altivec and the upcoming AMD SSE5.
- *Parallelization of Miller's Algorithm to compute pairings:* we develop a simple algorithm for parallel pairing computation which does not increase storage costs. Our parallelization is independent of the underlying pairing instantiation, allowing a parallel implementation to reach scalability in a variable number of processors unrelated to the pairing mathematical definition. This parallelization provides good scalability in fields of small characteristic.
- *Static load balancing technique:* we present a simple technique to balance the costs of parallel pairing computation between the available processing units. The technique is successfully applied for latency minimization, but its flexibility allows the implementation to determine controlled non-optimal partitions of the algorithm.
- *Experimental results:* speedups of parallel implementations over serial implementations are estimated and experimentally verified for platforms up to 8 processors. We also obtain an approximation of the performance up to 32 processing units and compare our serial and parallel execution times with the current state-of-the-art implementations with the same parameters.

The results of this work can improve serial and parallel implementations of pairings. The parallelization may be important to reduce the latency of pairing computation in two scenarios: (i) desktop-class processors running real-time applications with strict response time requirements; (ii) embedded multiprocessor architectures with weak processing units. The availability of parallel algorithms for application in these scenarios is suggested as an open problem by [3] and [4]. Our features of flexible load balancing and small storage overhead are critical for the second scenario, because they can support static scheduling schemes for compromises between pairing computation time and power consumption; and memory capacity is commonly restricted in embedded devices.

2   Finite Field Arithmetic

In this section we will represent the elements of $\mathbb{F}_{2^m}$ using a polynomial basis. Let $f(z)$ be an irreducible binary polynomial of degree m. The elements of $\mathbb{F}_{2^m}$ are the binary polynomials of degree at most $m-1$. A field element $a(z) = \sum_{i=0}^{m-1} a_i z^i$ is associated with the binary vector $a = (a_{m-1}, \ldots, a_1, a_0)$ of length m. In a software implementation, these bit coefficients are packed and stored in an array $(a[0], \ldots, a[n-1])$ of n W-bit words, where W is the word size of the processor. For simplicity, we assume that n is always even.

2.1   Vector Instruction Sets

Vector instructions, also called SIMD (Single Instruction, Multiple Data) because they operate in several data objects simultaneously, are widely supported in recent families of processor architectures. The number, functionality and efficiency of these instructions have been improved with each new generation of processors, and natural applications include multimedia processing, scientific applications or any software with high arithmetic density. Some well-known SIMD instruction sets are the Intel MMX and SSE [5] families, the Altivec extensions introduced by Apple and IBM in the Power architecture specification and AMD 3DNow. Instruction sets supported by current technology are restricted to 128-bit registers and provide simple orthogonal operations across 8, 16, 32 or 64-bit data units stored inside these registers, but future extensions such as Intel AVX and AMD SSE5 will support 256-bits registers with the added inclusion of a heavily-anticipated carry-less multiplier [6].

The Intel Core microarchitecture is equipped with several vector instruction sets which operate in 16 architectural 128-bit registers. A small subset of these instructions can be used to implement binary field arithmetic, some found in the Streaming SIMD Extensions 2 (SSE2) and others in the Supplementary SSE3 instructions (SSSE3). The SSE2 instruction set is also supported by the recent VIA Nano processors, AMD processors since the K8 family and Intel processors since the Pentium 4.

A non-exhaustive list of SSE2 instructions relevant for our work is given below. Each instruction described will be referred in the algorithms by the short mnemonic which follows the instruction opcode:

- MOVDQU/MOVDQA (*load/store*): implements load/store between unaligned/aligned memory addresses and registers. In our implementation, all allocated memory is stored in 128-bit aligned base addresses so that the faster MOVDQA instruction can always be used.
- PSLLQ/PSRLQ ($\ll_{\dagger 8}, \gg_{\dagger 8}$): implements bitwise left/right shifts of a pair of 64-bit integers while shifting in zero bits. This instruction does not propagate bits from the lower 64-bit integer to the higher 64-bit integer, thus additional shifts and additions are required to implement bitwise shifts of 128-bit values.
- PSLLDQ/PRLLDQ ($\ll_8, \gg_8$): implements byte-wise left/right shifts of a 128-bit register. Since this instruction propagates bytes from the lower half to the higher half of a 128-bit register, this instruction is preferred over the previous one when the shift amount is a multiple of 8. Thus shifts by multiples of 8 bits should be used whenever possible. The latency of this instruction is 2 cycles in the first generation of Core 2 Conroe/Merom (65nm) processors and 1 cycle in the more recent Penryn/Wolfdale (45nm) microarchitecture.
- PXOR/PAND/POR ($\oplus, \wedge, \vee$): implements bitwise XOR/AND/OR of two 128-bit registers. These instructions have a high throughput, reaching 3 instructions per cycle when the operands are registers and there are no dependencies between consecutive operations.
- PUNPCKLBW/PUNPCKHBW (*interlo/interhi*): interleaves the lower/higher bytes in a register with the lower/higher bytes of another register.

We also find application for powerful but often-missed SSSE3 instructions:

- PALIGNR ($\triangleleft$): takes registers r_a and r_b, concatenate their values, and pull out a 128-bit section from an offset given by a constant immediate; in other words, implements a right byte-wise shift with propagation of shifted out bytes from r_a to r_b. This instruction can be used to implement a left shift by s bytes with the immediate $(16 - s)$.

- PSHUFB (*lookup* or *shuffle* depending on functionality): takes registers of bytes $r_a = a_0, a_1, \ldots, a_{16}$ and $r_b = b_0, b_1, \ldots, b_{16}$ and replaces r_a with the permutation $a_{b_0}, a_{b_1}, \ldots, a_{b_{16}}$; except that it replaces a_i with zero if the most significant bit of b_i is set. A powerful use of this instruction is to perform 16 simultaneous lookups in a 16-byte lookup table. This can be easily done by storing the lookup table in r_a and the lookup indexes in r_b. Intel introduced a specific *Super Shuffle Engine* in the latest microarchitecture to reduce the latency of this instruction from 3 cycles to 1 cycle.

Alternate vector instruction sets present functional analogues of these instructions. In particular, the PSHUFB permutation instruction is implemented as VPERM in Altivec and as PPERM in SSE5, although the PPERM instruction is reportedly more powerful as it can also operate at bit level. SIMD instructions are critical for the performance of binary field arithmetic and can be easily accessed with compiler intrinsics. In the remainder of this section, the optimization techniques applied during the implementation of each field operation are detailed. We will describe algorithms in terms of *vector operations* using the mnemonics defined above so that algorithms can be easily transcribed to other target platforms. Specific instruction choices based on latency or functionality will be focused on the SSE family.

2.2 Squaring

Since the square of a finite field element $a(z) \in \mathbb{F}_{2^m}$ is given by $a(z)^2 = \sum_{i=0}^{m-1} a_i z^{2i} = a_{m-1} z^{2m-2} + \cdots + a_2 z^4 + a_1 z^2 + a_0$, the binary representation of $a(z)^2$ can be computed by inserting a zero bit between each pair of consecutive bits on the binary representation of $a(z)$. This operation can be accelerated by introducing a lookup table as discussed in [7]. This method can be improved further if the table lookups can be executed simultaneously. This way, for an implementation which processes 4 bits per iteration, squaring can be implemented mainly in terms of permutation instructions which convert groups of 4 bits (*nibbles*) to the corresponding expanded bytes. The proposed optimization is shown in Algorithm 1. The algorithm receives a field element a stored in a vector of n 64-bit words (or $\frac{n}{2}$ 128-bit values) and expands the input into a double-precision vector t which can be reduced modulo $f(z)$. At each iteration of this algorithm, a 128-bit value $a[2i]$ is loaded from memory and separated by a bit mask into two registers containing the low nibbles (a_L) and the high nibbles (a_H). Each group of nibbles is then expanded from 4 bits to 8 bits by a parallel table lookup. The proper order of bytes is restored by interleaving instructions which pick alternately the lower or higher bytes of a_L or a_H to form two consecutive 128-bit values ($t[2i], t[2i + 1]$) produced as the result.

Algorithm 1. Proposed implementation of squaring in $\mathbb{F}_{2^m}$.

Input: $a(z) = a[0..n-1]$.
Output: $c(z) = c[0..n-1] = a(z)^2 \bmod f(z)$.

1: ▷ Store in *table* the squares $u(z)^2$ of all 4-bit polynomials $u(z)$.
2:    *table* ← 0x5554515045444140,0x1514111005040100
3:    *mask* ← 0x0F0F0F0F0F0F0F0F,0x0F0F0F0F0F0F0F0F
4: **for** $i \leftarrow 0$ **to** $\frac{n}{2} - 1$ **do**
5:     $a_0 \leftarrow load(a[2i])$
6:     $a_L \leftarrow a_0 \wedge mask$, $a_L \leftarrow lookup(table, a_L)$
7:     $a_H \leftarrow a_0 \gg_{\dagger 8} 4$, $a_H \leftarrow a_H \wedge mask$, $a_H \leftarrow lookup(table, a_H)$
8:     $t[2i] \leftarrow store(interlo(a_L, a_H))$, $t[2i+1] \leftarrow store(interhi(a_L, a_H))$
9: **end for**
10: **return** $c = t \bmod f(z)$

2.3 Square Root

Given an element $a(z) \in \mathbb{F}_{2^m}$, the field element $c(z)$ such that $c(z)^2 = a(z) \bmod f(z)$ can be computed by $c(z) = a_{even} + \sqrt{z} \cdot a_{odd} \bmod f(z)$, where a_{even} represents the concatenation of even coefficients of $a(z)$, a_{odd} represents the concatenation of odd coefficients of $a(z)$ and $\sqrt{z}$ is a constant depending on the irreducible polynomial $f(z)$ [8]. When $f(z)$ is a suitable trinomial $f(z) = z^m + z^t + 1$ with odd exponents m, t, $\sqrt{z}$ has the sparse form $\sqrt{z} = z^{\frac{m+1}{2}} + z^{\frac{t+1}{2}}$ and multiplication by this constant can be computed with shifts and additions only.

This algorithm can also be implemented with simultaneous table lookups. Algorithm 2 presents our implementation of this method with vector instructions. The algorithm processes 128 bits of a in each iteration and progressively separates the coefficients of $a[2i]$ in even or odd coefficients. First, a permutation mask is used to divide $a[2i]$ in bytes of odd index and bytes of even index. The bytes with even indexes are stored in the lower 64-bit part of a_0 and the bytes with odd indexes are stored in the higher 64-bit part of a_0. The high and low nibbles of a_0 are then divided into a_L and a_H and additional lookup tables are applied to further separate the bits of a_L and a_H into bits with odd and even indexes. At the end of the 128-bit section, a_0 stores the interleaving of odd and even coefficients of a packed into groups of 4 bits. The remaining instructions in the 128-bit sections separate the even and odd coefficients into u and v, which can be reordered and multiplied by $\sqrt{z}$ inside the 64-bit section. We implement these final steps in 64-bit mode to avoid expensive shifts in 128-bit registers.

2.4 Multiplication

Two different strategies are commonly considered for the implementation of multiplication in $\mathbb{F}_{2^m}$. The first one consists in applying the Karatsuba algorithm [9] to divide the multiplication in sub-problems and solve each problem independently [7] (for $a(z) = A_1 z^{\lceil m/2 \rceil} + A_0$ and $b(z) = B_1 z^{\lceil m/2 \rceil} + B_0$):

$$c(z) = a(z) \cdot b(z) = A_1 B_1 z^m + [(A_1 + A_0)(B_1 + B_0) + A_1 B_1 + A_0 B_0] z^{\lceil m/2 \rceil} + A_0 B_0.$$

Algorithm 2. Proposed implementation of square root in $\mathbb{F}_{2^m}$.

Input: $a(z) = a[0..n-1]$, exponents m and t of trinomial $f(z)$.
Output: $c(z) = c[0..n-1] = a(z)^{\frac{1}{2}} \bmod f(z)$.

1: ▷ Permutation mask to divide a 128-bit value in bytes with odd and even indexes.
2: $perm \leftarrow$ 0x0F0D0B0907050301,0x0E0C0A0806040200
3: ▷ Tables to divide a low/high nibble in bits with odd and even indexes.
4: $sqrt_L \leftarrow$ 0x3332232231302120,0x1312030211100100
5: ▷ Table to divide a high nibble in bits with odd and even indexes ($sqrt_L \ll 2$).
6: $sqrt_H \leftarrow$ 0xCCC88C88C4C08480,0x4C480C0844400400
7: ▷ Bit masks to isolate bytes in lower or higher nibbles.
8: $mask_L \leftarrow$ 0x0F0F0F0F0F0F0F0F,0x0F0F0F0F0F0F0F0F
9: $mask_H \leftarrow$ 0xF0F0F0F0F0F0F0F0,0xF0F0F0F0F0F0F0F0
10: $c[0\ldots n-1] \leftarrow 0, h \leftarrow \frac{n+1}{2}, l \leftarrow \frac{t+1}{128}, s_1 \leftarrow \frac{m+1}{2} \bmod 64, s_2 \leftarrow \frac{t+1}{2} \bmod 64$
11: **for** $i \leftarrow 0$ **to** $\frac{n}{2} - 1$ **do**
12:     $a_0 \leftarrow load(a[2i]), \quad a_0 \leftarrow shuffle(a_0, perm)$
13:     $a_L \leftarrow a_0 \wedge mask_L, \; a_L \leftarrow lookup(sqrt_L, a_L)$,
14:     $a_H \leftarrow a_0 \wedge mask_H, \; a_H \leftarrow a_H \gg_{\dagger 8} 4, \quad a_H \leftarrow lookup(sqrt_H, a_H)$
15:     $a_0 \leftarrow a_L \vee a_H, \qquad a_L \leftarrow a_0 \wedge mask_L, \; a_H \leftarrow a_0 \wedge mask_H$
16:     $u \leftarrow store(a_L), v \leftarrow store(a_H)$
17:     ▷ From now on, operate in 64-bit registers.
18:     $a_{even} \leftarrow u[0] \vee u[1] \ll 4, a_{odd} \leftarrow v[1] \vee v[0] \gg 4$
19:     $c[i] \leftarrow c[i] \oplus a_{even}$
20:     $c[i+h-1] \leftarrow c[h+i-1] \oplus (a_{odd} \ll s_1)$
21:     $c[i+h] \leftarrow c[h+i] \oplus (a_{odd} \gg (64-s_1))$
22:     $c[i+l] \leftarrow c[i+l] \oplus (a_{odd} \ll s_2)$
23:     $c[i+l+1] \leftarrow c[i+l+1] \oplus (a_{odd} \gg (64-s_2))$
24: **end for**
25: **return** c

The second one consists in applying a direct algorithm like the *comb* method proposed by López and Dahab in [10]. Conventionally, the series of additions involved in this method are implemented through additions over sub parts of a double-precision vector. In order to reduce the number of memory accesses during these additions, we employ n registers. These registers simulate the series of memory additions by accumulating consecutive writes, allowing the implementation to reach maximum XOR throughput. We also employ an additional table T_1 analogue to T_0 which stores $u(z) \cdot (b(z) \ll 4)$ to eliminate shifts by 4, as discussed in [10]. Recall that shifts by multiples of 8 bits are faster in the target platform. We assume that the length of operand $b[0..n-1]$ is at most $64n-7$ bits; if necessary, terms of higher degree can be processed separately at relatively low cost. The implemented LD multiplication algorithm is shown as Algorithm 3. The element $a(z)$ is processed in groups of 8 bits separated by intervals of 128 bits. This avoids shifts of the register vector since a 128-bit shift can be emulated by referencing m_{i+1} instead of m_i. The multiple precision shift by 8 bits of the register vector ($\lhd 8$) is implemented with 15-byte shifts with carry propagation ($\lhd$) of register pairs.

Algorithm 3. LD multiplication implemented with n 128-bit registers.

Input: $a(z) = a[0..n-1], b(z) = b[0..n-1]$.
Output: $c(z) = c[0..n-1]$.
Note: m_i denotes the vector of $\frac{n}{2}$ 128-bit registers $(r_{(i-1+n/2)}, \ldots, r_i)$.

1: Compute $T_0(u) = u(z) \cdot b(z), T_1(u) = u(z) \cdot (b(z) \ll 4)$ for all $u(z)$ of degree < 4.
2: $(r_{n-1} \ldots, r_0) \leftarrow 0$
3: **for** $k \leftarrow 56$ **downto** 0 **by** 8 **do**
4:    **for** $j \leftarrow 1$ **to** $n-1$ **by** 2 **do**
5:       Let $u = (u_3, u_2, u_1, u_0)$, where u_t is bit $(k+t)$ of $a[j]$.
6:       Let $v = (v_3, v_2, v_1, v_0)$, where v_t is bit $(k+t+4)$ of $a[j]$.
7:       $m_{(j-1)/2} \leftarrow m_{(j-1)/2} \oplus T_0(u)$
8:       $m_{(j-1)/2} \leftarrow m_{(j-1)/2} \oplus T_1(v)$
9:    **end for**
10:   $(r_{n-1} \ldots, r_0) \leftarrow (r_{n-1} \ldots, r_0) \lhd 8$
11: **end for**
12: **for** $k \leftarrow 56$ **downto** 0 **by** 8 **do**
13:    **for** $j \leftarrow 0$ **to** $n-2$ **by** 2 **do**
14:       Let $u = (u_3, u_2, u_1, u_0)$, where u_t is bit $(k+t)$ of $a[j]$.
15:       Let $v = (v_3, v_2, v_1, v_0)$, where v_t is bit $(k+t+4)$ of $a[j]$.
16:       $m_{j/2} \leftarrow m_{j/2} \oplus T_0(u)$
17:       $m_{j/2} \leftarrow m_{j/2} \oplus T_1(v)$
18:    **end for**
19:    **if** $k > 0$ **then** $(r_{n-1} \ldots, r_0) \leftarrow (r_{n-1} \ldots, r_0) \lhd 8$
20: **end for**
21: **return** $c = (r_{n-1} \ldots, r_0) \bmod f(z)$

2.5 Modular Reduction

Efficient modular reduction depends on the format of the trinomial or pentanomial $f(z)$. In general, it's better to choose $f(z)$ such that bitwise shifts amounts are multiples of 8 bits. If the non-null coefficients of $f(z)$ are located in the lower words of the array representation of $f(z)$, consecutive writes into memory can also be accumulated into registers to avoid redundant memory writes. We illustrate these optimizations with modular reduction by $f(z) = z^{1223} + z^{255} + 1$ in Algorithm 4. The algorithm receives as input a vector of n 128-bit elements and reduces this vector by accumulating four memory writes at a time in registers. Note also that shifts by multiples of 8 bits are used whenever possible.

2.6 Inversion

For inversion in $\mathbb{F}_{2^m}$ we implemented a variant of the Extended Euclidean Algorithm for polynomials [7] where the length of each temporary vector is tracked. Since this algorithm requires flexible left shifts by arbitrary amounts, we implemented the full algorithm in 64-bit mode. Some Assembly in the form of a compiler intrinsic was used to efficiently count the number of leading 0 bits to determine the highest set bit.

Algorithm 4. Proposed modular reduction by $f(z) = z^{1223} + z^{255} + 1$.

Input: $t(z) = t[0..n-1]$ (vector of 128-bit elements).
Output: $c(z) \bmod f(z) = c[0..n-1]$.
Note: The accumulate function $R(r_3, r_2, r_1, r_0, t)$ executes:

$$s \leftarrow t \gg_{\dagger 8} 7, r_3 \leftarrow t \ll_{\dagger 8} 57$$
$$r_3 \leftarrow r_3 \oplus (s \ll_8 64)$$
$$r_2 \leftarrow r_2 \oplus (s \gg_8 64)$$
$$r_1 \leftarrow r_1 \oplus (t \ll_8 56)$$
$$r_0 \leftarrow r_0 \oplus (t \gg_8 72)$$

1: $r_0, r_1, r_2, r_3 \leftarrow 0$
2: **for** $i \leftarrow 19$ **downto** 15 **by** 4 **do**
3:     $R(r_3, r_2, r_1, r_0, t[i])$, $t[i-7] \leftarrow t[i-7] \oplus r_0$
4:     $R(r_0, r_3, r_2, r_1, t[i-1])$, $t[i-8] \leftarrow t[i-8] \oplus r_1$
5:     $R(r_1, r_0, r_3, r_2, t[i-2])$, $t[i-9] \leftarrow t[i-9] \oplus r_2$
6:     $R(r_2, r_1, r_0, r_3, t[i-3])$, $t[i-10] \leftarrow t[i-10] \oplus r_3$
7: **end for**
8: $R(r_3, r_2, r_1, r_0, t[11])$, $t[4] \leftarrow t[4] \oplus r_0$
9: $R(r_0, r_3, r_2, r_1, t[10])$, $t[3] \leftarrow t[3] \oplus r_1$
10: $t[2] \leftarrow t[2] \oplus r_2$, $t[1] \leftarrow t[1] \oplus r_3$, $t[0] \leftarrow t[0] \oplus r_0$
11: $r_0 \leftarrow m[9] \gg_8 64$, $r_0 \leftarrow r_0 \gg_{\dagger 8} 7$, $t[0] \leftarrow t[0] \oplus r_0$
12: $r_1 \leftarrow r_0 \ll_8 64$, $r_1 \leftarrow r_1 \ll_{\dagger 8} 63$, $t[1] \leftarrow t[1] \oplus r_1$
13: $r_1 \leftarrow r_0 \gg_{\dagger 8} 1$, $t[2] \leftarrow t[2] \oplus r_1$
14: **for** $i \leftarrow 0$ **to** 9 **do** $c[2i] \leftarrow store(t[i])$
15: $c[19] \leftarrow c[19] \wedge \texttt{0x7F}$
16: **return** c

2.7   Implementation Timings

In this section, we present our timings for finite field arithmetic. We implemented arithmetic in $\mathbb{F}_{2^{1223}}$ with irreducible trinomial $f(z) = z^{1223} + z^{255} + 1$. This field is suitable for instantiations of the η_T pairing over supersingular binary curves at the 128-bit security level [4]. The C programming language was used in conjunction with compiler intrinsics for accessing vector instructions. The chosen compiler was GCC version 4.1.2 because it generated the fastest code from vector intrinsics, as already observed by [4]. The differences between our implementations in the 65nm and 45nm processors can be explained by the lower cost of the PSLLDQ and PSHUFB instructions in the newer generation after the introduction of the *Super Shuffle Engine* by Intel.

Field multiplication was implemented by a combination of one instance of Karatsuba and the LD method depicted as Algorithm 3. Karatsuba's splitting point was at 632 bits and the divide-and-conquer steps were also implemented with vector instructions. Note that our binary field multiplier precomputes two tables of 16 rows, while the multiplier implemented in [4] precomputes a single table. This increase in memory consumption is negligible when compared to the total memory capacity of the target platform.

Table 1. Comparison of different software implementations of finite field arithmetic in two Intel Core 2 platforms. All timings are reported in cycles. Improvements are computed in comparison with the previous fastest result in a 65nm platform, since the related works do not present timings for field operations in a 45nm platform.

Implementation	Operation			
	$a^2 \bmod f$	$a^{\frac{1}{2}} \bmod f$	$a \cdot b \bmod f$	$a^{-1} \bmod f$
Hankerson et al. [4]	600	500	8200	162000
Beuchat et al. [11]	480	749	5438	–
This work (Core 2 65nm)	160	166	4030	149763
Improvement	66.7%	66.8%	25.9%	7.6%
This work (Core 2 45nm)	108	140	3785	149589

3 Pairing Computation

Miller's Algorithm for pairing computation requires a rich mathematical framework. We briefly present some definitions and point the reader to more complete treatments of the subject presented in [12,13].

3.1 Preliminary Definitions

An *admissible bilinear pairing* is an efficiently computable map $e : \mathbb{G}_1 \times \mathbb{G}_2 \to \mathbb{G}_T$, where $\mathbb{G}_1$ and $\mathbb{G}_2$ are additive groups of points in an elliptic curve E and $\mathbb{G}_T$ is a related multiplicative group. Let P, Q be r-torsion points. The computation of a bilinear pairing $e(P, Q)$ requires the construction and evaluation of a function $f_{r,P}$ such that $div(f_{r,P}) = r(P) - r(\mathcal{O})$ at a divisor $\mathcal{D}$ which is equivalent to $(Q) - (\mathcal{O})$. Miller constructs $f_{r,P}$ in stages by using a double-and-add method [14]. Let $g_{U,V} : E(\mathbb{F}_{q^k}) \to \mathbb{F}_{q^k}$ be the line equation through points U and V. If $U = V$, the line $g_{U,V}$ is the tangent to the curve at U. If $V = -U$, the line g_U is the shorthand for $g_{U,-U}$. A *Miller function* is any function $f_{c,P}$ with divisor $div(f_{c,P}) = c(P) - (cP) - (c-1)(\mathcal{O})$, $c \in \mathbb{Z}$. The following property is true for all integers $a, b \in \mathbb{Z}$ [13, Theorem 2]:

$$f_{a+b,P}(\mathcal{D}) = f_{a,P}(\mathcal{D}) \cdot f_{b,P}(\mathcal{D}) \cdot \frac{g_{aP,bP}(\mathcal{D})}{g_{(a+b)P}(\mathcal{D})}. \tag{1}$$

Direct corollaries are:

(i) $f_{1,P}(\mathcal{D}) = 1$.

(ii) $f_{a,P}(\mathcal{D}) = f_{a-1,P}(\mathcal{D}) \cdot \frac{g_{(a-1)P,P}(\mathcal{D})}{g_{aP}(\mathcal{D})}$.

(iii) $f_{2a,P}(\mathcal{D}) = f_{a,P}(\mathcal{D})^2 \cdot \frac{g_{aP,aP}(\mathcal{D})}{g_{2aP}(\mathcal{D})}$.

Miller's Algorithm is depicted in Algorithm 5. The work by Barreto et al. [13] later showed how to use the final exponentiation of the Tate pairing to eliminate the denominators involved in the algorithm and to evaluate $f_{r,P}$ at Q instead of the divisor $\mathcal{D}$. Additional optimizations published in the literature focus on minimizing the latency of the Miller loop, that is, reduce the length of r while keeping its low Hamming weight [1,15,16].

Algorithm 5. Miller's Algorithm [14].

Input: $r = \sum_{i=0}^{\log_2(r)} r_i 2^i$, P, $\mathcal{D} = (Q + R) - (R)$
Output: $f_{r,P}(\mathcal{D})$.

1: $T \leftarrow P$, $f \leftarrow 1$
2: **for** $i = \lfloor \log_2(r) \rfloor - 1$ **downto** 0 **do**
3:   $f \leftarrow f^2 \cdot \frac{g_{T,T}(Q+R)g_{2T}(R)}{g_{2T}(Q+R)g_{T,T}(R)}$
4:   $T \leftarrow 2T$
5:   **if** $r_i = 1$ **then**
6:     $f \leftarrow f \cdot \frac{g_{T,P}(Q+R)g_{T+P}(R)}{g_{T+P}(Q+R)g_{T,P}(R)}$
7:     $T \leftarrow T + P$
8:   **end if**
9: **end for**
10: **return** f

3.2 Related Work

In this work, we are interested in parallel algorithms for pairing computation with no static limits on scalability, or more precisely, algorithms in which the scalability is not restricted by the mathematical definition of the pairing. Practical limits will always exist when: (i) the communication cost is dominant; (ii) the cost of parallelization is higher than the cost of computation.

Several works already developed parallel strategies for the computation of pairings achieving mixed results. Grabher et al. [3] analyzes two approaches: parallel extension field arithmetic, which gives good results but has a clear limit on scalability; a parallel Miller loop strategy for two processors, where lines 3-4 for all iterations in Miller's Algorithm are precomputed by one processor and both processors compute in parallel the iterations where $r_i = 1$. Because r frequently has a low Hamming weight, this strategy results in performance losses due to unbalanced computational costs between the processors.

Mitsunari [17] observes that the different iterations of the algorithm can be computed in parallel if the points T of different iterations are available and proposes a specialized version of the η_T pairing over $\mathbb{F}_{3^m}$ for parallel execution in 2 processors. In this version, all the values $(x_P^{\frac{1}{3}^i}, y_P^{\frac{1}{3}^i}, x_Q^{3^i}, y_Q^{3^i})$ used for line evaluation in the i-th iteration of the algorithm are precomputed and the Miller loop iterations are divided in sets of the same size. Hence load balancing is trivially achieved. Since the cost of cubing and cube root computation is small, this approach achieves good speedups ranging from 1.61 to 1.76 at two different security levels. However, it requires significant storage overhead, since $4 \cdot (\frac{m+1}{2})$ field elements must be precomputed and stored. This approach is generalized and extended in the work by Beuchat et al. [11], where results are presented for fields of characteristic 2 and 3 at the 128-bit security level. For characteristic 2, the speedups achieved by parallel execution reach 1.75, 2.53 and 2.57 for 2, 4, and 8 processors, respectively. For characteristic 3, the speedups reach 1.65, 2.26 and 2.79, respectively. This parallelization represents the current state-of-the-art in parallel implementations of cryptographic pairings.

Cesena and Avanzi [18,19] propose a technique to compute pairings over trace zero varieties constructed from supersingular elliptic curves and extensions with degrees $a = 3$ or $a = 5$. This approach allows a pairing computation to be packed in a short parallel Miller loops by the action of the a-th power of Frobenius. The problem with this approach is again the scalability limit (restricted by the extension degree a). The speedup achieved with parallel execution in 3 processors is 1.11 over a serial implementation of the η_T pairing at the same security level [19].

3.3 Parallelization

In this section, a parallelization of Miller's Algorithm is derived. This parallelization can be used to accelerate serial pairing implementations or improve the scalability of parallel approaches restricted by the pairing definition. This formulation is similar to the parallelization presented by [17] and [11], but our method focuses on minimizing the number of points needed for parallel executions of different iterations of the algorithm. This allows us to eliminate the overhead of storing $4(\frac{m+1}{2})$ precomputed field elements.

Miller's Algorithm computes $f_{r,P}$ in $\log_2(r)$ iterations. For a parallel algorithm, we must divide these $\log_2(r)$ iterations between some number π of processors. To achieve this, first we need a simple property of Miller functions [16,20].

Lemma 1. *Let P, Q be points on $E(\mathbb{F}_q)$, $\mathcal{D} \sim (Q) - (\infty)$ and $f_{c,P}$ denote a Miller function. For all integers $a, b \in \mathbb{Z}$, $f_{a \cdot b, P}(\mathcal{D}) = f_{b,P}(\mathcal{D})^a \cdot f_{a,bP}(\mathcal{D})$.*

We need this property because Equation (1) just divides a Miller's Algorithm instance computed in $\log_2(r)$ iterations in two instances computed in at least $\log_2(r) - 1$ iterations. If we could represent r as a product $r_0 \cdot r_1$, it would be possible to compute $f_{r,P}$ in two instances of $\frac{\log_2(r)}{2}$ iterations. Since for some pairing instantiations, r is a prime group order, we write r in the simple and flexible form $2^w r_1 + r_0$, with $w \sim \frac{\log_2(r)}{2}$. This way, we can compute:

$$f_{r,P}(\mathcal{D}) = f_{2^w r_1 + r_0, P}(\mathcal{D}) = f_{2^w r_1, P}(\mathcal{D}) \cdot f_{r_0, P}(\mathcal{D}) \cdot \frac{g_{(2^w r_1)P, r_0 P}(\mathcal{D})}{g_{rP}(\mathcal{D})}. \qquad (2)$$

The previous Lemma provides two choices to further develop $f_{2^w r_1, P}(\mathcal{D})$:

(i) $f_{2^w r_1, P}(\mathcal{D}) = f_{r_1, P}(\mathcal{D})^{2^w} \cdot f_{2^w, r_1 P}(\mathcal{D})$.
(ii) $f_{2^w r_1, P}(\mathcal{D}) = f_{2^w, P}(\mathcal{D})^{r_1} \cdot f_{r_1, 2^w P}(\mathcal{D})$.

The choice can be made based on efficiency: (i) compute w squarings in the extension field $\mathbb{F}_{q^k}^*$ and a point multiplication by r_1; (ii) compute an exponentiation to r_1 in the extension field and a point multiplication by 2^w (or w repeated point doublings). In the general case, the most efficient strategy will depend on the curve and embedding degree. The higher the embedding degree, the higher the cost of exponentiation in the extension field in comparison with point multiplication in the elliptic curve. If r has low Hamming weight, the two strategies should have similar costs. We adopt the first strategy:

$$f_{r,P}(\mathcal{D}) = f_{r_1, P}(\mathcal{D})^{2^w} \cdot f_{2^w, r_1 P}(\mathcal{D}) \cdot f_{r_0, P}(\mathcal{D}) \cdot \frac{g_{(2^w r_1)P, r_0 P}(\mathcal{D})}{g_{rP}(\mathcal{D})}. \qquad (3)$$

This formula is clearly suitable for parallel execution in $\pi = 3$ processors, since each Miller function can be computed in $\frac{\log_2(r)}{2}$ iterations. For our purposes, however, r will have low Hamming weight and r_0 will be very small. In this case, $f_{r,P}$ can be computed by two Miller functions of approximately $\frac{\log_2(r)}{2}$ iterations. The parameter w can be adjusted to balance the costs in both processors (w extension field squarings with a point multiplication by r_1).

This formula can also be applied recursively for $f_{r_1,P}$ and f_{2^w,r_1P} to develop a parallelization suitable for any number of processors. Observe that π also does not have to be a power of 2, because of the flexible way we write r to exploit parallelism. An important detail is that a parallel implementation will only have significant speedups if the cost of the Miller loop is dominant over the communication overhead or the parallelization overhead. It is also important to note that the higher the number of processors, the higher the number of squarings and the smaller the constants r_i involved in point multiplication. However, applying the formula recursively can increase the size of the integers which multiply P, because they will be a product of r_i constants. Thus, the scalability of this algorithm for π processors depends on the cost of squarings in the extension field, the cost of point multiplications by r_i in the elliptic curve and the actual length of the Miller loop. Fortunately, these parameters are constant and can be statically determined. If P is fixed (a private key, for example), the multiples r_iP can also be precomputed and stored with low storage overhead.

3.4   Parallel η_T Pairing

In this section, the performance gain of a parallel implementation of the η_T pairing over a serial implementation is investigated following the analysis by [4].

Let E be a supersingular curve with embedding degree $k = 4$ defined over $\mathbb{F}_{2^m}$ with equation $E/\mathbb{F}_{2^m} : y^2 + y = x^3 + x + b$. The order of E is $2^m + 1 \pm 2^{\frac{m+1}{2}}$. A quartic extension is built over $\mathbb{F}_{2^m}$ with basis $\{1, s, t, st\}$, where $s^2 = s + 1$ and $t^2 = t + s$. Let $P, Q \in E(\mathbb{F}_{2^m})$ be r-torsion points. An associated distortion map ψ from $E(\mathbb{F}_{2^m})[r]$ to $E(\mathbb{F}_{2^{4m}})$ is defined by $\psi : (x, y) \rightarrow (x + s^2, y + sx + t)$. For this family of curves, Barreto et al. [1] defined the optimized η_T pairing:

$$\eta_T : E(\mathbb{F}_{2^m})[r] \times E(\mathbb{F}_{2^{4m}})[r] \rightarrow \mathbb{F}^*_{2^{4m}},$$
$$\eta_T(P, Q) = f_{T',P'}(Q')^M, \tag{4}$$

with $Q' = \psi(Q)$, $T' = (-v)(2^m - \#E(\mathbb{F}_{2^m}))$, $P' = (-v)P$, $M = (2^{2m} - 1)(2^m + 1 \pm 2^{\frac{m+1}{2}})$ for a curve-dependent parameter $v \in \{-1, 1\}$.

At the 128-bit security level, the base field must have $m = 1223$ bits [4]. Let E_1 be the supersingular curve with embedding degree $k = 4$ defined over $\mathbb{F}_{2^{1223}}$ with equation $E_1(\mathbb{F}_{2^{1223}}) : y^2 + y = x^3 + x$. The order of E_1 is $5r = 2^{1223} + 2^{612} + 1$, where r is a 1221-bit prime number. Applying the parallel form developed in Section 3.3, the pairing computation can be decomposed in:

$$f_{T',P'}(Q')^M = \left(f_{2^{612-w},P'}(Q')^{2^w} \cdot f_{2^w,2^{612-w}P'}(Q') \cdot \frac{g_{2^{612-w}P',P'}(Q')}{g_{T'P'}(Q')} \right)^M.$$

Since squarings in $\mathbb{F}_{2^{4m}}$ and point duplication in supersingular curves require only binary field squarings and these can be efficiently computed, the cost of parallelization is low, but further improvements are possible. Barreto *et al.* [1] proposed a closed formula for this pairing based on a reversed-loop approach with square roots which eliminates the extension field squarings in Miller's Algorithm. Beuchat et al. [21] encountered further algorithmic improvements and proposed a slightly faster formula for the η_T pairing computation. We can obtain a parallel algorithm directly from the parallel formula derived above by excluding the involved extension field squarings and simply dividing the loop iterations between the processors. This algorithm is shown as Algorithm 6. In this algorithm, each processor i starts the loop from the w_i counter, computing w_i squarings/square roots of overhead. Without extension field squarings to offset these operations, it makes sense to assign processor 1 the first line evaluation and to increase the loop parts executed by processors with small w_i. The total overhead is smaller because extension field squarings are not needed and point arithmetic in binary supersingular curves can be computed with inexpensive squarings and square roots. Observe that the combining step can be implemented in at least two different ways: (i) serial combining of results with $(\pi - 1)$ serial extension field multiplications executed in one processor; (ii) parallel logarithmic combining of results with latency of $\lceil \log_2(\pi) \rceil$ extension field multiplications. We adopt the parallel strategy for efficiency.

3.5   Performance Analysis

Now we proceed with performance analysis of Algorithm 6. Processor 1 has an initialization cost of 3 multiplications and 2 squarings. Processor i has a parallelization cost of $2w_i$ squarings and $2w_i$ square roots. Additional parallelization overhead is $\lceil \log_2(\pi) \rceil$ extension field multiplications to combine the results. A full extension field multiplication costs 9 field multiplications. Each iteration of the algorithm executes 2 square roots, 2 squarings, 1 field multiplication and 1 extension field multiplication. Exploring the sparsity of G_i, this extension field multiplication costs 6 field multiplications. The final exponentiation has a cost of 26 multiplications, 7 finite field squarings, 612 extension field squarings and 1 inversion. Each extension field squaring costs 4 finite field squarings [21].

Let $\widetilde{m}, \widetilde{s}, \widetilde{r}, \widetilde{i}$ be the cost of finite field operations: multiplication, squaring, square root and inversion, respectively. For our efficient implementation of finite field $\mathbb{F}_{2^{1223}}$ in an Intel Core 2 65nm processor, we have $\widetilde{r} \approx \widetilde{s}$, $\widetilde{m} \approx 25\widetilde{s}$ and $\widetilde{i} \approx 37\widetilde{m}$. From these ratios, we will illustrate how to compute the optimal w_i values which balance the computational cost between processors. Let $c_\pi(i)$ be the computational cost of a processor $0 < i \leq \pi$ while executing its portion of the parallel algorithm. For $\pi = 2$ processors:

$$c_2(1) = (3\widetilde{m} + 2\widetilde{s}) + (7\widetilde{m} + 4\widetilde{s})w_2 = 80\widetilde{s} + (186\widetilde{s})w_2$$
$$c_2(2) = (4\widetilde{s})w_2 + (7\widetilde{m} + 4\widetilde{s})(611 - w_2).$$

Naturally, we always have $w_1 = 0$ and $w_{\pi+1} = 611$. Solving $c_2(1) = c_2(2)$ for w_2, we can obtain the optimal $w_2 = 309$. For $\pi = 4$ processors, we solve

---

Algorithm 6. Proposed parallelization of the η_T pairing (π processors).

Input: $P = (x_P, y_P), Q = (x_Q, y_Q) \in E(\mathbb{F}_{2^m}[r])$, starting point w_i for processor i.
Output: $\eta_T(P, Q) \in \mathbb{F}_{2^{4m}}^*$.

1: $y_P \leftarrow y_P + 1 - \delta$
2: **parallel section**(processor i)
3: **if** $i = 0$ **then**
4:     $u_i \leftarrow x_P + \alpha, v_i \leftarrow x_Q + \alpha$
5:     $g_{0_i} \leftarrow u_i \cdot v_i + y_P + y_Q + \beta$
6:     $g_{1_i} \leftarrow u_i + x_Q, g_{2_i} \leftarrow v_i + x_P^2$
7:     $G_i \leftarrow g_{0_i} + g_{1_i}s + t$
8:     $L_i \leftarrow (g_{0_i} + g_{2_i}) + (g_{1_i} + 1)s + t$
9:     $F_i \leftarrow L_i \cdot G_i$
10: **else**
11:     $F_i \leftarrow 1$
12: **end if**
13: $x_{Q_i} \leftarrow (x_Q)^{2^{w_i}}, y_{Q_i} \leftarrow (y_Q)^{2^{w_i}}$
14: $x_{P_i} \leftarrow (x_P)^{\frac{1}{2^{w_i}}}, y_{P_i} \leftarrow (y_P)^{\frac{1}{2^{w_i}}}$
15: **for** $j \leftarrow w_i$ **to** $w_{i+1} - 1$ **do**
16:     $x_{P_i} \leftarrow \sqrt{x_{P_i}}, y_{P_i} \leftarrow \sqrt{y_{P_i}}, x_{Q_i} \leftarrow x_{Q_i}^2, y_{Q_i} \leftarrow y_{Q_i}^2$
17:     $u_i \leftarrow x_{P_i} + \alpha, v_i \leftarrow x_{Q_i} + \alpha$
18:     $g_{0_i} \leftarrow u_i \cdot v_i + y_{P_i} + y_{Q_i} + \beta$
19:     $g_{1_i} \leftarrow u_i + x_{Q_i}$
20:     $G_i \leftarrow g_{0_i} + g_{1_i}s + t$
21:     $F_i \leftarrow F_i \cdot G_i$
22: **end for**
23: $F \leftarrow \prod_{i=0}^{\pi} F_i$
24: **end parallel**
25: **return** F^M

$c_4(1) = c_4(2) = c_4(3) = c_4(4)$ to obtain $w_2 = 158, w_3 = 312, w_4 = 463$. Observe that by solving a simple system of equations it is always possible to balance the computational cost between the processors. Furthermore, the latency of the Miller loop will always be equal to $c_\pi(1)$. Let $c_1(1)$ be the cost of a serial implementation of the main loop, par be the parallelization overhead and exp be the cost of final exponentiation. Considering the additional $\lceil \log_2(\pi) \rceil$ extension field multiplications as parallelization overhead and $26\tilde{m} + (7 + 2446)\tilde{s} + \tilde{i}$ as the cost of final exponentiation, the speedup for π processors is the ratio between the cost of the serial implementation over the cost of the parallel implementation:

$$s(\pi) = \frac{c_1(1) + exp}{c_\pi(1) + par + exp} = \frac{77 + 179 \cdot 611 + 3978}{c_\pi(1) + 225\lceil \log_2(\pi) \rceil + 3978}.$$

Table 2 presents speedups estimated by our performance analysis. Note that our efficient implementation of binary field arithmetic in a 45nm processor has a bigger multiplication-to-squaring ratio, concentrating higher computational costs in the main loop of the algorithm. This explains why the speedups should be higher in the 45nm processor.

Table 2. Estimated speedups for our parallelization of the η_T pairing over supersingular binary curves at the 128-bit security level. The optimal partitions were computed by a Sage[1] script.

Estimated speedup $s(\pi)$	Number π of processors					
	1	2	4	8	16	32
Core 2 65nm	1.00	1.90	3.45	5.83	8.69	11.48
Core 2 45nm	1.00	1.92	3.54	6.11	9.34	12.66

4   Experimental Results

We implemented the parallel algorithm for the η_T pairing over our efficient binary field arithmetic in two Intel Core platforms: an Intel Core 2 Quad 65nm platform running at 2.4GHz (Platform 1) and a dual quad-core Intel Xeon 45nm processor running at 2.0GHz (Platform 2). The parallel sections were implemented with OpenMP[2] constructs. OpenMP is an application programming interface that supports multi-platform shared memory multiprocessing programming in C, C++ and Fortran. We used a special version of the GCC 4.1.2 compiler included in Fedora Linux 8 with OpenMP support backported from GCC 4.2 and SSSE3 support backported from GCC 4.3. This way, we could use both multiprocessing support and fast code generation for SSE intrinsics.

The timings and speedups presented in Table 3 were measured on 10^4 executions of each algorithm. We present timings in millions of cycles to ignore differences in clock frequency between the target platforms. From the table, we can observe that real implementations can obtain speedups close to the estimated speedups derived in the previous section. We verified that threading creation and synchronization overhead stayed in the order of microseconds, being negligible compared to the pairing computation time. Timings for $\pi > 4$ processors in Platform 1 and $\pi > 8$ processors in Platform 2 were measured through a high-precision per-thread counter measured by the main thread. These timings might be an accurate approximation of future real implementations, but memory effects (such as cache locality) or scheduling influence may impose penalties.

Table 3 shows that the proposed parallelization presents good scalability. We improve the state-of-the-art serial and parallel execution times significantly. The fastest timing for computing the η_T pairing obtained by our implementation was 1.51 milliseconds using all 8 cores of Platform 2. The work by Beuchat et al. [11] reports a timing of 3.08 milliseconds in a Intel Core i7 45nm processor clocked at 2.9GHz. Note that we obtain a much faster timing with a lower clock frequency and without requiring the storage overhead of $4 \cdot (\frac{m+1}{2})$ field elements present in [11], which may reach 365KB for these parameters and be prohibitive in resource-constrained embedded devices.

---

[1] SAGE: Software for Algebra and Geometry Experimentation,
    http://www.sagemath.org
[2] Open Multi-Processing, http://www.openmp.org

Table 3. Experimental results for serial/parallel executions of the η_T pairing. Times are presented in millions of cycles and the speedups are computed by the ratio between execution times of serial implementations over execution times of parallel implementations. The columns marked with (*) present estimates based on per-thread data.

	Number of threads					
Platform 1 – Intel Core 2 65nm	**1**	**2**	**4**	**8***	**16***	**32***
Hankerson et al. [4] – latency	39	–	–	–	–	–
Beuchat et al. [11] – latency	26.86	16.13	10.13	–	–	–
Beuchat et al. [11] – speedup	1.00	1.67	2.65	–	–	–
This work – latency	18.76	10.08	5.72	3.55	2.51	2.14
This work – speedup	1.00	1.86	3.28	5.28	7.47	8.76
Improvement	30.2%	32.9%	39.9%	–	–	–
Platform 2 – Intel Core 2 45nm	**1**	**2**	**4**	**8**	**16***	**32***
Beuchat et al. [11] – latency	23.03	13.14	9.08	8.93	–	–
Beuchat et al. [11] – speedup	1.00	1.77	2.54	2.58	–	–
This work – latency	17.40	9.34	5.08	3.02	2.03	1.62
This work – speedup	1.00	1.86	3.42	5.76	8.57	10.74
Improvement	24.4%	28.9%	44.0%	66.2%	–	–

5   Conclusion and Future Work

In this work, we proposed novel techniques for exploring parallelism during the implementation of the η_T pairing over supersingular binary curves in modern multi-core computers. Powerful vector instructions of the SSE family were shown to accelerate considerably the arithmetic in binary fields. We obtained significant performance in computing the η_T pairing, using an efficient implementation of field multiplication, squaring and square root computation. The optimizations improved the state-of-the-art timings of this pairing instantiation at the 128-bit security level by 24% and 30% in two different Intel Core processors.

We also derived a parallelization of Miller's Algorithm to compute pairings. This parallelization is generic and can be applied to any pairing algorithm or instantiation. The construction also achieves good scalability in the symmetric case and this scalability is not restricted by the definition of the pairing. We illustrated the formulation when applied to the η_T pairing over supersingular binary curves and validated our performance analysis with a real implementation. The experimental results show that the actual implementation could sustain performance gains close to the estimated speedups. Parallel execution of the η_T pairing improved the state-of-the-art timings by at least 28%, 44% and 66% in 2, 4 and 8 cores respectively. This parallelization is suitable for embedded platforms and can be applied to reduce computation latency when response time is critical.

Future work can adapt the introduced techniques for the case $\mathbb{F}_{3^m}$. Improvements to the parallelization should focus on minimizing the serial region and parallelization cost. The proposed parallelization should also be applied to an optimal asymmetric pairing setting, where parallelization costs are clearly higher. Preliminary data for the R-ate pairing [16] over Barreto-Naehrig curves at the 128-bit security level points to a 10% speedup using 2 processor cores.

References

1. Barreto, P.S.L.M., Gailbraith, S., Ó hÉigeartaigh, C., Scott, M.: Efficient Pairing Computation on Supersingular Abelian Varieties. Design, Codes and Cryptography 42(3), 239–271 (2007)
2. Wechsler, O.: Inside Intel Core Microarchitecture: Setting new standards for energy-efficient performance. Technology@Intel Magazine (2006)
3. Grabher, P., Groszschaedl, J., Page, D.: On Software Parallel Implementation of Cryptographic Pairings. In: Avanzi, R.M., Keliher, L., Sica, F. (eds.) Selected Areas in Cryptography. LNCS, vol. 5381, pp. 34–49. Springer, Heidelberg (2009)
4. Hankerson, D., Menezes, A., Scott, M.: Identity-Based Cryptography, ch. 12, pp. 188–206. IOS Press, Amsterdam (2008)
5. Intel 64 and IA-32 Architectures Software Developer's Manual Volume 2: Instruction Set Reference, http://www.intel.com/Assets/PDF/manual/253666.pdf
6. Gueron, S., Kounavis, M.E.: Carry-Less Multiplication and Its Usage for Computing The GCM Mode. White paper, http://software.intel.com/
7. Hankerson, D., Menezes, A., Vanstone, S.: Guide to Elliptic Curve Cryptography. Springer, Secaucus (2003)
8. Fong, K., Hankerson, D., López, J., Menezes, A.: Field inversion and point halving revisited. IEEE Transactions on Computers 53(8), 1047–1059 (2004)
9. Karatsuba, A., Ofman, Y.: Multiplication of many-digital numbers by automatic computers (in Russian). Doklady Akad. Nauk SSSR 145, 293–294 (1962)
10. López, J., Dahab, R.: High-speed software multiplication in GF(2^m). In: Roy, B., Okamoto, E. (eds.) INDOCRYPT 2000. LNCS, vol. 1977, pp. 203–212. Springer, Heidelberg (2000)
11. Beuchat, J., López-Trejo, E., Martínez-Ramos, L., Mitsunari, S., Rodríguez-Henríquez, F.: Multi-core implementation of the Tate pairing over supersingular elliptic curves. In: Garay, J.A., Miyaji, A., Otsuka, A. (eds.) CANS 2009. LNCS, vol. 5888, pp. 413–432. Springer, Heidelberg (2009)
12. Barreto, P.S.L.M., Lynn, B., Scott, M.: Efficient Implementation of Pairing-Based Cryptosystems. Journal of Cryptology 17(4), 321–334 (2004)
13. Barreto, P.S.L.M., Kim, H.Y., Lynn, B., Scott, M.: Efficient algorithms for pairing-based cryptosystems. In: Yung, M. (ed.) CRYPTO 2002. LNCS, vol. 2442, pp. 354–368. Springer, Heidelberg (2002)
14. Miller, V.S.: The Weil Pairing, and Its Efficient Calculation. Journal of Cryptology 17(4), 235–261 (2004)
15. Hess, F., Smart, N.P., Vercauteren, F.: The eta pairing revisited. IEEE Trans. on Information Theory 52, 4595–4602 (2006)
16. Lee, H., Lee, E., Park, C.: Efficient and Generalized Pairing Computation on Abelian Varieties. IEEE Trans. on Information Theory 55(4), 1793–1803 (2009)
17. Mitsunari, S.: A Fast Implementation of η_T Pairing in Characteristic Three on Intel Core 2 Duo Processor. Cryptology ePrint Archive, Report 2009/032 (2009)
18. Cesena, E.: Pairing with Supersingular Trace Zero Varieties Revisited. Cryptology ePrint Archive, Report 2008/404 (2008)
19. Cesena, E., Avanzi, R.: Trace Zero Varieties in Pairing-based Cryptography. In: Conference on Hyperelliptic curves, discrete Logarithms, Encryption, etc. (2009), http://inst-mat.utalca.cl/chile2009/Slides/Roberto_Avanzi_2.pdf
20. Vercauteren, F.: Optimal pairings. Cryptology ePrint Archive, Report 2008/096 (2008)
21. Beuchat, J., Brisebarre, N., Detrey, J., Okamoto, E., Rodríguez-Henríquez, F.: A Comparison Between Hardware Accelerators for the Modified Tate Pairing over $\mathbb{F}_{2^m}$ and $\mathbb{F}_{3^m}$. In: Galbraith, S.D., Paterson, K.G. (eds.) Pairing 2008. LNCS, vol. 5209, pp. 297–315. Springer, Heidelberg (2008)

Refinement of Miller's Algorithm Over Edwards Curves

Lei Xu[1,2] and Dongdai Lin[1]

[1] State Key Laboratory of Information Security, Institute of Software,
Chinese Academy of Sciences, Beijing, China
[2] Graduate University of Chinese Academy of Sciences, Beijing, China

Abstract. Edwards gave a new form of elliptic curves in [1], and these curves were introduced to cryptography by Bernstein and Lange in [2]. The Edwards curves enjoy faster addition and doubling operations, so they are very attractive for elliptic curve cryptography.

In 2006, Blake, Murty and Xu proposed three refinements to Millers algorithm for computing Weil/Tate pairings over Weierstraß curves. In this paper we extend their method to Edwards curve and propose a faster algorithm for computing pairings with Edwards coordinates, which comes from the analysis of divisors of rational functions.

Keywords: Cryptography, bilinear pairing, Miller algorithm, twisted Edwards curve.

1 Introduction

Bilinear pairing on elliptic curves are of great interests due to their application in cryptography. It was first introduced by Alfred J.Menezes, Tatsuaki Okamoto and Scott A.Vanstone to reduce the discrete logarithms of elliptic curves to fintie fields([3]), which is known as the MOV attack. Frey and Rück([4]) also consider this situation using Tate pairing instead of Weil pairing.

Recent work on bilinear pairing has considered their positive applications. Dan Boneh and Matt Franklin proposed the first practical identity based encryption scheme([5]), which was first described by Shamir([6]). And many interesting applications of bilinear pairing are developed. Such as a one round protocol for tripartite Diffie-Hellman key exchange by Antoine Joux([7]), a short signature from Weil pairing by D.Boneh, B.Lynn and H.Shacham([8]) and so on.

Due to their various applications, a lot of effort has gone into efficient computing of bilinear pairing. Miller propose the first effective algorithm to calculate the bilinear pairing, which works in double-and-add manner([9]). And many improvements had been done to accelerate Miller's algorithm, see [10] for a survey.

In 2007, Edwards generalized an example from Euler and Gauss and introduced a new form of elliptic curve([1]). He showed that all elliptic curves over number fields could be transformed to the shape $x^2 + y^2 = c^2(1 + x^2y^2)$, with $(0, c)$ as neutral element and with simple and symmetric addition law. Bernstein and Lange in [2] presented fast explicit formulas for addition and doubling in

J. Pieprzyk (Ed.): CT-RSA 2010, LNCS 5985, pp. 106–118, 2010.

projective coordinates on an Edwards curve. They also generalize the addition law to the curve $x^2 + y^2 = c^2(1 + dx^2y^2)$, which covers more elliptic curves over finite field. In [11], Bernstein, Birkner, Joye, Lange, and Peters further generalized the Edwards curve to cover all curves $ax^2 + y^2 = 1 + dx^2y^2$.

Compared with Weierstraß curves, Edwards curves enjoy more efficient addition and double operation. So discrete logarithm based systems such as Diffie-Hellman key exchange or digital signatures that require efficient computation of scalar multiples benefit from Edwards curve.

However, the situation becomes complicated when Edwards curve is in the world of pairing based cryptography, where Miller's algorithm needs a function whose divisor is $(P) + (Q) - (P + Q) - (\mathcal{O})$, for input points P, Q and their sum $P + Q$.

Until recently, little work was dedicated to the improvement of pairing computation over Edwards curves. M. Prem Laxman Das and Palash Sarkar([12]) used birational equivalence to Weierstraß curves to calculate the bilinear pairing, Sorina Ionica and Antoine Joux([13]) used a different map to curve of degree 3 and compute the 4-th power of the Tate pairing, which is faster than Das and Sarkar's algorithm. Christophe Aréne, Tanja Lange, Michael Naehrig, and Christophe Ritzenthaler([14]) improved the computation of bilinear pairing on Edwards curve and twisted Edwards curve in a way that is similar to [15]. [14] also gives a geometric interpretation of the group law on Edwards curves and concise formulas for the coefficients of the conic.

In this paper, we propose an efficient algorithm to compute bilinear pairing over twisted Edwards curves. Our improvement comes from the consideration of the different combinations of the divisors and is different from the previous effort on pairing computation on Edwards curves.

The remainder of this paper is organized as follows: Section 2 recalls basic properties of bilinear pairing and Edwards curves. Section 3 presents our improvements to the original Miller's algorithm for twisted Edwards curves. In Section 4 we compare the improved algorithms with the original algorithm and give some detailed analysis. Section 5 gives the conclusion and some comments.

2   Background on Pairing and Twisted Edwards Curves

2.1   Bilinear Pairing and Miller's Algorithm

Let E/K be an elliptic curve. Weil pairing and Tate pairing are the two most important bilinear pairings.

Definition 1 (Divisor). *A divisor is an element of the free abelian group (Denoted by $Div(E)$) generated by the set of points of $E(\overline{K})$.*

Given a divisor $D = \sum_{P \in E} n_P(P)$, the degree of D is defined by $deg(D) = \sum_{P \in E} n_P$. The sum of divisor D is defined by $sum(D) = \sum_{P \in E} n_P P$.

The divisor of degree 0 is a subgroup of $Div(E)$ and is denoted by $Div^0(E)$.

The support of divisor D is the set of points P with $n_P \neq 0$.

For a nonzero rational function f over E, the corresponding divisor is defined to be $div(f) = ord_P(f)(P)$. It can be proved that $div(f) \in Div^0(E)$, and $div(f)$ is called principal divisor. A characterization of principal divisors is : $D = \sum_{P \in E} n_P(P)$ is principal iff $deg(D) = 0$ and $sum(D) = \mathcal{O}$, where $\mathcal{O}$ is the neutral element of the points group. The relation $\sim$ on $Div^0(E)$ is defined to be $D_1 \sim D_2$ iff $D_1 - D_2$ is principal.

If f is a nonzero rational function such that $div(f)$ and D have disjoint supports, then the evaluation of f at D is defined by $f(D) = \prod_{P \in E} f(P)^{n_P}$.

For more information on rational functions, divisors, and their relations, we refer the readers to [16].

Let n be an integer which is prime to $p = char(K)$ and $E[n] = \{P \in E(\overline{K}) | nP = \mathcal{O}\}$. Take $P, Q \in E[n]$, there exist $D_P, D_Q \in Div^0(E)$ s.t. $D_P \sim (P) - (\mathcal{O}), D_Q \sim (Q) - (\mathcal{O})$. Let $div(f_P) = nD_P$, $div(f_Q) = nD_Q$, and suppose that D_P, D_Q have disjoint supports, the *Weil pairing* is defined to be:

$$e(P, Q) = \frac{f_P(D_Q)}{f_Q(D_P)}$$

Take a point $S \in E$ s.t. $D_Q = (Q + S) - (S)$ and $div(f_P)$ have disjoint supports. Then the *Tate pairing* is defined to be:

$$\phi_n : E(K)[n] \times (E(K)/nE(K)) \mapsto K^*/(K^*)^n$$
$$\phi_n(P, \overline{Q}) = \overline{f_P(D_Q)}$$

Here $\overline{Q}$ is the equivalence class in $E(K)/nE(K)$ containing Q and $\overline{f_P(D_Q)}$ is the equivalence class in $K^*/(K^*)^n$.

An essential part in computing the Weil/Tate pairing is the evaluation of rational function f_P at some divisor D. Miller gave an efficient algorithm for this calculation.

The main idea of Miller's algorithm is to calculate $f_P(D_Q)$ recursively. Specifically, pick a random point S, and let $D_P = (P + S) - (S)$. Then $div(f_P) \sim nD_P$. For each integer k, there is a rational function f_k s.t.

$$div(f_k) = k(P + S) - k(S) - (kP) + (\mathcal{O}).$$

In particular, $f_n = f_P$.

Let $L_{P,Q}$ be the line passing through points P, Q and L_P be the vertical line passing through point P. Then we have

$$div(L_{k_1 P, k_2 P}) = (k_1 P) + (k_2 P) + (-(k_1 + k_2)P)) - 3(\mathcal{O})$$
$$div(L_{kP}) = (kP) + (-kP) - 2(\mathcal{O}).$$

So

$$div(f_{k_1 + k_2}) = div(f_{k_1}) + div(f_{k_2}) + div(L_{k_1 P, k_2 P}) - div(L_{(k_1 + k_2)P}).$$

In other words,

$$f_{k_1 + k_2} = f_{k_1} f_{f_2} L_{k_1 P, k_2 P} / H_{(k_1 + k_2)P}.$$

The initial values are: $f_0 = 1$ and $f_1 = L_{P,R}/L_{P+R}$.

Algorithm 1 describes Miller's method(see [9] for details).

Algorithm 1. Miller's Algorithm

Input: Elliptic curve E, integer $n = \sum_{i=0}^{t} b_i 2^i, b_t \neq 0$, points $P, Q \in E, order(P) = n$
Output: $f = f_n(S)$
    $f \leftarrow f_1; Q \leftarrow P;$
    for $j \leftarrow t - 1$ down to 0 **do**
        $f \leftarrow f^2 \frac{L_{Q,Q}(S)}{L_{2Q}(S)}; Q \leftarrow 2Q;$
        if $b_j = 1$ **then**
            $f \leftarrow f_1 f \frac{L_{Q,P}(S)}{L_{Q+P}(S)}; Q \leftarrow Q + P;$
    end if
    end for
    return f

2.2 Twisted Edwards Curves

Bernstein et al. introduced the twisted Edwards curve in [11]. Here we give a brief description.

For finite field K with character different from 2, the twisted Edwards curve is defined as:

$$E_{a,d} : ax^2 + y^2 = 1 + dx^2 y^2, \text{ where } a, d \in K^* \text{ and } a \neq d$$

The neutral element is $\mathcal{O} = (0, 1)$ and element $\mathcal{O}' = (0, -1)$ has order two. It also has two points at infinity, denoted by $\Omega_1 = (1 : 0 : 0), \Omega_2 = (0 : 1 : 0)$. Notice these two points are singular and have multiplicity two.

The addition law on points of the curve $E_{a,d}$ is

$$(x_1, y_1) + (x_2, y_2) = (\frac{x_1 y_2 + y_1 x_2}{1 + dx_1 x_2 y_1 y_2}, \frac{y_1 y_2 - ax_1 x_2}{1 - dx_1 x_2 y_1 y_2}) \tag{1}$$

It is proved in [11] that if a is a square and d is not a square, then formula (1) is complete.

[11] also gives explicit formulae for twisted Edwards curves in projective coordinates. In projective coordinates, twisted Edwards curve is defined as:

$$(aX^2 + Y^2)Z^2 = Z^4 + dX^2 Y^2$$

For $Z_1 \neq 0$ the homogeneous point $(X_1 : Y_1 : Z_1)$ represents the affine point $(X_1/Z_1, Y_1/Z_1)$ on $E_{a,d}$.

Addition in Projective Twisted Coordinates. The following formulae compute $(X_3 : Y_3 : Z_3) = (X_1 : Y_1 : Z_1) + (X_2 : Y_2 : Z_2)$ in $10\mathbf{M} + 1\mathbf{S} + 2\mathbf{D} + 7\mathbf{add}$, where the $2\mathbf{D}$ are one multiplication by a and one by d:

$A = Z_1 \cdot Z_2; B = A^2; C = X_1 \cdot X_2; D = Y_1 \cdot Y_2; E = dC \cdot D;$
$F = B \cdot E; G = B + E; X_3 = A \cdot F \cdot ((X_1 + Y_1) \cdot (X_2 + Y_2) \cdot C \cdot D);$
$Y_3 = A \cdot G \cdot (D \cdot aC); Z_3 = F \cdot G.$

Doubling in Projective Twisted Coordinates. The following formulae compute $(X3 : Y3 : Z3) = 2(X1 : Y1 : Z1)$ in $3\mathbf{M} + 4\mathbf{S} + 1\mathbf{D} + 7\mathbf{add}$, where the $1\mathbf{D}$ is a multiplication by a:

$$B = (X_1 + Y_1)^2; C = X_1^2; D = Y_1^2; E = aC; F := E + D; H = Z_1^2;$$
$$J = F \cdot 2H; X_3 = (B \cdot C \cdot D) \cdot J; Y_3 = F \cdot (E \cdot D); Z_3 = F \cdot J.$$

2.3   Bilinear Pairing Over Edwards Curves

In [14], the authors gave a geometry explanation for the addition law over twisted Edwards curves and a method to construct rational function with divisor $(P_1) + (P_2) - (P_3) - (\mathcal{O})$, which is essential for Miller's algorithm.

Let h_{P_1,P_2} be the conic passing through $\Omega_1, \Omega_2, \mathcal{O}', P_1, P_2; \ell_{1,P_3}$ be the horizontal line passing through $P_3; \ell_{2,\mathcal{O}}$ be the vertical line passing through $\mathcal{O}$.

From [14], we have the following lemma:

Lemma 1. *Let ℓ_1, ℓ_2, h be defined as above. Then*

$$div(\ell_{1,P_3}) = (P_3) + (-P_3) - 2(\Omega_2) \tag{2}$$
$$div(\ell_{2,\mathcal{O}}) = (\mathcal{O}) + (\mathcal{O}') - 2(\Omega_1) \tag{3}$$
$$div(h_{P_1,P_2}) = (P_1) + (P_2) + (\mathcal{O}') + (-P_3) - 2(\Omega_1) - 2(\Omega_2) \tag{4}$$

It is easy to prove Corollary 1 with Lemma 1.

Corollary 1. *Let ℓ_1, ℓ_2, h be the same as in Lemma 1. Then*

$$\frac{h_{P_1,P_2}}{\ell_{1,P_3}\ell_{2,\mathcal{O}}} = (P_2) + (P_2) - (P_3) - (\mathcal{O}) \tag{5}$$

We get the Edwards edition of Miller's algorithm(Algorithm 2).

Algorithm 2. Miller's Algorithm for Twisted Edwards Curve

Input: Twisted Edwards curve $E_{a,d}$, integer $n = \sum_{i=0}^{t} b_i 2^i, b_t \neq 0$, points $P, S \in$
  $E, order(P) = n$
Output: $f = f_n(S)$
1: $f \leftarrow f_1; Q \leftarrow P;$
2: **for** $j \leftarrow t - 1$ down to 0 **do**
3:    $f \leftarrow f^2 \frac{h_{Q,Q}(S)}{\ell_{2,2Q}(S)\ell_{1,\mathcal{O}}(S)}$
4:    $Q \leftarrow 2Q$
5:    **if** $b_j = 1$ **then**
6:       $f \leftarrow f_1 f \frac{h_{Q,P}(S)}{\ell_{2,Q+P}(S)\ell_{1,\mathcal{O}(S)}}$
7:       $Q \leftarrow Q + P$
8:    **end if**
9: **end for**
10: **return** f

---

3   Our Improvements

The main loop of Miller's algorithm (i.e from line 2 to line 8 in Algorithm 2) takes most of the running time. So we focus on the improvements of the operations in the loop.

In 2006 Blake, Murty and Xu([17]) proposed a method to reduce the total number of lines in Miller's algorithm. Though this concept does not dramatically decrease the cost of points adding, it is novel and can be applied to decrease the number of field multiplications.

In this paper we extend their technique to twisted Edwards curves and achieve some improvements.

Specifically, notice that in Miller's algorithm for twisted Edwards curves (Algorithm 2), only one bit of the integer n is considered in one iteration. If we consider two consecutive bits at a time, we can achieve some improvements.

First we give a theorem which is fundamental to the improvements.

Theorem 1. *Let $E_{a,d}$ be a twisted Edwards curve and $Q \in E_{a,d}$ with $order(Q) = n$. Then[1]*

1.

$$\left(\frac{h_{Q,Q}}{\ell_{1,2Q}\ell_{2,\mathcal{O}}}\right)^2 \frac{h_{2Q,2Q}}{\ell_{2,4Q}\ell_{1,\mathcal{O}}} = \frac{h_{Q,Q}^2}{h_{-2Q,-2Q}h_{\mathcal{O},\mathcal{O}}}$$

2.

$$\frac{h_{Q,Q}}{\ell_{2,2Q}\ell_{1,\mathcal{O}}} \frac{h_{2Q,P}}{\ell_{2,2Q+P}\ell_{1,\mathcal{O}}} = \frac{h_{Q,Q}\ell_{2,P}}{h_{2Q+P,-P}\ell_{1,\mathcal{O}}}$$

Proof.   1. The divisor of the rational function

$$\left(\frac{h_{Q,Q}}{\ell_{1,2Q}\ell_{2,\mathcal{O}}}\right)^2 \frac{h_{2Q,2Q}}{\ell_{1,4Q}\ell_{2,\mathcal{O}}}$$

is

$$\frac{4(Q) + 2(-2Q) - 4(\Omega_1) - 4(\Omega_2) + 2(\mathcal{O}')}{2(2Q) + 2(-2Q) - 4(\Omega_2) + 2(\mathcal{O}) + 2(\mathcal{O}') - 4(\Omega_1)}$$
$$+ \frac{2(2Q) + (-4Q) - 2(\Omega_1) - 2(\Omega_2) + (\mathcal{O}')}{(4Q) + (-4Q) - 2(\Omega_2) + (\mathcal{O}) + (\mathcal{O}') - 2(\Omega_1)}$$
$$= \frac{4(Q) + 2(-2Q) - 4(\Omega_1) - 4(\Omega_2) + 2(\mathcal{O}')}{2(-2Q) - 2(\Omega_2) + 2(\mathcal{O}) + 2(\mathcal{O}') - 2(\Omega_1) + (4Q) - 2(\Omega_2) + (\mathcal{O}) - 2(\Omega_1)}$$
$$= 4(Q) + 2(-2Q) - 4(\Omega_1) - 4(\Omega_2) + 2(\mathcal{O}') +$$
$$\frac{1}{2(-2Q) + (4Q) - 2(\Omega_1) - 2(\Omega_2) + (\mathcal{O}')} +$$
$$\frac{1}{3(\mathcal{O}) + (\mathcal{O}') - 2(\Omega_2) - 2(\Omega_1)}$$

---

[1] Notice that the fraction of divisor a/b means $a - b$.

The divisor $4(Q) + 2(-2Q) - 4(\Omega_1) - 4(\Omega_2) + 2(\mathcal{O}')$ corresponds to $h_{Q,Q}^2$, $2(-2Q) + (4Q) - 2(\Omega_1) - 2(\Omega_2) + (\mathcal{O}')$ corresponds to $h_{-2Q,-2Q}$ and $3(\mathcal{O}) + (\mathcal{O}') - 2(\Omega_2) - 2(\Omega_1)$ corresponds to $h_{\mathcal{O},\mathcal{O}}$.
So we have

$$\left(\frac{h_{Q,Q}}{\ell_{1,2Q}\ell_{2,\mathcal{O}}}\right)^2 \frac{h_{2Q,2Q}}{\ell_{2,4Q}\ell_{1,\mathcal{O}}} = \frac{h_{Q,Q}^2}{h_{-2Q,-2Q}h_{\mathcal{O},\mathcal{O}}}.$$

2. The divisor of the rational function

$$\frac{h_{Q,Q}}{\ell_{2,2Q}\ell_{1,\mathcal{O}}} \frac{h_{2Q,P}}{\ell_{2,2Q+P}\ell_{1,\mathcal{O}}}$$

is

$$\frac{2(Q) + (-2Q) - 2(\Omega_1) - 2(\Omega_2) + (\mathcal{O}')}{(2Q) + (-2Q) - 2(\Omega_2) + (\mathcal{O}) + (\mathcal{O}') - 2(\Omega_1)}$$
$$+ \frac{(2Q) + (P) + (-(2Q+P)) - 2(\Omega_1) - 2(\Omega_2) + (\mathcal{O}')}{(2Q+P) + (-(2Q+P)) - 2(\Omega_2) + (\mathcal{O}) + (\mathcal{O}') - 2(\Omega_1)}$$
$$= 2(Q) + (-2Q) - 2(\Omega_1) - 2(\Omega_2) + (\mathcal{O}') +$$
$$\frac{(P)}{(2Q+P) + (-2Q) - 2(\Omega_1) - 2(\Omega_2) + (\mathcal{O}') + 2(\mathcal{O})}$$
$$= 2(Q) + (-2Q) - 2(\Omega_1) - 2(\Omega_2) + (\mathcal{O}') +$$
$$\frac{(P) + (-P) - 2(\Omega_2)}{(2Q+P) + (-2Q) + (-P) - 2(\Omega_1) - 2(\Omega_2) + (\mathcal{O}') + 2(\mathcal{O}) - 2(\Omega_2)}$$

The divisor $2(Q) + (-2Q) - 2(\Omega_1) - 2(\Omega_2) + (\mathcal{O}')$ corresponds to $h_{Q,Q}$, $(P) + (-P) - 2(\Omega_2)$ corresponds to $\ell_{1,P}$, $(2Q+P) + (-2Q) + (-P) - 2(\Omega_1) - 2(\Omega_2) + (\mathcal{O}')$ correponds to $h_{2Q+P,-P}$, and $2(\mathcal{O}) - 2(\Omega_2)$ corresponds to $\ell_{1,\mathcal{O}}$.
So we have

$$\frac{h_{Q,Q}}{\ell_{2,2Q}\ell_{1,\mathcal{O}}} \frac{h_{2Q,P}}{\ell_{2,2Q+P}\ell_{1,\mathcal{O}}} = \frac{h_{Q,Q}\ell_{2,P}}{h_{2Q+P,-P}\ell_{1,\mathcal{O}}}.$$

Next we describe the improvements using Theorem 1 in four different cases.

1. If two consecutive bits of n are "00", then according to Algorithm 2, line 3 ~ 4 are executed twice.
   The result of the execution is

$$f \leftarrow (f^2 \frac{h_{Q,Q}(S)}{\ell_{2,2Q}(S)\ell_{1,\mathcal{O}}(S)})^2 \frac{h_{2Q,2Q}(S)}{\ell_{2,4Q}(S)\ell_{1,\mathcal{O}}(S)}$$
$$Q \leftarrow 4Q$$

Using the first formula of Theorem 1, the above operations are equal to

$$f \leftarrow f^4 \frac{h_{Q,Q}^2(S)}{h_{-2Q,-2Q}(S)h_{\mathcal{O},\mathcal{O}}(S)}$$
$$Q \leftarrow 4Q$$

2. If two consecutive bits of n are "01", then according to Algorithm 2, line 3 $\sim$ 4 are executed twice, and line 6 $\sim$ 7 are executed.
   The result of the execution is

   $$f \leftarrow f_1(f^2 \frac{h_{Q,Q}(S)}{\ell_{2,2Q}(S)\ell_{1,O}(S)})^2 \frac{h_{2Q,2Q}(S)}{\ell_{2,4Q}(S)\ell_{1,O}(S)} \frac{h_{4Q,P}(S)}{\ell_{2,4Q+P}(S)\ell_{1,O(S)}}$$
   $$Q \leftarrow 4Q + P$$

   In this case we have two ways to combine the divisor of the result rational function f.

   (a) Using the first formula of Theorem 1, the above operations are equal to

   $$f \leftarrow f_1 f^4 \frac{h_{Q,Q}^2(S)}{h_{-2Q,-2Q}(S)h_{O,O}(S)} \frac{h_{4Q,P}(S)}{\ell_{2,4Q+P}\ell_{1,O}(S)}$$
   $$Q \leftarrow 4Q + P$$

   (b) We can also use the second formula of Theorem 1, then the above operations are equal to

   $$f \leftarrow f_1(f^2 \frac{h_{Q,Q}(S)}{\ell_{2,2Q}(S)\ell_{1,O}(S)})^2 \frac{h_{2Q,2Q}(S)\ell_{2,P}(S)}{h_{4Q+P,-P}(S)h_{O,O}(S)}$$
   $$Q \leftarrow 4Q + P$$

3. If two consecutive bits of n are "10", then according to Algorithm 2, line 3 $\sim$ 7 are executed, and line 3 $\sim$ 4 are executed.
   The result of the execution is

   $$f \leftarrow (f_1 f^2 \frac{h_{Q,Q}(S)}{\ell_{2,2Q}(S)\ell_{1,O}(S)} \frac{h_{2Q,P}(S)}{\ell_{2,2Q+P}(S)\ell_{1,O(S)}})^2 \frac{h_{2Q+P,2Q+P}(S)}{\ell_{2,4Q+2P}(S)\ell_{1,O}(S)}$$
   $$Q \leftarrow 4Q + 2P$$

   In this case we have two ways to combine the divisor of the result rational function f.

   (a) Using the first formula of Theorem 1, the above operations are equal to

   $$f \leftarrow f_1^2 f^4 \frac{h_{Q,Q}^2(S)}{\ell_{2,2Q}^2(S)\ell_{1,O}^2(S)} \frac{h_{2Q,P}^2(S)}{h_{-(2Q+P),-(2Q+P)}(S)h_{O,O}(S)}$$
   $$Q \leftarrow 4Q + 2P$$

   (b) Using the second formula of Theorem 1, the above operations are equal to

   $$f \leftarrow (f1f^2 \frac{h_{Q,Q}(S)\ell_{2,P}(S)}{h_{2Q+P,-P}(S)\ell_{1,O}(S)})^2 \frac{h_{2Q+P,2Q+P}(S)}{\ell_{2,4Q+2P}(S)\ell_{1,O}(S)}$$
   $$Q \leftarrow 4Q + 2P$$

4. If two consecutive bits of n are "11", then according to Algorithm 2, line 3 ~ 7 are executed twice.

The result of the execution is

$$f \leftarrow f_1(f_1 f^2 \frac{h_{Q,Q}(S)}{\ell_{2,2Q}(S)\ell_{1,O}(S)} \frac{h_{2Q,P}(S)}{\ell_{2,2Q+P}(S)\ell_{1,O(S)}})^2 \frac{h_{2Q+P,2Q+P}(S)}{\ell_{2,4Q+2P}(S)\ell_{1,O}(S)} \frac{h_{4Q+2P,P}(S)}{\ell_{2,4Q+3P}(S)\ell_{1,O(S)}}$$

$$Q \leftarrow 4Q + 3P$$

As in the cases "01" and "10", we have two ways to combine the divisor of the result rational function f.

(a) Using the first formula of Theorem 1, the above operations are equal to

$$f \leftarrow f_1^3 f^4 \frac{h_{Q,Q}^2(S)}{\ell_{2,2Q}^2(S)\ell_{1,O}^2(S)} \frac{h_{2Q,P}^2(S)}{h_{-(2Q+P),-(2Q+P)}(S)h_{O,O}(S)} \frac{h_{4Q+2P,P}(S)}{\ell_{2,4Q+3P}(S)\ell_{1,O}(S)}$$

$$Q \leftarrow 4Q + 3$$

(b) Using the second formula of Theorem 1, the above operations are equal to

$$f \leftarrow f_1(f_1 f^2 \frac{h_{Q,Q}(S)\ell_{2,P}(S)}{h_{2Q+P,-P}(S)\ell_{1,O}(S)})^2 \frac{h_{2Q+P,2Q+P}(S)\ell_{2,P}(S)}{h_{4Q+3P,-P}(S)h_{O,O(S)}}$$

$$Q \leftarrow 4Q + 3$$

It is easy to derive the concrete algorithm from the above description(Algorithm 3). To consider two bits of n at a time, we represent n in 4-base. If the number of bits of the integer n is odd, we initialize f with the first bit, otherwise we use the first two bits to initialize f. And we use different method to combine the divisor of f in different cases. The reasons to use different combinations are shown in Section 4.

4   Analysis and Comparison

The cost of the algorithms calculating bilinear pairing over Edwards curves consists of three parts: the cost of updating f, the cost of updating Q, and the cost of evaluating $h_{Q,R}, \ell_{1,Q}, \ell_{2,Q}$ at some point S. Note that here we mainly make comparison with the original Miller's algorithm(Algorithm 2).

Without special treatment, the cost of updating Q in Algorithm 3 is no more than that of Algorithm 2. And the cost of evaluating $h_{Q,R}, \ell_{1,Q}, \ell_{2,Q}$ at some point S is also the same for the two algorithms[2]. So we focus on the cost of updating f.

Let $\mathbf{M}$ denote the cost of finite field multiplication, $\mathbf{S}$ denote the cost of finite field square and $\mathbf{I}$ denote the finite field inversion. Notice that there is always a multiplication following an inversion.

Field inversion is much more expensive compared with multiplication. And square is cheaper than multiplication. We set $\mathbf{S} = 0.8\mathbf{M}$ and $\mathbf{I} > 8\mathbf{M}$. Because

---

[2] If we take these costs in consideration, there may be room for further improvements.

Algorithm 3. Improved Miller's Algorithm for Edwards Curve

Input: Twisted Edwards curve $E_{a,d}$, integer $n = \sum_{i=0}^{t} q_i 4^i, q_t \neq 0$, points $P, S \in E_{a,d}, order(P) = n$

Output: $f = f_n(S)$

$\quad f \leftarrow f_1, Q \leftarrow P$

$\quad$ if number of bits of n is even **then**

$\qquad$ if $q_r = 2$ **then**

$\qquad\quad f \leftarrow f^2 \dfrac{h_{P,P}(S)}{\ell_{2,2P}(S)\ell_{1,O}(S)}, Q \leftarrow 2P$

$\qquad$ end if

$\qquad$ if $q_r = 3$ **then**

$\qquad\quad f \leftarrow f^2 \dfrac{h_{P,P}(S)}{\ell_{2,2P}(S)\ell_{1,O}(S)} \dfrac{h_{2P,P}(S)}{\ell_{2,3P}(S)\ell_{1,O}(S)}, Q \leftarrow 3P$

$\qquad$ end if

$\quad$ end if

$\quad$ for $j = t - 1$ down to 0 **do**

$\qquad$ if $q_j = 0$ **then**

$\qquad\quad f \leftarrow f^4 \dfrac{h_{Q,Q}^2(S)}{h_{-2Q,-2Q}(S)h_{O,O}(S)}$

$\qquad\quad Q \leftarrow 4Q$

$\qquad$ end if

$\qquad$ if $q_j = 1$ **then**

$\qquad\quad f \leftarrow f_1 f^4 \dfrac{h_{Q,Q}^2(S)}{h_{-2Q,-2Q}(S)h_{O,O}(S)} \dfrac{h_{4Q,P}(S)}{\ell_{2,4Q+P}\ell_{1,O}(S)}$

$\qquad\quad Q \leftarrow 4Q + P$

$\qquad$ end if

$\qquad$ if $q_j = 2$ **then**

$\qquad\quad f \leftarrow f_1^2 f^4 \dfrac{h_{Q,Q}^2(S)}{\ell_{2,2Q}^2(S)\ell_{1,O}^2(S)} \dfrac{h_{2Q,P}^2(S)}{h_{-(2Q+P),-(2Q+P)}(S)h_{O,O}(S)}$

$\qquad\quad Q \leftarrow 4Q + 2P$

$\qquad$ end if

$\qquad$ if $q_j = 3$ **then**

$\qquad\quad f \leftarrow f_1(f_1 f^2 \dfrac{h_{Q,Q}(S)\ell_{2,P}(S)}{h_{2Q+P,-P}(S)\ell_{1,O}(S)})^2 \dfrac{h_{2Q+P,2Q+P}(S)\ell_{2,P}(S)}{h_{4Q+3P,-P}(S)h_{O,O(S)}}$

$\qquad\quad Q \leftarrow 4Q + 3P$

$\qquad$ end if

$\quad$ end for

$\quad$ return f

---

the inversion operation is so expensive, we keep the middle result in fraction form($f = \frac{a}{b}$) to avoid the inversions.

First, we give a comparison of the effects of different combinations of divisor. The result of different combinations is given in Section 3, and the cost of these combinations is shown in Table 1.

From Table 1, it is easy to see that in the cases "01" and "10", the first combination save 1 **M** than the second. And in the case "11", the second method is faster than the first one(also save 1 **M**). There is only one way to combine the divisor in the case "00".

The construction of Algorithm 3 follows the above observations.

Table 1. Comparison of Different Combinations of Divisor of f

Method to combine the divisor	case "00"	case "01"	case "10"	case "11"
Using the first formula of Theorem 1	5S + 3M = 7M	4S + 7M = 10.2M	4S + 7M = 10.2M	4S + 11M = 14.2M
Using the second formula of Theorem 1	5S + 3M = 7M	4S + 8M = 11.2M	4S + 8M = 11.2M	4S + 10M = 13.2M

Specifically,

1. In the case "01" and "10", use the first formula of Theorem 1 to combine the divisor.
2. In the case "11", use the second formula of Theorem 1 to combine the divisor.

Next, We compare Algorithm 2 and the original Miller's algorithm(Algorithm 3). The result is showed in Table 2.

We remind that Algorithm 2 is slower than that of [14], where denominator elimination was used. And we include the result of [14] in the last row of Table 2.

Table 2. Comparison of the Algorithms

	case "00"	case "01"	case "10"	case "11"
Algorithm 2	4S + 6M = 9.2M	4S + 10M = 13.2M	4S + 10M = 13.2M	4S + 14M = 17.2M
Algorithm 3	5S + 3M = 7M	4S + 7M = 10.2M	4S + 7M = 10.2M	4S + 10M = 13.2M
Method of [14]	2S + 2M = 3.6M	2S + 3M = 4.6M	2S + 3M = 4.6M	2S + 4M = 5.6M

From Table 2, we can see that the improved Algorithm 3 is more efficient than the original Algorithm 2 in all the four cases. And in the case "11", the new algorithm can save at most 4M per iteration. So if there are more "11"s in the binary representation of integer n, Algorithm 3 will save more time.

Let $n = \sum_{i=0}^{t} 2^i b_i$, suppose

$$Prob[b_i = 1] = Prob[b_i = 0] = 1/2, where\ 0 \leq i < t,$$

and b_i are mutually independent.

Then

$$Prob[b_i b_{i+1} = 00] = Prob[b_i b_{i+1} = 01] = Prob[b_i b_{i+1} = 10] =$$

$$Prob[b_i b_{i+1} = 11] = 1/4$$

So the total cost of Algorithm 3 is about 76.8% of that of Algorithm 2.

5   Conclusion

In this paper we propose an improved Miller's algorithm for twisted Edwards curves. And we give a detailed analysis of the improvement. The savings in the number of multiplication in the updating of f noted is important for the performance of algorithms in the pairing based cryptosystems.

At the same time, we pay little attention to the calculation of $4Q+P, 4Q+2P$, and $4Q+3P$. There may be faster methods to calculate these values than simple additions and doubles. So it is possible to make our analysis result better.

Acknowledgements. We thank Professor Tanja Lange and the anonymous reviewers for their precious comments.

References

1. Edwards, H.M.: A Normal Form for Elliptic Curves. Bulletin of the American Mathematical Society 44, 393–442 (2007)
2. Bernstein, D.J., Lange, T.: Faster Addition and Doubleing on Elliptic Curves. In: Kurosawa, K. (ed.) ASIACRYPT 2007. LNCS, vol. 4833, pp. 29–50. Springer, Heidelberg (2007)
3. Menezes, A.J., Okamoto, T., Vanstone, S.A.: Reducing Elliptic Curve Logarithms to Logarithms in a Finite Field. IEEE Transactions on Information Theory (1993)
4. Frey, G., Rück, H.G.: A Remark Concerning m-divisibility and the Discrete Logarithm in the Divisor Class Group of Curves. Mathematics of Computation 62, 865–874 (1994)
5. Boneh, D., Franklin, M.: Identity-based Encryption from the Weil Pairing. In: Kilian, J. (ed.) CRYPTO 2001. LNCS, vol. 2139, pp. 213–229. Springer, Heidelberg (2001)
6. Shamir, A.: Identity-based cryptosystems and signature schemes. In: Blakely, G.R., Chaum, D. (eds.) CRYPTO 1984. LNCS, vol. 196, pp. 47–53. Springer, Heidelberg (1985)
7. Joux, A.: A One Round Protocol for Tripartite Diffie-Hellman. In: Bosma, W. (ed.) ANTS 2000. LNCS, vol. 1838, pp. 385–393. Springer, Heidelberg (2000)
8. Boneh, D., Lynn, B., Shacham, H.: Short Signature from the Weil Pairing. In: Boyd, C. (ed.) ASIACRYPT 2001. LNCS, vol. 2248, pp. 514–532. Springer, Heidelberg (2001)
9. Miller, V.S.: The Weil Pairing, and its Efficient Calculation. Journal of Cryptology 17(4), 235–261 (2004)
10. Blake, I.F., Sroussi, G., Smart, P.N.: Advances in Elliptic Curve Cryptography. Cambridge University Press, Cambridge (2005)
11. Bernstein, D.J., Birkner, P., Joye, M., Lange, T., Peters, C.: Twisted Edwards Curves. In: Vaudenay, S. (ed.) AFRICACRYPT 2008. LNCS, vol. 5023, pp. 389–405. Springer, Heidelberg (2008)
12. Das, M.P.L., Sarkar, P.: Pairing Computation on Twisted Edwards form Elliptic Curves. In: Galbraith, S.D., Paterson, K.G. (eds.) Pairing 2008. LNCS, vol. 5209, pp. 192–210. Springer, Heidelberg (2008)
13. Ionica, S., Joux, A.: Another Approach to Pairing Computation in Edwards Coordinates. In: Chowdhury, D.R., Rijmen, V., Das, A. (eds.) INDOCRYPT 2008. LNCS, vol. 5365, pp. 400–413. Springer, Heidelberg (2008)

14. Aréne, C., Lange, T., Naehrig, M., Ritzenthaler, C.: Faster Pairing Computation. Cryptology ePrint Archive, Report 2009/155 (2009)
15. Barreto, P.S., Lynn, B., Scott, M.: Efficient Implementation of Pairing-based Cryptosystems. Journal of Cryptology 17, 321–334 (2004)
16. Hartshorne, R.: Algebraic Geometry. Graduate Texts in Mathematics. Springer, Heidelberg (1977)
17. Blake, I.F., Murty, V.K., Xu, G.: Refinements of Miller's Algorithm for Computing the Weil/Tate pairing. Journal of Algorithms 58, 134–149 (2006)

Probabilistic Public Key Encryption with Equality Test

Guomin Yang[1], Chik How Tan[1], Qiong Huang[2], and Duncan S. Wong[2]

[1] Temasek Laboratories
National University of Singapore
{tslyg,tsltch}@nus.edu.sg
[2] Department of Computer Science
City University of Hong Kong
duncan@cityu.edu.hk, csqhuang@student.cityu.edu.hk

Abstract. We present a (probabilistic) public key encryption (PKE) scheme such that when being implemented in a bilinear group, anyone is able to check whether two ciphertexts are encryptions of the same message. Interestingly, bilinear map operations are not required in key generation, encryption or decryption procedures of the PKE scheme, but is only required when people want to do an equality test (on the encrypted messages) between two ciphertexts that may be generated using different public keys. We show that our PKE scheme can be used in different applications such as searchable encryption and partitioning encrypted data. Moreover, we show that when being implemented in a non-bilinear group, the security of our PKE scheme can be strengthened from One-Way CCA to a weak form of IND-CCA.

Keywords: Public Key Encryption, Adaptive Chosen Ciphertext Attacks, Ciphertext Comparability, Searchable Encryption, Bilinear Map.

1 Introduction

Consider an outsourced database, data are stored in encrypted form. In order to maintain a good data structure, or extract some statistical information of the data, data may need to be partitioned. However, classical encryption schemes are not suitable for this purpose, since given a pile of ciphertexts, no one is able to tell the relationships among encrypted messages without knowing the decryption keys.

Searchable encryption (SE) schemes, introduced by Boneh et al. [7], and intensively studied in [3,1,21] may be one candidate to solve the problem. Informally speaking, in a searchable encryption scheme, a Tag T_M can be generated with respect to a message M and a key pair (PK, SK) (given T_M and PK, one should not be able to derive the message M). There is also a function Test' such that Test'(T_M, C) returns 1 if and only if the ciphertext C is an encryption of M under public key PK. So using the tag, one should be able to categorize the ciphertexts according to the encrypted messages. However, this method has one

J. Pieprzyk (Ed.): CT-RSA 2010, LNCS 5985, pp. 119–131, 2010.

shortcoming: the same message encrypted under different public keys cannot be categorized into one cluster.

Another possible approach is to use *deterministic* encryption schemes, which has been studied in recent years [3,5,6]. A PKE scheme is deterministic if the encryption algorithm is deterministic, namely given the same public key and the message, the encryption algorithm always outputs the same ciphertext. However, deterministic encryption suffers from the same problem that searchable encryption does for our purpose.

In this paper, we formalize the notion of Ciphertext Comparability for public key encryption schemes. We introduce a (probabilistic) public key encryption scheme such that when being implemented in a bilinear group its ciphertexts are publicly comparable. That is, given two ciphertexts C_1 and C_2 generated under public keys PK and PK', respectively, there is a function Test such that $\text{Test}(C_1, C_2)$ returns 1 if and only if C_1 and C_2 are encryptions of the same message, no matter $PK = PK'$ or not. Our encryption scheme itself does not invoke any bilinear map operation in key generation, encryption or decryption procedures, only the Test function does. Further more, we show that when being implemented in a non-bilinear group, our PKE scheme can achieve a higher level of confidentiality, but at the cost of losing ciphertext comparability.

Related Work. Since the introduction of public key cryptography due to the seminal paper of Diffie and Hellman [14], public key encryption schemes are in the center of modern cryptography. Different types of PKE schemes with different security goals have been constructed. For a long time, people were searching for PKE schemes providing strong confidentiality, namely indistinguishability or semantic security under chosen-ciphertext attacks (IND-CCA2) [24,15]. Nowadays there are many PKE schemes (e.g. [11,12,23,22,9,2,10,19,18,20]) achieving IND-CCA2 security. A historical survey on PKE can be found in [13].

In [7], Boneh et al. presented the notion of public key encryption with keyword search (PEKS) and several constructions that achieve semantic security. In a PEKS scheme, Alice, with a key pair (pk, sk), can provide Bob with a trapdoor T_W, which is computed as a function of her secret key sk and any keyword W of her choice. Using T_W, it is possible to check, using a function Test′, whether an arbitrary given ciphertext c is an encryption of W or not. The consistency condition is that $\text{Test}'(T_W, c)$ returns 1 if and only if c is an encryption of W. Otherwise, the trapdoor T_W should not give any information about the real encrypted message W' (besides $W' \neq W$). Later, a general connection between PEKS and (anonymous) IBE was given by Abdalla et al. [1]. Recently, Hofheinz and Weinreb [20] presented a searchable public key encryption with decryption (PEKSD) in the standard model.

In [3], Bellare et al. initiated the notion of deterministic public key encryption where the encryption algorithm is deterministic. Deterministic encryption has been shown to be a useful tool in many applications, e.g. fast (i.e. logarithmic time) searching on encrypted data. This topic is further studied in [5,6].

As discussed in the introduction, if we are concerning encryptions generated under one public key, both searchable encryption and deterministic encryption can be used to do equality tests among ciphertexts. However, we are interested in a more general multi-key setting.

This work is different from the *verifiable* encryption considered by Camenisch and Shoup in [8]. In a verifiable encryption scheme, the encrypter, who knows the plaintext, is able to prove (in zero-knowledge) to a third party that the message "encapsulated" in the ciphertext satisfies some property, e.g. the plaintext is the discrete log of an element with respect to a base in a group. So after generating the ciphertext, additional work is required from the encrypter in order to conduct the proof. Our work does not require any extra effort from the encrypters after generating the ciphertexts, and equality test can be performed just using the ciphertexts.

Our Contributions. In this paper, we give the notion of ciphertext comparability for public key encryption schemes, and present such a scheme. Our scheme allows anyone to compare two ciphertexts and check if they are encryptions of the same message, even though the ciphertexts may be generated using different public keys. We show that when being implemented in a bilinear group, our encryption scheme is one-way under chosen ciphertext attacks (as we will see shortly, it is impossible to achieve IND-ATK type of security for encryption schemes with ciphertext comparability) in the random oracle model. We show that PKE schemes with ciphertext comparability can be used in many applications, such as constructing searchable encryption, and partitioning encrypted data.

We then analyze the security of our encryption scheme when being implemented in a non-bilinear group. We show that under the DDH assumption our scheme achieves *Weak* Indistinguishability under Chosen Ciphertext Attacks (W-IND-CCA2) in the random oracle model.

Paper Organization. In the next section, we give the definitions and security models for public key encryption schemes with ciphertext comparability. Then we give our construction of a PKE scheme with ciphertext comparability in Sec. 3 and show some of its applications in Sec. 4. In Sec. 5, we give the definition of W-IND-ATK and prove that when being implemented in a non-bilinear group, our PKE scheme can achieve W-IND-CCA2 security.

2   Definitions

A public key encryption scheme $\Pi = (\mathcal{G}, \mathcal{E}, \mathcal{D})$ consists of a triple of algorithms. The key generation algorithm $\mathcal{G}$ takes a security parameter $k \in \mathbb{N}$ and outputs a public/private key pair (pk, sk). The encryption algorithm $\mathcal{E}$ takes a message m and the public key pk, and outputs a ciphertext c. The decryption algorithm $\mathcal{D}$ takes sk and c as input, and outputs m or $\bot$ (which indicates decryption failure). The correctness requirement is that $\forall k \in \mathbb{N}$ and $\forall m \in \mathsf{PtSp}(k)$, $(pk, sk) \leftarrow \mathcal{G}(1^k)$, $m \leftarrow \mathcal{D}(sk, \mathcal{E}(pk, m))$ where $\mathsf{PtSp}(k)$ is the plaintext space associated to Π.

We say that Π has *Ciphertext Comparability* with error ϵ for some function $\epsilon(\cdot)$ if there exists an efficiently computable deterministic function $\mathsf{Test}(\cdot, \cdot)$ such that for every k we have

1. Perfect Consistency: for every $x \in \mathsf{PtSp}(k)$

$$\Pr\left[\begin{array}{l} (pk, sk) \leftarrow \mathcal{G}(1^k), (pk', sk') \leftarrow \mathcal{G}(1^k), C \leftarrow \mathcal{E}(pk, x) \\ C' \leftarrow \mathcal{E}(pk', x): \mathsf{Test}(C, C') = 1 \end{array}\right] = 1$$

2. Soundness: for every polynomial time algorithm $\mathcal{M}$

$$\Pr\left[\begin{array}{l} (C, C', sk, sk') \leftarrow \mathcal{M}(1^k), x \leftarrow \mathcal{D}(sk, C), x' \leftarrow \mathcal{D}(sk', C') : \\ x \neq \bot \wedge x' \neq \bot \wedge x \neq x' \wedge \mathsf{Test}(C, C') = 1 \end{array}\right] \leq \epsilon(k)$$

In the above definition, consistency ensures that encryptions (even under different public keys) of the same message can be recognized. Soundness measures the probability of false-hits (i.e. $\mathsf{Test}(C, C') = 1$ but C and C' are encryptions of different messages).

Sanity Check. In our definition for ciphertext comparability, the Test function does not perform any sanity check on the ciphertexts, namely, we don't specify the output of Test when its input cannot be decrypted. We only require that when the ciphertexts are real encryptions of messages, the function works properly. On the other hand, checking whether a ciphertext is in the correct form without using the private key is another problem out of the scope of this paper.

In the following, we review the classical notions of privacy for public key encryption schemes, namely, indistinguishability under chosen plaintext and chosen ciphertext attacks [17,24,15].

Definition 1 (IND-ATK [4]). *Let $\Pi = (\mathcal{G}, \mathcal{E}, \mathcal{D})$ be a public key encryption scheme and let $\mathcal{A} = (\mathcal{A}_1, \mathcal{A}_2)$ be a polynomial-time adversary. For atk $\in$ $\{cpa, cca1, cca2\}$ and $k \in \mathbb{N}$ let*

$$\mathrm{Adv}_{\mathcal{A}, \Pi}^{\text{ind}-\text{atk}} \stackrel{\text{def}}{=} \Pr\left[\begin{array}{l} (pk, sk) \leftarrow \mathcal{G}(1^k), (x_0, x_1, \delta) \leftarrow \mathcal{A}_1^{\mathcal{O}_1}(pk), b \leftarrow \{0, 1\}, \\ y \leftarrow \mathcal{E}(pk, x_b), b' \leftarrow \mathcal{A}_2^{\mathcal{O}_2}(pk, x_0, x_1, \delta, y) : b' = b \end{array}\right] - \frac{1}{2}$$

where $x_0 \neq x_1 \wedge |x_0| = |x_1|$ and

If atk = cpa   then $\mathcal{O}_1(\cdot) = \varepsilon$ and $\mathcal{O}_2(\cdot) = \varepsilon$
If atk = cca1   then $\mathcal{O}_1(\cdot) = \mathcal{D}_{sk}(\cdot)$ and $\mathcal{O}_2(\cdot) = \varepsilon$
If atk = cca2   then $\mathcal{O}_1(\cdot) = \mathcal{D}_{sk}(\cdot)$ and $\mathcal{O}_2(\cdot) = \mathcal{D}_{sk}(\cdot)$

In the case of CCA2, we insist that $\mathcal{A}_2$ does not ask its oracle for decrypting y. We say that Π is secure in the sense of IND-ATK if $\mathrm{Adv}_{\mathcal{A}, \Pi}^{\text{ind}-\text{atk}}$ is negligible for any $\mathcal{A}$.

Unfortunately, indistinguishability based security notions are not applicable to PKE schemes with ciphertext comparability. Given the challenge ciphertext y and plaintexts x_0, x_1, an adversary $\mathcal{A}$ can compute another ciphertext

$y' = \mathcal{E}(pk, x_1)$, and then return $\mathsf{Test}(y, y')$ as her guess of the value b in the IND-ATK games. The advantage of the adversary $\mathcal{A}$ is

$$
\begin{aligned}
\mathrm{Adv}_{\mathcal{A},\Pi}^{\mathrm{ind-atk}} &= \frac{1}{2}\Pr[\mathsf{Test}(y, y') = 1 | b = 1] + \frac{1}{2}\Pr[\mathsf{Test}(y, y') = 0 | b = 0] - \frac{1}{2} \\
&= \frac{1}{2} + \frac{1}{2}\Pr[\mathsf{Test}(y, y') = 0 | b = 0] - \frac{1}{2} \\
&= \frac{1}{2}(1 - \Pr[\mathsf{Test}(y, y') = 1 | b = 0]) \\
&\geq \frac{1}{2}(1 - \epsilon(k))
\end{aligned}
$$

As ciphertext comparability and indistinguishability are irreconcilable, we go back to the one-way definition of privacy for public key encryption schemes.

Definition 2 (OW-ATK). *Let* $\Pi = (\mathcal{G}, \mathcal{E}, \mathcal{D})$ *be a public key encryption scheme and let* $\mathcal{A} = (\mathcal{A}_1, \mathcal{A}_2)$ *be a polynomial-time adversary. For atk* $\in \{cpa, cca1, cca2\}$ *and* $k \in \mathbb{N}$ *let*

$$
\mathrm{Adv}_{\mathcal{A},\Pi}^{\mathrm{ow-atk}} \stackrel{\mathrm{def}}{=} \Pr\left[
\begin{array}{l}
(pk, sk) \leftarrow \mathcal{G}(1^k), \delta \leftarrow \mathcal{A}_1^{\mathcal{O}_1}(pk), x \leftarrow \mathsf{PtSp}(k) \\
y \leftarrow \mathcal{E}(pk, x), x' \leftarrow \mathcal{A}_2^{\mathcal{O}_2}(pk, \delta, y) : x' = x
\end{array}
\right]
$$

where

$$
\begin{array}{llll}
\textit{If atk} = cpa & \textit{then} & \mathcal{O}_1(\cdot) = \varepsilon & \textit{and} \quad \mathcal{O}_2(\cdot) = \varepsilon \\
\textit{If atk} = cca1 & \textit{then} & \mathcal{O}_1(\cdot) = \mathcal{D}_{sk}(\cdot) & \textit{and} \quad \mathcal{O}_2(\cdot) = \varepsilon \\
\textit{If atk} = cca2 & \textit{then} & \mathcal{O}_1(\cdot) = \mathcal{D}_{sk}(\cdot) & \textit{and} \quad \mathcal{O}_2(\cdot) = \mathcal{D}_{sk}(\cdot)
\end{array}
$$

In the case of CCA2, we insist that $\mathcal{A}_2$ *does not ask its oracle to decrypt* y. *We say that* Π *is secure in the sense of OW-ATK if* $\mathrm{Adv}_{\mathcal{A},\Pi}^{\mathrm{ow-atk}}$ *is negligible for any* $\mathcal{A}$.

A Simpler Definition of OW-CCA2. The following theorem states that the OW-CCA2 definition can be simplified.

Theorem 1. *In the case of OW-CCA2, the definition with* $\mathcal{O}_1 = \mathcal{D}_{sk}(\cdot)$ *(denoted Def1) is equivalent to that with* $\mathcal{O}_1 = \varepsilon$ *(denoted Def2).*

Proof. 1. Def1 $\Rightarrow$ Def2: trivial.

2. Def2 $\Rightarrow$ Def1: we prove that if a PKE scheme Π is secure under definition Def2, it is also secure under definition Def1. First of all, an obvious fact is that if Π is secure under definition Def2 then $|\mathsf{PtSp}(k)| > p(k)$ for any polynomial p.

The proof is by contradiction. Suppose there exists a polynomial-time adversary $\mathcal{A}$ that breaks Π in Def1 with a non-negligible advantage, we construct another polynomial-time adversary $\mathcal{B}$ that breaks Π in Def2 also with a non-negligible advantage.

$\mathcal{B}$ is given (pk, y), where $(pk, sk) \leftarrow \mathcal{G}(1^k), x \leftarrow \mathsf{PtSp}(k), y \leftarrow \mathcal{E}(pk, x)$. $\mathcal{B}$ simulates the experiment of Def1 as follows: $\mathcal{B}$ gives pk to $\mathcal{A}$ as the public key, and simulates $\mathcal{O}_1 = \mathcal{D}_{sk}(\cdot)$ by asking its own decryption oracle. If $\mathcal{O}_1$ is queried with y, $\mathcal{B}$ makes a random guess on x and aborts the game. Otherwise, $\mathcal{B}$ gives y to $\mathcal{A}$ after $\mathcal{A}$ finishes to query $\mathcal{O}_1$, and continues to simulate $\mathcal{O}_2 = \mathcal{D}_{sk}(\cdot)$ in Def1 by using its own decryption oracle. Finally, $\mathcal{B}$ outputs whatever $\mathcal{A}$ outputs.

Denote E the event $\mathcal{A}$ queries $\mathcal{O}_1$ with y. If E does not occur, then the simulation is perfect. Since $|\mathsf{PtSp}(k)| > p(k)$ for any polynomial p, for any $\tilde{y}$ $\mathcal{A}$ queried to $\mathcal{O}_1$, the probability that $\mathcal{D}_{sk}(\tilde{y}) = x$ is negligible as x is randomly selected from $\mathsf{PtSp}(k)$. So the probability that E occurs is negligible. $\qquad\square$

The above theorem shows that the simplification does not weaken the definition of security. On the other hand, it helps simplify the security proofs.

3   PKE with Ciphertext Comparability in Bilinear Groups

Our construction is based on the Computational Diffie-Hellman (CDH) assumption in bilinear groups. Let $\mathbb{G}_1, \mathbb{G}_2$ denote two groups of prime order q, and $e : \mathbb{G}_1 \times \mathbb{G}_1 \to \mathbb{G}_2$ a bilinear map between them. The map satisfies the following properties:

1. Bilinear: For any $U, V \in \mathbb{G}_1$, and $a, b \in \mathbb{Z}_q$, we have $e(U^a, V^b) = e(U, V)^{ab}$;
2. Non-degenerate: If g is a generator of $\mathbb{G}_1$, then $e(g, g)$ is a generator of $\mathbb{G}_2$;
3. Computable: there exists an efficient algorithm to compute $e(U, V)$ for any $U, V \in \mathbb{G}_1$.

Computational Diffie-Hellman (CDH) Problem: Fix a generator g of $\mathbb{G}_1$. The CDH problem is as follows: given g, g^a, g^b as input where a, b are randomly selected from $\mathbb{Z}_q$, compute g^{ab}. We say that CDH is intractable if all polynomial time algorithms have a negligible advantage in solving CDH.

We build a public key encryption scheme with ciphertext comparability in a bilinear group where CDH is intractable. Our construction uses a hash function $H : \mathbb{G}_1^3 \to \{0,1\}^{k+\ell}$, where k and ℓ are security parameters such that elements of $\mathbb{G}_1$ are represented in k bits and elements of $\mathbb{Z}_q$ are represented in ℓ bits. Our PKE with ciphertext comparability works as follows:

- $\mathcal{G}(1^k)$: Select $x \leftarrow \mathbb{Z}_q^*$ and compute $y = g^x$. Set $pk = y$ and $sk = x$.
- $\mathcal{E}(pk, m)$: To encrypt a plaintext $m \in \mathbb{G}_1^*$ ($\stackrel{\text{def}}{=} \mathbb{G}_1 \backslash \{1\}$), select $r \leftarrow \mathbb{Z}_q^*$, compute $U = g^r$, $V = m^r$, $W = H(U, V, y^r) \oplus m\|r$. The ciphertext is $C = (U, V, W)$.
- $\mathcal{D}(sk, C)$: To decrypt a ciphertext $C = (U, V, W)$, compute $m\|r \leftarrow H(U, V, U^x) \oplus W$. If ($m \in \mathbb{G}_1^* \wedge r \in \mathbb{Z}_q^* \wedge U = g^r \wedge V = m^r$), return m; otherwise, return $\bot$.
- $\mathsf{Test}(C_1, C_2)$: Given two ciphertexts $C_1 = (U_1, V_1, W_1)$ and $C_2 = (U_2, V_2, W_2)$, if $e(U_1, V_2) = e(U_2, V_1)$, return 1; otherwise, return 0.

Theorem 2. *The above PKE scheme has perfect consistency and perfect soundness.*

Proof. The proof is straightforward, as follows:

1. Perfect Consistency. It is easy to see that for any $(pk, sk) \leftarrow \mathcal{G}(1^k)$, $(pk', sk') \leftarrow \mathcal{G}(1^k)$ and $C \leftarrow \mathcal{E}(pk, m), C' \leftarrow \mathcal{E}(pk', m)$ where $C = (g^r, m^r, W)$, $C' = (g^{r'}, m^{r'}, W')$, we have

$$e(g^r, m^{r'}) = e(g^{r'}, m^r) = e(g, m)^{rr'}$$

for any $m \in \mathbb{G}_1^*$ and $(r, r') \in \mathbb{Z}_q^{*2}$.

2. Perfect Soundness. Given two ciphertexts $C = (g^r, m^r, W), C' = (g^{r'}, m'^{r'}, W')$, we have

$$e(g^r, m'^{r'}) = e(g, m')^{rr'}, e(g^{r'}, m^r) = e(g, m)^{rr'}$$

then it must be true that $e(g, m')^{rr'} \neq e(g, m)^{rr'}$ for any $m \neq m'$ and $(r, r') \in \mathbb{Z}_q^{*2}$. □

Theorem 3. *The PKE scheme above with message space $\mathbb{G}_1^*$ is OW-CCA2 secure in the random oracle model assuming CDH is intractable.*

Proof. Let $\mathcal{A}$ be a PPT adversary attacking the OW-CCA2 security of the above PKE scheme. Suppose that $\mathcal{A}$ runs in time t and makes at most q_H hash queries and q_D decryption queries. Let $\mathsf{Adv}_{\mathcal{A}}^{\mathsf{OW\text{-}CCA2}}(t, q_H, q_D)$ denote the advantage of $\mathcal{A}$ in the OW-CCA2 experiment. We first consider the original game:

Game G_0

1. $x \leftarrow \mathbb{Z}_q^*$, $y = g^x$
2. $m \leftarrow \mathbb{G}_1^*$, $r \leftarrow \mathbb{Z}_q^*$, $U^* = g^r$, $V^* = m^r$, $W^* = H(U^*, V^*, y^r) \oplus (m\|r)$
3. $m' \leftarrow \mathcal{A}^{\mathcal{O}_H, \mathcal{O}_2}(y, U^*, V^*, W^*)$, where the oracles work as follows.
   - $\mathcal{O}_H$: On input a triple $(U, V, Y) \in \mathbb{G}_1^3$, a *compatible* random value is returned, where by '*compatible*' we mean that if the same input is asked multiple times, the same answer will be returned. Note that a query to this oracle is also issued when computing the challenge ciphertext or simulating the decryption oracle.
   - $\mathcal{O}_2$: On input a ciphertext (U, V, W), it runs the decryption algorithm $\mathcal{D}$ to decrypt it using the secret key x.

Let $\mathbf{X}_0$ be the event that $m' = m$ in **Game G_0**. In the following, let $\mathbf{X}_i$ be the event that $m' = m$ in **Game G_i** for $i = 1, 2, \cdots$. $\mathcal{A}$'s winning probability in **Game G_i** is $\Pr[\mathbf{X}_i]$. Next we modify **Game G_0** and obtain the following game.

Game G_1

1. $x \leftarrow \mathbb{Z}_q^*$, $y = g^x$, $T = \emptyset$
2. $m \leftarrow \mathbb{G}_1^*$, $r \leftarrow \mathbb{Z}_q^*$, $U^* = g^r$, $V^* = m^r$, $R^* \leftarrow \{0,1\}^{k+\ell}$, $W^* = R^* \oplus (m\|r)$,
   $T = T \cup \{(U^*, V^*, (U^*)^x, R^*)\}$
3. $m' \leftarrow \mathcal{A}^{\mathcal{O}_H, \mathcal{O}_2}(y, U^*, V^*, W^*)$, where the oracles work as follows.
   - $\mathcal{O}_H$: On input $(U, V, Y) \in \mathbb{G}_1^3$, if there is an entry (U, V, Y, h) in the hash table T, h is returned; otherwise, a random value h is selected and returned, and (U, V, Y, h) is added into T.

– $\mathcal{O}_2$: On input a ciphertext (U, V, W), a hash query on (U, V, U^x) is issued. Suppose the answer is $h \in \{0, 1\}^{k+\ell}$. Then $m\|r$ is computed as $h \oplus W$, and the validity check on whether $U = g^r$ and $V = m^r$ is performed. If the check fails, $\perp$ is returned; otherwise, m is returned.

Due to the idealness of the random oracle, **Game G_1** is identical to **Game G_0**. Thus $\Pr[X_1] = \Pr[X_0]$. In the next game, we further modify the simulation in an indistinguishable way.

Game G_2

1. $x \leftarrow \mathbb{Z}_q^*$, $y = g^x$, $T = \emptyset$
2. $m \leftarrow \mathbb{G}_1^*$, $r \leftarrow \mathbb{Z}_q^*$, $U^* = g^r$, $V^* = m^r$, $W^* \leftarrow \{0,1\}^{k+\ell}$,
   $T = T \cup \{(U^*, V^*, (U^*)^x, W^* \oplus (m\|r))\}$
3. $m' \leftarrow \mathcal{A}^{\mathcal{O}_H, \mathcal{O}_2}(y, U^*, V^*, W^*)$, where the oracles work as follows.
   – $\mathcal{O}_H$: It is simulated in the same way as that in **Game G_1** except that if $\mathcal{A}$ asks $(U^*, \cdot, (U^*)^x)$, the game is aborted. Let this event be **E**.
   – $\mathcal{O}_2$: The same as that in **Game G_1** except that if $\mathcal{A}$ asks for decryption of (U^*, V^*, W') where $W' \neq W^*$, $\perp$ is returned.

The challenge ciphertext generated in this game is identically distributed to that in **Game G_1**, as W^* is a random value in both **Game G_1** and **Game G_2**. Also, the simulation of $\mathcal{O}_2$ is perfect since W^* is uniquely determined by U^* and V^*. Therefore, if event **E** does not occur, **Game G_2** is identical to **Game G_1**. Next, we show that event **E** occurs with negligible probability.

Lemma 1. *Event* **E** *happens in* **Game G_2** *with negligible probability if CDH is intractable.*

Proof. Suppose that $\Pr[E]$ is non-negligible. We construct a PPT algorithm $\mathcal{B}$ to break the CDH assumption. Given a tuple $(g, g^a, g^c) \in \mathbb{G}_1^3$, $\mathcal{B}$ randomly selects $\alpha \leftarrow \mathbb{Z}_q^*$ and $R^* \leftarrow \{0,1\}^{k+\ell}$, then sets the public key $y = g^a$ and $U^* = g^c$, and computes $m = g^\alpha$. It then adds $(U^*, V^* = (U^*)^\alpha, \top, \top)$ into table T which is initially empty, where $\top$ represents that the value is unknown. $\mathcal{B}$ invokes adversary $\mathcal{A}$ on input (y, U^*, V^*, R^*), where the challenge ciphertext (U^*, V^*, R^*) has the same distribution as that in **Game G_2**. The oracles for $\mathcal{A}$ are simulated as follows.

– $\mathcal{O}_H$: $\mathcal{B}$ simulates the oracle as described in **Game G_1**, except that if $\mathcal{A}$ makes a query on $(U^*, \cdot, Z)$, $\mathcal{B}$ checks if $e(g, Z) = e(y, U^*)$. If the equation holds, $\mathcal{B}$ outputs Z and aborts the game.
– $\mathcal{O}_2$: On input (U, V, W), if the input is $U = U^*$, $V = V^*$ and $W \neq R^*$, $\mathcal{B}$ returns $\perp$. Otherwise, $\mathcal{B}$ searches T for an entry of the form $(U, V, \cdot, \cdot)$. For each item (U, V, Y, h), $\mathcal{B}$ computes $m\|r = h \oplus W$ and proceeds as follows.
   1. If $U = g^r$, $V = m^r$ and $Y = y^r$, m is returned;
   2. Otherwise, $\mathcal{B}$ continues to search T for the next entry of the form $(U, V, \cdot, \cdot)$. If nothing is returned to $\mathcal{A}$ in the above loop for all entries $(U, V, \cdot, \cdot)$ in T, $\mathcal{B}$ returns $\perp$.

Denote $\mathbf{E}'$ the event that $\mathcal{A}$ queries $\mathcal{O}_H$ on input $(U^*, \cdot, g^{ac})$. At the end of the simulation, if $\mathbf{E}'$ does not occur, $\mathcal{B}$ aborts with failure.

(*Analysis*): We first show that the decryption queries are simulated indistinguishably from **Game $\mathbf{G_2}$**. We separate all the decryption queries into two types:

1. Type 1: (U, V, U^a) has been queried to $\mathcal{O}_H$ before a decryption query (U, V, W) is issued. In this case, W is uniquely determined after (U, V, U^a) is queried to $\mathcal{O}_H$. So the decryption oracle is simulated perfectly.
2. Type 2: (U, V, U^a) has never been queried to $\mathcal{O}_H$ when a decryption query (U, V, W) is issued. In this case, $\perp$ is returned by the decryption oracle. The simulation fails if (U, V, W) is a valid ciphertext. However, due to the idealness of the random oracle, this happens with probability $1/2^{k+\ell}$.

Denote $\mathbf{E_2}$ the event that a valid ciphertext is rejected in the simulation. Then we have

$$\Pr[\mathbf{E_2}] \leq \frac{q_D}{2^{k+\ell}}.$$

If $\mathbf{E_2}$ does not happen, then the simulation is identical to **Game $\mathbf{G_2}$**, so $\Pr[\mathbf{E}'|\neg\mathbf{E_2}] = \Pr[\mathbf{E}]$. Then we have

$$\begin{aligned}
\Pr[\mathbf{E}'] &= \Pr[\mathbf{E}'|\mathbf{E_2}]\Pr[\mathbf{E_2}] + \Pr[\mathbf{E}'|\neg\mathbf{E_2}]\Pr[\neg\mathbf{E_2}] \\
&\geq \Pr[\mathbf{E}'|\neg\mathbf{E_2}]\Pr[\neg\mathbf{E_2}] \\
&= \Pr[\mathbf{E}](1 - \Pr[\mathbf{E_2}]) \\
&\geq \Pr[\mathbf{E}] - \Pr[\mathbf{E_2}]
\end{aligned}$$

Therefore,

$$\mathsf{Adv}_{\mathcal{B}}^{\mathsf{CDH}} \geq \Pr[\mathbf{E}] - \frac{q_D}{2^{k+\ell}}$$

which is non-negligible. This completes the proof of Lemma 1.   □

Since **Game $\mathbf{G_1}$** and **Game $\mathbf{G_2}$** are the same if event $\mathbf{E}$ does not occur, we have,

$$|\Pr[\mathbf{X_1}] - \Pr[\mathbf{X_2}]| \leq \Pr[\mathbf{E}].$$

Lemma 2. $\Pr[\mathbf{X_2}]$ *is negligible under the CDH assumption.*

Proof. Suppose that $\Pr[\mathbf{X_2}]$ is non-negligible. We construct a PPT algorithm $\mathcal{B}$ to break the CDH assumption. Given a tuple $(g, \hat{U} = g^r, \hat{V} = m^r) \in \mathbb{G}_1^3$, where $r \leftarrow \mathbb{Z}_q$ and $m \leftarrow \mathbb{G}_1^*$, $\mathcal{B}$'s goal is to compute m. $\mathcal{B}$ randomly selects $x \leftarrow \mathbb{Z}_q^*$ and sets the public key $y = g^x$. It then adds $(U^* = \hat{U}, V^* = \hat{V}, (U^*)^x, \top)$ into table T which is initially empty, where $\top$ represents that the value is unknown. $\mathcal{B}$ randomly selects $R^* \leftarrow \{0,1\}^{k+\ell}$ and invokes adversary $\mathcal{A}$ on input (y, U^*, V^*, R^*). $\mathcal{B}$ simulates the game by following the description of **Game $\mathbf{G_2}$**. Finally $\mathcal{B}$ outputs whatever $\mathcal{A}$ outputs. So we have

$$\Pr[\mathbf{X_2}] \leq \mathsf{Adv}_{\mathcal{B}}^{\mathsf{CDH}}.$$

□

Therefore, we have

$$\mathsf{Adv}_{\mathcal{A}}^{\mathsf{OW\text{-}CCA2}}(t, q_H, q_D) = \Pr[\mathbf{X}_0]$$
$$= \Pr[\mathbf{X}_1]$$
$$\leq \Pr[\mathbf{X}_2] + \mathsf{Adv}^{\mathsf{CDH}} + \frac{q_D}{2^{k+\ell}}$$
$$\leq 2\mathsf{Adv}^{\mathsf{CDH}} + \frac{q_D}{2^{k+\ell}}.$$

This completes the proof of Theorem 3.   □

4   Variants and Applications

Encrypting Long Messages. In our PKE scheme above, we assume that messages are elements of group $\mathbb{G}_1^*$. To encrypt long messages, we can use a collision resistant hash function $H' : \{0,1\}^* \rightarrow \mathbb{G}_1^*$ and a pseudo-random bit generator PRG [16]. We modify the scheme as follows:

- $\mathcal{G}(1^k)$: Unchanged.
- $\mathcal{E}(pk, m)$: To encrypt a plaintext $M \in \{0,1\}^*$, compute $m \leftarrow H'(M)$. Select $r \leftarrow \mathbb{Z}_q^*$, compute $U \leftarrow g^r$, $V \leftarrow m^r$, $K \leftarrow H(U, V, pk^r)$ and $W \leftarrow \mathsf{PRG}(K) \oplus M\|r$. The ciphertext is $C = (U, V, W)$.
- $\mathcal{D}(sk, C)$: To decrypt a ciphertext $C = (U, V, W)$, compute $K \leftarrow H(U, V, U^{sk})$, $M\|r \leftarrow \mathsf{PRG}(K) \oplus W$, and $m \leftarrow H'(M)$. If $(r \in \mathbb{Z}_q^* \wedge U = g^r \wedge V = m^r)$, return M; otherwise, return $\perp$.
- $\mathsf{Test}(C_1, C_2)$: Unchanged

It follows that the above (hybrid) encryption scheme also has ciphertext comparability. The perfect consistency is still maintained, however, the soundness is no longer perfect, but is bounded by the collision probability of H'.

Theorem 4. *The modified PKE scheme above is OW-CCA2 secure assuming H, H' are random oracles, PRG is a secure pseudo-random bit generator, and CDH is intractable.*

The proof essentially follows that of Theorem 3, we can replace the pseudo-random bit string with a truly random string when generating the challenge ciphertext. Since the adversary does not know the random seed of the PRG (due to the CDH assumption and H is a random oracle), the difference between the games is negligible provided PRG is a secure pseudo-random bit generator.

Searchable Encryption. A PKE scheme with ciphertext comparability is naturally searchable. To generate a tag T_M for message M, one can simply encrypt M under any valid public key to generate a ciphertext C, and set $T_M = C$. Then using the Test function, anyone is able to search encryptions of the message M, even if they are generated using different public keys.

Partitioning Encrypted Data. In applications such as outsourced databases, by using a PKE scheme with ciphertext comparability, the database administrator is able to do a partition of the database according to the encrypted messages without any help from the message owners. These schemes may also be useful in other similar applications such as collection and categorization of confidential data through an agent.

5   Weak IND-CCA2 vs Ciphertext Comparability

In Sec. 2, we have shown that ciphertext comparability and indistinguishability are irreconcilable. In this section, we are interested in the following question: if we don't need ciphertext comparability, or when being implemented in a non-bilinear group, what kind of security level can our PKE scheme in Sec. 3 achieve? The first security model we'd like to try is of course IND-CCA. Unfortunately, our scheme is even not IND-CPA secure, as shown below.

Theorem 5. *The PKE scheme in Sec. 3 with message space $\mathbb{G}_1^*$ is not IND-CPA secure.*

Proof. We construct a PPT adversary $\mathcal{A}$ as follows. Given public key y, $\mathcal{A}$ computes $m_0 = g^{r_0}$ and $m_1 = g^{r_1}$ for any two distinct r_0 and r_1 chosen arbitrarily from $\mathbb{Z}_q^*$. $\mathcal{A}$ sends (m_0, m_1) to the game simulator. After receiving the challenge ciphertext $c^* = (U, V, W)$, $\mathcal{A}$ checks if $V = U^{r_0}$. If yes, $\mathcal{A}$ returns 0; otherwise $\mathcal{A}$ returns 1. The probability that $\mathcal{A}$ guesses correctly the value of b is 1. $\qquad\square$

The above attack demonstrates the advantage the adversary can get from selecting the challenge plaintexts. In the next, we define a different set of indistinguishability games where the adversary has no such power.

Definition 3 (W-IND-ATK). *Let $\Pi = (\mathcal{G}, \mathcal{E}, \mathcal{D})$ be a public key encryption scheme and let $\mathcal{A} = (\mathcal{A}_1, \mathcal{A}_2)$ be a polynomial-time adversary. For atk $\in \{cpa, cca1, cca2\}$ and $k \in \mathbb{N}$ let*

$$\mathrm{Adv}_{\mathcal{A},\Pi}^{\mathrm{w-ind-atk}} \overset{\mathrm{def}}{=} \Pr \left[\begin{array}{l} (pk, sk) \leftarrow \mathcal{G}(1^k), \delta \leftarrow \mathcal{A}_1^{\mathcal{O}_1}(pk), \\ (x_0, x_1) \leftarrow \mathsf{PtSp}(k), b \leftarrow \{0,1\}, y \leftarrow \mathcal{E}(pk, x_b), \\ b' \leftarrow \mathcal{A}_2^{\mathcal{O}_2}(pk, x_0, x_1, \delta, y) : b' = b \end{array} \right] - \frac{1}{2}$$

where $x_0 \neq x_1 \wedge |x_0| = |x_1|$ and

If atk = cpa   then $\mathcal{O}_1(\cdot) = \varepsilon$ and $\mathcal{O}_2(\cdot) = \varepsilon$
If atk = cca1   then $\mathcal{O}_1(\cdot) = \mathcal{D}_{sk}(\cdot)$ and $\mathcal{O}_2(\cdot) = \varepsilon$
If atk = cca2   then $\mathcal{O}_1(\cdot) = \mathcal{D}_{sk}(\cdot)$ and $\mathcal{O}_2(\cdot) = \mathcal{D}_{sk}(\cdot)$

In the case of CCA2, we insist that $\mathcal{A}_2$ does not ask its oracle for decrypting y. We say that Π is secure in the sense of W-IND-ATK if $\mathrm{Adv}_{\mathcal{A},\Pi}^{\mathrm{w-ind-atk}}$ is negligible for any $\mathcal{A}$.

W-IND-CCA2 Security of Our PKE. Interestingly, we can show that when being implemented in a non-bilinear group, our PKE scheme given in Sec. 3 can achieve W-IND-CCA2 security under the DDH assumption which is described below.

Decisional Diffie-Hellman (DDH) Problem. Fix a generator g of $\mathbb{G}_1$. The DDH assumption claims that $\{g, g^a, g^b, Z\}$ and $\{g, g^a, g^b, g^{ab}\}$ are computationally indistinguishable where a, b are randomly selected from $\mathbb{Z}_q$ and Z is a random element of $\mathbb{G}_1$.

Theorem 6. *The PKE scheme in Sec. 3 with message space $\mathbb{G}_1^*$ is W-IND-CCA2 secure in the random oracle model under the DDH assumption.*

The proof is by contradiction. Suppose there exists an adversary who can break the encryption scheme, we plant the DDH problem $(g, m, U = g^r, V = Z)$ into the challenge ciphertext to the adversary, and simulate the decryption oracle in a similar way as in the proof of Theorem 3. Then depending on $Z = m^r$ (i.e. Z is in the "right" form) or $Z \leftarrow \mathbb{G}_1$ (Z is independent of m), the adversary would have different probability in winning the game, so we can use the adversary to solve the DDH problem. The detailed proof is deferred to the full version of the paper.

Acknowledgement. We would like to thank the anonymous reviewers for their comments and suggestions.

References

1. Abdalla, M., Bellare, M., Catalano, D., Kiltz, E., Kohno, T., Lange, T., Malone-Lee, J., Neven, G., Paillier, P., Shi, H.: Searchable encryption revisited: Consistency properties, relation to anonymous IBE, and extensions. J. Cryptology 21(3), 350–391 (2008)

2. Abe, M., Gennaro, R., Kurosawa, K., Shoup, V.: Tag-KEM/DEM: A new framework for hybrid encryption and a new analysis of Kurosawa-Desmedt KEM. In: Cramer, R. (ed.) EUROCRYPT 2005. LNCS, vol. 3494, pp. 128–146. Springer, Heidelberg (2005)

3. Bellare, M., Boldyreva, A., O'Neill, A.: Deterministic and efficiently searchable encryption. In: Menezes, A. (ed.) CRYPTO 2007. LNCS, vol. 4622, pp. 535–552. Springer, Heidelberg (2007)

4. Bellare, M., Desai, A., Pointcheval, D., Rogaway, P.: Relations among notions of security for public-key encryption schemes. In: Krawczyk, H. (ed.) CRYPTO 1998. LNCS, vol. 1462, pp. 26–45. Springer, Heidelberg (1998)

5. Bellare, M., Fischlin, M., O'Neill, A., Ristenpart, T.: Deterministic encryption: Definitional equivalences and constructions without random oracles. In: Wagner, D. (ed.) CRYPTO 2008. LNCS, vol. 5157, pp. 360–378. Springer, Heidelberg (2008)

6. Boldyreva, A., Fehr, S., O'Neill, A.: On notions of security for deterministic encryption, and efficient constructions without random oracles. In: Wagner, D. (ed.) CRYPTO 2008. LNCS, vol. 5157, pp. 335–359. Springer, Heidelberg (2008)

7. Boneh, D., Crescenzo, G.D., Ostrovsky, R., Persiano, G.: Public key encryption with keyword search. In: Cachin, C., Camenisch, J.L. (eds.) EUROCRYPT 2004. LNCS, vol. 3027, pp. 506–522. Springer, Heidelberg (2004)

8. Camenisch, J., Shoup, V.: Practical verifiable encryption and decryption of discrete logarithms. In: Boneh, D. (ed.) CRYPTO 2003. LNCS, vol. 2729, pp. 126–144. Springer, Heidelberg (2003)

9. Canetti, R., Halevi, S., Katz, J.: Chosen-ciphertext security from identity-based encryption. In: Cachin, C., Camenisch, J.L. (eds.) EUROCRYPT 2004. LNCS, vol. 3027, pp. 207–222. Springer, Heidelberg (2004)

10. Canetti, R., Halevi, S., Katz, J.: A forward-secure public-key encryption scheme. J. Cryptology 20(3), 265–294 (2007)

11. Cramer, R., Shoup, V.: A practical public key cryptosystem provably secure against adaptive chosen ciphertext attack. In: Krawczyk, H. (ed.) CRYPTO 1998. LNCS, vol. 1462, pp. 13–25. Springer, Heidelberg (1998)

12. Cramer, R., Shoup, V.: Universal hash proofs and a paradigm for adaptive chosen ciphertext secure public-key encryption. In: Knudsen, L.R. (ed.) EUROCRYPT 2002. LNCS, vol. 2332, pp. 45–64. Springer, Heidelberg (2002)

13. Dent, A.W.: A brief history of provably-secure public-key encryption. In: Vaudenay, S. (ed.) AFRICACRYPT 2008. LNCS, vol. 5023, pp. 357–370. Springer, Heidelberg (2008)

14. Diffie, W., Hellman, M.E.: New directions in cryptography. IEEE Transactions on Information Theory 22, 644–654 (1978)

15. Dolev, D., Dwork, C., Naor, M.: Nonmalleable cryptography. SIAM J. Comput. 30(2), 391–437 (2000)

16. Goldreich, O.: Foundations of Cryptography: Basic Tools. Cambridge University Press, Cambridge (2001)

17. Goldwasser, S., Micali, S.: Probabilistic encryption. J. Comput. Syst. Sci. 28(2), 270–299 (1984)

18. Hanaoka, G., Kurosawa, K.: Efficient chosen ciphertext secure public key encryption under the computational diffie-hellman assumption. In: Pieprzyk, J. (ed.) ASIACRYPT 2008. LNCS, vol. 5350, pp. 308–325. Springer, Heidelberg (2008)

19. Hofheinz, D., Kiltz, E.: Secure hybrid encryption from weakened key encapsulation. In: Menezes, A. (ed.) CRYPTO 2007. LNCS, vol. 4622, pp. 553–571. Springer, Heidelberg (2007)

20. Hofheinz, D., Kiltz, E.: Practical chosen ciphertext secure encryption from factoring. In: Joux, A. (ed.) EUROCRYPT 2009. LNCS, vol. 5479, pp. 313–332. Springer, Heidelberg (2009)

21. Hofheinz, D., Weinreb, E.: Searchable encryption with decryption in the standard model. Cryptology ePrint Archive, Report 2008/423 (2008),
http://eprint.iacr.org/

22. Kurosawa, K., Desmedt, Y.: A new paradigm of hybrid encryption scheme. In: Franklin, M. (ed.) CRYPTO 2004. LNCS, vol. 3152, pp. 426–442. Springer, Heidelberg (2004)

23. Lucks, S.: A variant of the cramer-shoup cryptosystem for groups of unknown order. In: Zheng, Y. (ed.) ASIACRYPT 2002. LNCS, vol. 2501, pp. 27–45. Springer, Heidelberg (2002)

24. Rackoff, C., Simon, D.R.: Non-interactive zero-knowledge proof of knowledge and chosen ciphertext attack. In: Feigenbaum, J. (ed.) CRYPTO 1991. LNCS, vol. 576, pp. 433–444. Springer, Heidelberg (1992)

Efficient CCA-Secure PKE from Identity-Based Techniques

Junzuo Lai[1], Robert H. Deng[2], Shengli Liu[1], and Weidong Kou[3]

[1] Department of Computer Science and Engineering
Shanghai Jiao Tong University, Shanghai 200030, China
{laijunzuo,slliu}@sjtu.edu.cn
[2] School of Information Systems,
Singapore Management University, Singapore 178902
robertdeng@smu.edu.sg
[3] School of Computer Science and Technology
Xi Dian University, Xi'an 710071, China
kou_weidong@yahoo.com.cn

Abstract. Boneh, Canetti, Halevi, and Katz showed a general method for constructing CCA-secure public key encryption (PKE) from any selective-ID CPA-secure identity-based encryption (IBE) schemes. Their approach treated IBE as a black box. Subsequently, Boyen, Mei, and Waters demonstrated how to build a direct CCA-secure PKE scheme from the Waters IBE scheme, which is adaptive-ID CPA secure. They made direct use of the underlying IBE structure, and required no cryptographic primitive other than the IBE scheme itself. However, their scheme requires long public key and the security reduction is loose. In this paper, we propose an efficient PKE scheme employing identity-based techniques. Our scheme requires short public key and is proven CCA-secure in the standard model (without random oracles) with a tight security reduction, under the Decisional Bilinear Diffie-Hellman (DBDH) assumption. In addition, we show how to use our scheme to construct an efficient threshold public key encryption scheme and a public key encryption with non-interactive opening (PKENO) scheme.

Keywords: Chosen Ciphertext Security, Public Key Encryption, Identity-Based Encryption.

1 Introduction

Chosen-ciphertext security (CCA-security, for short) [31,16] is now considered as a standard notion of security for public key encryption (PKE) in practice. There have been several efficient PKE schemes shown to be secure in the random oracle (RO) model [3]. Unfortunately, the RO model is heuristic, and a proof of security in the RO model does not directly imply anything about the security of a PKE scheme in the real world. In fact, it has been demonstrated that there exist cryptographic schemes which are secure in the RO model but which are inherently insecure when the random oracle is instantiated with any real hash

J. Pieprzyk (Ed.): CT-RSA 2010, LNCS 5985, pp. 132–147, 2010.

function [8,29,19,2]. Throughout this paper, we focus on PKE schemes whose security are proven in the standard model (without random oracles).

Dolev, Dwork, and Naor [16] were the first to come up with a CCA-secure PKE scheme in the standard model. Later Cramer and Shoup [11] proposed the first practical CCA-secure PKE scheme in the standard model, under the Decisional Diffie-Hellman (DDH) assumption. Interestingly, Elkind and Sahai [17] showed that both techniques can be viewed as special cases of a single paradigm.

Canetti, Halevi, and Katz [9] presented a new paradigm for constructing CCA-secure PKE schemes using IBE as a building block. The idea is to use, for each encryption, a fresh random verification key of a one-time signature scheme as the "identity" for IBE encryption. In order to tie the IBE ciphertext to this verification key, the ciphertext is signed using the corresponding signing key. If the IBE scheme is selective-ID CPA secure then the resulting PKE scheme is CCA secure. Boneh and Katz [6] further improved the efficiency of the scheme by using a MAC instead of a one-time signature. Kiltz [25] showed that a tag-based encryption (TBE) scheme is sufficient for the transformation in [9] to obtain a CCA-secure PKE scheme.

Boyen, Mei, and Waters [7] showed how to build a direct CCA-secure PKE scheme from the Waters IBE scheme [33]. Unlike the Canetti-Halevi-Katz (CHK) scheme [9] and the Boneh-Katz (BK) scheme [6] that use IBE as a black box, their approach is endogenous, very simple, and compact. They constructed a CCA-secure PKE scheme, referred to as the BMW scheme, in which a ciphertext consists of just three group elements without attached signature or MAC. Compared with the CHK scheme and the BK scheme, the main difference is to use the first two elements of the ciphertext to determine a one-time "identity", instead of a fresh random "identity" generated by a one-time signature as in the CHK scheme or encapsulation as in the BK scheme. When proving security of the scheme, they took advantage of the CPA security of the Waters IBE scheme in the adaptive-ID security model (as opposed to the weaker selective-ID model). The drawback of the BMW scheme, however, is that the user needs *long* public key and the security is reduced only *loosely* to the Decisional Bilinear Diffie-Hellman (DBDH) assumption. Note that, an inefficient security reduction would imply either a lower security level or the requirement of larger key and ciphertext sizes to obtain the same security level.

1.1 Hybrid Encryption

Cramer and Shoup [12,32] formalized the notion of *hybrid encryption*, where a public key cryptosystem is used to encapsulate the (session) key of a symmetric cipher which is subsequently used to conceal the data. This is also known as the KEM/DEM approach. A folklore composition theorem (formalized in [12]) shows that if both KEM and DEM are CCA-secure then the hybrid encryption is CCA-secure. Kurosawa and Desmedt [27] came up with a hybrid encryption scheme improving the performance of the Cramer-Shoup scheme both in computational efficiency and in ciphertext length. Abe, Gennaro, Kurosawa and Desmedt [1] established the Tag-KEM/DEM framework, and explained the security of Kurosawa-Desmedt scheme in this framework. Hofheinz and Kiltz [21] presented another paradigm for constructing hybrid encryption with strictly weakened KEM. The DDH assumption still is required for these extensions except for one of Hofheinz and Kiltz's schemes which depends on the *n-linear* DDH assumption.

Kiltz [26] presented a practical CCA-secure KEM scheme whose security is proven under the gap hashed Diffie-Hellman (GHDH) assumption. Cash, Kiltz and Shoup [10] proposed CCA-secure hybrid encryption schemes under the computational Diffie-Hellman (CDH) or hashed Diffie-Hellman (HDH) assumption by using the twin DH problem (which is also applicable to a wide range of cryptographic primitives). Note that the CDH and HDH assumptions are weaker than the DDH assumption. Based on Naor-Pinkas broadcast encryption (BE) scheme [28], Hanaoka and Kurosawa [20] proposed more efficient CCA-secure hybrid encryption schemes under the CDH or HDH assumption. Recently, Hofheinz and Kiltz [22] proposed a practical CCA-secure hybrid encryption scheme whose security can be reduced to the assumption that factoring is intractable. However, all these hybrid encryption schemes are not suited for constructing threshold public key encryption and public key encryption with non-interactive opening schemes.

1.2 Our Contribution

In this paper, we propose a more efficient PKE scheme employing identity-based techniques. The proposed scheme has small public key size and is proven CCA-secure in the standard model with a tight security reduction, under the DBDH assumption. We follow a similar method in the proof simulation as that in the CHK, BK and BMW schemes. After the step phase there is a certain set of well-formed ciphertexts that the simulator can decrypt corresponding to "identities" that the simulator knows the private keys. The remainder of the well-formed ciphertexts, that the simulator cannot decrypt corresponding to "identities" for which the simulator does not know the private keys, can be used as challenge ciphertexts in the simulation.

Our scheme has the desirable property that it allows the validity of ciphertexts to be checked publicly. Using this property, we extend our scheme to an efficient threshold public key encryption scheme and an efficient PKE with non-interactive opening (PKENO) scheme. An overview comparing the efficiency of our PKE scheme to those of other PKE schemes employing identity-based techniques is given in Table 1.

1.3 Organization

The rest of the paper is organized as follows. In Section 2, we present some definitions and a related complexity assumption. We describe and analysis our PKE scheme in Section 3. In Section 4, we introduce two extensions of practical interest to our PKE scheme. Finally, we state our conclusion in Section 5.

Table 1. Comparison of public key encryption schemes employing identity-based techniques. "exp" denotes an exponentiation operation. (Some of the exponentiations are actually multi-exponentiations.) "pr" denotes a pairing operation. The CHK [9] and BK [6] schemes are instantiated with the first Boneh-Boyen IBE scheme from [4]. (Kiltz [24] showed that the CHK transformation maps the first and second Boneh-Boyen IBE schemes from [4] to nearly one single encryption scheme.)

Scheme	PK size	Encryption	Decryption	Ciphertext size	TPKE	PKENO
CHK[9]	$3|\mathbb{G}|+1|\mathbb{G}_T|$	3 exp + Sig	1 exp + 1 pr + Ver	$2|\mathbb{G}|+1|\mathbb{G}_T|$+vk+sig	√	√
BK[6]	$3|\mathbb{G}|+1|\mathbb{G}_T|$	3 exp	1 exp + 1 pr	$2|\mathbb{G}|+1|\mathbb{G}_T|$+com+tag	×	×
BMW[7]	$162|\mathbb{G}|+1|\mathbb{G}_T|$	3 exp	1 exp + 1 pr	$2|\mathbb{G}|+1|\mathbb{G}_T|$	√	√
Ours	$4|\mathbb{G}|+1|\mathbb{G}_T|$	3 exp	1 exp + 1 pr	$2|\mathbb{G}|+1|\mathbb{G}_T|+1|\mathbb{Z}_p|$	√	√

2 Preliminaries

For a group $\mathbb{G}$, we denote the size of a group-element representation as $|\mathbb{G}|$. We say that a function $f(\lambda)$ is *negligible* if for every $c > 0$ there exists an λ_c such that $f(\lambda) < 1/\lambda^c$ for all $\lambda > \lambda_c$.

2.1 Bilinear Pairings

Let $\mathbb{G}$ be a cyclic multiplicative group of prime order p and $\mathbb{G}_T$ be a cyclic multiplicative group of the same order p. A bilinear pairing is a map $e : \mathbb{G} \times \mathbb{G} \to \mathbb{G}_T$ with the following properties:

- Bilinearity: $\forall g_1, g_2 \in \mathbb{G}, \forall a, b \in \mathbb{Z}_p^*$, we have $e(g_1^a, g_2^b) = e(g_1, g_2)^{ab}$;
- Non-degeneracy: There exist $g_1, g_2 \in \mathbb{G}$ such that $e(g_1, g_2) \neq 1$;
- Computability: There exists an efficient algorithm to compute $e(g_1, g_2)$ for $\forall g_1, g_2 \in \mathbb{G}$.

2.2 Complexity Assumption

Definition 1 (DBDH Problem). *Given a group $\mathbb{G}$ of prime order p with generator g and elements $g^a, g^b, g^c \in \mathbb{G}$, $e(g, g)^z \in \mathbb{G}_T$ where a, b, c, z are selected uniformly at random from $\mathbb{Z}_p^*$. A fair binary coin $\beta \in \{0, 1\}$ is flipped. If $\beta = 1$, it outputs the tuple $(g, g^a, g^b, g^c, T = e(g, g)^{abc})$. If $\beta = 0$, it outputs the tuple $(g, g^a, g^b, g^c, T = e(g, g)^z)$. The Decisional Bilinear Diffie-Hellman (DBDH) problem is to guess the value of β.*

An adversary $\mathcal{A}$ has at least an ϵ advantage in solving the DBDH problem if

$$|\mathsf{Pr}[\mathcal{A}(g^a, g^b, g^c, T = e(g, g)^{abc}) = 1] - \mathsf{Pr}[\mathcal{A}(g^a, g^b, g^c, T = e(g, g)^z) = 1]| \geq \epsilon$$

where the probability is over the randomly chosen a, b, c, z and the random bits consumed by $\mathcal{A}$. We refer to the distribution on the left as $\mathcal{P}_{BDH}$ and the one on the right as $\mathcal{R}_{BDH}$.

Definition 2 (DBDH assumption). *We say that the (ϵ, t)-DBDH assumption holds in a group $\mathbb{G}$ if no algorithm running in time at most t can solve the DBDH problem in $\mathbb{G}$ with advantage at least ϵ.*

2.3 Collision-Resistant Hashing

Formally, we say that a function $H : X \rightarrow Y$ is a target-collision resistant (CR) hash function, if for all PPT $\mathcal{A}$, $\mathsf{Adv}_{\mathcal{A}}^{\mathsf{CR}}(\lambda)$ is negligible in λ, where $\mathsf{Adv}_{\mathcal{A}}^{\mathsf{CR}}(\lambda) = \Pr[x, x' \leftarrow \mathcal{A}(H) : x' \neq x \wedge H(x') = H(x)]$.

2.4 Public Key Encryption

A public key encryption scheme is a tuple of algorithms described as follows:

KeyGen(λ). Takes as input a security parameter λ. It outputs a public/private key pair (PK, SK).

Encrypt(PK, m). Takes as input a public key PK and a message m. It outputs a ciphertext.

Decrypt(SK, C). Takes as input a private key SK and a ciphertext C. It outputs a plaintext message or the special symbol $\perp$ meaning that the ciphertext is invalid.

We insist that all public key encryption schemes satisfy the obvious correctness condition (that decryption "undoes" encryption).

The strongest and commonly accepted notion of security for a public key encryption scheme is that of indistinguishability against an adaptive chosen ciphertext attack (IND-CCA). It is defined using the following game between an attack algorithm $\mathcal{A}$ and a challenger.

Setup. The challenger runs KeyGen(λ) to obtain a public/private key pair (PK, SK). It gives the public key PK to the adversary.

Query phase 1. The adversary $\mathcal{A}$ adaptively issues decryption queries C. The challenger responds with Decrypt(SK, C).

Challenge. The adversary $\mathcal{A}$ submits two (equal length) messages m_0, m_1. The challenger selects a random bit $\beta \in \{0, 1\}$, sets $C^* = \mathsf{Encrypt}(\mathsf{PK}, m_\beta)$ and sends C^* to the adversary as its challenge ciphertext.

Query phase 2. The adversary continues to adaptively issue decryption queries C, as in Query phase 1, but with the natural constraint that the adversary does not request the decryption of C^*.

Guess. The adversary $\mathcal{A}$ outputs its guess $\beta' \in \{0, 1\}$ for β and wins the game if $\beta = \beta'$.

We define $\mathcal{A}$'s advantage in attacking the public key encryption scheme PKE with the security parameter λ as $\mathsf{Adv}_{\mathcal{A}}^{\mathsf{PKE}}(\lambda) = |\Pr[\beta = \beta'] - \frac{1}{2}|$.

Definition 3. *We say that a public key encryption scheme PKE is (t, q, ϵ)-IND-CCA secure, if for all t-time algorithms $\mathcal{A}$ making at most q decryption queries have advantage at most ϵ in winning the above game.*

2.5   Public Key Encryption with Non-interactive Opening

A public key encryption with non-interactive opening (PKENO) scheme is a tuple of algorithms described as follows:

KeyGen(λ). Takes as input a security parameter λ. It outputs a public/private key pair (PK, SK).

Encrypt(PK, m). Takes as input a public key PK and a message m. It outputs a ciphertext.

Decrypt(SK, C). Takes as input a private key SK and a ciphertext C. It outputs a plaintext message or the special symbol $\perp$ meaning that the ciphertext is invalid.

Prove(SK, C). Takes as input a private key SK and a ciphertext C. It outputs a proof π or the special symbol $\perp$ meaning that the ciphertext is invalid.

Ver(PK, C, m, π). Takes as input a public key PK, a ciphertext C, a message m and a proof π. It outputs a result $res \in \{0, 1\}$ meaning accepted and rejected proof, respectively. In particular $1 \leftarrow$ Ver(PK, $C, \perp, \pi$) must be interpreted as the verifier being convinced that C is an invalid ciphertext.

We insist that all PKENO schemes satisfy the obvious correctness condition (that decryption "undoes" encryption). In addition, we require, for a honestly generated key pair (PK, SK) and all ciphertexts C, $1 \leftarrow$ Ver(PK, C, Decrypt(SK, C), Prove(SK, C)).

The notion of security for a PKENO scheme is indistinguishability against chosen-ciphertext and prove attacks (IND-CCPA) and satisfies computational proof soundness. IND-CCPA is defined using the following game between an attack algorithm $\mathcal{A}$ and a challenger.

Setup. The challenger runs KeyGen(λ) to obtain a public/private key pair (PK, SK). It gives the public key PK to the adversary.

Query phase 1. The adversary $\mathcal{A}$ adaptively issues decryption or proof queries on C. The challenger responds with Decrypt(SK, C) or Prove(SK, C).

Challenge. The adversary $\mathcal{A}$ submits two (equal length) messages m_0, m_1. The challenger selects a random bit $\beta \in \{0, 1\}$, sets $C^* =$ Encrypt(PK, m_β) and sends C^* to the adversary as its challenge ciphertext.

Query phase 2. The adversary continues to adaptively issue decryption or proof queries C, as in Query phase 1, but with the natural constraint that decryption or proof queries on C^* are not allowed.

Guess. The adversary $\mathcal{A}$ outputs its guess $\beta' \in \{0, 1\}$ for b and wins the game if $\beta = \beta'$.

We define $\mathcal{A}$'s advantage as $\mathsf{Adv}_{\mathsf{PKENO},\mathcal{A}}^{\mathsf{IND\text{-}CCPA}}(\lambda) = |\Pr[\beta = \beta'] - \frac{1}{2}|$.

Definition 4. *We say that a PKENO scheme is IND-CCPA secure, if for every adversary $\mathcal{A}$, the advantage $\mathsf{Adv}_{\mathsf{PKENO},\mathcal{A}}^{\mathsf{IND\text{-}CCPA}}(\cdot)$ is negligible.*

Computational proof soundness is defined using the following game between an attack algorithm $\mathcal{A}$ and a challenger.

Setup. The challenger runs KeyGen(λ) to obtain a public/private key pair (PK, SK). It gives the key pair (PK, SK) to the adversary.

Challenge. The adversary $\mathcal{A}$ submits a message m. The challenger sends $C =$ Encrypt(PK, m) to the adversary.

Output. The adversary $\mathcal{A}$ outputs (m', π').

We define $\mathcal{A}$'s advantage in forging proof by $\mathsf{Adv}^{\mathsf{snd}}_{\mathsf{PKENO}, \mathcal{A}}(\lambda) = \Pr[1 \leftarrow \mathsf{Ver}(\mathsf{PK}, C, m', \pi') \wedge m' \neq m]$.

Definition 5. *We say that a PKENO scheme satisfies computational proof soundness, if for every adversary $\mathcal{A}$, the advantage $\mathsf{Adv}^{\mathsf{snd}}_{\mathsf{PKENO}, \mathcal{A}}(\cdot)$ is negligible.*

Definition 6. *We say that a PKENO scheme is secure, if it is IND-CCPA secure and satisfies computational proof soundness.*

2.6   Threshold Public Key Encryption

A threshold public key encryption (TPKE) scheme is a tuple of algorithms described as follows:

Setup(n, k, λ). Takes as input the number of decryption servers n, a threshold k where $1 \leq k \leq n$ and a security parameter λ. It outputs a public key PK, a verification key VK and private key SK = $(\mathsf{SK}_1, \ldots, \mathsf{SK}_n)$ which is a vector of n private key shares. The verification key VK is used to check validity of responses from decryption servers.

Encrypt(PK, m). Takes as input a public key PK and a message m. It outputs a ciphertext.

ShareDecrypt(SK_i, C). Takes as input a private key share SK_i and a ciphertext C. It outputs a decryption share $\mu_i = (i, d_{C,i})$ or the special symbol $(i, \bot)$.

ShareVerify(VK, C, μ_i). Takes as input the verification key VK, a ciphertext C and a decryption share μ_i. It outputs valid meaning that μ_i is a valid decryption share of C or invalid.

Combine(PK, VK, $C, \{\mu_1, \ldots, \mu_k\}$). Takes as input the public key PK, the verification key VK, a ciphertext C and k decryption shares $\mu_1, \ldots, \mu_k$. It outputs a plaintext message or the special symbol $\bot$.

We require, for all ciphertext C, ShareVerify(VK, C, ShareDecrypt(SK_i, C)) = valid. In addition, let $\mu_1, \ldots, \mu_k$ are k distinct valid decryption shares of C, where $C =$ Encrypt(PK, m), then we require Combine(PK, VK, $C, \{\mu_1, \ldots, \mu_k\}$) = m.

Security against chosen ciphertext attack is defined using the following game between an attack algorithm $\mathcal{A}$ and a challenger.

Init. The adversary outputs a set $S \subset \{1, \ldots, n\}$ of $k - 1$ decryption servers to corrupt.

Setup. The challenger runs Setup(n, k, λ) to obtain a triple (PK, VK, SK). It gives PK, VK, and all (j, SK_j) for $j \in S$ to the adversary.

Query phase 1. The adversary $\mathcal{A}$ adaptively issues decryption queries (C, i). The challenger responds with $\mathsf{ShareDecrypt}(\mathsf{SK}_i, C)$.

Challenge. The adversary $\mathcal{A}$ submits two (equal length) messages m_0, m_1. The challenger selects a random bit $\beta \in \{0, 1\}$, sets $C^* = \mathsf{Encrypt}(\mathsf{PK}, m_\beta)$ and sends C^* to the adversary as its challenge ciphertext.

Query phase 2. The adversary continues to adaptively issue decryption queries (C, i), as in Query phase 1, but with the natural constraint that the adversary may not request the decryption of C^*.

Guess. The adversary $\mathcal{A}$ outputs its guess $\beta' \in \{0, 1\}$ for b and wins the game if $\beta = \beta'$.

We define $\mathcal{A}$'s advantage as $\mathsf{Adv}_{\mathcal{A}}^{\mathsf{TPKE}}(\lambda) = |\mathsf{Pr}[\beta = \beta'] - \frac{1}{2}|$.

Definition 7. *We say that a threshold public key encryption scheme TPKE is secure, if for every adversary, the advantage $Adv_{\mathcal{A}}^{TPKE}(\cdot)$ is negligible.*

3   The Proposed PKE Scheme

Our scheme is motivated by the recent signature scheme by Hohenberger and Waters [23]. Recall that in the CHK, BK and BMW schemes, for each encryption, the encryptor first generates a one-time "identity", and then encrypts the message with respect to the "identity". In the CHK and BK schemes, the one-time "identity" is generated randomly by the encryptor; in the BMW scheme, the first two elements of a ciphertext are hashed to form the one-time "identity". In our proposed PKE scheme, we make use of two "identities". One "identity" is generated randomly as in the CHK and BK schemes, while the other "identity" is generated based on the approach in the BMW scheme. The benefit of doing this is twofold. Compared with the CHK and BK schemes, our ciphertexts are short without attached signature or MAC; and compared with the BMW scheme, our scheme has small public key size and is proven secure with a tight security reduction.

Our scheme consists of the following algorithms:

$\mathsf{KeyGen}(\lambda)$. Given the security parameter λ, a bilinear map group system $\langle p, \mathbb{G}, \mathbb{G}_T, e \rangle$ is constructed. Pick a generator g of $\mathbb{G}$, select random $\alpha, x, y, z \in \mathbb{Z}_p$ and set $g_1 = g^\alpha, u = g^x, v = g^y, d = g^z$. Next, choose random element $g_2 \in \mathbb{G}$. Finally, choose a collision-resistant hash function $H : \mathbb{G}_T \times \mathbb{G} \to \mathbb{Z}_p$. The published public key is

$$\mathsf{PK} = (p, \mathbb{G}, \mathbb{G}_T, e, g, H, \ Z = e(g_1, g_2), \ u, v, d),$$

and the private key is $\mathsf{SK} = (g_2^\alpha, x, y, z)$.

$\mathsf{Encrypt}(\mathsf{PK}, m)$. Given PK and a message $m \in \mathbb{G}_T$, randomly choose $s, r \in \mathbb{Z}_p$ and compute

$$C_0 = m \cdot Z^s, \ C_1 = g^s, \ C_2 = (u^t v^r d)^s,$$

where $t = H(C_0, C_1)$. Finally, output the ciphertext $C = (C_0, C_1, C_2, r) \in \mathbb{G}_T \times \mathbb{G}^2 \times \mathbb{Z}_p$.

Decrypt(SK, C). Given SK $= (g_2^\alpha, x, y, z)$ and a ciphertext $C = (C_0, C_1, C_2, r)$, compute $t = H(C_0, C_1)$. Then check whether

$$(C_1)^{tx+ry+z} = C_2.$$

If not, output $\perp$, else output

$$C_0/e(C_1, g_2^\alpha).$$

Theorem 1. *The above public key encryption scheme is (t, q, ϵ) IND-CCA secure, assuming the (t', ϵ')-DBDH assumption holds in $\mathbb{G}$ (the multiplicative group of prime order p), where*

$$t' = t + O(q), \quad \epsilon' \geq \epsilon - \mathsf{Adv}_\mathcal{A}^{CR} - q/p.$$

Proof. Suppose there exists a (t, q, ϵ)-IND-CCA adversary $\mathcal{A}$ against our public key encryption scheme. We are going to construct another PPT $\mathcal{B}$ that makes use of $\mathcal{A}$ to solve the DBDH problem with probability at least ϵ' and in the time at most t'.

$\mathcal{B}$ is given as input a random 5-tuple (g, g^a, g^b, g^c, T) that is either sampled from $\mathcal{P}_{BDH}$ (where $T = e(g, g)^{abc}$) or from $\mathcal{R}_{BDH}$ (where T is uniform and independent in $\mathbb{G}_T$). Algorithm $\mathcal{B}$'s goal is to output 1 if $T = e(g, g)^{abc}$ and 0 otherwise. Algorithm $\mathcal{B}$ runs $\mathcal{A}$ executing the following steps.

Setup. $\mathcal{B}$ chooses random $x_v, x_d, y_u, y_v, y_d \in \mathbb{Z}_p$ and sets $g_1 = g^a, g_2 = g^b, u = g^b g^{y_u}, v = g^{bx_v} g^{y_v}, d = g^{bx_d} g^{y_d}$. Then, choose a target-collision resistant hash function $H : \mathbb{G}_T \times \mathbb{G} \to \mathbb{Z}_p$. The public key

$$\mathsf{PK} = (p, \mathbb{G}, \mathbb{G}_T, e, g, H, \ Z = e(g_1, g_2), \ u, v, d)$$

is passed to $\mathcal{A}$. The private key is SK $= (g_2^\alpha = g_2^a = g^{ab}, x = b + y_u, y = bx_v + y_v, z = bx_d + y_d)$ which is unknown to $\mathcal{B}$.

Query phase 1. When $\mathcal{A}$ issues decryption query on a ciphertext $C = (C_0, C_1, C_2, r)$, $\mathcal{B}$ first computes $t = H(C_0, C_1)$ and checks whether

$$e(C_1, u^t v^r d) = e(g, C_2).$$

If not, output $\perp$. Check whether $t + rx_v + x_d = 0$. If so, $\mathcal{B}$ aborts and randomly outputs a bit, else chooses random $\gamma \in \mathbb{Z}_p$ and computes

$$d_C^1 = g_1^{-(ty_u + ry_v + y_d)/(t+rx_v+x_d)}(u^t v^r d)^\gamma,$$
$$d_C^2 = g_1^{-1/(t+rx_v+x_d)} g^\gamma.$$

Let $\tilde{\gamma} = \gamma - \frac{a}{(t+rx_v+x_d)}$. Then we have

$$d_C^1 = g_2^a (u^t v^r d)^{\tilde{\gamma}}, \ d_C^2 = g^{\tilde{\gamma}}.$$

Finally, $\mathcal{B}$ outputs

$$C_0 \cdot e(C_2, d_C^2)/e(C_1, d_C^1).$$

Challenge. The adversary $\mathcal{A}$ outputs two equal-length plaintexts (m_0, m_1). $\mathcal{B}$ flips a fair coin, $\beta \in \{0, 1\}$ and constructs the ciphertext as follows:

1. It computes

$$C_0^* = m_\beta \cdot T, \ C_1^* = g^c, t^* = H(C_0^*, C_1^*).$$

2. Then, it sets $r^* = -(t^* + x_d)/x_v$ and computes $C_2^* = (g^c)^{(t^* y_u + r^* y_v + y_d)}$.
3. Finally, return the ciphertext $C^* = (C_0^*, C_1^*, C_2^*, r^*)$.

Since $C^* = (m_\beta \cdot T, \ g^c, \ (u^{t^*} v^{r^*} d)^c, \ r^*)$, the challenge ciphertext is a valid encryption of m_β with the correct distribution whenever $T = e(g, g)^{abc} = e(g_1, g_2)^c$ (as is the case when the input 5-tuple is sampled from $\mathcal{P}_{BDH}$). On the other hand, when T is uniform and independent in $\mathbb{G}_T$ (which occurs when the input 5-tuple is sampled from $\mathcal{R}_{BDH}$) the challenge ciphertext C^* is independent of β in the adversary's view.

Query phase 2. $\mathcal{A}$ continues to adaptively issue decryption query $C = (C_0, C_1, C_2, r)$, $\mathcal{B}$ performs the following steps:

1. Check if $C = C^*$. If so, output $\perp$.
2. Check if $C = (C_0, C_1^*, C_2^*, r^*)$ and $H(C_0, C_1) = t^*$. If so, $\mathcal{B}$ aborts and randomly outputs a bit.

   Note that, if $\mathcal{A}$ were able to produce such a ciphertext, this would represent a collision in the hash function H, and so the probability that this event occurs is negligible.
3. Check if $t + rx_v + x_d = 0$ where $t = H(C_0, C_1)$. If so, $\mathcal{B}$ aborts and randomly outputs a bit, else $\mathcal{B}$ responds as in Query phase 1.

   Observe that the values x_v and x_d are initially hidden by blinding factors y_v and y_d, respectively.

   When the adversary $\mathcal{A}$ issues decryption query $C = (C_0, C_1, C_2, r)$:
   - if $e(C_1, u^t v^r d) \neq e(g, C_2)$, $\mathcal{B}$ outputs $\perp$ and do not leak any information about either x_v or x_d.
   - else $e(C_1, u^t v^r d) = e(g, C_2)$, $\mathcal{B}$ computes $(d_C^1 = g_2^a (u^t v^r d)^{\tilde{\gamma}}, d_C^2 = g^{\tilde{\gamma}})$ and outputs

$$C_0 \cdot \frac{e(C_2, d_C^2)}{e(C_1, d_C^1)} = C_0 \cdot \frac{e(C_2, g^{\tilde{\gamma}})}{e(C_1, g_2^a (u^t v^r d)^{\tilde{\gamma}})}$$

$$= C_0 \cdot \frac{e(C_2, g)^{\tilde{\gamma}}}{e(C_1, g_2^a) \cdot e(C_1, u^t v^r d)^{\tilde{\gamma}}} = \frac{C_0}{e(C_1, g_2^a)}.$$

So, the adversary could not obtain any information about either x_v or x_d from the decryption queries.

For the challenge ciphertext, the adversary could obtain the information that $t^* + r^* x_v + x_d = 0$. However, there are exactly p possible (x_v, x_d) pairs that satisfy this equation and each of them are equally likely. Thus, information-theoretically, the probability that $t + rx_v + x_d = 0$ is at most $1/p$.

Guess. The adversary $\mathcal{A}$ outputs a bit β'. $\mathcal{B}$ concludes its own game by outputting a guess as follows. If $\beta' = \beta$ then $\mathcal{B}$ outputs 1 meaning $T = e(g, g)^{abc}$. Otherwise, it outputs 0 meaning $T \neq e(g, g)^{abc}$.

The probability that $\mathcal{B}$ does not abort during the simulation is at most $\mathsf{Adv}_{\mathcal{A}}^{\mathsf{CR}} + q/p$. When the input 5-tuple is sampled from $\mathcal{P}_{BDH}$ (where $T = e(g,g)^{abc}$) and $\mathcal{B}$ does not abort then $\mathcal{A}$'s view is identical to its view in a real attack game. On the other hand, when the input 5-tuple is sampled from $\mathcal{R}_{BDH}$ (where T is uniform in $\mathbb{G}_T$) and $\mathcal{B}$ does not abort then the advantage that $\mathcal{A}$ wins the attack game is $1/2$. The running time of $\mathcal{A}$ is dominated by the paring computation in response to $\mathcal{A}$'s decryption queries.

This concludes the proof of Theorem 1.

4   Practical Extensions

In this section, we describe two interesting extensions to our PKE scheme. In the following, we only present the extended schemes. Their security proofs can be performed in a similar manner as in Section 3 and are therefore omitted.

4.1   Public Key Encryption with Non-interactive Opening

Public key encryption with non-interactive opening (PKENO) was recently introduced in [13,14] as a means to enable publicly-verifiable decryption. In a PKENO scheme, the receiver of a ciphertext C can, convincingly and without interaction, reveal what the result was of decryption C, without compromising the confidentiality of non-opened ciphertexts. The construction of PKENO can be obtained by using public key encryption with witness-recovering decryption (PKEWR) [30]. Here the receiver can efficiently reconstruct the "randomness" that was used for encryption. This randomness then serves as the proof. Verification performs re-encrypting using the randomness and the message. The proof is valid if the result equals the ciphertext. The existing constructions of PKEWR [30] in the standard model, however, are relatively inefficient since the ciphertext size is linear in the message length.

Damgård, Hofheinz, Kiltz and Thorbek [14] proposed two efficient constructions of PKENO schemes. The first proposal is a generic construction and resembles the CHK transformation [9]. The idea is to use, for each PKENO encryption, a fresh random verification key of a one-time signature scheme as the "identity" for IBE encryption. The private key corresponding to the "identity" serves as the proof. Verification performs decryption using the private key. The second proposal is a concrete scheme based on the CCA-secure key encapsulation mechanism by Boyen, Mei and Waters [7]. Recently, Galindo [18] showed the second scheme in [14] is insecure and proposed a fix based on direct CCA-secure PKE from identity-based techniques by Boyen, Mei and Waters [7]. Their scheme needs *long* public keys. Based on our PKE scheme, we propose a more efficient PKENO scheme as detailed in the following.

KeyGen(λ). Given the security parameter λ, a bilinear map group system $\langle p, \mathbb{G}, \mathbb{G}_T, e\rangle$ is constructed. Pick a generator g of $\mathbb{G}$, select random $\alpha, x, y, z \in \mathbb{Z}_p$ and set $g_1 = g^\alpha, u = g^x, v = g^y, d = g^z$. Next, choose random element

$g_2 \in \mathbb{G}$. Finally, choose a collision-resistant hash function $H : \mathbb{G}_T \times \mathbb{G} \to \mathbb{Z}_p$. The published public key is

$$\mathsf{PK} = (p, \mathbb{G}, \mathbb{G}_T, e, g, H,\ Z = e(g_1, g_2),\ u, v, d),$$

and the private key is $\mathsf{SK} = (g_2^\alpha, x, y, z)$.

Encrypt(PK, m). Given PK and a message $m \in \mathbb{G}_T$, randomly choose $s, r \in \mathbb{Z}_p$ and compute

$$C_0 = m \cdot Z^s = m \cdot e(g_1, g_2)^s,\ C_1 = g^s,\ C_2 = (u^t v^r d)^s,$$

where $t = H(C_0, C_1)$. Finally, output the ciphertext $C = (C_0, C_1, C_2, r) \in \mathbb{G}_T \times \mathbb{G}^2 \times \mathbb{Z}_p$.

Decrypt(SK, C). Given $\mathsf{SK} = (g_2^\alpha, x, y, z)$ and a ciphertext $C = (C_0, C_1, C_2, r)$, compute $t = H(C_0, C_1)$. Then check whether

$$(C_1)^{tx+ry+z} = C_2.$$

If not, output $\bot$, else output

$$C_0 / e(C_1, g_2^\alpha).$$

Prove(SK, C). Given $\mathsf{SK} = (g_2^\alpha, x, y, z)$ and a ciphertext $C = (C_0, C_1, C_2, r)$, compute $t = H(C_0, C_1)$. Then check whether

$$(C_1)^{tx+ry+z} = C_2.$$

If not, output $\bot$, else randomly choose $\gamma \in \mathbb{Z}_p$ and output $\pi = (d_C^1, d_C^2) \in \mathbb{G}^2$, where

$$d_C^1 = g_2^\alpha (u^t v^r d)^\gamma,\ d_C^2 = g^\gamma.$$

Ver(PK, C, m, π). Given PK, a ciphertext $C = (C_0, C_1, C_2, r)$, a message m and a proof $\pi = (d_C^1, d_C^2)$, compute $t = H(C_0, C_1)$. Then check whether

$$e(C_1, u^t v^r d) = e(g, C_2),\ e(g, d_C^1) = Z \cdot e(u^t v^r d, d_C^2) \text{ and}$$
$$m = C_0 \cdot e(C_2, d_C^2) / e(C_1, d_C^1).$$

If not, output 0, else output 1.

4.2 Threshold Public Key Encryption

In a threshold public key encryption (TPKE) scheme [15], the private key corresponding to a public key is shared among a set of n decryption servers. In such a scheme, a message is encrypted and sent to a group of decryption servers, in such a way that the cooperation of at least k of them (where k is the threshold) is necessary in order to recover the original message. In a non-interactive threshold scheme, no communication is needed amongst the decryption servers performing the partial decryptions. Such schemes have many applications in situations where one cannot fully trust a unique person, but possibly a pool of individuals.

Recall that the Cramer-Shoup scheme [11] provides efficient CCA-secure encryption without random oracles. The scheme requires that the private key be used to check ciphertext validity during decryption. In a threshold environment none of the decryption servers possess the private key needed to perform this validity check. Consequently, constructing a threshold version of the Cramer-Shoup scheme is non-trivial.

Boneh, Boyen and Halevi [5] showed that CCA-secure threshold public key encryption schemes (without random oracles) are easier to derive from selective-ID CPA secure identity based encryption than from the Cramer-Shoup paradigm. The main reason is that in the IBE-to-CCA transformation [9], the validity check performed during decryption requires only the public key. Consequently, each decryption server can check ciphertext validity on its own and only release a partial decryption of valid ciphertexts. Note that the more efficient transformation of Boneh and Katz [6] does not have this property and is thus less suitable for threshold decryption.

Boyen, Mei and Waters [7] gave a very simple and efficient CCA-secure threshold key encapsulation mechanism (KEM) based on the Boneh-Boyen IBE framework. However, designing a full threshold PKE from a threshold KEM is not an easy task. Let us have a glimpse on it. A standard (hybrid) PKE scheme can be obtained by using the KEM to securely transport a random session key that is fed into a symmetric encryption scheme to encrypt the plaintext message. If both the KEM and the symmetric encryption scheme are chosen-ciphertext secure, then the resulting hybrid PKE is also chosen-ciphertext secure. A symmetric encryption scheme secure against chosen-ciphertext attacks can be built from relatively weak primitives, i.e. from any (one-time) symmetric encryption scheme by essentially adding a MAC. Unfortunately, sharing a MAC is not trivial in general, and will often lead to costly computations.

In our PKE scheme, the decryptor needs to verify the ciphertext before attempting to decrypt it. This check is efficiently performed using a single exponentiation in $\mathbb{G}$, but requires knowledge of the private key (the exponents x, y, z). In fact, the validity check could have been performed publicly, using additional application of the bilinear map, by checking whether $e(C_1, u^t v^r d) = e(g, C_2)$. Since under such a modification the ciphertext validity check no longer requires the private key, our PKE scheme is suitable for non-interactive threshold decryption. The following is the detailed construction of the threshold version of our PKE scheme. It bears some resemblance to the threshold schemes in [5] due to its roots in identity-based techniques.

Setup(n, k, λ). Given the security parameter λ, a bilinear map group system $\langle p, \mathbb{G}, \mathbb{G}_T, e \rangle$ is constructed. Select random generators g, g_2, u, v, d of $\mathbb{G}$ and a random degree $k - 1$ polynomial $f \in \mathbb{Z}_p[X]$. Set $\alpha = f(0) \in \mathbb{Z}_p$ and $g_1 = g^\alpha$. Choose a collision-resistant hash function $H : \mathbb{G}_T \times \mathbb{G} \to \mathbb{Z}_p$. The published public key is

$$\mathsf{PK} = (p, \mathbb{G}, \mathbb{G}_T, e, g, H,\ Z = e(g_1, g_2),\ g_2, u, v, d).$$

For $i = 1, \ldots, n$ the secret key SK_i of server i is defined as $\mathsf{SK}_i = g_2^{f(i)}$. The public verification key VK is the n-tuple $(g^{f(1)}, \ldots, g^{f(n)})$.

Encrypt(PK, m). Given PK and a message $m \in \mathbb{G}_T$, randomly choose $s, r \in \mathbb{Z}_p$ and compute

$$C_0 = m \cdot Z^s = m \cdot e(g_1, g_2)^s, \; C_1 = g^s, \; C_2 = (u^t v^r d)^s,$$

where $t = H(C_0, C_1)$. Finally, output the ciphertext $C = (C_0, C_1, C_2, r) \in \mathbb{G}_T \times \mathbb{G}^2 \times \mathbb{Z}_p$.

ShareDecrypt(SK_i, C). Given SK_i and a ciphertext $C = (C_0, C_1, C_2, r)$, decryption server i computes $t = H(C_0, C_1)$. Then check whether

$$e(C_1, u^t v^r d) = e(g, C_2).$$

If not, output $\mu_i = (i, \bot)$, else randomly choose $\gamma \in \mathbb{Z}_p$ and output the decryption share $\mu_i = (i, (d_{C,i}^1, d_{C,i}^2))$, where

$$d_{C,i}^1 = \mathsf{SK}_i \cdot (u^t v^r d)^\gamma, \; d_{C,i}^2 = g^\gamma.$$

ShareVerify(VK, C, μ_i). Given $\mathsf{VK} = (h_1, \ldots, h_n)$ where $h_i = g^{f(i)}$, a ciphertext $C = (C_0, C_1, C_2, r)$ and a decryption share $\mu_i = (i, (d_{C,i}^1, d_{C,i}^2))$ of the ciphertext C, compute $t = H(C_0, C_1)$. Then check whether

$$e(C_1, u^t v^r d) = e(g, C_2) \text{ and } e(g, d_{C,i}^1) = e(h_i, g_2) \cdot e(u^t v^r d, d_{C,i}^2).$$

If not, output `invalid`, else output `valid`.

Combine(PK, VK, C, $\{\mu_1, \ldots, \mu_k\}$). Given PK, VK, a ciphertext $C = (C_0, C_1, C_2, r)$ and the partial decryptions $\mu_1, \ldots, \mu_k$, first check that all decryption shares $\mu_i = (i, (d_{C,i}^1, d_{C,i}^2))$ bear distinct server indices i, and that they are all *valid*, i.e., that all ShareVerify(VK, C, μ_i) = `valid`; otherwise output $\bot$. Without loss of generality, assume that the shares $\mu_1, \ldots, \mu_k$ were generated by the decryption servers $i = 1, \ldots, k$, respectively. Then compute the Lagrange coefficients $\lambda_1, \ldots, \lambda_k \in \mathbb{Z}_p$ so that $\alpha = f(0) = \sum_{i=1}^k \lambda_i f(i)$, and set

$$d_C^1 = \prod_{i=1}^k (d_{C,i}^1)^{\lambda_i}, \; d_C^2 = \prod_{i=1}^k (d_{C,i}^2)^{\lambda_i}.$$

Finally, output

$$C_0 \cdot e(C_2, d_C^2) / e(C_1, d_C^1).$$

5   Conclusions

We described an efficient CCA-secure public key encryption scheme whose performance is competitive with previous CCA-secure public key encryption schemes employing identity-based techniques. Our scheme is based on the identity-based encryption schemes of Boneh and Boyen [4], and the signature scheme of Hohenberger and Waters [23]. In addition, we showed that our scheme is well suited for constructing TPKE and PKENO schemes. In fact, our approach can be applied to obtain more efficient CCA-secure hierarchical identity based encryption (HIBE) scheme based on the Waters CPA-secure HIBE scheme [33].

Acknowledgement

We are grateful to the anonymous reviewers for their helpful comments. This work is partially funded by National Natural Science Foundation of China (No. 60873229) and Shanghai Rising-star Program (No. 09QA1403000), and also supported in part by the Office of Research, Singapore Management University.

References

1. Abe, M., Gennaro, R., Kurosawa, K., Shoup, V.: Tag-KEM/DEM: A new framework for hybrid encryption and a new analysis of Kurosawa-Desmedt KEM. In: Cramer, R. (ed.) EUROCRYPT 2005. LNCS, vol. 3494, pp. 128–146. Springer, Heidelberg (2005)
2. Bellare, M., Boldyreva, A., Palacio, A.: An Uninstantiable Random-Oracle-Model Scheme for a Hybrid-Encryption Problem. In: Cachin, C., Camenisch, J.L. (eds.) EUROCRYPT 2004. LNCS, vol. 3027, pp. 171–188. Springer, Heidelberg (2004)
3. Bellare, M., Rogaway, P.: Random oracles are practical: a paradigm for designing efficient protocols. In: Proc. of ACM CCS 1993, pp. 62–73. ACM Press, New York (1993)
4. Boneh, D., Boyen, X.: Efficient selective-ID secure identity-based encryption without random oracles. In: Cachin, C., Camenisch, J.L. (eds.) EUROCRYPT 2004. LNCS, vol. 3027, pp. 223–238. Springer, Heidelberg (2004)
5. Boneh, D., Boyen, X., Halevi, S.: Chosen ciphertext secure public key threshold encryption without random oracles. In: Pointcheval, D. (ed.) CT-RSA 2006. LNCS, vol. 3860, pp. 226–243. Springer, Heidelberg (2006)
6. Boneh, D., Katz, J.: Improved efficiency for CCA-secure cryptosystems built using identity-based encryption. In: Menezes, A. (ed.) CT-RSA 2005. LNCS, vol. 3376, pp. 87–103. Springer, Heidelberg (2005)
7. Boyen, X., Mei, Q., Waters, B.: Direct chosen ciphertext security from identity-based techniques. In: Proc. of ACM CCS 2005, pp. 320–329. ACM Press, New-York (2005)
8. Canetti, R., Goldreich, O., Halevi, S.: The Random Oracle Model Revisited. In: Proceedings of STOC 1998. ACM, New York (1998)
9. Canetti, R., Halevi, S., Katz, J.: Chosen-ciphertext security from identity-based encryption. In: Cachin, C., Camenisch, J.L. (eds.) EUROCRYPT 2004. LNCS, vol. 3027, pp. 207–222. Springer, Heidelberg (2004)
10. Cash, D., Kiltz, E., Shoup, V.: The twin Diffie-Hellman problem and applications. In: Smart, N.P. (ed.) EUROCRYPT 2008. LNCS, vol. 4965, pp. 127–145. Springer, Heidelberg (2008)
11. Cramer, R., Shoup, V.: A practical public key cryptosystem provably secure against adaptive chosen ciphertext attack. In: Krawczyk, H. (ed.) CRYPTO 1998. LNCS, vol. 1462, pp. 13–25. Springer, Heidelberg (1998)
12. Cramer, R., Shoup, V.: Universal hash proofs and a paradigm for adaptive chosen ciphertext secure public-key encryption. In: Knudsen, L.R. (ed.) EUROCRYPT 2002. LNCS, vol. 2332, pp. 45–64. Springer, Heidelberg (2002)
13. Damgård, I., Thorbek, R.: Non-interactive proofs for integer multiplication. In: Naor, M. (ed.) EUROCRYPT 2007. LNCS, vol. 4515, pp. 412–429. Springer, Heidelberg (2007)
14. Damgård, I., Hofheinz, D., Kiltz, E., Thorbek, R.: Public-key encryption with non-interactive opening. In: Malkin, T.G. (ed.) CT-RSA 2008. LNCS, vol. 4964, pp. 239–255. Springer, Heidelberg (2008)

15. Desmedt, Y., Frankel, Y.: Threshold cryptosystems. In: Brassard, G. (ed.) CRYPTO 1989. LNCS, vol. 435, pp. 307–315. Springer, Heidelberg (1990)
16. Dolev, D., Dwork, C., Naor, M.: Non-malleable cryptography. In: Proc. of STOC 1991, pp. 542–552 (1991)
17. Elkind, E., Sahai, A.: A unified methodology for constructing public-key encryption schemes secure against adaptive chosen-ciphertext attack. Cryptology ePrint Archive, Report 2002/042 (2002), http://eprint.iacr.org/
18. Galindo, D.: Breaking and Repairing Damgård et al. Public Key Encryption Scheme with Non-interactive Opening. In: Fischlin, M. (ed.) CT-RSA 2009. LNCS, vol. 5473, pp. 389–398. Springer, Heidelberg (2009)
19. Goldwasser, S., Tauman, Y.: On the (In)security of the Fiat-Shamir Paradigm. In: Proc. of FOCS. IEEE, Los Alamitos (2003)
20. Hanaoka, G., Kurosawa, K.: Efficient Chosen Ciphertext Secure Public Key Encryption under the Computational Diffie-Hellman Assumption. In: Pieprzyk, J. (ed.) ASIACRYPT 2008. LNCS, vol. 5350, pp. 308–325. Springer, Heidelberg (2008)
21. Hofheinz, D., Kiltz, E.: Secure hybrid encryption from weakened key encapsulation. In: Menezes, A. (ed.) CRYPTO 2007. LNCS, vol. 4622, pp. 553–571. Springer, Heidelberg (2007)
22. Hofheinz, D., Kiltz, E.: Practical Chosen Ciphertext Secure Encryption from Factoring. In: Joux, A. (ed.) EUROCRYPT 2009. LNCS, vol. 5479, pp. 313–332. Springer, Heidelberg (2009)
23. Hohenberger, S., Waters, B.: Realizing Hash-and-Sign Signatures under Standard Assumptions. In: Joux, A. (ed.) EUROCRYPT 2009. LNCS, vol. 5479, pp. 333–350. Springer, Heidelberg (2009)
24. Kiltz, E.: On the Limitations of the Spread of an IBE-to-PKE Transformation. In: Yung, M., Dodis, Y., Kiayias, A., Malkin, T.G. (eds.) PKC 2006. LNCS, vol. 3958, pp. 274–289. Springer, Heidelberg (2006)
25. Kiltz, E.: Chosen-ciphertext security from tag-based encryption. In: Halevi, S., Rabin, T. (eds.) TCC 2006. LNCS, vol. 3876, pp. 581–600. Springer, Heidelberg (2006)
26. Kiltz, E.: Chosen-ciphertext secure key-encapsulation based on gap hashed Diffie-Hellman. In: Okamoto, T., Wang, X. (eds.) PKC 2007. LNCS, vol. 4450, pp. 282–297. Springer, Heidelberg (2007)
27. Kurosawa, K., Desmedt, Y.: A new paradigm of hybrid encryption scheme. In: Franklin, M. (ed.) CRYPTO 2004. LNCS, vol. 3152, pp. 426–442. Springer, Heidelberg (2004)
28. Naor, M., Pinkas, B.: Efficient trace and revoke schemes. In: Frankel, Y. (ed.) FC 2000. LNCS, vol. 1962, pp. 1–20. Springer, Heidelberg (2001)
29. Nielsen, J.B.: Separating Random Oracle Proofs from Complexity Theoretic Proofs: The Non-committing Encryption Case. In: Yung, M. (ed.) CRYPTO 2002. LNCS, vol. 2442, pp. 111–126. Springer, Heidelberg (2002)
30. Peikert, C., Waters, B.: Lossy Trapdoor Functions and Their Applications. In: STOC 2008, pp. 187–196. ACM, New York (2008)
31. Rackoff, C., Simon, D.R.: Non-interactive zero-knowledge proof of knowledge and chosen ciphertext attack. In: Feigenbaum, J. (ed.) CRYPTO 1991. LNCS, vol. 576, pp. 433–444. Springer, Heidelberg (1992)
32. Shoup, V.: Using hash functions as a hedge against chosen ciphertext attack. In: Preneel, B. (ed.) EUROCRYPT 2000. LNCS, vol. 1807, pp. 275–288. Springer, Heidelberg (2000)
33. Waters, B.: Efficient identity-based encryption without random oracles. In: Cramer, R. (ed.) EUROCRYPT 2005. LNCS, vol. 3494, pp. 114–127. Springer, Heidelberg (2005)

Anonymity from Asymmetry:
New Constructions for Anonymous HIBE

Léo Ducas

Ecole Normale Superieure, Paris
leo.ducas@ens.fr

Abstract. A Hierarchical Identity Based Encryption (HIBE) system is anonymous if the ciphertext reveals no information about the recipient's identity. create it. While there are multiple constructions for secure HIBE, far fewer constructions exist for *anonymous* HIBE. In this paper we show how to use asymmetric pairings to convert a large family of IBE and HIBE constructions into anonymous IBE and HIBE systems. We also obtain a delegatable-HVE which is a generalization of anonymous HIBE.

Keywords: Anonymity, Identity Based Encryption, HIBE, delegatable Hidden Vector Encryption.

1 Introduction

In an Identity Based Encryption system (IBE) [Sha85, BF03] any string can function as a public key. A master secret is used to generate private keys for any public-key of interest. An extension of IBE, called Hierarchical-IBE [HL02, GS02], allows for a hierarchy of identities where any path from the root to a node can function as a public-key. An IBE or HIBE is said to be *recipient anonymous* or simply *anonymous* if the ciphertext leaks no information about the recipient's identity. Both anonymous IBE and HIBE are building blocks for encryption systems supporting searching on encrypted data [BCOP04, ABC+05, SBC+07, BW07].

While there are several approaches to constructing an IBE using bilinear maps [BF03], most constructions in the standard model are not recipient anonymous [CHK03, BB04, Wat05, BBG05] — there is a simple attack that can tell if a given ciphertext is encrypted for a specific identity (the system in [Gen06] is an exception). Oddly, by changing the type of pairing used, the anonymity attack goes away. In particular, if one uses an *asymmetric* pairing $e : \mathbb{G} \times \hat{\mathbb{G}} \to \mathbb{G}_t$ (as discussed in the next section) then it is no longer known how to break anonymity of the systems in [BB04, Wat05, BBG05]. However, it is not known how to prove anonymous security of these systems from simple assumptions.

In this paper we resolve this issue and show that a small tweak to the systems in [BB04, Wat05, BBG05] can make them provably anonymous when using *asymmetric* pairings. All our proofs are set in the standard model (*i.e.* without

J. Pieprzyk (Ed.): CT-RSA 2010, LNCS 5985, pp. 148–164, 2010.
© Springer-Verlag Berlin Heidelberg 2010

relying on random oracles). In addition to hiding the recipient's identity, ciphertexts in our system also hide the public parameters under which the ciphertext
was created.

We are certainly not the first to construct an anonymous IBE or HIBE without
random oracles. Boyen and Waters [BW06] gave the first construction based on
the decision linear assumption. Other anonymous IBE systems were presented
in [Gen06, BW07, IP08, SW08]. Our IBE system is a little simpler than [BW06]
and relies on a weaker assumption than [Gen06]. A more substantial advantage
comes up in our HIBE.

The HIBE case. Consider a hierarchy of depth ℓ. The anonymous HIBE system
in [BW06] generates secret keys of size proportional to ℓ^2 and ciphertext of
size proportional to ℓ. Our system, which is derived from [BBG05], has secret
keys of size $O(\ell)$ and constant size ciphertext. As in [BBG05], security relies
on a complexity assumption whose size is proportional to ℓ. We note that very
recently Shi and Waters [SW08] used composite order groups to construct an
HIBE where private key size is $O(\ell)$ and ciphertext size linear in ℓ. Seo et al.
in [SKOS09] also used composite order groups to construct a compact anonymous
HIBE, with constant size ciphertext and linear size private keys. Ciphertexts in
our system are considerably shorter.

We note that our construction also gives a delegatable-HVE [BW07, SW08],
which is a generalization of anonymous HIBE. We present the system in
Section 5.1.

2   Anonymous IBE and HIBE: Definitions

We briefly review the definition of anonymous IBE and HIBE. A more detailed
definition can be found in [ABC+05]. In this paper we only consider CPA attacks,
and define both the selective-ID and adaptive-ID security games.

A public-key in an HIBE is a vector $\mathsf{ID} = (I_1, \ldots, I_k)$ representing an identity
at depth k of the hierarchy. We use ℓ to denote the maximum hierarchy depth.
The case $\ell = 1$ is an IBE. An HIBE system consists of five algorithms: *Setup*
to generate public parameters PP and a master secret mk; *Extract* to generate
a secret key for an identity ID using the master key; *Derive* to generate a secret
for an identity $\mathsf{ID} = (I_1, \ldots, I_k)$ given a secret key for $\mathsf{ID} = (I_1, \ldots, I_{k-1})$; and
Encrypt and *Decrypt* to encrypt and decrypt messages for identity ID.

We use $\mathcal{C}_{\lambda,\ell}$ to denote the finite set of all possible ciphertexts for a given
security parameters λ and maximum hierarchy depth ℓ.

The following security games capture both semantic security and recipient
anonymity properties of the system. We begin with the selective-ID game between a challenger and adversary $\mathcal{A}$. Both are given the security parameter λ as
input. For $b = 0, 1$ game $\Gamma^{(b)}(\lambda)$ is defined as follows:

- Initialization: The adversary sends to the challenger (ℓ, ID^*) where $\ell > 0$ is
  the maximal hierarchy depth ℓ and $\mathsf{ID}^* = (I_1^*, \ldots, I_k^*)$ is an identity that
  $\mathcal{A}$ intends to attack (where $k \leq \ell$). When defining IBE security we require
  $\ell = 1$.

- Setup: The challenger runs $Setup(\lambda, \ell)$ and sends PP to $\mathcal{A}$.
- Phase 1: $\mathcal{A}$ issues up to q_s private key queries where no query is a prefix of ID^*. The challenger responds to the ith query ID_i by sending $\mathcal{A}$ the output of $Extract(mk, \mathsf{ID}_i)$.
- Challenge: $\mathcal{A}$ outputs a message m^*. The challenger responds by choosing a random ciphertext $c \xleftarrow{R} \mathcal{C}_{\lambda, \ell}$ and sending c^* to the challenger, where c^* is defined as

$$c^* \leftarrow \begin{cases} Encrypt(PP, \mathsf{ID}^*, m^*) & \text{if } b = 0 \quad (\text{game } \Gamma^{(0)}) \\ c & \text{if } b = 1 \quad (\text{game } \Gamma^{(1)}) \end{cases}$$

- Phase 2: The adversary continues to issue private-key queries subject to the same restriction as in phase 1.
- Guess: Finally, $\mathcal{A}$ outputs a guess $b' \in \{0, 1\}$ for b.

For $b = 0, 1$ let W_b be the event that $b = b'$ in Game $\Gamma^{(b)}$ and define $\mathcal{A}$'s advantage as
$$\mathsf{Adv}^{\mathsf{aIND\text{-}sIDCPA}}(\lambda) := \big| \Pr[W_0] - \Pr[W_1] \big|.$$

Definition 1. *We say that an HIBE system is selective-ID anonymous if for all PPT $\mathcal{A}$ we have that $\mathsf{Adv}^{\mathsf{aIND\text{-}sIDCPA}}(\lambda)$ is a negligible function of λ.*

Note that our definition of anonymity is a little stronger than usual [ABC+05]. We require that an encryption of m^* for ID^* under PP is indistinguishable from a random ciphertext in $\mathcal{C}_{\lambda, \ell}$. Consequently, not only does the ciphertext hide m^* and ID^*, it also hides PP. Moreover, our definition implies that ciphertext length must be independent of the depth of ID^* in the hierarchy.

As usual, to define full-IBE security one modifies the game to allow the adversary to specify ID^* in the challenge step instead of at initialization.

3 Complexity Assumptions

3.1 Asymmetric Pairings

Let p be a prime and let $\mathbb{G}$, $\hat{\mathbb{G}}$, and $\mathbb{G}_t$ be groups of order p. An asymmetric pairing is a map $e : \mathbb{G} \times \hat{\mathbb{G}} \to \mathbb{G}_t$ that is bilinear, non-degenerate, and efficiently computable. The term *asymmetric* refers to the fact that the groups $\mathbb{G}$ and $\hat{\mathbb{G}}$ need not be the same.

It is well known that when $\mathbb{G} = \hat{\mathbb{G}}$ then the Decision Diffie-Hellman problem in $\mathbb{G}$ is easy [Jou00]. However, when $\mathbb{G}$ and $\hat{\mathbb{G}}$ are distinct and there is no efficiently computable map from $\mathbb{G}$ to $\hat{\mathbb{G}}$ then Decision Diffie-Hellman in $\mathbb{G}$ can still be hard. This is partially the reason why the anonymity attacks on [BB04, Wat05, BBG05] do not seem to apply when using an asymmetric pairing.

The assumption that DDH is hard in $\mathbb{G}$ is sometimes called the XDH assumption. This assumption and its variants have been used in [Sco02, CHL05, ACd05, BKM05]. In Sections 3.2 and 3.3 we state the specific assumptions we will use.

3.2   The Bilinear Diffie-Hellman Assumption

The BDH problem for a symmetric pairing $e : \mathbb{G} \times \mathbb{G} \to \mathbb{G}_t$ is stated as follows [Jou00, BF03]:

Given a tuple $(g, g^a, g^b, g^c) \in \mathbb{G}^4$ as input, output $e(g, g)^{abc} \in \mathbb{G}_t$.

We extend the BDH problem to asymmetric bilinear groups by giving $(g, g^a, g^c, \hat{g}, \hat{g}^a, \hat{g}^b) \in \mathbb{G}^3 \times \hat{\mathbb{G}}^3$ as input and asking for $e(g, \hat{g})^{abc} \in \mathbb{G}_t$.

Asymmetric Decision BDH. Consider the following two distributions: For $g \in \mathbb{G}$, $\hat{g} \in \hat{\mathbb{G}}$, $a, b, c \in \mathbb{Z}_p$, and $T \in \mathbb{G}_t$ chosen uniformly at random, define:

- $\mathcal{P}_{BDH} := \left(g, g^a, g^c, \hat{g}, \hat{g}^a, \hat{g}^b, e(g, \hat{g})^{abc} \right) \in \mathbb{G}^3 \times \hat{\mathbb{G}}^3 \times \mathbb{G}_t$
- $\mathcal{R}_{BDH} := \left(g, g^a, g^c, \hat{g}, \hat{g}^a, \hat{g}^b, T \right) \in \mathbb{G}^3 \times \hat{\mathbb{G}}^3 \times \mathbb{G}_t$

For an algorithm $\mathcal{A}$ we let $\mathsf{Adv}_{\mathcal{A}}^{\text{D-BDH}}$ be the advantage of $\mathcal{A}$ is distinguishing these two distributions. That is,

$$\mathsf{Adv}_{\mathcal{A}}^{\text{D-BDH}} = \big| \Pr[\mathcal{A}(D) = 1] - \Pr[\mathcal{A}(R) = 1] \big|$$

where D is sampled from $\mathcal{P}_{BDH}$ and R is sampled from $\mathcal{R}_{BDH}$.

We say that an algorithm $\mathcal{B}$ that outputs a bit in $\{0, 1\}$ has advantage $\mathsf{Adv}_{\mathcal{B}}^{\text{D-BDH}} = \epsilon$ in solving the *Decision*-BDH problem in $(\mathbb{G}, \hat{\mathbb{G}})$ if

$$\Big| \Pr\left[\mathcal{B}(g, g^a, g^c, \hat{g}, \hat{g}^a, \hat{g}^b, e(g, \hat{g})^{abc}) = 0 \right] - \Pr\left[\mathcal{B}(g, g^a, g^c, \hat{g}, \hat{g}^a, \hat{g}^b, T) = 0 \right] \Big| \geq \epsilon$$

where the probability is over the random choice of generator $g \in \mathbb{G}$ and $\hat{g} \in \hat{\mathbb{G}}$, exponents a, b, c in $\mathbb{Z}_p$, $T \in \mathbb{G}_t$, and the random bits used by $\mathcal{B}$.

As usual, to state the assumption asymptotically we rely on a bilinear group generator $\mathcal{G}$ that takes a security parameter λ as input and outputs the description of a bilinear group.

Definition 2. *Let $\mathcal{G}$ be a bilinear group generator. We say that the Decision BDH holds for $\mathcal{G}$ if, for all PPT algorithms $\mathcal{A}$, the function $\mathsf{Adv}_{\mathcal{A}}^{D\text{-}BDH}(\lambda)$ is a negligible function of λ.*

3.3   Additional Assumptions

To prove the anonymity property of our systems, we will need a slightly stronger assumption. Consider the following two distributions: For $g \in \mathbb{G}$, $\hat{g} \in \hat{\mathbb{G}}$, $a, b, c \in \mathbb{Z}_p$, and $T \in \mathbb{G}$ chosen uniformly at random, define:

- $\mathcal{D}_N := \left(g, g^a, g^{ab}, g^c, \hat{g}, \hat{g}^a, \hat{g}^b, g^{abc} \right) \in \mathbb{G}^4 \times \hat{\mathbb{G}}^3 \times \mathbb{G}$
- $\mathcal{D}_R := \left(g, g^a, g^{ab}, g^c, \hat{g}, \hat{g}^a, \hat{g}^b, T \right) \in \mathbb{G}^4 \times \hat{\mathbb{G}}^3 \times \mathbb{G}$

For an algorithm $\mathcal{A}$ we let $\mathsf{Adv}_{\mathcal{A}}^{\text{P-BDH}}$ be the advantage of $\mathcal{A}$ is distinguishing these two distributions.

Definition 3. *Let $\mathcal{G}$ be a bilinear group generator. We say that the Decision $\mathcal{P}$-BDH holds for $\mathcal{G}$ if, for all PPT algorithms $\mathcal{A}$, the function $\mathsf{Adv}_{\mathcal{A}}^{P\text{-}BDH}(\lambda)$ is a negligible function of λ.*

3.4 Discussion About the Assumptions

Intuitively, the Decision $\mathcal{P}$-BDH assumption is a combination of the Decision BDH-assumption used by the BB_1 HIBE system, and the XDH-assumption (stating that Decision Diffie-Hellman problem is hard in one of the groups despite the existence of the pairing). Indeed, the Decision $\mathcal{P}$-BDH assumption implies both assumptions via the following simple reductions: A Decision $\mathcal{P}$-BDH instance $\left(g, g^a, g^{ab}, g^c, \hat{g}, \hat{g}^a, \hat{g}^b, T\right)$ can be solved using:

- a Decision BDH adversary $\mathcal{A}$: run $\mathcal{A}\left(g, g^a, g^c, \hat{g}, \hat{g}^a, \hat{g}^b, e(T, \hat{g})\right)$;
- a Decision DH adversary $\mathcal{A}$: run $\mathcal{A}\left(g, g^{ab}, g^c, T\right)$.

The existence of an efficiently computable homomorphism from $\hat{\mathbb{G}}$ to $\mathbb{G}$ does not contradict with our assumption, but is not required either for the construction nor the security proof. Thus, the bilinear group of our system may be instantiated by either type 2 or type 3 pairing on elliptic curves, as defined in [GPS06].

4   An Efficient Anonymous IBE

We first construct an IBE system that is anonymous under the $\mathcal{P}$-BDH assumption in asymmetric bilinear groups.

4.1 IBE Construction

We are given a bilinear map $e : \mathbb{G} \times \hat{\mathbb{G}} \to \mathbb{G}_t$ over a bilinear group pair $(\mathbb{G}, \hat{\mathbb{G}})$ of prime order p, with respective generators $g \in \mathbb{G}^*$ and $\hat{g} \in \hat{\mathbb{G}}^*$. The size of p is determined by the security parameter.

Our IBE system works as follows:

Setup: To generate system parameters for an IBE, given bilinear groups $(\mathbb{G}, \hat{\mathbb{G}})$ with generators $(g, \hat{g})$, the setup algorithm first selects a random $(\alpha, \beta, \gamma, \delta, \eta) \in \mathbb{Z}_p^5$, and sets: $g_1 = g^\alpha$, $g_2 = g^\beta$, $h = g^\gamma$, $f = g^\delta$, $t = g^\eta$, and their analogues: $\hat{g}_1 = \hat{g}^\alpha$, $\hat{g}_2 = \hat{g}^\beta$, $\hat{h} = \hat{g}^\gamma$, $\hat{f} = \hat{g}^\delta$, $\hat{t} = \hat{g}^\eta$. The public parameters PP and the master secret mk are given by

$$PP = \left(g, g_1, h, f, t, \hat{g}, \hat{g}_2, \hat{h}\right) \in \mathbb{G}^5 \times \hat{\mathbb{G}}^3$$
$$mk = (\hat{g}_0 = \hat{g}^{\alpha\beta}, \hat{f}, \hat{t}) \in \hat{\mathbb{G}}^3$$

Extract(mk, ID): To extract a private key d_{ID} for an identity $\mathsf{ID} = I \in \mathbb{Z}_p^*$ the authority holding the master key picks random $r, R \in \mathbb{Z}_p$ and outputs

$$d_{\mathsf{ID}} = \left(\hat{g}_0\,(\hat{h}^I\,\hat{f})^r\,\hat{t}^R,\ \ \hat{g}^r,\ \ \hat{g}^R\right) \in \hat{\mathbb{G}}^3$$

Encrypt(PP, ID, M). To encrypt a message $M \in \mathbb{G}_t$ under the public key $\mathsf{ID} = I \in \mathbb{Z}_p^*$, pick a random $s \in \mathbb{Z}_p$ and output

$$C = \left(M \cdot e(g_1, \hat{g}_2)^s, \ g^s, \ (h^I f)^s, \ t^s \right) \in \mathbb{G}_t \times \mathbb{G}^3$$

Decrypt(d_{ID}, C). To decrypt a given ciphertext $C = (A, B, C_1, Z) \in \mathbb{G}_t \times \mathbb{G}^3$ using the private key $d_{\mathsf{ID}} = (d_0, d_1, d_2) \in \hat{\mathbb{G}}^3$, output

$$A \cdot e(C_1, d_1) \cdot e(Z, d_2) \, \big/ \, e(B, d_0) \ \in \mathbb{G}_t$$

The system in consistent. Indeed, for a valid ciphertext encrypted under the identity $\mathsf{ID} = I$ to which the private key d_{ID} belongs, we have

$$A \cdot \frac{e(C_1, d_1) \cdot e(Z, d_2)}{e(B, d_0)} = A \cdot \frac{e(h^I f, \hat{g})^{sr} \cdot e(t, \hat{g})^{sR}}{e(g, \hat{g}_0)^s \cdot e(g, \hat{h}^I \hat{f})^{sr} \cdot e(g, \hat{t})^{sR}} = A \cdot \frac{1}{e(g_1, \hat{g}_2)^s} = M$$

The system is closely related to the BB_1 IBE system from [BB04]. The only difference is the additional element t^s in the ciphertext and the additional blinding value $\hat{t}^R$ in the private key.

4.2 Security Reduction

The following theorem proves security of our system under the Decision $\mathcal{P}$-BDH assumption.

Theorem 1 (IBE security). *Our IBE is selective-ID anonymous assuming the Decision $\mathcal{P}$-BDH assumption holds for the bilinear group generator $\mathcal{G}$. In particular, for all PPT algorithms $\mathcal{B}$, the function $\mathrm{Adv}_{\mathcal{B}}^{\mathsf{aIND\text{-}sIDCPA}}(\lambda)$ is a negligible function of λ.*

The proof proceeds by a hybrid argument across a number of games. Let $\mathsf{CT} = (A, B, C_1, Z) \in \mathbb{G}_t \times \mathbb{G}^3$ denote the challenge ciphertext given to the adversary during a real attack (game $\Gamma^{(0)}$ in Definition 1). Additionally, let R be a random element of $\mathbb{G}_t$ and R', R_1 be random elements of $\mathbb{G}$. We define the following hybrid experiments, which differ in how the challenge ciphertext is generated:

- **Game Γ:** The challenge ciphertext is $\mathsf{CT} = (A, B, C_1, Z)$
- **Game Γ':** The challenge ciphertext is $\mathsf{CT}' = (R, B, C_1, Z)$
- **Game Γ_0:** The challenge ciphertext is $\mathsf{CT}_0 = (R, B, C_1, R')$
- **Game Γ_1:** The challenge ciphertext is $\mathsf{CT}_1 = (R, B, R_1, R')$

Game Γ is the same as game $\Gamma^{(0)}$ in Definition 1. Game Γ_1 is the same as game $\Gamma^{(1)}$ in Definition 1, where the adversary is given a random ciphertext. Therefore,

$$\mathrm{Adv}_{\mathcal{B}}^{\mathsf{aIND\text{-}sIDCPA}} \leq \left| \Pr\left[\mathcal{A}^{\Gamma} = 0 \right] - \Pr\left[\mathcal{A}^{\Gamma_1} = 0 \right] \right|.$$

To prove that Γ is indistinguishable from Γ_1 we prove that each step of the hybrid is indistinguishable from the next. We do so in a sequence of lemmas whose proofs are given in the full version of this paper [Duc09].

Lemma 1 (semantic security). *Let $\mathcal{A}$ be an adversary playing the* aIND-sIDCPA *attack game. Then, there exist an algorithm $\mathcal{B}$ solving the Decision $\mathcal{P}$-BDH problem such that:*

$$\left| \Pr\left[\mathcal{A}^{\Gamma} = 0 \right] - \Pr\left[\mathcal{A}^{\Gamma'} = 0 \right] \right| \leq \mathsf{Adv}_{\mathcal{B}}^{\mathcal{P}\text{-}BDH}$$

This proof is just the adaptation of the original BB$_1$ security proof. The lemma is stated using decision $\mathcal{P}$-BDH problem, but, the proof is using decision BDH problem, which is a weaker assumption.

Lemma 2 (anonymity, part 1). *Let $\mathcal{A}$ be an adversary playing the* aIND-sIDCPA *attack game. Then, there exist an algorithm $\mathcal{B}$ solving the Decision $\mathcal{P}$-BDH problem such that:*

$$\left| \Pr\left[\mathcal{A}^{\Gamma'} = 0 \right] - \Pr\left[\mathcal{A}^{\Gamma_0} = 0 \right] \right| \leq \mathsf{Adv}_{\mathcal{B}}^{\mathcal{P}\text{-}BDH}$$

Lemma 3 (anonymity, part 2). *Let $\mathcal{A}$ be an adversary playing the* aIND-sIDCPA *attack game. Then, there exist an algorithm $\mathcal{B}$ solving the Decision $\mathcal{P}$-BDH problem such that:*

$$\left| \Pr\left[\mathcal{A}^{\Gamma_0} = 0 \right] - \Pr\left[\mathcal{A}^{\Gamma_1} = 0 \right] \right| \leq \mathsf{Adv}_{\mathcal{B}}^{\mathcal{P}\text{-}BDH}$$

Thus, if there is no algorithm $\mathcal{B}$ that solve $\mathcal{P}$-BDH problem with an advantage better than ϵ, then, for all adversary $\mathcal{A}$

$$
\begin{aligned}
\left| \Pr\left[\mathcal{A}^{\Gamma} = 0 \right] - \Pr\left[\mathcal{A}^{\Gamma_1} = 0 \right] \right| \leq{} & \left| \Pr\left[\mathcal{A}^{\Gamma} = 0 \right] - \Pr\left[\mathcal{A}^{\Gamma'} = 0 \right] \right| \\
& + \left| \Pr\left[\mathcal{A}^{\Gamma'} = 0 \right] - \Pr\left[\mathcal{A}^{\Gamma_0} = 0 \right] \right| \\
& + \left| \Pr\left[\mathcal{A}^{\Gamma_0} = 0 \right] - \Pr\left[\mathcal{A}^{\Gamma_1} = 0 \right] \right| \\
\leq{} & 3\epsilon
\end{aligned}
$$

Full IBE security. We proved that our system is anonymous under a CPA and selective-ID attacks. It can be made fully secure using known tools.

First, to make the system chosen ciphertext secure one can use the results of Canetti et al. [CHK04,BCHK07]. To construct an anonymous chosen-ciphertext secure IBE we need a 2-level HIBE where the first level is anonymous, but the second need not be. Following the BB$_1$ HIBE construction, we can build a 2-level HIBE which is anonymous relative to the first level, but not the second. We thus obtain an anonymous chosen ciphertext secure IBE.

Second, to obtain full security against adaptive attacks (rather than selective security), one can use random oracles or inefficient reductions. It is also possible to apply the technique of Waters [Wat05] to our system to obtain an anonymous fully secure IBE without random oracles.

HIBE. The original BB_1 system extends to an HIBE by expanding f to a vector $f_1, \ldots, f_\ell$, one f_i per level of the hierarchy. Unfortunately, we cannot use the same method to extend our IBE to an anonymous HIBE. The problem is that to enable key delegation, we must include the values $\hat{f}_2, \ldots, \hat{f}_\ell$ in private keys. But providing these values breaks anonymity for all levels except the first one. A different approach is needed to extend our anonymous IBE to an anonymous HIBE. We develop this in the next section.

5   Anonymous Hierarchical IBE and Delegatable HVE

As before, we assume a bilinear group $\mathbb{G}$ and a map $e : \mathbb{G} \times \hat{\mathbb{G}} \to \mathbb{G}_t$, where $\mathbb{G}$, $\hat{\mathbb{G}}$ and $\mathbb{G}_t$ have prime order p.

To extend our anonymous IBE to an anonymous HIBE we add $\hat{f}_i$ terms to the private keys to enable key delegation, but we blind them so as not to break anonymity. We can then build an anonymous HIBE under the same compact assumption as before, but the private key size now becomes quadratic in the depth of the hierarchy. A similar problem was encountered in [BW06].

Another approach to making BB_1 a (non-anonymous) HIBE was proposed in [BBG05]. The construction uses a stronger assumption, but provides constant size ciphertext, and constant pairing steps during decryption. Using this construction, we can extend the anonymous IBE of the previous section to a very efficient anonymous HIBE with short ciphertext, fast decryption, and *linear* size private keys.

The two approaches we outlined above (linear keys with a strong assumption and quadratic keys with a compact assumption) can be done simultaneously to obtain a hybrid anonymous HIBE with good performance and relying on a semi-compact assumption. This system is described in the full version of this paper [Duc09], and we discuss several instantiations of this general construction in Section 5.3.

From anonymous HIBE to delegatable HVE. A delegatable Hidden Vector Encryption system (dHVE) [BW07,SW08] can be viewed as an extension of anonymous HIBE. Messages in a dHVE of depth ℓ are encrypted depending on a property vector $v \in \mathbb{S}^\ell$. We define tokens as being vectors in $\mathbb{S}^{*\ell}$, with $\mathbb{S}^* = \mathbb{S} \cup \{*\}$, the $*$ symbol being used as a wildcard. We also set a partial order on those tokens: $\mathbf{v} \geq \mathbf{w}$ iff $\forall 1 \leq i \leq l, v_i = w_i \vee v_i = *$.

A master authority provide public parameters PP allowing anyone to encrypt with any property vector $v \in \mathbb{S}^\ell$. With its master key mk, this authority must also be able to extract keys for any token $\mathbf{w} \in \mathbb{S}^{*\ell}$. Knowing a key k for token $\mathbf{w}$ should allow one to decrypt all messages encoded with property vectors $\mathbf{v} \leq \mathbf{w}$, otherwise no information should be leaked about the message m or the property vector $\mathbf{v}$ used to encrypt it. Furthermore, anyone with a key k for a token $\mathbf{w}$ should be able to delegate keys for any vectors $\mathbf{w}' \leq \mathbf{w}$. We refer to [SW08] for the definition of security.

Any anonymous HIBE is also a dHVE where property vectors are used as identities and we only require keys for vectors of the form $\mathbf{w} = (v_1, \ldots, v_k, *, \ldots, *)$

with $v_1 \ldots v_k \neq *$. To encrypt for an identity of depth $k < \ell$, the encryptor pads the property vector with random values of $\mathbb{S}$ (or with a special token in $\mathbb{S}$ that is not used in identity vectors).

Unfortunately, the construction from [BBG05] doesn't give efficient way to decrypt messages without knowing the exact identity of the recipient, even with a private key for a higher identity in the hierarchy. Thus, our anonymizing tweak on this system doesn't lead to a proper dHVE.

5.1 A Delegatable HVE

Our system uses the property set $\mathbb{S} = \mathbb{Z}_p^*$ and encode the wildcard $*$ by $0 \in \mathbb{Z}_p$.

Convention and reindexation. For a token $\mathbf{w}$, we note $S_{\mathbf{w}}$ (resp $\bar{S}_{\mathbf{w}}$) the subset of indexes such that $\mathbf{w}_i \neq *$ (resp $\mathbf{w}_i = *$), and $k_{\mathbf{w}}$ denotes $|S_{\mathbf{w}}|$. Keys will be seen as matrices of size $(l+2) \times (k_{\mathbf{w}} + 2)$, indexed the following way:

$$
\begin{pmatrix}
 & 0 & s_1 & \cdots & s_{k_{\mathbf{w}}} & -1 \\
\hline
-1 & \cdot & \cdot & \cdots & \cdot & \cdot \\
0 & \cdot & \cdot & \cdots & \cdot & \cdot \\
1 & \cdot & \cdot & \cdots & \cdot & \cdot \\
\vdots & \vdots & \vdots & & \vdots & \vdots \\
\ell & \cdot & \cdot & \cdots & \cdot & \cdot
\end{pmatrix}
$$

where $s_1 \ldots s_{k_{\mathbf{w}}}$ are the naturally ordered elements of $S_{\mathbf{w}}$. But for readability, we can always consider for a given $\mathbf{w}$ that $S_{\mathbf{w}} = 1 \ldots k_{\mathbf{w}}$, by reordering rows and columns of the considered matrices. With this new indexation, row -1 will be called the decryption part (of the key), rows $0 \ldots k_{\mathbf{w}}$ the rerandomization part, and rows $(k_{\mathbf{w}} + 1) \ldots \ell$ the delegation part.

Linear algebra notation. The description of our system is greatly simplified by the use of notation from linear algebra. We will be using vectors and matrices whose components are elements in the groups $\mathbb{G}$ or $\hat{\mathbb{G}}$. The sum of two such matrices is defined by doing a cell by cell group product. We define the product of a vector $\mathbf{x} = (x_1 \ \cdots \ x_n)$ in $(\mathbb{Z}_p)^n$ with a vector $\mathbf{g} = (\hat{g}_1 \ \cdots \ \hat{g}_n)$ of $\hat{\mathbb{G}}^n$ by:

$$
(x_1 \ \cdots \ x_n) \cdot (g_1 \ \cdots \ g_n) = \prod_{i=1}^{n} g_i^{x_i} \quad \in \hat{\mathbb{G}}
$$

This definition extends naturally to the product of matrices over $\mathbb{Z}_p$ by matrices over $\hat{\mathbb{G}}$. When we write those products, we will always place the matrices over $\hat{\mathbb{G}}$ on the right. Vector will be written in bold, and if $\mathbf{v}$ is a vector, then v_i is its ith component of $\mathbf{v}$, and $\mathbf{v}_{|S}$ is the restriction of $\mathbf{v}$ to its components of index in S: $[v_i]_{i \in S}$ (in the natural order). We will also use the plus sign over group matrices:

$A + B$ corresponds the component by component group product. Last, when writing block matrices, **Id** refers to the square identity matrix of the needed dimension.

Using this notation, we can now describe the delegatable HVE system as follows.

Setup(ℓ): To generate system parameters for a dHVE of depth ℓ, given bilinear groups $(\mathbb{G}, \hat{\mathbb{G}})$ with generators $(g, \hat{g})$, the setup algorithm first selects a randoms $\alpha, \beta, \gamma, \eta \in \mathbb{Z}_p, \boldsymbol{\delta} \in (\mathbb{Z}_p)^\ell$, and set: $g_1 = g^\alpha$, $g_2 = g^\beta$, $h = g^\gamma$, $\mathbf{f} = \boldsymbol{\delta} \cdot (g)$, $t = g^\eta$, and their analogues: $\hat{g}_1 = \hat{g}^\alpha$, $\hat{g}_2 = \hat{g}^\beta$, $\hat{h} = \hat{g}^\gamma$, $\hat{\mathbf{f}} = \boldsymbol{\delta} \cdot (\hat{g})$, $\hat{t} = \hat{g}^\eta$. The public parameters PP and the master secret mk are:

$$PP = (\, g, g_1, h, \mathbf{f}, t, \hat{g}, \hat{g}_2, \hat{h} \,) \in \mathbb{G}^{4+l} \times \hat{\mathbb{G}}^3, \quad mk = (\hat{g}_0 = \hat{g}^{\alpha\beta}, \hat{\mathbf{f}}, \hat{t}) \in \hat{\mathbb{G}}^{2+l}$$

Extract($mk, \mathbf{w}$): We first suppose, using reindexation if necessary, that $\mathbf{w}$ is of the form $(\mathbf{w}_1, \ldots, \mathbf{w}_k, *, \ldots, *)$. To generate private key $d_{\mathbf{w}}$, output the matrix:

$$d_{\mathbf{w}} = \begin{pmatrix} d_{\mathbf{w}}^{\text{dec}} \\ d_{\mathbf{w}}^{\text{rer}} \\ d_{\mathbf{w}}^{\text{del}} \end{pmatrix} = \begin{pmatrix} 1 & R_1 & 0 \\ 0 & R_2 & 0 \\ 0 & R_3 & \text{Id} \end{pmatrix} \cdot M_{\mathbf{w}} = \begin{pmatrix} M_{\mathbf{w}}^{\text{dec}} + R_1 \cdot M_{\mathbf{w}}^{\text{rer}} \\ R_2 \cdot M_{\mathbf{w}}^{\text{rer}} \\ M_{\mathbf{w}}^{\text{del}} + R_3 \cdot M_{\mathbf{w}}^{\text{rer}} \end{pmatrix} \in \hat{\mathbb{G}}^{(l+2)\times(2+k_{\mathbf{w}})}$$

where $R_1 \in (\mathbb{Z}_p)^{1\times(k+1)}, R_2 \in (\mathbb{Z}_p)^{(k+1)\times(k+1)}$, and $R_3 \in (\mathbb{Z}_p)^{(\ell-k)\times(k+1)}$ are random matrices, and

$$M_{\mathbf{w}} = \begin{pmatrix} M_{\mathbf{w}}^{\text{dec}} \\ M_{\mathbf{w}}^{\text{rer}} \\ M_{\mathbf{w}}^{\text{del}} \end{pmatrix} = \begin{pmatrix} \hat{g}_0 & & & \hat{g} \\ \hat{t} & & & \\ \hat{h}^{\mathbf{w}_1}\hat{f}_1 & \hat{g} & & \\ \vdots & & \ddots & \\ \hat{h}^{\mathbf{w}_k}\hat{f}_k & & \hat{g} & \\ \hat{f}_{k+1} & & & \\ \vdots & & & \\ \hat{f}_\ell & & & \end{pmatrix}$$

In this matrix blanks correspond to cells containing the group identity $\hat{g}^0$. Unrolling the definition, and without reindexation we thus have:

$$d_{\mathbf{w}}^{\text{dec}} = (\, \hat{g}_0 \, (\mathbf{r}_{-1} \cdot (\mathbf{w} \cdot \hat{h} + \hat{\mathbf{f}})_{|S_{\mathbf{w}}}) \, t^{R_{-1}}, \, \mathbf{r}_{-1} \cdot (\hat{g}), \, \hat{g}^{R_{-1}}) \qquad \in \hat{\mathbb{G}}^{1\times(2+k_{\mathbf{w}})}$$

$$d_{\mathbf{w}}^{\text{rer}} = \left[(\mathbf{r}_j \cdot (\mathbf{w} \cdot \hat{h} + \hat{\mathbf{f}})_{|S_{\mathbf{w}}}) \, t^{R_j}, \quad \mathbf{r}_j \cdot (\hat{g}), \, \hat{g}^{R_j} \right]_{j \in S_{\mathbf{w}} \cup \{0\}} \in \hat{\mathbb{G}}^{(1+k_{\mathbf{w}})\times(2+k_{\mathbf{w}})}$$

$$d_{\mathbf{w}}^{\text{del}} = \left[\hat{f}_j \, (\mathbf{r}_j \cdot (\mathbf{w} \cdot \hat{h} + \hat{\mathbf{f}})_{|S_{\mathbf{w}}}) \, t^{R_j}, \quad \mathbf{r}_j \cdot (\hat{g}), \, \hat{g}^{R_j} \right]_{j \in \bar{S}_{\mathbf{w}}} \in \hat{\mathbb{G}}^{(\ell-k_{\mathbf{w}})\times(2+k_{\mathbf{w}})}$$

for random $R_{-1} \ldots R_u$ in $\mathbb{Z}_p$ and $\mathbf{r}_{-1} \ldots \mathbf{r}_\ell$ in $(\mathbb{Z}_p)^{k_{\mathbf{w}}}$.

The idea behind this definition of $d_{\mathbf{w}}^{\text{del}}$ is to embed the $\hat{f}_i$ needed for delegation in the private key, but blinded in a way that maintains their utility.

***Rerand*(w, $d_\mathbf{w}$):** We present a helper algorithm that will be useful for key delegation.

If we have one valid key $d_\mathbf{w}$ for a given token $\mathbf{w}$, we can build another key $d'_\mathbf{w}$ for the same token $\mathbf{w}$, that has the same distribution as the output of algorithm *Extract*(mk, $\mathbf{w}$), and independent from $d_\mathbf{w}$ by re-randomization. Once again, we invoke reindexation and suppose that $\mathbf{w}$ is of the form $(\mathbf{w}_1, \ldots, \mathbf{w}_k, *, \ldots, *)$. Let $R'_1 \in (\mathbb{Z}_p)^{1\times(k+1)}, R'_2 \in (\mathbb{Z}_p)^{(k+1)\times(k+1)}$, and $R'_3 \in (\mathbb{Z}_p)^{(u-k-1)\times(k+1)}$ be random matrices. Note that this corresponds to $(k+1)(\ell+2)$ random values in $\mathbb{Z}_p$, as in *Extract*. We build $d'_\mathbf{w}$ by:

$$d'_\mathbf{w} = \left(\begin{array}{c|c|c} 1 & R'_1 & 0 \\ \hline 0 & R'_2 & 0 \\ \hline 0 & R'_3 & \mathbf{Id} \end{array}\right) \cdot d_\mathbf{w} = \left(\begin{array}{c} d_\mathbf{w}^{\mathrm{dec}} + R'_1 \cdot d_\mathbf{w}^{\mathrm{rer}} \\ R'_2 \cdot d_\mathbf{w}^{\mathrm{rer}} \\ d_\mathbf{w}^{\mathrm{del}} + R'_3 \cdot d_\mathbf{w}^{\mathrm{rer}} \end{array}\right)$$

Let R_1, R_2, R_3 be the matrices giving the previous decomposition of $d_\mathbf{w}$:

$$d_\mathbf{w} = \left(\begin{array}{c|c|c} 1 & R_1 & 0 \\ \hline 0 & R_2 & 0 \\ \hline 0 & R_3 & \mathbf{Id} \end{array}\right) \cdot M_\mathbf{w}$$

thus we have:

$$d'_\mathbf{w} = \left(\begin{array}{c|c|c} 1 & R'_1 & 0 \\ \hline 0 & R'_2 & 0 \\ \hline 0 & R'_3 & \mathbf{Id} \end{array}\right) \cdot \left(\begin{array}{c|c|c} 1 & R_1 & 0 \\ \hline 0 & R_2 & 0 \\ \hline 0 & R_3 & \mathbf{Id} \end{array}\right) \cdot M_\mathbf{w} = \left(\begin{array}{c} M_\mathbf{w}^{\mathrm{dec}} + (R'_1 \cdot R_2 + R_1) \cdot M_\mathbf{w}^{\mathrm{rer}} \\ R'_2 \cdot R_2 \cdot M_\mathbf{w}^{\mathrm{rer}} \\ M_\mathbf{w}^{\mathrm{del}} + (R'_3 \cdot R_2 + R_3) \cdot M_\mathbf{w}^{\mathrm{del}} \end{array}\right)$$

It is not hard to see that if R_2 is full rank, then $d'_\mathbf{w}$ is distributed as the output of algorithm *Extract*(mk, $\mathbf{w}$). If R_2 is not full ranked then $d'_\mathbf{w}$ is distributed differently (we then say that $d'_\mathbf{w}$ is ill-formed), but since this happens with probability about $1/p$ (which is negligible), $d'_\mathbf{w}$ is then distributed statically close to the distribution of *Extract*(mk, $\mathbf{w}$).

Note that this difference arise between the real-world and the security model, thus it will not appear in the security reduction.

***Derive*(w, $d_\mathbf{w}$, w′):** The derivation algorithm only needs to answer when $\mathbf{w} \geq \mathbf{w}'$ and can assume that $d_\mathbf{w}$ is indeed a valid key for the token $\mathbf{w}$.

It is sufficient to show a correct algorithm working only when $k_{\mathbf{w}'} = k_\mathbf{w} + 1$, and then do delegation step-by-step. Using reindexation, we can also assume that $\mathbf{w} = (\mathbf{w}_1, \ldots, \mathbf{w}_k, *, \ldots, *)$ and $\mathbf{w}' = (\mathbf{w}_1, \ldots, \mathbf{w}_{k+1}, *, \ldots, *)$

Lets write the private key for $d_\mathbf{w}$ as

$$d_\mathbf{w} = \left(\begin{array}{ccc|c} D & G_1 & \cdots & G_{k-1} & T \\ D_0 & G_{0,1} & \cdots & G_{0,k-1} & T_0 \\ \vdots & \vdots & \ddots & \vdots & \vdots \\ D_\ell & G_{\ell,1} & \cdots & G_{\ell,k-1} & T_\ell \end{array}\right)$$

First, we define $d_{\mathbf{w}}^{\text{temp}}$ (with one more column):

$$d_{\mathbf{w}'}^{\text{temp}} = \left(\begin{array}{c|ccc|c|c} D & G_1 & \cdots & G_{k-1} & & T \\ D_0 & G_{0,1} & \cdots & G_{0,k-1} & & T_0 \\ \vdots & \vdots & \ddots & \vdots & & \vdots \\ D_{k-1} & G_{k-1,1} & \cdots & G_{k-1,k-1} & & T_{k-1} \\ \hline \hat{h}^{\mathbf{w}_k} D_k & G_{k,1} & \cdots & G_{k,k-1} & \hat{g} & T_k \\ \hline D_{k+1} & G_{k+1,1} & \cdots & G_{k+1,k-1} & & T_{k+1} \\ \vdots & \vdots & \ddots & \vdots & & \vdots \\ D_\ell & G_{\ell,1} & \cdots & G_{\ell,k-1} & & T_\ell \end{array}\right)$$

Also in this matrix, blanks correspond to cells containing the group identity $\hat{g}^0$.

Intuitively, we moved the highlighted line from the delegation part to the re-randomization part of the key. Let R_1, R_2, R_3 be the matrices giving the previous decomposition of $d_{\mathbf{w}}$, and R_1', R_2', R_3' for $d_{\mathbf{w}'}^{\text{temp}}$, we have:

$$R_1' = \left(\, R_1 \mid 0 \,\right), \quad R_2' = \left(\begin{array}{c|c} & 0 \\ R_2 & \vdots \\ & 0 \\ \hline (R_3)_1 & 1 \end{array}\right), \quad R_3' = \left(\begin{array}{c|c} (R_3)_2 & 0 \\ \vdots & \vdots \\ (R_3)_{k-2} & 0 \end{array}\right)$$

Note that if R_2 rank is $k+1$, then rank of R_2' is $k+2$, so if $d_{\mathbf{w}}$ isn't an ill-formed key, $d_{\mathbf{w}'}^{\text{temp}}$ isn't ill-formed either. We can thus output $Rerand(d_{\mathbf{w}'}^{\text{temp}})$, which is a valid key following the same distribution as if it was directly generated by $Extract(mk, \mathbf{w}')$.

Encrypt($PP, \mathbf{v}, M$). To encrypt a message $M \in \mathbb{G}_T$ under the property vector $\mathbf{v} = (v_1, \ldots, v_\ell) \in \mathbb{Z}_p^*[\ell]$, pick a random $s \in \mathbb{Z}_p$ and output

$$\mathsf{CT} = \left(\, e(g_1, \hat{g}_2)^s \cdot M, \quad g^s, \quad (s) \cdot (\mathbf{v} \cdot (h) + \mathbf{f}), \quad t^s \,\right) \in \mathbb{G}_t \times \mathbb{G}^{2+\ell}.$$

Decrypt($d_{\mathbf{w}}^{\text{dec}}, \mathbf{w}, \mathsf{CT}$). Consider the token $\mathbf{w} = (\mathbf{w}_1, \ldots, \mathbf{w}_\ell)$ associated to the key $d_{\mathbf{w}}$. To decrypt a ciphertext $\mathsf{CT} = (A, B, C_1, \ldots, C_\ell, Z)$ using the decryption part of the private key $d_{\mathbf{w}}^{\text{dec}} = (a_0, [b_i]_{i \in S_{\mathbf{w}}}, z)$, output

$$M' = A \cdot e(Z, z) \cdot \prod_{i \in S_{\mathbf{w}}} e(C_i, b_i) \Big/ e(B, a_0).$$

Correctness. We briefly check that decryption is correct. With the same notations as in *Decrypt*. Assuming $\mathbf{w} \geq \mathbf{v}$, we have $\forall i \in S_{\mathbf{w}}, w_i = v_i$.

$$
\begin{aligned}
e(B, a_0) &= e(g^s, \hat{g}_0 \; \mathbf{r}_{-1} \cdot (\mathbf{w} \cdot \hat{h} + \hat{\mathbf{f}})_{|S_{\mathbf{w}}}) t^{R_{-1}}) \\
&= e(g^s, \hat{g}_0) \cdot e(g^s, \hat{t}^{R_{-1}}) \cdot \prod_{i \in S_{\mathbf{w}}} e(g^s, (\hat{h}^{w_i} \hat{\mathbf{f}}_i)^{r_{-1,i}}) \\
&= e(g_0, \hat{g})^s \cdot e(t^s, \hat{g}^{R_{-1}}) \cdot \prod_{i \in S_{\mathbf{w}}} e((h^{w_i} \mathbf{f}_i)^s, \hat{g}^{r_{-1,i}}) \\
&= e(g_0, \hat{g})^s \cdot e(Z, z) \cdot \prod_{i \in S_{\mathbf{w}}} e(C_i, b_i)
\end{aligned}
$$

If CT is the encryption of M under the public key ID, then $A = e(g_0, \hat{g})^s \cdot M$, and hence we have $M' = M$.

5.2 Security Reduction

The following theorem proves security of our system under the Decision $\mathcal{P}$-BDH assumption.

Theorem 2 (dHVE security). *Our dHVE is selective-ID anonymous assuming the Decision $\mathcal{P}$-BDH assumption holds for the bilinear group generator $\mathcal{G}$. In particular, for all PPT algorithms $\mathcal{B}$ and all $\ell > 0$, the function $\mathsf{Adv}_{\mathcal{B}}^{\mathsf{aIND\text{-}sIDCPA}}(\lambda)$ is a negligible function of λ.*

The proof proceeds by a hybrid argument across a number of games. Let $\mathsf{CT} = (A, B, [C_i]_{i=1}^{\ell}, Z)$ in $\mathbb{G}_t \times \mathbb{G}^3$ denote the challenge ciphertext given to the adversary during the selective-ID game $\Gamma^{(0)}$ of definition 1. Additionally, let R be a random element of $\mathbb{G}_t$, and R' and $[R_i]_{i=1}^{\ell}$ be random elements of $\mathbb{G}$. We define the following hybrid games, which differ on what challenge ciphertext is given to the adversary:

- **Game Γ:** The challenge ciphertext is $\mathsf{CT} = \left(A, B, [C_i]_{i=1}^{\ell}, Z\right)$
- **Game Γ':** The challenge ciphertext is $\mathsf{CT}' = \left(R, B, [C_i]_{i=1}^{\ell}, Z\right)$
- **Game Γ_n $(n = 0 \ldots \ell)$:** The challenge is $\mathsf{CT}_n = \left(R, B, [R_i]_{i=1}^{n}, [C_i]_{i=n+1}^{\ell}, R'\right)$

Game Γ is the same as game $\Gamma^{(0)}$ in Definition 1. Game Γ_ℓ is the same as game $\Gamma^{(1)}$ in Definition 1, where the adversary is given a random ciphertext. Therefore,

$$
\mathsf{Adv}_{\mathcal{B}}^{\mathsf{aIND\text{-}sIDCPA}} \leq \left| \Pr\left[\mathcal{A}^{\Gamma} = 0 \right] - \Pr\left[\mathcal{A}^{\Gamma_\ell} = 0 \right] \right|.
$$

To prove that Γ is indistinguishable from Γ_ℓ we prove that each step of the hybrid is indistinguishable from the next. We do so in a sequence of lemmas.

Lemma 4 (semantic security). *Let $\mathcal{A}$ be an adversary playing the aIND-sIDCPA attack game. Then, there exist an algorithm $\mathcal{B}$ solving the Decision $\mathcal{P}$-BDH problem such that:*

$$
\left| \Pr\left[\mathcal{A}^{\Gamma} = 0 \right] - \Pr\left[\mathcal{A}^{\Gamma'} = 0 \right] \right| \leq \mathsf{Adv}_{\mathcal{B}}^{\mathcal{P}\text{-}BDH}
$$

Lemma 5 (anonymity, part 1). *Let $\mathcal{A}$ be an adversary playing the aIND-sIDCPA attack game. Then, there exist an algorithm $\mathcal{B}$ solving the Decision $\mathcal{P}$-BDH problem such that:*

$$\left| \Pr\left[\mathcal{A}^{\Gamma'} = 0 \right] - \Pr\left[\mathcal{A}^{\Gamma_0} = 0 \right] \right| \leq \mathsf{Adv}_{\mathcal{B}}^{\mathcal{P}\text{-}BDH}$$

Lemma 6 (anonymity, part 2). *Let $\mathcal{A}$ be an adversary playing the aIND-sIDCPA attack game. Then, for all $n = 1 \ldots \ell$, there exist an algorithm $\mathcal{B}$ solving the Decision $\mathcal{P}$-BDH problem such that:*

$$\left| \Pr\left[\mathcal{A}^{\Gamma_{n-1}} = 0 \right] - \Pr\left[\mathcal{A}^{\Gamma_n} = 0 \right] \right| \leq \mathsf{Adv}_{\mathcal{B}}^{\mathcal{P}\text{-}BDH}$$

Thus, if there is no algorithm $\mathcal{B}$ that solve $\mathcal{P}$-BDH problem with an advantage better than ϵ, there, for all adversary $\mathcal{A}$ making at most:

$$
\begin{aligned}
\left| \Pr\left[\mathcal{A}^{\Gamma} = 0 \right] - \Pr\left[\mathcal{A}^{\Gamma_\ell} = 0 \right] \right| &\leq \left| \Pr\left[\mathcal{A}^{\Gamma} = 0 \right] - \Pr\left[\mathcal{A}^{\Gamma'} = 0 \right] \right| \\
&+ \left| \Pr\left[\mathcal{A}^{\Gamma'} = 0 \right] - \Pr\left[\mathcal{A}^{\Gamma_0} = 0 \right] \right| \\
&+ \sum_{n=1}^{\ell} \left| \Pr\left[\mathcal{A}^{\Gamma_{n-1}} = 0 \right] - \Pr\left[\mathcal{A}^{\Gamma_n} = 0 \right] \right| \\
&\leq (2 + \ell)\,\epsilon
\end{aligned}
$$

Consequently, under the $\mathcal{P}$-BDH assumption, game Γ is indistinguishable from Γ_ℓ.

The proofs of those lemmas are given in the full version [Duc09].   □

5.3 Instantiations

We now compare several instantiations of the anonymous HIBE system and the dHVE system that can be build using our techniques with other similar anonymous constructions. In the Hybrid system described in the full version [Duc09], there is a parameter $\omega \in [0, 1]$ that we can also adjust to get different trade-off in terms of hypothesis strength and performances. Its security rely on a stronger assumption: $\mathcal{P}^n$-BDH, which is also described in the full version. There are 3 noticeable values for ω: $\omega = 1$ is not using mechanics of [BBG05] and can be extended to our dHVE, $\omega = 0$ provides constant size ciphertexts and linear keys, last $\omega = 1/2$ gives optimal key size and sublinear ciphertexts.

	group order	assumption	key size	ciphertext size
BB$_1$ based IBE	prime	$\mathcal{P}$-BDH	3	4
BB$_1$ based dHVE ($\omega = 1$)	prime	$\mathcal{P}$-BDH	$\sim \ell^2$	$\ell + 3$
Hybrid BBG-based HIBE	prime	$\mathcal{P}^{\lceil \ell^{1-\omega} \rceil}$-BDH	$\leq 3(\ell + \ell^{2\omega})$	$\sim \ell^\omega$
BBG based HIBE ($\omega = 0$)	prime	$\mathcal{P}^\ell$-BDH	$\sim 3\ell$	4
Hybrid BBG-based HIBE ($\omega = 1/2$)	prime	$\mathcal{P}^{\lceil \sqrt{\ell} \rceil}$-BDH	$\sim \ell$	$\sim \sqrt{\ell}$
dHVE from [SW08]	composite	composite-BDH	$\sim \ell^2$	$\sim \ell$
HIBE from [SKOS09]	composite	composite-BDH	$\sim 3\ell$	4

Sizes are expressed in group elements. Number of pairings for decryption is always bound by the ciphertext size. Our construction offers better efficiency by a constant factor than previous results, depending on the chosen trade-off. Using asymmetric pairing let us avoid composite groups, which are inevitably larger. For elements of $\mathbb{G}$ (used in the ciphertext) the gain can be very substantial (1024 bits versus 170 bits to achieve common concrete security). However, the use of type 2 or type 3 pairing doesn't allow compact representation of $\hat{\mathbb{G}}$ elements, so that the improvement in private key size is not as significant (see [GPS06]).

6   Conclusions

We presented a technique for using asymmetric bilinear groups to add anonymity to a family of non-anonymous HIBE systems. One of those HIBE naturally extend to a delegatable HVE system. The resulting systems are more efficient than several existing constructions for anonymous systems.

Acknowledgments

This work was done while the author was visiting Stanford University. I would like to express my gratitude to Dan Boneh, who gave me precious advice all along this work. Stanford University staff also deserves some thanks for welcoming me at the Computer Science lab for this internship. Finally, I would like to thank the anonymous reviewers for their wise comments on this paper.

References

[ABC+05] Abdalla, M., Bellare, M., Catalano, D., Kiltz, E., Kohno, T., Lange, T., Malone-Lee, J., Neven, G., Paillier, P., Shi, H.: Searchable encryption revisited: Consistency properties, relation to anonymous IBE, and extensions. In: Shoup, V. (ed.) CRYPTO 2005. LNCS, vol. 3621, pp. 205–222. Springer, Heidelberg (2005)

[ACd05]   Ateniese, G., Camenisch, J., deMedeiros, B.: Untraceable rfid tags via insubvertible encryption. In: Proceedings of the 12th ACM conference on Computer and communications security (2005)

[BB04]   Boneh, D., Boyen, X.: Efficient selective-ID identity based encryption without random oracles. In: Cachin, C., Camenisch, J.L. (eds.) EUROCRYPT 2004. LNCS, vol. 3027, pp. 223–238. Springer, Heidelberg (2004)

[BBG05]   Boneh, D., Boyen, X., Goh, E.-J.: Hierarchical identity based encryption with constant size ciphertext. In: Cramer, R. (ed.) EUROCRYPT 2005. LNCS, vol. 3494, pp. 440–456. Springer, Heidelberg (2005)

[BCHK07] Boneh, D., Canetti, R., Halevi, S., Katz, J.: Chosen-ciphertext security from identity-based encryption. SIAM Journal of Computing 36(5) (2007)

[BCOP04] Boneh, D., Crescenzo, G.D., Ostrovsky, R., Persiano, G.: Public key encryption with keyword search. In: Cachin, C., Camenisch, J.L. (eds.) EUROCRYPT 2004. LNCS, vol. 3027, pp. 506–522. Springer, Heidelberg (2004)

[BF03]   Boneh, D., Franklin, M.: Identity-based encryption from the Weil pairing.
           SIAM Journal of Computing 32(3) (2003); Preliminary version In: Kilian, J.
           (ed.) CRYPTO 2001. LNCS, vol. 2139, p. 213. Springer, Heidelberg (2001)
[BKM05]   Ballard, L., Kamara, S., Monrose, F.: Achieving efficient conjunctive key-
           word searches over encrypted data. In: Qing, S., Mao, W., López, J.,
           Wang, G. (eds.) ICICS 2005. LNCS, vol. 3783, pp. 414–426. Springer,
           Heidelberg (2005)
[BW06]   Boyen, X., Waters, B.: Anonymous hierarchical identity-based encryption
           (without random oracles). In: Dwork, C. (ed.) CRYPTO 2006. LNCS,
           vol. 4117, pp. 290–307. Springer, Heidelberg (2006)
[BW07]   Boneh, D., Waters, B.: Conjunctive, subset, and range queries on encrypted
           data. In: Vadhan, S.P. (ed.) TCC 2007. LNCS, vol. 4392, pp. 535–554.
           Springer, Heidelberg (2007)
[CHK03]   Canetti, R., Halevi, S., Katz, J.: A forward-secure public-key encryption
           scheme. In: Biham, E. (ed.) EUROCRYPT 2003. LNCS, vol. 2656. Springer,
           Heidelberg (2003)
[CHK04]   Canetti, R., Halevi, S., Katz, J.: Chosen-ciphertext security from identity-
           based encryption. In: Cachin, C., Camenisch, J.L. (eds.) EUROCRYPT
           2004. LNCS, vol. 3027, pp. 207–222. Springer, Heidelberg (2004)
[CHL05]   Camenisch, J., Hohenberger, S., Lysyanskaya, A.: Compact e-cash. ePrint
           Report 2005/060 (2005)
[Duc09]   Ducas, L.: Anonymity from asymmetry: New constructions for anonymous
           hibe (2009), http://www.eleves.ens.fr/home/ducas/publi/ahibe10/
[Gen06]   Gentry, C.: Practical identity-based encryption without random oracles.
           In: Vaudenay, S. (ed.) EUROCRYPT 2006. LNCS, vol. 4004, pp. 445–464.
           Springer, Heidelberg (2006)
[GPS06]   Galbraith, S.D., Paterson, K.G., Smart, N.P.: Pairings for cryptographers.
           Cryptology ePrint Archive, Report 2006/165 (2006),
           http://eprint.iacr.org/
[GS02]   Gentry, C., Silverberg, A.: Hierarchical ID-based cryptography. In: Zheng,
           Y. (ed.) ASIACRYPT 2002. LNCS, vol. 2501, pp. 548–566. Springer,
           Heidelberg (2002)
[HL02]   Horwitz, J., Lynn, B.: Towards hierarchical identity-based encryption. In:
           Knudsen, L.R. (ed.) EUROCRYPT 2002. LNCS, vol. 2332, p. 466. Springer,
           Heidelberg (2002)
[IP08]   Iovino, V., Persiano, G.: Hidden-vector encryption with groups of prime
           order. In: Galbraith, S.D., Paterson, K.G. (eds.) Pairing 2008. LNCS,
           vol. 5209, pp. 75–88. Springer, Heidelberg (2008)
[Jou00]   Joux, A.: A one round protocol for tripartite Diffie-Hellman. In: Bosma,
           W. (ed.) ANTS 2000. LNCS, vol. 1838. Springer, Heidelberg (2000)
[SBC+07]   Shi, E., Bethencourt, J., Chan, H.T.-H., Song, D.X., Perrig, A.: Multi-
           dimensional range query over encrypted data. In: SP 2007: Proceedings of
           the 2007 IEEE Symposium on Security and Privacy (2007)
[Sco02]   Scott, M.: Authenticated id-based key exchange and remote log-in with
           simple token and pin number. ePrint Report 2002/164 (2002)
[Sha85]   Shamir, A.: Identity-based cryptosystems and signature schemes. In:
           Blakely, G.R., Chaum, D. (eds.) CRYPTO 1984. LNCS, vol. 196, pp. 47–53.
           Springer, Heidelberg (1985)

[SKOS09]   Seo, J.H., Kobayashi, T., Ohkubo, M., Suzuki, K.: Anonymous hierarchical identity-based encryption with constant size ciphertexts. In: Irvine: Proceedings of the 12th International Conference on Practice and Theory in Public Key Cryptography (2009)

[SW08]   Shi, E., Waters, B.: Delegating capabilities in predicate encryption systems. In: Aceto, L., Damgård, I., Goldberg, L.A., Halldórsson, M.M., Ingólfsdóttir, A., Walukiewicz, I. (eds.) ICALP 2008, Part II. LNCS, vol. 5126, pp. 560–578. Springer, Heidelberg (2008)

[Wat05]   Waters, B.: Efficient identity-based encryption without random oracles. In: Cramer, R. (ed.) EUROCRYPT 2005. LNCS, vol. 3494, pp. 114–127. Springer, Heidelberg (2005)

Making the Diffie-Hellman Protocol Identity-Based*

Dario Fiore[1],** and Rosario Gennaro[2]

[1] Dipartimento di Matematica ed Informatica – Università di Catania, Italy
fiore@dmi.unict.it
[2] IBM T. J. Watson Research Center – Hawthorne, New York 10532
rosario@us.ibm.com

Abstract. This paper presents a new identity based key agreement protocol. In id-based cryptography (introduced by Adi Shamir in [29]) each party uses its own identity as public key and receives his secret key from a master Key Generation Center, whose public parameters are publicly known.

The novelty of our protocol is that it can be implemented over any cyclic group of prime order, where the Diffie-Hellman problem is supposed to be hard. It does not require the computation of expensive bilinear maps, or additional assumptions such as factoring or RSA.

The protocol is extremely efficient, requiring only twice the amount of bandwith and computation of the *unauthenticated* basic Diffie-Hellman protocol. The design of our protocol was inspired by MQV (the most efficient authenticated Diffie-Hellman based protocol in the public-key model) and indeed its performance is competitive with respect to MQV (especially when one includes the transmission and verification of certificates in the MQV protocol, which are not required in an id-based scheme). Our protocol requires a single round of communication in which each party sends only 2 group elements: a very short message, especially when the protocol is implemented over elliptic curves.

We provide a full proof of security in the Canetti-Krawczyk security model for key exchange, including a proof that our protocol satisfies additional security properties such as forward secrecy, and resistance to reflection and key-compromise impersonation attacks.

1 Introduction

Identity-based cryptography was introduced in 1984 by Adi Shamir [29]. The goal was to simplify the management of public keys and in particular the association of a public key to the identity of its holder. Usually such binding of a public key to an identity is achieved by means of *certificates* which are signed statements by trusted third parties that a given public key belongs to a user. This requires users

---

* A full version of this paper is available at http://eprint.iacr.org/2009/174
** Work done while visiting NYU and IBM Research.

J. Pieprzyk (Ed.): CT-RSA 2010, LNCS 5985, pp. 165–178, 2010.
© Springer-Verlag Berlin Heidelberg 2010

to obtain and verify certificates whenever they want to use a specific public key, and the management of public key certificates remains a technically challenging problem.

Shamir's idea was to allow parties to use their identities as public keys. An id-based scheme works as follows. A trusted *Key Generation Center* (KGC) generates a master public/secret key pair, which is known to all the users. A user with identity ID receives from the KGC a secret key S_{ID} which is a function of the string ID and the KGC's secret key (one can think of S_{ID} as a signature by the KGC on the string ID). Using S_{ID} the user can then perform cryptographic tasks. For example in the case of *id-based encryption* any party can send an encrypted message to the user with identity ID using the string ID as a public key and the user (and only the user and the KGC) will be able to decrypt it using S_{ID}. Note that the sender can do this even if the recipient has not obtained yet his secret key from the KGC. All the sender needs to know is the recipient's identity and the public parameters of the KGC. This is the major advantage of id-based encryption.

ID-BASED KEY AGREEMENT AND ITS MOTIVATIONS. This paper is concerned with the task of *id-based key agreement*. Here two parties Alice and Bob, with identities A, B and secret keys S_A, S_B respectively, want to agree on a common shared key, in an *authenticated* manner (i.e. Alice must be sure that once the key is established, only Bob knows it – and viceversa). Since key agreement is inherently an interactive protocol (both parties are "live" and ready to establish a session) there is a smaller gain in using an id-based solution: indeed certificates and public keys can be easily sent as part of the protocol communication.

Yet the ability to avoid sending and verifying public key certificates is a significant practical advantage (see e.g. [32]). Indeed known shortcomings of the public key setting are the requirement of centralized certification authorities, the need for parties to cross-certify each other (via possibly long certificate chains), and the management of some form of large-scale coordination and communication (possibly on-line) to propagate certificate revocation information. Identity-based schemes significantly simplify identity management by bypassing the certification issues. All a party needs to know in order to generate a shared key is its own secret key, the public information of the KGC, and the identity of the communication peer (clearly, the need to know the peer's identity exists in any scheme including a certificate-based one).

Another advantage of identity-based systems is the versatility with which identities may be chosen. Since identities can be arbitrary string, they can be selected according to the function and attributes of the parties (rather than its actual "name"). For example in vehicular networks a party may be identified by its location ("the checkpoint at the intersection of a and b") or in military applications a party can be identified by its role ("platoon x commander"). This allows parties to communicate securely with the intended recipient even without knowing its "true" identity but simply by the definition of its function in the network.

Finally, identities can also include additional attributes which are temporal in nature: in particular an "expiration date" for an identity makes revocation of the corresponding secret key much easier to achieve.

For the reasons described above, id-based KA protocols are very useful in many systems where bandwith and computation are at a premium (e.g. sensor networks), and also in ad-hoc networks where large scale coordination is undesirable, if not outright impossible. Therefore it is an important question to come up with very efficient and secure id-based KA protocols.

PREVIOUS WORK ON ID-BASED KEY AGREEMENT. Following Shamir's proposal of the concept of id-based cryptography, some early proposals for id-based key agreement appeared in the literature: we refer in particular to the works of Okamoto [24] (later improved in [25]) and Gunther [18]. A new impetus to this research area came with the breakthrough discovery of bilinear maps and their application to id-based encryption in [4]: starting with the work of Sakai *et al.* [28] a large number of id-based KA protocols were designed that use pairings as tool. We refer the readers to [5] and [10] for surveys of these pairing-based protocols.

The main problem with the current state of the art is that many of these protocols lack a proof of security, and some have even been broken. Indeed only a few (e.g., [7,33]) have been proven according to a formal definition of security.

OUR CONTRIBUTION. By looking at prior work we see that provably secure id-based KAs require either groups that admit bilinear maps [7,33], or to work over a composite RSA modulus [25].

This motivated us to ask the following question: can we find an efficient and provably secure id-based KA protocol such that:

1. It that can be implemented over *any* cyclic group in which the Diffie-Hellman problem is supposed to be hard. The advantages of such a KA protocol would be several, in particular: (i) it would avoid the use of computationally expensive pairing computations; (ii) it could be implemented over much smaller groups (since we could use 'regular' elliptic curves, rather than the ones that admit efficient pairings computations for high security levels, or the group Z_N^* for a composite N needed for Okamoto-Tanaka).
2. It is more efficient than any KA protocols in the public key model (such as MQV [22]), when one includes the transmission and verification of certificates which are not required in an id-based scheme. This is a very important point since, as we pointed out earlier in this Section, id-based KA protocols are only relevant if they outperform PKI based ones in efficiency.

Our new protocol presented in this paper, achieves all these features. It can be implemented over any cyclic group over which the Diffie-Hellman problem is assumed to be hard. In addition it requires an amount of bandwith and computation similar to the *unauthenticated* basic Diffie-Hellman protocol. Indeed our new protocol requires a single round of communication in which each party sends just two group elements (as opposed to one in the Diffie-Hellman protocol). Each party must compute four exponentiations to compute the session key (as opposed to two in the Diffie-Hellman protocol).

A similar favorable comparison holds with the Okamoto-Tanaka protocol in [25]. While that protocol requires only two exponentiations, it works over Z_N^* therefore requiring the use of a larger group size, which almost totally absorbs the computational advantage, and immediately implies a much larger bandwith requirement. Detailed efficiency comparisons to other protocols in the literature are discussed in Section 5.

We present a full proof of security of our protocol in the Canetti-Krawczyk security model. Our results hold in the random oracle model, under the Strong Diffie-Hellman Assumption. We also present some variations of our protocol that can be proven secure under the basic Computational Diffie-Hellman Assumption. Our protocol can be proven to satisfy additional desirable security properties such as perfect forward secrecy[1], and resistance to reflection and key-compromise impersonation attacks.

OUR APPROACH. The first direction we took in our approach was to attempt to analyze the id-based KA protocols by Gunther [18] and Saeednia [27]. They also work over any cyclic group where the Diffie-Hellman problem is assumed to be hard, but lack a formal proof of security. While the original protocols cannot be shown to be secure, we were able to prove the security of modified versions of them. Nevertheless these two protocols were not very satisfactory solutions for the problem we had set out to solve, particularly for reasons of efficiency since they required a large number of exponentiations, which made them less efficient than say MQV with certificates. The security analysis of these modified Gunther and Saeednia protocols will be included in the final version.

Our protocol improves over these two protocols by using Schnorr's signatures [30], rather than ElGamal, to issue secret keys to the users. The simpler structure of Schnorr's signatures permits a much more efficient computation of the session key, resulting in less exponentiations and a single round protocol. Our approach was inspired by the way the MQV protocol [22] achieves *implicit authentication* of the session key. Indeed our protocol can be seen as an id-based version of the MQV protocol.

2   Preliminaries

In this section we present some standard definitions needed in the rest of the paper.

Let $\mathbb{N}$ the set of natural numbers. We will denote with $\ell \in \mathbb{N}$ the security parameter. The partecipants to our protocols are modeled as probabilistic Turing machines whose running time is bounded by some polynomial in ℓ. If S is a set, we

---

[1] We can prove PFS only in the case the adversary was passive in the session that he is attacking – though he can be active in other sessions. As proven by Krawczyk in [21], this is the best that can be achieved for 1-round protocols with *implicit* authentication, such as ours.

denote with $s \xleftarrow{\$} S$ the process of selecting an element uniformly at random from S. A function is said to be *negligible* if it vanishes faster than any polynomial.

The security of our protocol is based on the Strong Diffie-Hellman Assumption (SDH) [1], which is a variant of the standard Computational Diffie-Hellman (CDH) [13] where the adversary is provided with an oracle that solves the decisional problem.

Our new protocol is proven secure in the Canetti-Krawczyk (CK) [8,9] model for key agreement, adapted to the identity-based setting. For lack of space we defer the description of the assumptions and the model to the full version of the paper.

3   The New Protocol IB-KA

Protocol setup. The Key Generation Center (KGC) chooses a group $\mathbb{G}$ of prime order q (where q is ℓ-bits long), a random generator $g \in \mathbb{G}$ and two hash functions $H_1 : \{0,1\}^* \to \mathbb{Z}_q$ and $H_2 : \mathbb{Z}_q \times \mathbb{Z}_q \to \{0,1\}^\ell$. Then it picks a random $x \xleftarrow{\$} \mathbb{Z}_q$ and sets $y = g^x$. Finally the KGC outputs the public parameters $MPK = (\mathbb{G}, g, y, H_1, H_2)$ and keeps the master secret key $MSK = x$ for itself.

Key Derivation. A user with identity ID receives, as its secret key, a Schnorr's signature [30] of the message $m = ID$ under public key y. More specifically, the KGC after verifying the user's identity, creates the associated secret key as follows. First it picks a random $k \xleftarrow{\$} \mathbb{Z}_q$ and sets $r_{ID} = g^k$. Then it uses the master secret key x to compute $s_{ID} = k + H_1(ID, r_{ID})x$. (r_{ID}, s_{ID}) is the secret key returned to the user. The user can verify the correctness of its secret key by using the public key y and checking the equation $g^{s_{ID}} \stackrel{?}{=} r_{ID} \cdot y^{H_1(ID, r_{ID})}$.

A protocol session. Let's assume that Alice wants to establish a session key with Bob. Alice owns secret key (r_A, s_A) and identity A while Bob has secret key (r_B, s_B) and identity B.

Alice selects a random $t_A \xleftarrow{\$} \mathbb{Z}_q$, computes $u_A = g^{t_A}$ and sends the message $\langle A, r_A, u_A \rangle$ to Bob. Analogously Bob picks a random $t_B \xleftarrow{\$} \mathbb{Z}_q$, computes $u_B = g^{t_B}$ and sends $\langle B, r_B, u_B \rangle$ to Alice. After the parties have exchanged these two messages, they are able to compute the same session key $Z = H_2(z_1, z_2)$. In particular Alice computes

$$z_1 = (u_B r_B y^{H_1(B, r_B)})^{t_A + s_A} \quad \text{and} \quad z_2 = u_B^{t_A}.$$

On the other hand Bob computes

$$z_1 = (u_A r_A y^{H_1(A, r_A)})^{t_B + s_B} \quad \text{and} \quad z_2 = u_A^{t_B}.$$

It is easy to see that both the parties are computing the same values $z_1 = g^{(t_A + s_A)(t_B + s_B)}$ and $z_2 = g^{t_A t_B}$. The state of a user ID during a protocol session contains only the fresh random exponent t_{ID}. We assume that after a session is completed, the parties erase their state and keep only the session key.

Remark: In the next section we show that protocol IB-KA is secure under the Strong Diffie-Hellman Assumption. However, in the full version of the paper we show how to modify IB-KA to obtain security under the basic CDH Assumption, at the cost of a slight degradation in efficiency.

4 Security Proof

We prove the security of the protocol by a usual reduction argument. More precisely we show how to reduce the existence of an adversary breaking the protocol into an algorithm that is able to break the SDH Assumption with non-negligible probability. The adversary is modeled as a CK attacker: in particular it will choose a test session among the complete and unexposed sessions and will try to distinguish between its real session key and a random one.

In our reduction we will make use of the General Forking Lemma, stated by Bellare and Neven in [2]. It follows the original forking lemma of Pointcheval and Stern [26], but, unlike that, it makes no mention of signature schemes and random oracles. In this sense it is more general and it can be used to prove the security of our protocol. We briefly recall it in the following.

Lemma 1 (General Forking Lemma [2]). *Fix an integer $Q \geq 1$ and a set H of size $|H| \geq 2$. Let $\mathcal{B}$ be a randomized algorithm that on input $x, h_1, \ldots, h_Q$ returns a pair (J, σ) where $J \in \{0, \ldots, Q\}$ and σ is referred as side output. Let IG be a randomized algorithm called the input generator. Let $acc_{\mathcal{B}} = \Pr[J \geq 1 : x \xleftarrow{\$} IG, h_1, \ldots, h_Q \xleftarrow{\$} H; (J, \sigma) \xleftarrow{\$} \mathcal{B}(x, h_1, \ldots, h_Q)]$ be the accepting probability of $\mathcal{B}$.*

The forking algorithm $F_{\mathcal{B}}$ associated to $\mathcal{B}$ is the randomized algorithm that takes in input x and proceeds as follows:

> *Algorithm $F_{\mathcal{B}}(x)$*
> *Pick random coins ρ for $\mathcal{B}$*
> $h_1, \ldots, h_Q \xleftarrow{\$} H$
> $(J, \sigma) \xleftarrow{\$} \mathcal{B}(x, h_1, \ldots, h_Q; \rho)$
> *If $J = 0$ then return $(0, \bot, \bot)$*
> $h'_1, \ldots, h'_Q \xleftarrow{\$} H$
> $(J', \sigma') \xleftarrow{\$} \mathcal{B}(x, h_1, \ldots, h_{J-1}, h'_J, \ldots, h'_Q, ; \rho)$
> *If $(J = J'$ and $h_J \neq h'_J)$ then return $(1, \sigma, \sigma')$*
> *Else return $(0, \bot, \bot)$.*

Let $frk = \Pr[b = 1 : x \xleftarrow{\$} IG; (b, \sigma, \sigma') \xleftarrow{\$} F_{\mathcal{B}}(x)]$. Then $frk \geq acc_{\mathcal{B}}(\frac{acc_{\mathcal{B}}}{Q} - \frac{1}{|H|})$.

Theorem 1. *Under the Strong-DH Assumption, if we model H_1 and H_2 as random oracles, then protocol IB-KA is a secure identity-based key agreement protocol.*

Proof. For sake of contradiction let us suppose there exists a PPT adversary $\mathcal{A}$ that has non-negligible advantage ϵ into breaking the protocol IB-KA , then we show how to build a solver algorithm S for the CDH problem.

In our reduction we will proceed into two steps. First, we describe an intermediate algorithm $\mathcal{B}$ (i.e. the simulator) that interacts with the IB-KA adversary $\mathcal{A}$ and returns a side output σ. Second, we will show how to build an algorithm S that exploits $F_\mathcal{B}$, the forking algorithm associated with $\mathcal{B}$, to solve the CDH problem under the Strong-DH Assumption.

$\mathcal{B}$ receives in input a tuple $(\mathbb{G}, g, U, V)$, where $U = g^u, V = g^v$ and u, v are random exponents in $\mathbb{Z}_q$, and a set of random elements $h_1, \ldots, h_Q \in \mathbb{Z}_q$. The simulator is also given access to a DH oracle $\mathsf{DH}(U, \cdot, \cdot)$ that on input $(\hat{V}, \hat{W})$ answers "yes" if $(U, \hat{V}, \hat{W})$ is a valid DDH tuple . The side output of $\mathcal{B}$ is $\sigma \in \mathbb{G} \times \mathbb{Z}_q$ or $\perp$. Let n be an upper bound to the number of sessions of the protocol run by the adversary $\mathcal{A}$ and Q_1 and Q_2 be the number of queries made by $\mathcal{A}$ to the random oracles H_1, H_2 respectively. Moreover, let Q_c be the number of users corrupted by $\mathcal{A}$ and $Q = Q_1 + Q_c + 1$.

Algorithm $\mathcal{B}^{\mathsf{DH}(U, \cdot, \cdot)}(\mathbb{G}, g, U, V, h_1, \ldots, h_Q)$
  Initialize $ctr \leftarrow 0; bad \leftarrow false$; empty tables $\overline{H_1}, \overline{H_2}$;
  Run $\mathcal{A}$ on input $(\mathbb{G}, g, y = U)$ as the public parameters of the protocol and simulates the protocol's environment for $\mathcal{A}$ as follows:
    Guess the test session by choosing at random the user (let us call him Bob) and the order number of the test session. If n is an upper bound to the number of all the sessions initiated by $\mathcal{A}$ then the guess is right with probability at least $1/n$.
  H_2 **queries** On input a pair (z_1, z_2):
    If $\overline{H_2}[z_1, z_2] = \perp$: choose a random string $Z \in \{0, 1\}^\ell$ and store $\overline{H_2}[z_1, z_2] = Z$
    Return $\overline{H_2}[z_1, z_2]$ to $\mathcal{A}$
  H_1 **queries** On input (ID, r):
    If $\overline{H_1}[ID, r] = \perp$, then $ctr \leftarrow ctr + 1; \overline{H_1}[ID, r] = h_{ctr}$
    Return $\overline{H_1}[ID, r]$ to $\mathcal{A}$
  Party Corruption. When $\mathcal{A}$ asks to corrupt party $ID \neq B$, then:
    $ctr \leftarrow ctr + 1; s \xleftarrow{\$} \mathbb{Z}_q; r = g^s y^{-h_{ctr}}$
    If $\overline{H_1}[ID, r] \neq \perp$ then $bad \leftarrow true$
    Store $\overline{H_1}[ID, r] = h_{ctr}$ and return (r, s) as ID's private key.
    For the case of Bob, the simulator simply chooses the r_B component of Bob's private key by picking a random $k_B \xleftarrow{\$} \mathbb{Z}_q$ and setting $r_B = g^{k_B}$. We observe that in this case $\mathcal{B}$ is not able to compute the corresponding s_B. However, since Bob is the user guessed for the test session, we can assume that the adversary will not ask for his secret key.
  Simulating sessions. First we describe how to simulate sessions different from the test session. Here the main point is that the adversary is allowed to ask session-key queries and thus the simulator must be able to produce the correct session key for each of these sessions. The simulator has full information about all the users' secret keys except Bob. Therefore $\mathcal{B}$ can easily simulate all the protocol sessions that do not include Bob, and answer any of the attacker's queries about these sessions. Hence we concentrate on describing how $\mathcal{B}$ simulates interactions with Bob.

Assume that Bob has a session with Charlie (whose identity is the string C). If Charlie is an uncorrupted party this means that $\mathcal{B}$ will generate the messages on behalf of him. In this case $\mathcal{B}$ knows Charlie's secret key and also has chosen his ephemeral exponent t_C. Thus it is trivial to see that $\mathcal{B}$ has enough information to compute the correct session key. The case when the adversary presents a message $\langle C, r_C, u_C \rangle$ to Bob as coming from Charlie is more complicated. Here is where $\mathcal{B}$ makes use of the oracle $\mathsf{DH}(y, \cdot, \cdot)$ to answer a session-key query about this session. The simulator replies with a message $\langle B, r_B, u_B = g^{t_B} \rangle$ where t_B is chosen by $\mathcal{B}$. Recall that the session key is $H_2(z_1, z_2)$ with $z_1 = g^{(s_C + t_C)(s_B + t_B)}$ and $z_2 = u_C^{t_B}$. So z_1 is the Diffie-Hellman result of the values $u_C g^{s_C}$ and $u_B g^{s_B}$, where $g^{s_C} = r_C y^{H_1(C, r_C)}$ and $g^{s_B} = r_B y^{H_1(B, r_B)}$ can be computed by the simulator. Notice also that the simulator knows t_B and k_B (the discrete log of r_B in base g). Therefore it checks if $\overline{H_2}[z_1, z_2] = Z$ where $z_2 = u_C^{t_B}$ and $\mathsf{DH}(y, u_C g^{s_C}, \overline{z_1}) = $ "yes" where $\overline{z_1} = \frac{z_1}{(u_C g^{s_C})(k_B + t_B) H_1(B, r_B) - 1}$. If $\mathcal{B}$ finds a match then it outputs the corresponding Z as session key for Bob. Otherwise it generates a random $\zeta \xleftarrow{\$} \{0,1\}^\ell$ and gives it as response to the adversary. Later, for each query (z_1, z_2) to H_2, if (z_1, z_2) satisfies the equation above it sets $\overline{H_2}[z_1, z_2] = \zeta$ and answers with ζ. This makes oracle's answers consistent.

In addition observe that the simulator can easily answer to state reveal queries as it chooses the fresh exponents on its own.

Simulating the test session. Let $\langle B, \rho_B, u_B = g^{t_B} \rangle$ be the message from Bob to Alice sent in the test session. We notice that such message may be sent by the adversary who is trying to impersonate Bob. In this case $\mathcal{A}$ may use a value $\rho_B = g^{\lambda_B}$ of its choice as the public component of Bob's private key (i.e. different than $r_B = g^{k_B}$ which $\mathcal{B}$ simulated and for which it knows k_B). $\mathcal{B}$ responds with the message $\langle A, r_A, u_A = V \rangle$ as coming from Alice. Finally $\mathcal{B}$ provides $\mathcal{A}$ with a random session key.

Run until $\mathcal{A}$ halts and outputs its decision bit

If $\overline{H_1}[B, \rho_B] = \bot$ then set $ctr \leftarrow ctr + 1$ and $\overline{H_1}[B, \rho_B] = h_{ctr}$

If $bad = true$ then return $(0, \bot)$

Let $i \in \{1, \ldots, Q\}$ such that $H_1(B, \rho_B) = h_i$

Let $Z = H_2(z_1, z_2)$ be the correct session key for the test session where
$z_1 = (u_A r_A y^{H_1(A, r_A)})^{(t_B + \lambda_B + x h_i)}$ and $z_2 = u_A^{t_B}$.

If $\mathcal{A}$ has success into distinguishing Z from a random value it must necessarily query the correct pair (z_1, z_2) to the random oracle H_2. This means that $\mathcal{B}$ can efficiently find the pair (z_1, z_2) in the table $\overline{H_2}$ using the Strong-DH oracle.

Compute $\tau = \frac{z_1}{z_2 (u_B \rho_B y^{h_i})^{s_A}} = \rho_B^v W^{h_i}$

Return $(i, (\tau, h_i))$

Let IG be the algorithm that generates a random Diffie-Hellman tuple $(\mathbb{G}, g, U, V)$ and acc_B be the accepting probability of B.[2] Then we have that:

$$acc_B \geq \frac{\epsilon}{n} - \Pr[bad = true].$$

The probability that $bad = true$ is the probability that the adversary has guessed the "right" r for a corrupted party ID before corrupting it, in one of the H_1 oracle queries beforehand. Since r is uniformly distributed the probability of guessing it is $1/q$, and since the adversary makes at most Q queries to H_1 and corrupts at most Q_c parties (and $q > 2^\ell$) we have that

$$acc_B \geq \frac{\epsilon}{n} - \frac{Q_c(Q)}{2^\ell}.$$

which is still non-negligible, since ϵ is non-negligible.

Once we have described the algorithm B we can now show how to build a solver algorithm S that can exploit F_B, the forking algorithm associated with the above B.

The algorithm S plays the role of a CDH solver under the Strong-DH Assumption. It receives in input a CDH tuple $(\mathbb{G}, g, U, V)$ where $U = g^u, V = g^v$ and u, v are random exponents in $\mathbb{Z}_q$. S is also given access to a decision oracle $\mathsf{DH}(U, \cdot, \cdot)$ that on input $(\hat{V}, \hat{W})$ answers "yes" if $(U, \hat{V}, \hat{W})$ is a valid DH tuple.

Algorithm $S^{\mathsf{DH}(U,\cdot,\cdot)}(\mathbb{G}, g, U, V)$

$(b, \tau, \tau') \xleftarrow{\$} F_B^{\mathsf{DH}(U,\cdot,\cdot)}(\mathbb{G}, g, U, V)$
If $b = 0$ then return 0 and halt
Parse σ as (τ, h) and σ' as (τ', h')
Return $W = (\tau/\tau')^{(h-h')^{-1}}$

If the forking algorithm F_B has success, this means that there exist random coins ρ, an index $J \geq 1$ and $h_1, \ldots, h_Q, h'_J, \ldots, h'_Q \in \mathbb{Z}_q$ with $h = h_J \neq h'_J = h'$ such that: the first execution of $B(\mathbb{G}, g, U, V, h_1, \ldots, h_Q; \rho)$ outputs $\tau = \rho_B^v W^h$ where $\overline{H_1}[B, \rho_B] = h$; the second execution of $B(\mathbb{G}, g, U, V, h_1, \ldots, h_{J-1}, h'_J, \ldots, h'_Q; \rho)$ outputs $\tau' = (\rho'_{B'})^v W^{h'}$ where $\overline{H_1}[B', \rho'_{B'}] = h'$. Since the two executions of B are the same until the response to the J-th query to H_1, then we must have $B = B'$ and $\rho_B = \rho'_{B'}$. Thus it is easy to see that S achieves its goal by computing $W = (\tau/\tau')^{\frac{1}{h-h'}} = g^{uv}$.

Finally, by the General Forking Lemma, we have that if A has non-negligible advantage into breaking the security of IB-KA , then S's success probability is also non-negligible.

4.1   Additional Security Properties of IB-KA

In addition to the notion of session key security, any key-agreement protocol should satisfy other important properties. Below we describe the additional security properties enjoyed by IB-KA .

---

[2] We say that B accepts if it outputs (J, σ) such that $J \geq 1$.

Forward secrecy. We say that a KA protocol has forward secrecy, if after a session is completed and its session key erased, the adversary cannot learn it *even if* it corrupts the parties involved in that session. In other words, learning the private keys of parties should not jeopardize the security of past completed sessions.

A relaxed notion of forward secrecy (which we call *weak*) assumes that only past sessions in which the adversary was passive (i.e. did not choose the messages) are not jeopardized.

The following theorem (whose proof is deferred to the full version of the paper) shows that the protocol IB-KA satisfies this notion of *weak forward secrecy*.

Theorem 2. *Let $\mathcal{A}$ be a PPT adversary that is able to break the weak forward secrecy of the IB-KA protocol with advantage ϵ. Let n be the an upper bound to the number of sessions of the protocol run by $\mathcal{A}$ and Q_1 and Q_2 be the number of queries made by the adversary to the random oracles H_1, H_2 respectively. Then we can solve the CDH problem with probability at least $\epsilon/(nQ_2)$.*

Resistance to reflection attacks. A *reflection attack* occurs when an adversary can compromise a session in which the two parties have the same identity (and the same private key). Though, at first glance, this seems to be only of theoretical interest, there are real-life situations in which this scenario occurs. For example consider the case when Alice is at her office and wants to establish a secure connection with her PC at home, therefore running a session between two computers with the same identity and private key.

The current proof actually does not work when the adversary sends a message with the same value r_B provided by the KGC (for which the simulator knows the discrete logarithm k_B, but cannot compute the corresponding s_B). The issue is that the knowledge of s_B would be needed to extract the solution of the CDH problem. We point out that a reflection attack using a value $\rho_B \neq r_B$ is captured by the current proof. Moreover it is reasonable to assume that a honest party refuses connections from itself that use a "wrong" key.

However it is possible to adapt the proof in this specific case. In particular we can show that a successful run of the adversary enables the simulator to compute g^{u^2} instead of g^{uv}. As showed in [23] by Maurer and Wolf, such an algorithm can be easily turned into a solver for CDH. For lack of space this is deferred to the full version of the paper.

Resistance to Key Compromise Impersonation. Suppose that the adversary learns Alice's private key. Then, it is trivial to see that this knowledge enables the adversary to impersonate Alice to other parties. A *key compromise impersonation* (KCI) attack can be carried out when the knowledge of Alice's private key allows the adversary to impersonate another party to Alice.

To see that the protocol IB-KA is resistant to KCI attacks it suffices to observe that in the proof of security, when the adversary tries to impersonate Bob to Alice, we are able to output Alice's private key whenever it is asked by the adversary. It means that the proof continues to be valid even in this case.

Ephemeral Key Compromise Impersonation. A recent paper by Cheng and Ma [12] shows that our protocol is susceptible to an ephemeral key compromise attack. Roughly speaking this attack considers the case when the adversary can make state-reveal queries (in order to learn the ephemeral key of a user) even in the test session. Though the paper is correct, we point out that this kind of attack is not part of the standard Canetti-Krawczyk security model that is considered in this paper.

5   Comparisons with Other IB-KA Protocols

In this section we compare IB-KA with other id-based KA protocols from the literature. In particular, we consider the protocol by Chen and Kudla [11] (SCK-2) (which is a modification of Smart's [31]) and two protocols proposed very recently by Boyd et al. [6] (BCNP1, BCNP2).

For our efficiency comparisons we consider a security parameter of 128 and implementations of SCK-2, BCNP1 and BCNP2 with Type 3 pairings[3], which are the most efficient pairings for this kind of security level (higher than 80). Our protocol is assumed to be implemented in an elliptic curves group $\mathbb{G}$ with the same security parameter. In this scenario elements of $\mathbb{G}$ and $\mathbb{G}_1$ need 256 bit to be represented, while 512 bits are needed for $\mathbb{G}_2$ elements and 3072 bits for an element of $\mathbb{G}_T$. We estimate the computational cost of all the protocols using the costs per operation for Type 3 pairings given by Chen et al. in [10]. The bandwidth cost is expressed as the amount of data in bits sent by each party to complete a session of the protocol[4]. According to the work of Chen et al. [10] SCK-2 is the most efficient protocol with a proof of security in the CK model for all types of pairings. It is proved secure using random oracles under the Bilinear Diffie-Hellman Assumption and requires one round of communication with only one group element sent by each party. To be precise, we point out that the protocol of Boyd et al. (BMP) [7] would appear computationally more efficient than SCK-2, but unfortunately it works only in type 1 and type 4 pairings and is proven secure only in symmetric pairings. BCNP1 and BCNP2 are generic constructions based on any CCA-secure IB-KEM. When implemented (as suggested by the authors of [6]) using one of the IB-KEMs by Kiltz [19], Kiltz-Galindo [20] or Gentry [17] they lead to a two-pass single-round protocol with (CK) security in the standard model. BCNP2 provides weak FS and resistance to KCI attacks, while BCNP1 satisfies only the former property.

The results are summarized in Table 1 assuming protocols BCNP1 and BCNP2 to be implemented with Kiltz's IB-KEM (the most efficient for this application according to the work of Boyd et al. [6]). We defer to the original papers of SCK-2 [11] and BCNP1, BCNP2 [6] for more details about these costs. As described in the table, our protocol has a reasonable bandwidth requirement and achieves the best computational efficiency among the other id-based KA protocols.

---

[3] This classification of pairing groups into several types is provided by Galbraith et al. in [15].

[4] We do not consider the identity string sent with the messages as it can be implicit and, in any way, appears in all the protocols.

Table 1. Comparisons between IB-KA protocols

	weak FS	KCI	Standard model	Efficiency	
				Bandwidth	Cost per party
BCNP1	✗	✓	✓	768	56
BCNP2	✓	✓	✓	1024	59
SCK-2	✓	✓	✗	256	43
IB-KA	✓	✓	✗	512	6

COMPARISON WITH PKI-BASED PROTOCOLS. We also compare our protocol to MQV [22], and its provably secure version HMQV [21], which is the most efficient protocol in the public-key setting. When comparing our protocol to a PKI-based scheme, like MQV, we must also consider the additional cost of sending and verifying certificates.

We measure the computation costs of the protocols in terms of the number of exponentiations in the underlying group needed to compute the session key. If the exponentiations is done with an exponent that is half the length of the group size, then obviously we count it as $1/2$ exponentiation. Also if an exponentiation is done over a fixed basis, we apply precomputation schemes to speed up the computation, e.g. [16].

Our protocol requires each party to send a single message consisting of two group elements. To compute the session key, the parties perform 2 full exponentiations over variable basis, and one half exponentiation over a fixed basis[5]. For our security parameter, following [16], the latter half exponentiation can be computed with less than 20 group multiplications, with a precomputation table of moderate size.

In MQV, each party sends a single message consisiting of one group element, and performs 1.5 exponentiations to compute the session key. Moreover, in HMQV certificates are sent and verified. Here we distinguish two cases: the certificate is based either on an RSA signature, or on a discrete-log signature, e.g. Schnorr's.

In the RSA case, a short exponent e.g. $e = 2^{16} + 1$, is typically used, and the verification cost is basically equivalent to the cost of the half exponentiation with precomputation in our protocol above. Therefore in this case, MQV is faster, but by a mere half exponentiation. The price to pay however is a massive increase in bandwidth to send the RSA signature (i.e. 3072 bits), and the introduction of the RSA Assumption in order to prove security of the entire scheme. If we use a Schnorr signature for the certificate, then MQV require sending two more group elements, and therefore its bandwidth requirement is already worse than our protocol (by one group element). The parties then must compute one full and one half exponentiation, both with fixed basis[6] to verify the certificate. This extra computational cost can be compared to an additional half exponentiation, making the computation requirement of MQV with Schnorr certificates equivalent to that of our protocol.

---

[5] Indeed since the input to the hash function H_1 is randomized, we can set its output length to be half of the length of the group size.

[6] Though different basis, which means that in order to apply precomputation techniques, the parties need to maintain two tables.

In conclusion, when comparing our protocol with MQV with certificates we find that our protocol: (i) has comparable computational cost; (ii) has better bandiwdth (by far in the case of RSA certificates) and (iii) simplifies protocol implementation by removing entirely the need to manage certificates and to interact with a PKI[7].

References

1. Abdalla, M., Bellare, M., Rogaway, P.: The oracle Diffie-Hellman assumptions and an analysis of DHIES. In: Naccache, D. (ed.) CT-RSA 2001. LNCS, vol. 2020, pp. 143–158. Springer, Heidelberg (2001)
2. Bellare, M., Neven, G.: New Multi-Signature Schemes and a General Forking Lemma. In: Proceedings of the 13th Conference on Computer and Communications Security – ACM CCS 2006. ACM Press, New York (2006)
3. Boneh, D., Boyen, X.: Short Signatures without Random Oracles. In: Cachin, C., Camenisch, J.L. (eds.) EUROCRYPT 2004. LNCS, vol. 3027, pp. 56–73. Springer, Heidelberg (2004)
4. Boneh, D., Franklin, M.K.: Identity-Based Encryption from the Weil Pairing. SIAM J. Comput. 32(3), 586–615 (2003); In: Kilian, J. (ed.) CRYPTO 2001. LNCS, vol. 2139, pp. 213–615. Springer, Heidelberg (2001)
5. Boyd, C., Choo, K.-K.R.: Security of two-party identity-based key agreement. In: Dawson, E., Vaudenay, S. (eds.) Mycrypt 2005. LNCS, vol. 3715, pp. 229–243. Springer, Heidelberg (2005)
6. Boyd, C., Cliff, Y., Nieto, J.G., Paterson, K.G.: Efficient One-Round Key Exchange in the Standard Model. In: Mu, Y., Susilo, W., Seberry, J. (eds.) ACISP 2008. LNCS, vol. 5107, pp. 69–83. Springer, Heidelberg (2008)
7. Boyd, C., Mao, W., Paterson, K.G.: Key Agreement Using Statically Keyed Authenticators. In: Jakobsson, M., Yung, M., Zhou, J. (eds.) ACNS 2004. LNCS, vol. 3089, pp. 248–262. Springer, Heidelberg (2004)
8. Canetti, R., Krawczyk, H.: Analysis of Key-Exchange Protocols and Their Use for Building Secure Channels. In: Pfitzmann, B. (ed.) EUROCRYPT 2001. LNCS, vol. 2045, pp. 453–474. Springer, Heidelberg (2001)
9. Canetti, R., Krawczyk, H.: Universally Composable Notions of Key Exchange and Secure Channels. In: Knudsen, L.R. (ed.) EUROCRYPT 2002. LNCS, vol. 2332, pp. 337–351. Springer, Heidelberg (2002)
10. Chen, L., Cheng, Z., Nigel, P.: Smart. Identity-based key agreement protocols from pairings. Int. J. Inf. Sec. 6(4), 213–241 (2007)
11. Chen, L., Kudla, C.: Identity Based Authenticated Key Agreement Protocols from Pairings. In: 16th IEEE Computer Security Foundations Workshop - CSFW 2003, pp. 219–233. IEEE Computer Society Press, Los Alamitos (2003)
12. Cheng, Q., Ma, C.: Ephemeral Key Compromise Attack on the IB-KA protocol. Cryptology Eprint Archive, Report 2009/568 (2009), http://eprint.iacr.org/2009/568
13. Diffie, W., Hellman, M.: New Directions in Cryptography. IEEE Transactions on Information Theory 22(6), 644–654 (1976)

---

[7] In the above, we did not account for the cost of verifying group membership for the elements sent by the parties, which is necessary both in the case of MQV and our protocol, and is the same in both protocols.

14. Fiat, A., Shamir, A.: How to Prove Yourself: Practical Solutions of Identification and Signature Problems. In: Odlyzko, A.M. (ed.) CRYPTO 1986. LNCS, vol. 263, pp. 186–194. Springer, Heidelberg (1987)
15. Galbraith, S.D., Paterson, K.G., Smart, N.P.: Pairings for Cryptographers. Cryptology ePrint Archive, Report 2006/165 (2006), http://eprint.iacr.org
16. Lim, C.H., Lee, P.J.: More Flexible Exponentiation with Precomputation. In: Desmedt, Y.G. (ed.) CRYPTO 1994. LNCS, vol. 839, pp. 95–107. Springer, Heidelberg (1994)
17. Gentry, C.: Practical identity-based encryption without random oracles. In: Vaudenay, S. (ed.) EUROCRYPT 2006. LNCS, vol. 4004, pp. 445–464. Springer, Heidelberg (2006)
18. Gunther, C.G.: An Identity-Based Key-Exchange Protocol. In: Quisquater, J.-J., Vandewalle, J. (eds.) EUROCRYPT 1989. LNCS, vol. 434, pp. 29–37. Springer, Heidelberg (1990)
19. Kiltz, E.: Direct Chosen-Ciphertext Secure Identity-Based Encryption in the Standard Model with short Ciphertexts. Cryptology Eprint Archive, Report 2006/122 (2006), http://eprint.iacr.org/2006/122
20. Kiltz, E., Galindo, D.: Direct Chosen-Ciphertext Secure Identity-Based Key Encapsulation Without Random Oracles. Cryptology Eprint Archive, Report 2006/034 (2006), http://eprint.iacr.org/2006/034
21. Krawczyk, H.: HMQV: A High-Performance Secure Diffie-Hellman Protocol. In: Shoup, V. (ed.) CRYPTO 2005. LNCS, vol. 3621, pp. 546–566. Springer, Heidelberg (2005)
22. Law, L., Menezes, A., Qu, M., Solinas, J., Vanstone, S.: An efficient Protocol for Authenticated Key Agreement. Designs, Codes and Cryptography 28, 119–134 (2003)
23. Maurer, U., Wolf, S.: Diffie-Hellman oracles. In: Koblitz, N. (ed.) CRYPTO 1996. LNCS, vol. 1109, pp. 268–282. Springer, Heidelberg (1996)
24. Okamoto, E.: Key Distribution Systems Based on Identification Information. In: Pomerance, C. (ed.) CRYPTO 1987. LNCS, vol. 293, pp. 194–202. Springer, Heidelberg (1988)
25. Okamoto, E., Tanaka, K.: Key Distribution System Based on Identification. Information. IEEE Journal on Selected Areas in Communications 7(4), 481–485 (1989)
26. Pointcheval, D., Stern, J.: Security Arguments for Digital Signatures and Blind Signatures. Journal of Cryptology 13(3), 361–396 (2000)
27. Saeednia, S.: Improvement of Gunther's identity-based key exchange protocol. Electonics Letters 31(18), 1535–1536 (2000)
28. Sakai, R., Ohgishi, K., Kasahara, M.: Cryptosystems based on pairing. In: Symposium on Cryptography and Information Security, Okinawa, Japan (2000)
29. Shamir, A.: Identity-Based Cryptosystems and Signature Schemes. In: Blakely, G.R., Chaum, D. (eds.) CRYPTO 1984. LNCS, vol. 196, pp. 47–53. Springer, Heidelberg (1985)
30. Schnorr, C.P.: Efficient identification and signatures for smart cards. In: Brassard, G. (ed.) CRYPTO 1989. LNCS, vol. 435, pp. 239–252. Springer, Heidelberg (1990)
31. Smart, N.P.: An identity-based authenticated key-agreement protocol based on the Weil pairing. Electronics letters 38, 630–632 (2002)
32. Smetters, D.K., Durfee, G.: Domain-based Administration of Identity-Based Cryptosystems for Secure E-Mail and IPSEC. In: SSYM 2003: Proceedings of the 12th Conference on USENIX Security Symposium, p. 15. USENIX Association (2003)
33. Wang, Y.: Efficient Identity-Based and Authenticated Key Agreement Protocol. Cryptology ePrint Archive, Report 2005/108 (2005), http://eprint.iacr.org/2005/108/

On Extended Sanitizable Signature Schemes*

Sébastien Canard[1] and Amandine Jambert[1,2]

[1] Orange Labs, 42 rue des Coutures, BP6243, 14066 Caen Cedex, France
[2] IMB, Université Bordeaux 1, 351 cours de la Libération, 33405 Talence, France

Abstract. Sanitizable signature schemes allow a semi-trusted entity to modify some specific portions of a signed message while keeping a valid signature of the original off-line signer. In this paper, we give a new secure sanitizable signature scheme which is, to the best of our knowledge, the most efficient construction with such a high level of security. We also enhance the Brzuska *et al.* model on sanitizable signature schemes by adding new features. We thus model the way to limit the set of possible modifications on a single block, the way to force the same modifications on different admissible blocks, and the way to limit both the number of modifications of admissible blocks and the number of versions of a signed message. We finally present two cryptanalysis on proposals for two of these features due to Klonowski and Lauks at ICISC 2006 and propose some new practical constructions for two of them.

Keywords: Sanitizable signature, chameleon hash, accumulator scheme.

1 Introduction

Since the appearance of public key cryptography, signature schemes have been one of the most widely studied cryptographic tool. Among some others, one of the main security properties a signature scheme should verify is the integrity of the message. However, in some cases, such as medical applications, secure routing or content protection [1,5], it may be necessary for a designated semi-trusted entity to delete or modify some parts of the signed message.

In this paper, we focus on sanitizable signature schemes (introduced in [1] and later formalized in [2]) which permits a signer to produce a signature on a document, which can be further modified, in a limited and controlled fashion, by a designated semi-trusted "sanitizer", with no interaction with the original signer. Moreover, the signature on the resulting message should be verifiable as a signature from the original signer. On the other hand, the sanitizer should be able to modify only the sanitizable parts of the message, that is, the parts that have been stated as modifiable/admissible by the signer.

1.1 Related Work

The first sanitizable signature scheme [1] makes use of chameleon hash functions [8] and this is also the case for most of existing ones. The first problem with

---

* This work has been financially supported by the French Agence Nationale de la Recherche and the TES Cluster under the PACE project.

J. Pieprzyk (Ed.): CT-RSA 2010, LNCS 5985, pp. 179–194, 2010.

this construction is the possibility to obtain some new sanitized messages from two different ones, without the secret key of the sanitizer (i.e. it is forgeable). Moreover, a judge is unable to decide whether a signature has been sanitized or not. Thus, according to [2], the scheme is not accountable.

Canard et al. [5] fix both problems, in the context of trapdoor sanitizable signature, by adding an extra modifiable block corresponding to the whole message. The original message is used as a unique identifier to obtain accountability. However, as the original message is obviously recognizable from a sanitized one, this scheme is not transparent [2].

In [2], Brzuska et al. propose the first secure scheme (i.e. immutable, transparent and accountable). They propose to add a tag (verifiably and pseudo-randomly generated by the signer and randomly by the sanitizer) to each modifiable block. Thus, the signer can prove which one she constructed. Unforgeability (and thus accountability) is reached thanks to the computation of a new tag per message. This implies to compute a collision for each modifiable block, even if this block has not been modified. Thus, this solution lacks of efficiency.

Yuen et al.[10] also give a solution in the standard model but without accountability.

At ICISC'06, Klonowski and Lauks propose [7] several extensions of the sanitization signature paradigm: force the sanitizer to construct less than l versions of a message, modify at maximum k sanitizable blocks or limit the values available for some blocks. However neither security model nor proofs are given.

1.2 Our Contribution

In this paper, we provide several contributions to sanitizable signature schemes. We first extend in Section 4 the Brzuska et al. [2] model (see Section 2) by taking into account the way to (i) limit block modifications in a set, (ii) secretly force the same modifications on different admissible blocks (which can be different at the beginning), (iii) limit the number of admissible blocks modifiable and (iv) limit the number of versions of a signed message. We also give in Section 3 a new sanitizable signature scheme without additional features which is, to the best of our knowledge, the most efficient and secure construction. After that, we show in Section 5 a cryptanalysis on two proposals for additional features (ii) and (iii) due to Klonowski and Lauks [7]. Finally, we present in Section 6 practical constructions for the extensions (iii) and (iv) and show how the idea from Klonowski and Lauks for (i) can be made secure.

2   Initial Model for Sanitizable Signatures

In the following, the size of the message (in bits) is denoted ℓ and a message is divided (by the signer) into t blocks. The variable ADM includes, for each block m_i, $i \in [1,t]$, the length ℓ_i of the corresponding i-th block (thus $\ell = \sum_{i=1}^{t} \ell_i$) and a subset of $[1,t]$ corresponding to the ranks of the blocks modifiables by the sanitizer (i.e. admissible). The variable MOD is a set of elements of the form (i, m'_i). A value $(i, \mathsf{m}'_i) \in \mathsf{MOD}$ if and only if the i-th block is modified into m'_i during the sanitization. By misuse of notation, we denote $i \in \mathsf{MOD}$ if $\exists \mathsf{m}'_i / (i, \mathsf{m}'_i) \in \mathsf{MOD}$. We say that MOD *matches* ADM if $\forall i \in \mathsf{MOD}, i \in \mathsf{ADM}$.

2.1 Procedures and Correctness

A *sanitizable signature scheme* $\mathcal{SS}$ is composed of the following algorithms (each of them may output an error $\perp$), where λ is a security parameter.

- SETUP takes as input 1^λ and outputs the parameters param of the system. In the following, we consider that λ is included into param.
- SIGKEYGEN (resp. SANKEYGEN) on input param outputs the key pair $(\mathsf{pk}_{sig}, \mathsf{sk}_{sig})$ for the signer (resp $(\mathsf{pk}_{san}, \mathsf{sk}_{san})$ for the sanitizer).
- SIGN takes as input a message m of length ℓ divided into t blocks, the secret key sk_{sig}, the public key pk_{san} and the variable ADM. It outputs a sanitizable signature σ on the message m. In the following, ADM is included into σ.
- SANITIZE takes as input a message m, a sanitizable signature σ, the public key pk_{sig}, the secret key sk_{san} and the modifications MOD that the sanitizer wants to do on m. It outputs a new signature σ' and message m'.
- VERIFY permits to verify a signature σ on a message m with the public keys pk_{sig} and pk_{san}. It outputs true if the signature is correct and false otherwise.
- PROOF takes as input a signature σ on a given message m, the secret key sk_{sig}, the public key pk_{san} and the set of message-signature pairs she has produced $(\mathsf{m}_i, \sigma_i)_{i=1,2,\cdots,q}$. It outputs a proof π.
- JUDGE is a public algorithm which aims at deciding who has produced a given signature. It takes as input (m, σ), a proof π from PROOF and the public keys pk_{sig} and pk_{san} and outputs signer or sanitizer.

First, a sanitizable signature scheme needs to verify some correctness properties:

- **Signing correctness** says that a signature from SIGN with the secret key from SIGKEYGEN is accepted with an overwhelming probability by VERIFY.
- **Sanitizing correctness** says that a signature from SANITIZE from a valid signature with the secret key from SANKEYGEN is accepted with an overwhelming probability by VERIFY.
- **Proof correctness** says that for any sanitized message, the signer is able to output a proof π, using PROOF, such that JUDGE outputs sanitizer.

2.2 Security Requirements

According to Brzuska *et al.*, a sanitizable signature scheme is secure if it verifies the following security properties. Formal experiments are given in the table below.

- **Immutability.** It is not possible for the sanitizer to modify non admissible blocks of a signed message. In the corresponding experiment, the adversary impersonates the sanitizer.

- **Transparency.** Only the signer and the sanitizer are able to distinguish an original signature from a sanitized one. During this experiment, the adversary is given access to a SIGN/SANIT oracle which on input a bit b outputs either a sanitized signature if $b = 0$ (output by SANITIZE) or a signed message if $b = 1$ (output by SIGN).
- **Accountability.** In case of an argument about the origin of a signature and a message, the judge is able to correctly settle it.

Immutability: $Succ_{imm}^{SS} = Pr[1 \longleftarrow \text{EXP}_{imm}^{SS}]$ where EXP_{imm}^{SS} is as follows
- $(\mathsf{pk}_{sig}, \mathsf{sk}_{sig}) \longleftarrow \text{SIGKEYGEN}(1^\lambda)$
- $(\mathsf{pk}_{san}^*, m^*, \sigma^*) \longleftarrow \mathcal{A}^{\text{SIGN}(\cdot, \mathsf{sk}_{sig}, \cdot, \cdot), \text{PROOF}(\mathsf{sk}_{sig}, \cdot, \cdot, \cdot)}(\mathsf{pk}_{sig})$
- Let $(m_i, \mathsf{ADM}_i, \mathsf{pk}_{san,i})$ and σ_i for $i \in [1, q]$ be the queries related to oracle SIGN
- return 1 if $\text{VERIFY}(m^*, \sigma^*, \mathsf{pk}_{sig}, \mathsf{pk}_{san}^*) = \mathsf{true}$ and for all $i = 1, 2, \cdots, q$ we have
    - $\mathsf{pk}_{san}^* \neq \mathsf{pk}_{san,i}$ or
    - $\exists j_i \notin \mathsf{ADM}_i$ such that $m^*[j_i] \neq m_i[j_i]$

Transparency: $Adv_{trans}^{SS} = Pr[1 \longleftarrow \text{EXP}_{trans}^{SS}] - 1/2$ where EXP_{trans}^{SS} is as follows
- $(\mathsf{pk}_{sig}, \mathsf{sk}_{sig}) \longleftarrow \text{SIGKEYGEN}(1^\lambda)$
- $(\mathsf{pk}_{san}, \mathsf{sk}_{san}) \longleftarrow \text{SANKEYGEN}(1^\lambda)$
- $b \longleftarrow \{0, 1\}$
- $b' \longleftarrow \mathcal{A}^{\text{SIGN}(\cdot, \mathsf{sk}_{sig}, \cdot, \cdot), \text{SANIT}(\cdot, \cdot, \cdot, \cdot, \mathsf{sk}_{san}), \text{PROOF}(\mathsf{sk}_{sig}, \cdot, \cdot, \cdot), \text{SIGN}/\text{SANIT}(\cdot, \cdot, \cdot, \mathsf{sk}_{sig}, \mathsf{sk}_{san}, b)}(\mathsf{pk}_{sig}, \mathsf{pk}_{san})$
- return 1 if $b' = b$

Sanitizer Accountability: $Succ_{san-acc}^{SS} = Pr[1 \longleftarrow \text{EXP}_{san-acc}^{SS}]$ where $\text{EXP}_{san-acc}^{SS}$ is as follows
- $(\mathsf{pk}_{sig}, \mathsf{sk}_{sig}) \longleftarrow \text{SIGKEYGEN}(1^\lambda)$
- $(\mathsf{pk}_{san}^*, m^*, \sigma^*) \longleftarrow \mathcal{A}^{\text{SIGN}(\cdot, \mathsf{sk}_{sig}, \cdot, \cdot), \text{PROOF}(\mathsf{sk}_{sig}, \cdot, \cdot, \cdot)}(\mathsf{pk}_{sig})$
- Let $(m_i, \mathsf{ADM}_i, \mathsf{pk}_{san,i})$ and σ_i for $i \in [1, q]$ be the queries related to oracle SIGN
- $\pi \longleftarrow \text{PROOF}(sk_{sig}, m^*, \sigma^*, m_1, \sigma_1, \cdots, m_q, \sigma_q, pk_{san}^*)$
- return 1 if $\text{VERIFY}(m^*, \sigma^*, \mathsf{pk}_{sig}, \mathsf{pk}_{san}^*) = \mathsf{true}$ and
    - $(\mathsf{pk}_{san}^*, m^*) \neq (\mathsf{pk}_{san,i}, m_i)$ for all $i = 1, \cdots, q$ and
    - $\text{JUDGE}(m^*, \sigma^*, pk_{sig}, pk_{san}^*, \pi) = \mathsf{signer}$

Signer Accountability: $Succ_{sig-acc}^{SS} = Pr[1 \longleftarrow \text{EXP}_{sig-acc}^{SS}]$ where $\text{EXP}_{sig-acc}^{SS}$ is as follows
- $(\mathsf{pk}_{san}, \mathsf{sk}_{san}) \longleftarrow \text{SANKEYGEN}(1^\lambda)$
- $(\mathsf{pk}_{sig}^*, \pi^*, m^*, \sigma^*) \longleftarrow \mathcal{A}^{\text{SANIT}(\cdot, \cdot, \cdot, \cdot, \mathsf{sk}_{san})}(\mathsf{pk}_{san})$
- Let (m_i', σ_i') for $i = 1, 2, \cdots, q$ be the answers from oracle SANIT.
- return 1 if $\text{VERIFY}(m^*, \sigma^*, \mathsf{pk}_{sig}^*, \mathsf{pk}_{san}) = \mathsf{true}$ and
    - $(\mathsf{pk}_{sig,i}^*, m^*) \neq (\mathsf{pk}_{sig,i}^*, m_i')$ for all $i = 1, \cdots, q$ and
    - $\text{JUDGE}(m^*, \sigma^*, pk_{sig}^*, pk_{san}, \pi^*) = \mathsf{sanitizer}$

2.3   Useful Tools

Signature Schemes. We need a *signature scheme* $\mathcal{S} = (\text{KEYGEN}, \text{SIGN}, \text{VERIFY})$ using a security parameter λ. The secret key is denoted ssk and the corresponding public verification key spk. The verification algorithm outputs true if the signature is correct and false if not. The used signature scheme needs to be existentially unforgeable against chosen message attacks (EU-CMA), that is $Succ_{EU-CMA}^{\mathcal{S}}$ is negligible in the security parameter [6].

Chameleon Hash Schemes. We will use a *chameleon hash scheme* $\mathcal{CH} = (\text{SETUP}, \text{PROCEED}, \text{FORGE})$ using a security parameter λ. SETUP permits the generation of the key pairs $(\mathsf{chpk}, \mathsf{chsk})$ on input 1^λ. PROCEED takes as input the chameleon hash public key chpk, a message m and a random r and outputs the hash value h of the message m. FORGE, on input the chameleon hash secret key chsk, the message m, the random r, the hash value h and a new message m', outputs a new random r' such that $\mathsf{h} = \text{PROCEED}(\mathsf{chpk}, \mathsf{m}', \mathsf{r}')$.

A chameleon hash function is said strong (resp. weak) secure if it is both uniform (the distribution of the output of FORGE are indistinguishable from a random [8,5]) and strong (resp. weak) collision resistant (it is impossible to find a collision (m', r') on $h = \text{PROCEED}(\text{chpk}, m, r)$, only having access to h, m, r, chpk and an oracle FORGE (resp. to h, m, r, chpk)). We note $Succ^{\mathcal{CH}}_{SCollRes}$ (resp. $Succ^{\mathcal{CH}}_{WCollRes}$) the success of an adversary against strong (resp. weak) collision and $Adv^{\mathcal{CH}}_{Uni}$ the advantage of an adversary against uniformity.

Pseudorandom Generators. We use in the following a pseudorandom generator PRG mapping λ-bits to 2λ-bits and a pseudorandom function PRF mapping λ-bits to λ-bits. We note $Adv^{PRG}_{Pseudorand}$ and $Adv^{PRF}_{Pseudorand}$ the advantages of an adversary against pseudo-randomness and Adv^{PRG}_{OneWay} the one-wayness of PRG.

Accumulator Schemes. An accumulator scheme [4,9,3] ACC permits to accumulate a large set of objects in a single short value, called the accumulator and denoted Acc. Such scheme provides evidence that a given object belongs to the accumulator by producing a witness w related to Acc and x by the relation Acc = ACC(x, w). We denote $x \in$ Acc, or $(x, w) \in$ Acc, if x is accumulated in Acc with the witness w. If someone reveals a value x together with a witness w, she proves that the value x is truly accumulated in the accumulator Acc iff ACC(x, w) = Acc. Such scheme is divided into several procedures including the parameter generation which initializes the parameters, the accumulation phase to accumulate values in a new accumulator and the witnesses computation phase. Existing constructions provide the main security property of an accumulator scheme, named the collision resistant one, which says [9] that this is infeasible for an adversary, on input an accumulator Acc, to output a value and a witness that this value is accumulated in Acc, while this is not the case. In the following, we say that ACC is secure if $Succ^{\text{ACC}}_{CR}$ is negligible in the security parameter.

3   A New Construction in the Initial Model

3.1   High Level Description

Both [5] and [2] are based on Ateniese *et al.* [1], which works as follows. Let $m = m_1 \| m_2 \| \cdots \| m_t$ be the message to sign and ID_m a random unique identifier.

- The SIGN procedure consists in executing, for each admissible block m_i, $\mathcal{CH}.\text{PROCEED}$, with a random r_i, to obtain h_i. Then, the signer computes a modified version $\tilde{m}_i$ of each block as either h_i if $i \in$ ADM or $m_i \| i$ otherwise. Finally, she executes the signature algorithm on the message $\tilde{m} = \tilde{m}_1 \| \tilde{m}_2 \| \cdots \| \tilde{m}_t$, as $s = \mathcal{S}.\text{SIGN}(\text{ssk}, ID_m \| t \| \text{pk}_{san} \| \tilde{m})$. The final sanitizable signature σ on the message m is $(s, \mathcal{R}, \text{ADM})$ where $\mathcal{R} = \{r_i : i \in \text{ADM}\}$.
- The SANITIZE step from a message $m = m_1 \| m_2 \| \cdots \| m_t$ to a message $m' = m'_1 \| m'_2 \| \cdots \| m'_t$ consists in using $\mathcal{CH}.\text{FORGE}$ to obtain the new r'_i for all $i \in$ ADM, so that the value h_i (and thus the signature s) is unchanged after the modification of the block message (m_i to m'_i).

This solution has two main problems. First, it is not accountable since a judge can not decide whether a signature has been sanitized or not. We propose to use a pseudorandom number TAG, instead of ID_m, which is linked to the version of the message and generated using both a PRG and a PRF [2]. Thus, the signer can prove that she has correctly computed the tag, while being transparent.

Second, it is forgeable since it is possible to obtain a new sanitization from two versions of the same message, without chsk. Let $m = m_1\|m_2\|m_3\|m_4$, m_2 and m_4 are sanitized into $m' = m_1\|m_2'\|m_3\|m_4'$. We obtain $(s, \{r_2', r_4'\}, TAG)$ from $(s, \{r_2, r_4\}, TAG)$. Then everyone can obtain $(s, \{r_2', r_4'\}, ADM)$ on $m_1\|m_2'\|m_3\|m_4$. We add a final admissible block corresponding to the whole message [5]. As this block is updated at each version, the attack does not work any more.

3.2 Definition of Procedures

More formally, our sanitizable signature scheme works as follows.

- SIGKEYGEN. This step consists in executing $\mathcal{S}$.KEYGEN to obtain (ssk, spk) and in choosing randomly a secret key κ in $\{0,1\}^\lambda$ for the PRF.
- SANKEYGEN. This algorithm executes $\mathcal{CH}$.SETUP to obtain (sk_{san}, pk_{san}).
- SIGN. First, the signer generates the variable ADM as defined in the model. Let u be the number of modifiable parts in m. During this step, the signer generates the tag TAG by computing $x = PRF(\kappa, Nonce)$ where $Nonce \in \{0,1\}^\lambda$, and $TAG = PRG(x)$. In order to compute the chameleon hash function, she randomly chooses $r_1, \cdots, r_u, r_c$ in $\{0,1\}^\lambda$. Then, for each admissible block, she executes what we call the (public) "reconstruction procedure", which takes as input the message m, TAG, the r_i's and the public key pk_{san}. It is divided into several steps.
  1. Compute the values $\tilde{m}_i$ for each block:
  $$\forall i, \tilde{m}_i = \begin{cases} h_i = \mathcal{CH}.\text{PROCEED}(pk_{san}, m_i\|i, r_i) & \text{if } m_i \in ADM \\ m_i\|i & \text{else} \end{cases}$$
  2. Compute the final block : $h_c = \mathcal{CH}.\text{PROCEED}(pk_{san}, TAG\|m, r_c)$.

  After that, the signer signs the message $\tilde{m} = \tilde{m}_1\|\cdots\|\tilde{m}_t\|h_c\|pk_{san}$ as $s = \mathcal{S}.\text{SIGN}(sk_{sig}, \tilde{m})$. She finally obtains the sanitizable signature $\sigma = (s, TAG, Nonce, \mathcal{R}, ADM)$ with $\mathcal{R} = \{r_1, \cdots, r_u, r_c\}$. The signature and the message are added to the signer's database DB.
- SANITIZE. The sanitizer uses the reconstruction procedure to obtain the h_i's and h_c. For all $i \in MOD$, she finds a collision on h_i, using sk_{san}. She computes $\forall j \in MOD, r_j' = \mathcal{CH}.\text{FORGE}(sk_{san}, m_j\|j, m_j'\|j, h_j)$ and $r_c' = \mathcal{CH}.\text{FORGE}(sk_{san}, TAG\|m, TAG'\|m', h_c)$, where $Nonce'$ and TAG' are random values. The sanitized signature is $\sigma' = (s, TAG', Nonce', \mathcal{R}', ADM)$ where the set $\mathcal{R}' = \{r_1', \cdots, r_u', r_c'\}$ (with $r_j' = r_j$ if $j \notin MOD$).
- VERIFY. The verifier executes the reconstruction procedure as described above to obtain $\tilde{m} = \tilde{m}_1\|\cdots\|\tilde{m}_t\|h_c\|pk_{san}$. She finally returns the output of $\mathcal{S}.\text{VERIFY}(pk_{sig}, s, \tilde{m})$.
- PROOF. The signer searches in DB an integer $i \in [1, q]$ such that
$$\mathcal{CH}.\text{PROCEED}(pk_{san}, TAG\|m, r_c) = \mathcal{CH}.\text{PROCEED}(pk_{san}, TAG_i\|m_i, r_{c_i}) \quad (1)$$

with $\mathsf{TAG}_i = \mathrm{PRG}(x_i)$ for $x_i = \mathrm{PRF}(\kappa, \mathsf{Nonce}_i)$ and $\mathsf{m} \neq \mathsf{m}_i$. If it exists, it outputs $\pi = (\mathsf{pk}_{sig}, \mathsf{TAG}_i, \mathsf{m}_i, \mathsf{r}_{c_i}, x_i)$ else, it outputs $\bot$.

- JUDGE. If $\pi = \bot$, then it returns signer. Else, $\pi = \{\mathsf{pk}_{sig}, \mathsf{TAG}_i, \mathsf{m}_i, \mathsf{r}_{c_i}, x_i\}$ and the algorithm checks if Equation (1) holds, with $\mathsf{m} \neq \mathsf{m}_i$ and $\mathsf{TAG}_i = PRG(x_i)$. If so it outputs sanitizer. If not, it outputs signer.

3.3 Security Considerations

Theorem 1. *Our scheme is secure if the signature scheme is EU-CMA, PRG and PRF are pseudo-random and $\mathcal{CH}$ is strong secure.*

Proof. We prove that our scheme is immutable, transparent and accountable.

- The **Immutability** is reached thanks to the fact that non-admissible blocks are directly signed with the EU-CMA signature scheme $\mathcal{S}$. More formally let $\mathcal{A}^{SS}_{Imm}$ be an adversary against our scheme which, at the end of the experiment, outputs a message m^* and a public key pk^*_{san}.
  - It exists $\mathsf{m}^*_j \neq \mathsf{m}_{i_j}$ for some $j \notin$ ADM. We can use $\mathcal{A}^{SS}_{Imm}$ to break the EU-CMA property of $\mathcal{S}$. Each time $\mathcal{A}^{SS}_{Imm}$ queries a sanitizable signature from the signer, we use the signing oracle of $\mathcal{S}$. At the end, $\mathcal{A}^{SS}_{Imm}$ outputs a valid new pair m^*, σ^*. As it exists $\mathsf{m}^*_j \neq \mathsf{m}_{i_j}$ for some $j \notin$ ADM, the underlying signed message $\tilde{\mathsf{m}}^*$ and the corresponding signature s^* give us a forge on the signature scheme $\mathcal{S}$.
  - $\mathsf{pk}^*_{san} \neq \mathsf{pk}_{san_i}$ on all requests. We first recall that the message signed by the signer is $\tilde{\mathsf{m}} = \tilde{\mathsf{m}}_1 || \cdots || \tilde{\mathsf{m}}_t || \mathsf{h}_c || \mathsf{pk}_{san}$. By assumption, the underlying signed message $\tilde{\mathsf{m}}^*$ is different from all queried $\tilde{\mathsf{m}}_i$ on, at least, its last part corresponding to pk^*_{san}. All $\tilde{\mathsf{m}}_i$'s are signed using the signing oracle while the output signed message is a forge. We have thus broken the existential unforgeability of $\mathcal{S}$.

  As a consequence the probability of success of an adversary against the immutability of our scheme is $Succ^{SS}_{Imm} \leq Succ^{\mathcal{S}}_{EU-CMA}$.
- The **Transparency** is satisfied since the outputs of the signature and sanitization algorithms are similar, except in the construction of TAG and r_i.
  - Let us first focus on the r_i's. In this case, the transparency property results in the distributional property of the chameleon hash function. During a SIGN procedure, the r_i for the $\mathcal{CH}$.PROCEED algorithm are chosen at random while during the SANITIZE algorithm, the r_i's corresponds to the outputs of $\mathcal{CH}$.FORGE. Thus, the probability of success of the adversary in this case is the same as against the uniformity property of the chameleon hash function, $Adv^{\mathcal{CH}}_{Uni}$.
  - Regarding TAG, the signer chooses at each new signature a new pseudo-random value Nonce and uses it to compute the value TAG thanks to PRF and PRG. Thus, TAG is indistinguishable from a random value, under the pseudorandomness of functions PRF and PRG.

  As a conclusion, the advantage of an adversary against the transparency is $Adv^{SS}_{Trans} \leq Adv^{\mathcal{CH}}_{Uni} + Adv^{PRG}_{Pseudorand} + Adv^{PRF}_{Pseudorand}$.

- For the **Signer-Accountability**, there are two possibilities.
  1. The adversary uses a collision generated by the sanitizer, and thus successfully obtains a value x such that $\mathsf{TAG} = PRG(x)$. This is impossible under the one-wayness of the function PRG.
  2. The adversary uses a TAG she has constructed. To win the experiment, the adversary has to generate a collision on the chameleon hash function, which can happen with probability $Succ_{CollRes}^{CH}$.

  As a consequence, the probability of success of the adversary $\mathcal{A}_{sig-Acc}^{SS}$ is
  $$Succ_{sig-Acc}^{SS} \leq Succ_{OneWay}^{PRG} + Succ_{SCollRes}^{CH}.$$
- A successful adversary against the **Sanitizer-accountability** has to find a correct collision on a message m, using the PROOF algorithm. As m necessary respects ADM, the signature s in σ is necessary a forge of the classical signature scheme. As a consequence, the probability of success against the Sanitizer-accountability is $Succ_{san-Acc}^{SS} \leq Succ_{EU-CMA}^{S}$. $\square$

4   Model for Extended Sanitizable Signatures

4.1   Additional Features for Sanitizable Signatures

In this paper, we study in detail 4 additional features for sanitizable signature, from which 3 have been introduced in [7]. These restrictions are set by the signer and must be taken into account by the sanitizer.

- LimitSet: this feature permits the signer to force some admissible blocks of a signed message to be modified only into a predefined set of sub-messages. More precisely, during the SIGN algorithm the signer may define for each admissible block i a set V_i of available sub-messages. Then, the sanitizer must use one element $\mathsf{m}_i' \in V_i$ during her sanitization of this block. For this purpose, we introduce the set $\mathcal{V} = \{V_i \subset \{0,1\}^{\ell_i} : i \in \mathsf{ADM}\}$. Each V_i defines the set for the modifications of the block m_i. If the signer does not want to restrict the sanitizer in her modifications of the block m_i, then $V_i = \{0,1\}^{\ell_i}$.
- EnforceModif: with this feature, the signer forces the sanitizer to modify similarly several admissible blocks. If one is modified by the sanitizer during the SANITIZE procedure, she must use the same modification for the other admissible blocks designated by the signer. We introduce $\mathsf{cond}_m = \{\mathcal{S}_i \subset [1, 2^\ell] : \forall j \in \mathcal{S}_i, j \in \mathsf{ADM} \text{ and } \forall k \neq i, j \notin \mathcal{S}_k\}$. Each set in cond_m corresponds to a set of, at least, two admissible blocks which should be modified similarly. Note that an admissible block can only belong to one set $\mathcal{S}_i$.
- LimitNbModif: the sanitizer should modify less than a number k, fixed by the signer, out of the $|\mathsf{ADM}|$ admissible blocks. If the sanitizer modifies more than k blocks, one of her secret key becomes available. cond_k is a condition simply described by the integer $k \in [1, |\mathsf{ADM}|]$. In case this feature is not chosen by the signer, then $k = |\mathsf{ADM}|$.
- LimitNbSanit: this feature limits the number of versions one sanitizer can do from an original signed message. If the sanitizer does one extra sanitization, one of her secret key becomes available. We here define cond_l, which

corresponds to an integer l. If $l \neq \infty$, then the sanitizer can only sanitize the corresponding signed message l times. If $l = \infty$, then, there is no restriction on the number of sanitizations the sanitizer can do.

4.2   Modification of the Initial Model

We now modify the model of Brzuska et al. [2] to introduce the above new features. We first study the case of the classical procedures for sanitizable signature schemes SIGN and SANITIZE (as usual each of them output $\perp$ in case of error).

- SIGN takes as input a message m of length ℓ divided into t blocks, the secret key sk_{sig}, the public key pk_{san} and ADM. It outputs a sanitizable signature σ on the message m and the variables $\mathcal{V}$, cond_m, cond_k and cond_l as defined above. This procedure may also output some secret data denoted s that would be needed by the sanitizer. We denote s $=\perp$ if this is not relevant for the signer in the scheme. In the following, ADM is included into σ.
- SANITIZE takes as input a message m, a signature σ, the keys pk_{sig} and sk_{san}, the modifications MOD and furthermore the variables $\mathcal{V}$, cond_m, cond_k and cond_l, as defined above, and the secret data s. It outputs a new signature σ', the message m$'$ modified according to the different conditions and variables $\mathcal{V}$, cond_m, cond_k and cond_l defined by the signer.

Remark 1. Note that the different variables may not be used to verify the signature. In some cases, it may be necessary to keep these data secret. The verifier may e.g. not know how many times the sanitizer can sanitize a message. What is important is to detect a fraud, even if it may be simpler using the value l.

We now consider that the sanitization secret key sk_{san} is divided into two parts. The first one, usk_{san}, is considered as the user secret key and can be retrieved in case of fraud. It can be computed during a distinct procedure (e.g. some kind of USERKEYGEN procedure) or included into the SANKEYGEN phase. The second key, ssk_{san}, is used to sanitize messages, as the sanitization secret key in the initial model. We now introduce the new procedures.

- TESTFRAUD is a public algorithm which on input the public keys pk_{sig}, pk_{san} and a set DB of pairs (message, signature), checks if a fraud has been done on the number of admissible blocks and/or on the number of versions of message. It outputs either $\perp$ is everything is ok, or usk_{san} and a proof of guilt π otherwise. We consider that π includes DB .
- VERIFYFRAUD is a public algorithm which on input a proof π and a user key usk_{san} outputs either 1 if the proof π is valid, and 0 otherwise.

4.3   Relation between Security Properties

We now focus on the security properties that need to be modified to take into account the above features. First of all, the accountability and the transparency properties are not modified by the above additional features. However the oracles used in both experiment should be modified in order to consider the additional input. For example, in the transparency experiment, the SIGN/SANIT oracle generates a sanitized signature for all values of the challenge bit b and this oracle and the SIGN one should be honest together (i.e. they cannot make more than l sanitizations (for example) in total). Note that regarding the privacy property (implied by transparency [2]), we need to slightly modify the corresponding experiment to add in the LORSANIT oracle the choice of the $\mathcal{V}_i$'s: for each admissible part $k \in \mathsf{ADM}^+$, both initial and final messages should belong to the randomly chosen set $\mathcal{V}_k$. Thus, privacy is also induced by transparency in the extended model. We now concentrate on immutability.

Extended immutability. Let us focus on the two first additional extensions which modify the immutability property. In the new model, we add two new conditions to the classical experiment (i) one modifiable part m_i is not in the set of acceptable values for that part $(\mathsf{m}'_i \notin V_i)$ or (ii) two admissible blocks from the same set element $\mathcal{S}_{i_0}$ of cond_m (that is that are forced to be modified accordingly) are different. Note that we have the following lemma.

Lemma 1. *An Extended Immutable signature scheme is Immutable.*

Proof. A successful adversary $\mathcal{A}$ against the immutability experiment [2] outputs $(\mathsf{pk}_{san}{}^*, \mathsf{m}^*, \sigma^*)$ such that $\mathrm{VERIFY}(\mathsf{m}^*, \sigma^*, \mathsf{pk}_{sig}, \mathsf{pk}_{san}{}^*) = \mathsf{true}$ and for all $i = 1, 2, \cdots, q$ either $\mathsf{pk}_{san}{}^* \neq \mathsf{pk}_{san,i}$ or $\exists j_i \notin \mathsf{ADM}_i | \mathsf{m}^*[j_i] \neq \mathsf{m}_i[j_i]$, so she directly wins the Extended Immutability experiment above. $\square$

Extended traceability. We now introduce a new security property which we call "extended traceability". This ensures that an adversary is not able to do more modifications than stated by SIGN, or to execute more sanitizations of the same signed message than the sanitizer is allowed to, without being accused of.

In the corresponding experiment, the adversary outputs several valid pairs (message, signature) under the same sanitizer public key such that TESTFRAUD, with as input this set of pairs, detects a fraud i.e. returns a pair $(\mathsf{usk}_{san}, \pi)$. The adversary wins the game if usk_{san} is not part of the corresponding sanitizer secret key or if VERIFYFRAUD outputs 0.

Extended Immutability: $Succ_{ext-imm}^{SS} = Pr[1 \longleftarrow \mathrm{ExP}_{ext-imm}^{SS}]$ where $\mathrm{ExP}_{ext-imm}^{SS}$ is:
- $(\mathsf{pk}_{sig}, \mathsf{sk}_{sig}) \longleftarrow \mathrm{SIGKEYGEN}(1^\lambda)$
- $(\mathsf{pk}_{san}^*, \mathsf{m}^*, \sigma^*, l^*, k^*, \mathcal{V}^*) \longleftarrow \mathcal{A}^{\mathrm{SIGN}(\cdot, \mathsf{sk}_{sig}, \cdot, \cdot), \mathrm{PROOF}(\mathsf{sk}_{sig}, \cdot, \cdot, \cdot)}(\mathsf{pk}_{sig})$
- Let $(\mathsf{m}_i, \mathsf{ADM}_i, \mathsf{pk}_{san,i})$ and $(\sigma_i, \mathcal{V}_i, \mathsf{cond}_{m_i}, \mathsf{cond}_{k_i}, \mathsf{cond}_{l_i})$ for $i \in [1, q]$ be the queries and answers to and from oracle SIGN.
- return 1 if $\mathrm{VERIFY}(\mathsf{m}^*, \sigma^*, \mathsf{pk}_{sig}, \mathsf{pk}_{san}^*) = \mathsf{true}$ and for all $i = 1, 2, \cdots, q$ we have
    - $\mathsf{pk}_{san,i}^* \neq \mathsf{pk}_{san,i}$ or
    - $\exists j_i \notin \mathsf{ADM}_i$ such that $\mathsf{m}^*[j_i] \neq \mathsf{m}_i[j_i]$ or
    - $\exists j$ such that $\mathsf{m}^*[j] \notin V_j^*$ or
    - $\exists i_0$ such that $\exists j, j' \in \mathcal{S}_{i_0}$ such that $\mathsf{m}^*[j] \neq \mathsf{m}^*[j']$.

Extended Traceability: $Succ_{ext-tra}^{SS} = Pr[1 \longleftarrow \mathrm{ExP}_{ext-tra}^{SS}]$ where $\mathrm{ExP}_{ext-tra}^{SS}$ is:
- $(\mathsf{pk}_{sig}, \mathsf{sk}_{sig}) \longleftarrow \mathrm{SIGKEYGEN}(1^\lambda)$
- $(\mathsf{pk}_{san}^*, \mathsf{DB}^* = \{(m_p^*, \sigma_p^*), p = 1, \cdots, n\}) \longleftarrow \mathcal{A}^{\mathrm{SIGN}(\cdot, \mathsf{sk}_{sig}, \cdot, \cdot), \mathrm{PROOF}(\mathsf{sk}_{sig}, \cdot, \cdot, \cdot)}(\mathsf{pk}_{sig})$
- If it exists $p \in [1, n]$ such that $\mathrm{VERIFY}(m_p^*, \sigma_p^*, \mathsf{pk}_{sig}, \mathsf{pk}_{san}^*) = \mathsf{false}$, then outputs 0
- $(\mathsf{usk}_{san}, \pi) \longleftarrow \mathrm{TESTFRAUD}(\mathsf{pk}_{sig}, \mathsf{pk}_{san}^*, \mathsf{DB}^*)$
- return 1 if usk_{san} does not correspond to pk_{san}^*, or
    - $\mathrm{VERIFYFRAUD}(\pi, \mathsf{usk}_{san}) = 0$.

5   Cryptanalysis of Extended Sanitization Scheme

In this section, we review the paper of Klonowski and Lauks [7] and show that their EnforceModif and LimitNbModif extensions are insecure.

5.1   The EnforceModif Extension

We first recall the proposal in [7]. We assume that the signer signs $m = m_1 \| \cdots \| m_t$ with d blocks $m_{i_1}, \cdots, m_{i_d}$ such that the sanitizer can modify similarly.

- SIGN: the signer chooses at random $x_1, \cdots, x_t, r$ and computes, for all $i \in [1, t]$, $h_i = g^{x_i}$ with g a public value. Then, she computes $c = h_1^{m_1} \cdots h_t^{m_t} g^r$ and a classical signature s on c. Finally, the signature σ is $(c, r, s, h_1, \cdots, h_t)$ and the sanitizer is given a secret value $s = x_{i_1} + \cdots + x_{i_d}$.
- SANITIZE: the sanitizer wants to modify the signed message $m_1 \| \cdots \| m_t$ into the new one $m_1^* \| \cdots \| m_t^*$, with $m = m_{i_1} = \cdots = m_{i_d}$ and $m^* = m_{i_1}^* = \cdots = m_{i_d}^*$. On input $(c, r, s, h_1, \cdots, h_t)$ and s, the sanitized signature is simply $(c, r^*, s, h_1, \cdots, h_t)$ where $r^* = r + (m - m^*)s$.
- VERIFY: from the signature $\sigma = (c, r, s, h_1, \cdots, h_t)$ and the message $m = m_1 \| \cdots \| m_t$, one can verify that $c = h_1^{m_1} \cdots h_t^{m_t} g^r$ and that s on c is valid.

In fact, from two different versions, with m and m^* identically modified, one can retrieve s and thus sanitize any message. In fact, if these blocks are the only admissible ones, from the construction of the SANITIZE, we have $r^* = r + (m - m^*)s$. As r, m, r^* and m^* are included into the signatures or the messages and $m \neq m^*$, one can compute $s = \frac{r - r^*}{m^* - m}$. Thus this scheme is not accountable.

5.2   The LimitNbModif Extension

The solution proposed in [7] is based on polynomial interpolation: there is exactly one polynomial of degree at most k going through $k + 1$ fixed points. Then, the principle is to define a secret polynomial F of degree k, such that the sanitizer key $usk_{san} = F(0)$. Each time the sanitizer sanitizes a block, a point of the polynomial F leaks. Thus, when $k + 1$ blocks are modified, $k + 1$ points are available, the secret polynomial can be interpolated and the sanitizer's secret key is retrieved. In [7], the basic sanitizable signature scheme is the one of Ateniese *et al.* with a chameleon hash function not resistant to the key exposure attack. For each modified block during the SANITIZE procedure, a point on the polynomial is chosen as the used key. Thus, as soon as a block is modified, a collision is computed and the point leaks.

More precisely, their scheme works as follows. Let usk_{san} be the sanitizer secret key related to $upk_{san} = g^{usk_{san}}$. During the SIGN procedure, the sanitizer chooses at random k values $f_1, \cdots, f_k$ and constructs the polynomial $F(y) = usk_{san} + f_1 y + \cdots + f_k y^k$. She next computes $\{g_i = g^{f_i}\}_{i \in [1, k]}$, where g is a public generator, and sends it to the signer. The signer computes a sanitizable public key for each admissible block m_i as $z_i = g^{F(i)} = \mathsf{upk}_{san} \cdot g_1^i \cdot \ldots \cdot g_k^{i^k}$. She next chooses an identifier for the message ID_m and a random r_i for each admissible block. She uses the chameleon hash function and computes, for all i, $\tilde{\mathsf{m}}_i = \text{PROCEED}(z_i, \tilde{\mathsf{m}}_i = ID_\mathsf{m}\|\mathsf{m}_i\|i, r_i) = z_i^{\tilde{\mathsf{m}}_i} \cdot g^{r_i}(= g^{F(i)\cdot\tilde{\mathsf{m}}_i + r_i})$ if $i \in ADM$ and $\tilde{\mathsf{m}}_i = \mathsf{m}_i\|i$ otherwise. The signature is $\sigma = (ID_\mathsf{m}, t, \{r_i\}_{\forall i \in ADM}, \{z_i\}_{\forall i \in ADM}, s)$ with s a classical signature on $(ID_\mathsf{m}\|t\|\mathsf{upk}_{san}\|\tilde{\mathsf{m}}_1\|\cdots\|\tilde{\mathsf{m}}_t)$.

During the SANITIZE algorithm, for each block m_j the sanitizer wants to modify, she uses the secret key $F(j)$ and computes a collision on $\tilde{\mathsf{m}}_j$. As the function is weak against the key exposure attack, $F(j)$ necessary leaks. Thus if she modifies $k+1$ blocks, $k+1$ points of $F(y)$ leaks and anybody can find usk_{san}.

Again, this solution is not secure since, after one sanitization, the value $F(i)$ necessary leaks. As the knowledge of $F(i)$ is enough to construct any collision on $\tilde{\mathsf{m}}_i$, one can construct as many other sanitizations as she wants from only one sanitization: the scheme is not accountable.

6 Constructions in the Extended Model

6.1 The LimitSet Extension

This feature has been nicely solved in [7] by the use of accumulators. It consists in accumulating all possible modifications for a block into one accumulator. The sanitizer is given the accumulator, the accumulated values and the corresponding witnesses to prove that one value is truly accumulated. Then, the accumulator is signed by the signer as a non admissible part of the message. During the sanitization process, the sanitizer should have to give the accumulated value, which is the new message block, and the corresponding witness, so that the verifier can verify that the modified block is a valid message for the focused admissible block.

- SIGKEYGEN. This step executes the SIGKEYGEN procedure of the initial scheme, as described in Section 3. We thus obtain $(\mathsf{ssk}, \mathsf{spk})$ and a secret key κ in $\{0,1\}^\lambda$ for the PRF. Then it executes the initialisation algorithm of the chosen accumulator scheme ACC.
- SANKEYGEN. This algorithm is identical to the initial SANKEYGEN.
- SIGN. Let $\mathsf{m} = \mathsf{m}_1\|\cdots\|\mathsf{m}_t$ be the message to be signed. The signer first generates the variable ADM: she decides for each block $i \in [1, t]$ whether the block is admissible or not. There are then two cases:
    1. the i-th block is not admissible ($i \notin ADM$). The signer sets $\tilde{\mathsf{m}}_i = \mathsf{m}_i\|i$.
    2. the i-th block is admissible ($i \in ADM$). There are two new cases:
        (a) there are no restriction on the value for this block (we say that $i \in ADM^-$). The signer next chooses at random a value denoted $r_i \in \{0,1\}^\lambda$ and computes $\tilde{\mathsf{m}}_i = \mathcal{CH}.\text{PROCEED}(\mathsf{pk}_{san}, \mathsf{m}_i\|i, r_i)$.
        (b) the i-th message block should lie in a set of authorized values $\mathcal{V}_i$ defined by the signer (we say that $i \in ADM^+$). In this case, the signer first initializes an empty accumulator Acc_i and, for each element $a_{k,i} \in \mathcal{V}_i$, she accumulates it in Acc_i and computes the corresponding

witness $w_{k,i}$. The set of all witnesses for the block i is denoted $\mathcal{W}_i = \{w_{k,i} : k \in [1, |\mathcal{V}_i|]\}$. Note that the value m_i necessary lies in $\mathcal{V}_i$ for obvious reasons. That is, it exists k_0 such that $\mathsf{m}_i = a_{k_0,i}$. At the end of this step, the signer defines $\tilde{\mathsf{m}}_i = \mathrm{Acc}_i$.

In the following, we denote $W_0 = \{w_{k_0,i} : i \in \mathsf{ADM}^+\}$ the set of all witnesses used by the signer for the message m, $\mathcal{A} = \{\mathrm{Acc}_i : i \in \mathsf{ADM}^+\}$, $\mathcal{W} = \{\mathcal{W}_i : i \in \mathsf{ADM}^+\}$ and $\mathcal{R} = \{\mathsf{r}_i : i \in \mathsf{ADM}^-\}$.

The signer generates $\mathsf{TAG} = \mathrm{PRG}(x)$ where $x = \mathrm{PRF}(\kappa, \mathsf{Nonce})$ with $\mathsf{Nonce} \in \{0,1\}^\lambda$. She computes $(\mathsf{h}_c, \mathsf{r}_c) = \mathcal{CH}.\mathrm{PROCEED}(\mathsf{pk}_{san}, \mathsf{TAG}\|\mathsf{m}, \mathsf{r}_c)$ and signs the message $\tilde{\mathsf{m}} = \tilde{\mathsf{m}}_1\| \cdots \|\tilde{\mathsf{m}}_t\|\mathsf{h}_c\|\mathsf{pk}_{san}$ as $s = \mathcal{S}.\mathrm{SIGN}(\mathsf{sk}_{sig}, \tilde{\mathsf{m}})$. Finally, the signature is $\sigma = (s, \mathsf{TAG}, \mathsf{Nonce}, \mathcal{R} \cup \{\mathsf{r}_c\}, \mathsf{ADM}, W_0)$. The set of authorized values for each admissible block $\mathcal{V} = \{\mathcal{V}_i : i \in \mathsf{ADM}^+\}$ and the corresponding set of all witnesses $\mathcal{W}$ (if they are not publicly computable) are independently sent to the sanitizer.

- SANITIZE. The sanitizer wanting to modify the message m to the message m' performs the following actions, for each block $j \in [1, t]$:

  1. the j-th block is not admissible, the sanitizer does not do anything.
  2. the j-th block is admissible. There are two new cases:
     (a) if $j \in \mathsf{ADM}^-$, she computes $\mathsf{r}'_j = \mathcal{CH}.\mathrm{FORGE}(\mathsf{sk}_{san}, \mathsf{m}_j\|j, \mathsf{m}'_j\|j, \mathsf{h}_j)$.
     (b) if $j \in \mathsf{ADM}^+$, the sanitizer checks that the new block message $\mathsf{m}'_j \in \mathcal{V}_j$ and finds the corresponding $w_{k_0,j}$ in the set of all witnesses.

The sanitizer next sets $W'_0 = \{w_{k_0,j} : j \in \mathsf{ADM}^+\}$ the set of all used witnesses for the new message m' and by $\mathcal{R}' = \{\mathsf{r}'_i : i \in \mathsf{ADM}^-\}$.

She next chooses at random Nonce' and TAG' and uses sk_{san} to find a collision on the chameleon hash for the obtained message. That is, she recomputes h_c and computes $\mathsf{r}'_c = \mathcal{CH}.\mathrm{FORGE}(\mathsf{sk}_{san}, \mathsf{TAG}\|\mathsf{m}, \mathsf{TAG}'\|\mathsf{m}', \mathsf{h}_c)$. The new sanitize signature is finally $\sigma' = (s, \mathsf{TAG}', \mathsf{Nonce}', \mathcal{R}' \cup \{\mathsf{r}'_c\}, \mathsf{ADM}, W'_0)$.

- VERIFY. The verifier executes the reconstruction procedure. For all i, she defines $\tilde{\mathsf{m}}_i$ as (i) $\mathrm{ACC}(\mathsf{m}_i, w_i)$ if $\mathsf{m}_i \in \mathsf{ADM}^+$, (ii) h_i if $\mathsf{m}_i \in \mathsf{ADM}^-$ or (iii) $\mathsf{m}_i\|i$ otherwise. Then the verifier computes $\mathsf{h}_c = \mathcal{CH}.\mathrm{PROCEED}(\mathsf{pk}_{san}, \mathsf{TAG}\|\mathsf{m}, \mathsf{r}_c)$ and verifies whether $\mathcal{S}.\mathrm{VERIFY}(\mathsf{pk}_{sig}, s, \tilde{\mathsf{m}})$ returns true or false.

- PROOF and JUDGE are identical to our classical construction (cf. Section 3).

Theorem 2. *Our scheme is secure if the signature scheme is $EU-CMA$, PRG and PRF are pseudo-random, and $\mathcal{CH}$ is strong secure and ACC is secure.*

Proof. Our scheme is ext-immutable, transparent, accountable and ext-traceable.

- In the **Ext-Immutability** experiment, $\mathcal{A}$ outputs $(\mathsf{pk}^*_{san}, \mathsf{m}^*, \sigma^*, l^*, k^*, \mathcal{V}^*)$. Either $\mathsf{pk}^*_{san} \neq \mathsf{pk}_{san,i}$ or $\exists j_i \notin \mathsf{ADM}_i$ such that $\mathsf{m}^*[j_i] \neq \mathsf{m}_i[j_i]$. That cases lies on the chosen signature EU-CMA property, similarly as in the proof of Immutability of our classical scheme (cf. Section 3). Or $\exists j$ such that $\mathsf{m}^*[j] \notin V^*_j$. In that case, either $\mathsf{m}^*[j]$ has been added to the accumulator of the i-th block and we can construct an adversary $\mathcal{A}^{\mathcal{S}}_{EU-CMA}$ against the EU-CMA property in outputting the forgery on the message $\tilde{\mathsf{m}}^*$. Or the adversary has find a value which has not been accumulated and we are able to construct an adversary $\mathcal{A}_{CR}^{Acc}$ against the collision resistance of ACC. The success probability is finally $Succ_{Ext-Imm}^{set-SS} \leq Succ_{EU-CMA}^{S} + Succ_{CR}^{Acc}$.

- For the **Transparency** property, we remark that an original sanitizable signature is $(s, \mathsf{TAG}, \mathsf{Nonce}, \mathcal{R} \cup \{r_c\}, \mathsf{ADM}, W_0)$, while a sanitized one is $(s, \mathsf{TAG}', \mathsf{Nonce}', \mathcal{R}' \cup \{r'_c\}, \mathsf{ADM}, W'_0)$. As the used witnesses (in W_0 and W'_0) are constructed in the same way for an original or a sanitized signature, the transparency property relies on the classical parts of the signature and the proof is identical to the proof in the classical case. The advantage of an adversary is $Adv_{Trans}^{set-SS} \leq Adv_{Uni}^{\mathcal{CH}} + Adv_{Pseudorand}^{PRG} + Adv_{Pseudorand}^{PRF}$.

- Both **Accountability** properties rely on the construction of the last block of our construction h_c: it depends on the construction of TAG and on the incapability of the signer to obtain collisions on the chameleon hash. As accumulators are not implied on this part of the protocol, the proof is identical as in the classical case. Thus, $Succ_{sig-Acc}^{set-SS} \leq Succ_{OneWay}^{PRG} + Succ_{SCollRes}^{\mathcal{CH}}$ and $Succ_{san-Acc}^{set-SS} \leq Succ_{EU-CMA}^{S}$. $\square$

6.2   The LimitNbModif Extension

As in [7], our solution uses polynomial interpolation and a chameleon hash function CH weak against the key exposure attack. However, contrary to [7], we use a second *secure* chameleon hash function $\mathcal{CH}$. Thus, the sanitization phase on the message m_i requires the user to know both $\{F(i)\}_{i \in \mathsf{MOD}}$ (keys for CH) and the sanitizer secret key sk_{san} (for $\mathcal{CH}$). Thus, the leakage of $F(i)$ does not compromise the unforgeability any more.

More precisely, let usk_{san} be the sanitizer secret key, related to the public key $\mathsf{upk}_{san} = g^{\mathsf{usk}_{san}}$.

- SIGN. The signer executes the signature procedure as for our initial scheme and obtains ADM, the value $\tilde{m}$, TAG, Nonce and $\mathcal{R} = \{r_i\}_{i \in \mathsf{ADM}}$. Then she randomly chooses $\mathcal{R}^* = \{r^*_i\}_{i \in \mathsf{ADM}}$. Meanwhile, the sanitizer randomly chooses k values $f_1, \cdots, f_k$ and constructs $F(y) = \mathsf{usk}_{san} + f_1 y + \cdots + f_k y^k$. She next computes the set $\{g_i = g^{f_i}\}_{i \in [1,k]}$, where g is a public generator, and sends it to the signer. After that, the signer computes the set $\{z_i\}_{i \in \mathsf{ADM}}$ such that $z_i = \mathsf{upk}_{san} \cdot g_1^{i} \cdot \ldots \cdot g_k^{i^k} (= g^{F(i)})$ and uses each z_i as the public key of CH to hide the corresponding r_i: $\forall i \in \mathsf{ADM}, t_i = CH.\mathrm{PROCEED}(z_i, r_i, r^*_i) = z_i^{r_i} \cdot g^{r^*_i} (= g^{F(i) \cdot r_i + r^*_i})$. Finally, she classically signs the concatenation of $\tilde{m}$ and the $\{t_i\}_{i \in \mathsf{ADM}}$: $\bar{s} = \mathcal{S}.\mathrm{SIGN}(\mathsf{sk}_{sig}, \tilde{m} \| t_{i_1} \| \cdots \| t_{i_{|\mathsf{ADM}|}})$ and obtains the signature $\bar{\sigma} = \{\bar{s}, TAG, \mathsf{Nonce}, \mathcal{R}, \mathcal{R}^*, \{z_i\}_{i \in \mathsf{ADM}}, \mathsf{ADM}\}$.

- SANITIZE. The sanitizer computes, for each block m_j she wants to modify, a collision on $\tilde{m}_j$ thanks to sk_{san} and a collision on t_j thanks to $F(j)$.

Remark 2. Note that the SIGN process is suppose to be non-interactive. In the above description, we can imagine than the sanitizer regularly publishes some g_i's that can be used by the signer when necessary. In some other cases, such as for content protection [5], the sanitizer is considered as on line during this process, and thus can compute on line these g_i's.

With this method, the sanitizer can easily modify k blocks. As in [7], if she modifies $k + 1$ blocks, $k + 1$ points of $F(y)$ are available and usk_{san} leaks. But in our case, the collision resistance of $\mathcal{CH}$ already fixes the message. thus it does not impact the security of the scheme any more.

Theorem 3. *Our scheme is secure if the signature scheme is EU-CMA, PRG and PRF are pseudo-random, $\mathcal{CH}$ (resp. CH) is strong (resp. weak) secure.*

Proof. Our scheme is extended immutable, transparent and accountable for the same reasons than our main scheme. For the ext-traceability there are two cases. Either, the adversary outputs $(\mathsf{pk}^*_{san}, \mathsf{DB}^* = \{(m^*_p, \sigma^*_p), p = 1, \cdots, n\})$ such that pk^*_{san} does not correspond to usk_{san} output by TESTFRAUD. In our scheme, this is checked by TESTFRAUD. So the underlying success probability is 0. Or, $\mathcal{A}$ outputs $(\mathsf{pk}^*_{san}, \mathsf{DB}^* = \{(m^*_p, \sigma^*_p), p = 1, \cdots, n\})$ such that TESTFRAUD detects a fraud $(\mathsf{usk}_{san}, \pi)$ and that VERIFYFRAUD outputs 0. This may happens either if usk_{san} is not the secret key corresponding to the public key of the sanitizer, but this has already been studied, or if the polynomial $F(y)$ obtained through interpolation is such that $g^{F(0)} \neq \mathsf{upk}_{san}$. As only one polynomial of degree n goes through $n + 1$ given point, and considering that the public key is signed, an adversary able to modify this value can be used to construct an adversary against the EU-CMA of the chosen signature. So the success probability is $Succ^{SS}_{Ext-trans} \leq Succ^{S}_{EU-CMA}$. $\square$

6.3 The LimitNbSanit Extension

In a nutshell, our system is based on a method which has been first proposed for the e-cash purpose. It consists in using the soundness property of zero-knowledge proofs of knowledge of a secret. Honest-verifier proofs are three-move protocols: a commitment t based on random values, a question c and an answer s related to the above random values, the question and the secret. The soundness of these constructions ensures that given a single t, if someone is able to provide s and s' related to c and c' s.t. $c \neq c'$, then it is possible to retrieve the secret.

More precisely, our methodology works as follows. First, we use our main sanitizable signature scheme described in Section 3. Let usk_{san} be the sanitizer secret key, related to the public key $\mathsf{upk}_{san} = g^{\mathsf{usk}_{san}}$.

- SIGN. The sanitizer chooses at random l values $a_1, \cdots, a_l$. She next computes, for all $i \in [1, l]$, the value $t_i = g^{a_i}$, with g a public generator. Each value t_i corresponds to a version number authorized by the signer. The sanitizer next sends $\{t_i\}_{1 \leq i \leq l}$ to the signer. After that, the signer chooses two random values α and ρ and constructs its own version number: $t_0 = \mathsf{upk}^{\rho}_{san} g^{\alpha}$. Then the signer accumulates all the t_i's (including t_0) into one single accumulator Acc and executes the SIGN procedure of our main sanitizable signature scheme, using ρ as random for the final execution of the chameleon hash function $\mathcal{CH}$ on the whole message, and adding Acc in the final classical signature: she obtains σ. The signature is $\bar{\sigma} = \{\sigma, t_0, \alpha, \mathsf{Acc}, w_0\}$ with w_0 the witness for t_0. Finally, the witnesses of the accumulated values are given to the sanitizer as the secret s.

- SANITIZE. The sanitizer executes the SANITIZE procedure of main scheme. Then she reveals a new t_i and its corresponding witness, denoted w_i, her public key upk_{san} and the value $\alpha_i = a_i - \rho_i \mathsf{usk}_{san}$ with ρ_i the pseudorandom value output during the generation of the collision on the whole message.

With this method, the sanitizer can easily use the l different accumulated values. However, if the sanitizer executes $l + 1$ times this procedure, she has to use twice the same accumulated value with her secret key usk_{san} but with two different random values ρ_i and ρ_j. It is thus possible to retrieve usk_{san}.

Theorem 4. *Our scheme is secure if the signature scheme is EU-CMA, PRG and PRF are pseudo-random and $\mathcal{CH}$ is strong secure and ACC is secure.*

Proof. Our scheme is ext-immutable, transparent and accountable for the same reasons than our main scheme. The **Ext-Traceability** implies that $\mathcal{A}$ outputs DB^* under pk^*_{san} such that TESTFRAUD finds a fraud. In fact, $\mathcal{A}$ should either embed more values in Acc, which breaks the security of ACC, or is able to output a signature on a wrong accumulator and thus breaks the EU-CMA of the signature scheme. In conclusion, $Succ^{SS}_{Ext-trans} \leq Succ^{\text{Acc}}_{CR} + Succ^{S}_{EU-CMA}$. □

References

1. Ateniese, G., Chou, D.H., De Medeiros, B., Tsudik, G.: Sanitizable signatures. In: di Vimercati, S.d.C., Syverson, P.F., Gollmann, D. (eds.) ESORICS 2005. LNCS, vol. 3679, pp. 159–177. Springer, Heidelberg (2005)
2. Brzuska, C., Fischlin, M., Freudenreich, T., Lehmann, A., Page, M., Schelbert, J., Schröder, D., Volk, F.: Security of sanitizable signatures revisited. In: Jarecki, S., Tsudik, G. (eds.) PKC 2009. LNCS, vol. 5443, pp. 317–336. Springer, Heidelberg (2009)
3. Camenisch, J., Kohlweiss, M., Soriente, C.: An accumulator based on bilinear maps and efficient revocation for anonymous credential. In: Jarecki, S., Tsudik, G. (eds.) PKC 2009. LNCS, vol. 5443, pp. 481–500. Springer, Heidelberg (2009)
4. Camenisch, J., Lysyanskaya, A.: Dynamic accumulators and application to efficient revocation of anonymous credentials. In: Yung, M. (ed.) CRYPTO 2002. LNCS, vol. 2442, pp. 61–76. Springer, Heidelberg (2002)
5. Canard, S., Laguillaumie, F., Milhau, M.: Trapdoor sanitizable signatures and their application to content protection. In: Bellovin, S.M., Gennaro, R., Keromytis, A.D., Yung, M. (eds.) ACNS 2008. LNCS, vol. 5037, pp. 258–276. Springer, Heidelberg (2008)
6. Goldwasser, S., Micali, S., Rivest, R.L.: A digital signature scheme secure against adaptive chosen-message attacks. SIAM Journal on Computing 17, 281–308 (1988)
7. Klonowski, M., Lauks, A.: Extended sanitizable signatures. In: Rhee, M.S., Lee, B. (eds.) ICISC 2006. LNCS, vol. 4296, pp. 343–355. Springer, Heidelberg (2006)
8. Krawczyk, H., Rabin, T.: Chameleon signatures. In: NDSS 2000. The Internet Society (2000)
9. Nguyen, L.: Accumulators from bilinear pairings and applications. In: Menezes, A. (ed.) CT-RSA 2005. LNCS, vol. 3376, pp. 275–292. Springer, Heidelberg (2005)
10. Yuen, T.H., Susilo, W., Liu, J.K., Mu, Y.: Sanitizable signatures revisited. In: Franklin, M.K., Hui, L.C.K., Wong, D.S. (eds.) CANS 2008. LNCS, vol. 5339, pp. 80–97. Springer, Heidelberg (2008)

Unrolling Cryptographic Circuits: A Simple Countermeasure Against Side-Channel Attacks

Shivam Bhasin, Sylvain Guilley, Laurent Sauvage, and Jean-Luc Danger

Institut TELECOM / TELECOM ParisTech, CNRS LTCI (UMR 5141),
Département COMELEC, 46 rue Barrault, 75 634 PARIS Cedex 13, France

Abstract. Cryptographic cores are used to protect various devices but their physical implementation can be compromised by observing dynamic circuit emanations in order to derive information about the secrets it conceals. Protection against these attacks, also called side channel attacks are major concern of the cryptographic community. Masking and dual-rail precharge logic are promoted as its countermeasures but each has its own vulnerabilities. In this article, we propose a simple countermeasure which comprises unrolling rounds of a cryptographic algorithm such that multiple rounds are executed per clock cycle. This will require a stronger hypothesis on multiple bits due to deeper diffusion of the key. Results show that it resist against correlation power analysis on Hamming distance and Hamming weight model if the datapath is cleared after each operation. We also evaluated mutual information metric on the design and results show that unrolled DES is less vulnerable.

Keywords: Data encryption standard, side-channel attack, architectural countermeasure, mutual information metric.

1   Introduction

With the generalization of open networks, information society regards security as a critical factor. Modern cryptographic algorithms which ensure security are robust and free from practical cryptanalysis. However, other methods which target the physical implementation of an algorithm can be deployed to break the security. These attacks can be mounted by merely observing or perturbing the targeted system. Observing the activity of the system and its correlation with potential guesses can yield sensible information. Such attacks are better known as Side Channel Attacks (SCAs) [1]. When a device is perturbed such that it yields a non-nominal output, this together with expected output can lead to the secret key. Such attacks are called as Differential Fault Analyses (DFAs) [2]. The passive attacks that consist in observing the chip are difficult to protect since the chip is even not aware of the attack. Therefore these attacks are considered more critical.

SCAs try to recognize synchronous operations (rounds of cryptographic operations) in the leakage of a device. Then for a chosen round, the leakage is correlated with some guesses to reveal secret information. It is possible to guess some

J. Pieprzyk (Ed.): CT-RSA 2010, LNCS 5985, pp. 195–207, 2010.

key bits because the value of key remains same for one or a set of synchronous operations. For example if we consider DES, cryptanalysis is impractical as we need a huge number of plaintext or ciphertext. Whereas with power attacks only the power consumption of a few hundreds of encryption are needed to break a non-protected implementation. For instance in DPA contest [3], the participants have demonstrated that DES could be broken in 141 traces in average. Therefore it is essential to protect implementations against SCA.

State of the art countermeasures can be widely classified into two categories *i.e.* information making and information hiding. Masking [4] countermeasures rely on confusing the attacker. A random generated mask is used while running the algorithm such as the mask affects the intermediate states without affecting the end result. Owing to this technique, the attacker observes leakage corresponding to mask and not the actual key bits. Although a nicely masked circuit can resist first order SCA but higher order SCA can still compromise the security of the design.

Information hiding as the name suggests hides the information from attacker. The algorithm is implemented in such a way that leakage remains constant irrespective of the computations performed. Dual-rail precharge logic (DPL) [5] is a countermeasure based on information hiding. The principle of this countermeasure is to generate a design equivalent and with opposite behaviour of the target design such that every part of the circuit is perfectly balanced. This way the activity of the doubled design remains constant. There are some countermeasures which combine hiding and masking techniques in order to achieve higher level of security. The major problem of these countermeasures is that its hard to design a perfectly balanced circuit. Even minor imbalance in space (unbalanced dual nets) or time (early evaluation) can be exploited by sophisticated attacking techniques to reveal sensitive information.

In [6], the effect of pipelining on security is studied. In this article, we investigate the other trend, namely pipelining less; this way, all registers become unpredictable depending on the key (*i.e.* a hypothesis test involves too many key hypotheses). The idea is to implement the design in such a way that the key changes more than once during a synchronous operation. In other words, more than one round of a cryptographic algorithm are executed in one synchronous operation. The rest of the paper is organized as follows. Section 2 explains the theory of the proposed countermeasure. It also details the implementation details of a fully unrolled DES. Section 3 evaluates fully unrolled DES against the iterative DES using correlation power analysis (CPA [7]). Finally, section 4 concludes the paper.

2   Proposed Countermeasure

2.1   Rationale of the Countermeasure

In a cryptographic block product algorithm, data is ciphered by repeating a set of operations with a different key value each time generated from the previous key. These set of operations are called as rounds. The number of rounds are chosen such that linear and differential cryptanalysis are more difficult than an exhaustive key search. Normally, cryptographic circuits are designed to perform either some operations of a round or the whole round in one clock cycle. Thus the value of the key remains the same for one or more clock cycles. The attacker can guess some of the key bits and correlate it with leakage acquired. A correct guess will give a much higher correlation as compared to wrong guesses.

Most of the traditional SCA attacks target the registers where the result of each round is stored. This is because the leakage from the register is high due to its load and the leakage is synchronised to the clock. In combinatorial logic, the leakage is low and spread over time. If the result of a round is stored in the register at the end of each clock cycle, attacker can easily retrieve the subkey by guessing and correlating. Now, if the key is changed more than once during one clock cycle *i.e.* multiple rounds are executed per clock cycle the key used for one round is further diffused deeper into the design and mixed with the second key and so on. Thus exploiting this property we propose to design the cryptographic coprocessors in such a way that it executes multiple rounds in one clock cycle. We call this as unrolling the rounds of the algorithm. Also we define unrolling factor as the number of rounds unrolled. An implementation unrolled twice means that two rounds are performed at every clock cycle.

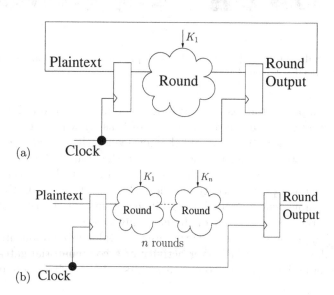

Fig. 1. (a) Architecture of a iterative cryptographic algorithm. (b) Architecture of a fully unrolled cryptographic algorithm.

Figure 1(a) shows the architecture of one round of a normal iterative cryptographic algorithm while figure 1(b) shows the architecture of an unrolled cryptographic algorithm. An idea of the difficulty to mount a side channel attack on the unrolled version can be estimated from the following discussion. Suppose,

(a)   (b)

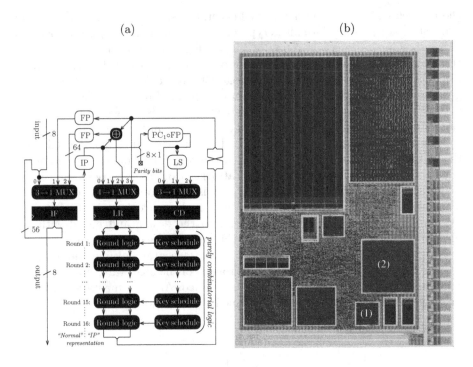

Fig. 2. (a) Unrolled DES Architecture. (b) Floorplan of the ASIC implementing DES iterative (1) and DES unrolled (2).

we have two implementations of a cryptographic algorithm: one iterative and the other unrolled with an unrolling factor of 2 as shown in fig 1(a) and (b) respectively. Let us see the signal and the noise when the attack is mounted on 1-bit. In the iterative design, the signal will be the sum of the power activity of all the combinatorial gates and flip-flop involved in calculating that bit. The noise shall be sum of power activity of other gates and flip-flops. In the unrolled design, if we implement an attack on 1-bit in the first of the two rounds, the signal will be the power activity of the gates involved only as the result is not memorised. The noise shall be twice the previous value as components are doubled. As explained before the power activity of a combinatorial gates is lesser than the power activity of a register. This results in SNR reduction of more than twice.

A rough evaluation of the theoretical complexity of this countermeasure in terms of area is given by the unrolling factor. Thus a design unrolled twice will have double the area of its original design as far as combinatorial part is concerned. In terms of performance, the trade-off is almost the same as original design. Unrolling factor of n will multiply the critical path by n times and thus maximum frequency is reduced $1/n$ times. Since n rounds are executed per clock cycle, N/n clock cycles are needed to execute the whole algorithm where N is the total number of rounds. Thus the throughput is approximately the same

```
set_current_module des_datapath_combi_wrapper; # Internal constraints
set_current_instance [find -hier -inst I_REG_LR];
# The following constraint (1+15 cycles allowed for the computation)
# concerns the whole bus:
set_cycle_addition -from [get_info [lindex [find -port q] 0] bus] 15;
set_current_instance [find -hier -inst I_REG_CD];
set_cycle_addition -from [get_info [lindex [find -port q] 0] bus] 15;
set_current_module des_datapath_combi; # External constraint
set_false_path -from [find -port sel_left_not_right]; # Encrypt/Decrypt
```

Fig. 3. TCL timing constraints crafted for the "multi-cycle" DES combinatorial datapath synthesis by Cadence **bgx_shell**

for original and unrolled design. The practical results are better than the one described below as some of the unnecessary components like multiplexers are removed while unrolling. Thus the area is less than n times and the operating frequency is more than $1/n$ times. We also point out that the unrolling does not impact the possibility of the encrypting block to be used in any mode of operation (CBC, CFB, OFB, *etc.*).

Fully unrolled DES implementation. An iterative architecture can be made combinatorial, by removing its register transfers occurring during the rounds [8]. In the case of DES, the algorithm combinatorial depth is thus roughly increased by a factor of sixteen, but the registers LR and CD remain frozen during sixteen clock cycles, which makes up for the delay through the gates. The architecture, based on that described in [9], and the floorplan are depicted in Fig. 2(a) and (b). It is a special case of the so called *brutal countermeasure* mentioned in [10], where the "glued blocks" actually make up the entire datapath. The inputs 1 of the LR multiplexer and 2 of the CD multiplexer play the role of enable for the corresponding registers. The key schedule consists in a sequence of pre-computed circular shifts which can be implemented just by switching wires and requires no logic. Such a technique is only valid for certain algorithms like DES and the absence of logic in key schedule avoids leakage. Thus attacks like [11] cannot be mounted anymore.

The synthesizers, in default mode, attempt to fit a timing path into one clock cycle. To synthesize such a design there is need to relax the timing constraints. In the combinatorial DES specific case, the logic driven by LR and CD has time equivalent to sixteen clock cycles to execute. This piece of information cannot be easily inferred, thus user constraints must be set. They basically consist in specifying spare clock cycles for some timing arcs. The timing paths that are concerned thus start at registers LR and CD, plus the Boolean signal originating from the control that tells whether the current operation is a ciphering or a deciphering , where the shifts can be interpreted left or right-wise. The "multi-cycle" constraints listed in Fig. 3 express the fact that outputs of LR and CD are sixteen times slower that the clock and that the signal to decide between ciphering and deciphering is a false timing path. This last path is indeed never critical because the choice between encryption and decryption is not modified during one computation. The key schedule can be implemented by mere routing of wires, with no logic usage. Indeed, every round key in DES is obtained by simply selecting the adequate bits from the 56 bit master key. However, this peculiar property applies to DES only and cannot be generalized for all the cryptographic algorithms.

3   Experimental Results

We implemented an iterative DES and a fully unrolled DES on SecMatV2: an academic ASIC for security evaluation of cryptoprocessors implemented in 130 nm technology from STMicroelectronics. The placement constraint used for both modules is that their placement density is 95%. Therefore we found that iterative DES consumes an area of 24787 μm^2 while the unrolled DES consumes an area of 139816 μm^2. The ratio in terms of surface is thus as low as 5.64 lower than expected *i.e.* 16, the unrolling factor which is due to removal of registers, removal of logic involved in the iteration management (multiplexers), round boundaries optimization. Also the key schedule is completely dissolved in mere routing which is a property specific to DES algorithm. In terms of performance for a nominal operating frequency, the iterative DES needs almost 5 times more time for single encryption. However, the operating frequency is not the maximal operating frequency in this case.

The average side-channel curves for one DES encryption are shown in Fig. 4(a) and 4(b) respectively for the iterative reference DES and the combinatorial instance. It clearly appears in Fig. 4 that the variations increase during the encryption.

Side-channel attacks can be roughly divided into two categories. On one hand correlation attacks make the assumption of a known leakage model; several

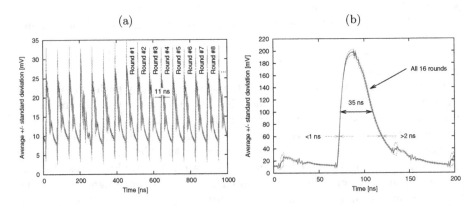

Fig. 4. (a) Sequential iterative DES encryption signature, with the average variation margin, for statistics collected on 10k measurements. (b) Average combinatorial DES encryption signature, with the average variation margin, for statistics collected on 100k measurements.

Table 1. Key recovery attack on the iterative reference DES using a CPA over 10K traces

Sbox index	Key Actual	Key Guessed	Lock_t $0 \leq \cdot \leq 10\,000$	SNR	Max CPA [%]
1	56	56	4 314	4.38603	8.40
2	11	11	7 848	3.94818	5.68
3	59	59	1 247	5.29027	6.81
4	38	38	3 555	5.09747	5.94
5	0	0	2 272	7.25941	8.86
6	13	13	3 868	4.52662	8.10
7	25	25	4 399	4.69634	6.28
8	55	55	273	6.81590	14.68

models corresponding to different values of the secret are devised. The model that correlate the better with the concrete measurements discloses the secret. On the other hand, template attacks divide into two steps. The first step is done off-line; it consists in pre-characterizing the circuit in an almost blind fashion, for as many representative values of the message and key inputs. Stochastic attacks are a variant where the pre-characterization is made more simple by injecting some partial knowledge about the target's leakage. The second step is the on-line attack proper. The attacker attempts to recognize the secret by matching measurements obtained from a fixed albeit unknown secret key.

We show that correlation attacks are made very implausible on a fully combinatorial implementation, due to the signal's desynchronization, even in the early rounds (represented in Fig. 5). First of all, we apply the same attack that is successful on the iterative reference implementation. It consists in a correlation of the measurements with the consecutive values of the right datapath register R_0, that leaks $\mathcal{L}(initial : R_0, final : L_0 \oplus f(R_0, K_1)) = |R_0 \oplus L_0 \oplus f(R_0, K_1)|$. The attack results on DES iterative and unrolled are shown in Tab. 1 and 2 respectively . Without any surprise, this attack completely fails on the combinatorial instance of DES, since the targeted transition has disappeared in the unrolled implementation. We would like to emphasize that each time a encryption is done, the datapath should be cleared. This can be done like precharge in DPL or by propagating random values without interference from the key. This is because, if two consecutive computations are done then some correlation can be found on the basis of previous computation.

3.1 Attack on the Unrolled DES

Now let us see a case when the previously described constraints are not respected i.e. two encryption are done without clearing the datapath. We explore two leakage models, namely the Hamming weight (HW) and the Hamming distance (HD), on two neuralgic positions of the algorithm, namely the Feistel function output (P1) and the round output right half (P2). We find that the HD on P1

Table 2. Key recovery attack on the unrolled DES using a CPA over 100K traces

Sbox index	Key		Lock_t $0 \le \cdot \le 10\,000$	SNR	Max CPA [%]
	Actual	Guessed			
1	56	58	87 976	1.83827	3.25
2	11	21	75 073	3.04394	1.52
3	59	17	97 462	2.07826	2.69
4	38	25	71 369	1.63005	4.85
5	0	53	70 590	3.45533	2.18
6	13	26	99 982	3.01725	1.18
7	25	22	70 433	2.07131	3.37
8	55	47	74 552	2.78395	3.26

Table 3. Key recovery attack using the a CPA with a Hamming distance model (with respect to the previous encryption) over 100K traces

Sbox index	Key		Lock_t $0 \le \cdot \le 100\,000$	SNR	Max CPA [%]
	Actual	Guessed			
1	56	56	16 557	2.20267	2.17
2	11	11	44 092	2.15008	2.09
3	59	59	36 090	2.50697	2.22
4	*38*	*9*	*3 291*	*3.73242*	*5.01*
5	0	0	27 164	1.96649	2.28
6	13	13	20 138	2.13591	2.65
7	25	25	17 862	2.11245	2.86
8	55	55	37 317	2.77701	2.75

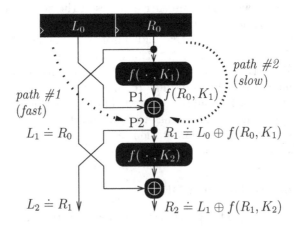

Fig. 5. Notations used to describe the combinatorial DES leakage functions

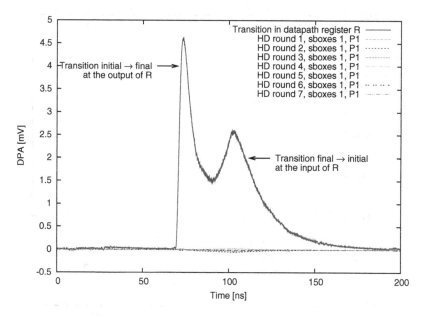

Fig. 6. DPA covariance for the register transfer R_0, and round correlations for the first sbox outputs

completely discloses the key. The results are given in Tab. 3. We can see that for all the eight broken substitution boxes, the signal-to-noise ratio (SNR) is much smaller than for the case of the reference circuit. The results for the sbox 4 are printed in italics, because actually two keys are guessed simultaneously in a unrolled implementation, due to a mathematical property of this sbox. The fourth sbox S_4 of DES has the following property: $\forall x, y \in \{0,1\}^6$, $S_4(x) \oplus S_4(y)$ and $S_4(x \oplus \texttt{0x2f}) \oplus S_4(y \oplus \texttt{0x2f})$ are palindromic. This fact can be shown by computing exhaustively the two expressions and comparing them.

Therefore, we have a remarkable Hamming distance conservation property: $\forall x, y \in \{0,1\}^6$, $|S_4(x) \oplus S_4(y)| = |S_4(x \oplus \texttt{0x2f}) \oplus S_4(y \oplus \texttt{0x2f})|$. As a conclusion, in a Hamming distance model, two keys are retrieved in pairs: the correct one and one another (false), equal to the correct key translated by $\texttt{0x2f}$.

To show that the correlations of the sboxes output (locus P1) are very disrupted due to their combinatorial nature, we have computed the DPA peaks, shown in Fig. 6. We favor DPA [12] over CPA [7], because, as explained in the technical article [13], the covariance used by DPA extracts the activity of some nets in the netlist, which is interesting for leakage characterization. As for the CPA, it is more suitable for attacks, because the normalization by the trace standard deviation corrects the fact that the leakage is not necessarily maximum at the times where the side-channel is [14]. The DPA covariance $|f(R_r^{-1}, K_{r+1}) \oplus f(R_r, K_{r+1})|$ for all $r \in [0,6]$ are plotted in Fig. 6. We have also added the transition in R_0 between two consecutive messages, because it indicates the computation beginning and its end. The beginning consists of the

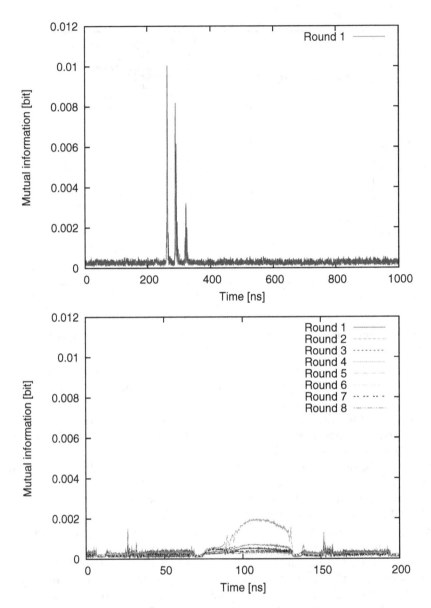

Fig. 7. Mutual information metric for sequential (*top*) and combinatorial (*bottom*) DES

R_0 register sampling at the rising edge of the clock. The end corresponds to the other transition (final → initial), in the R_0 register input latches, that are transparent, and that dissipate even in the absence of a clock event. We observe that the DPA covariances do not especially show peaks ordered in time. This indicates the link between the data and the side-channel measurement is destroyed as early as the first couple of rounds.

To conclude with the security analysis, we discuss briefly on the unsuitability of other SCAs. Template attacks are expected to become less a concern as technology typical feature sizes shrink and characteristics dispersion increases [15]. Preliminary works on 130 nm technologies [16] suggest that the intra-die technological mismatches are the preponderant source of variation, surpassing the imperfections of the logic style.

3.2   Evaluation Based on Mutual Information Metric

Mutual information analysis (MIA) has been introduced in [17] and further discussed in [18]. This analysis captures whatsoever dependence between measurements and a leakage model. It is thus a tool suited for an information leakage evaluation, as pointed out in [19]. The default leakage model does not assume any device-specific knowledge. Therefore it considers plain dependency with one sensitive and predicable word within the device. The notions of sensitivity and predictability have been defined in [20]. Basically, a variable is sensitive if it depends on one secret, and predictable if testing all the hypotheses for this variable is computationally tractable. The leakage-agnostic approach is the one employed in template attacks [21].

We have computed the mutual information (MI) between the right half of the datapath for sbox #1 and each point of our experimental traces. The results are plotted in Fig. 7 for the 80k traces of the iterative DES module and the 100k traces of the unrolled one. In the iterative circuit, the MI is roughly the same for each round. However, it depends on the round index for the combinatorial circuit; therefore we represent a couple of them in Fig. 7. It appears clearly that the sequential circuit is leaking more information about the first round than the combinatorial. Hence the vulnerability is less significant for our proposed countermeasure.

4   Conclusion and Perspectives

Information masking and hiding are two protection techniques against side-channel attacks. We propose a new countermeasure which comprises unrolling of rounds of a cryptographic algorithm to execute during a single clock. Results show that unrolling is secure against power attacks with a constraint of clearing the datapath after each encryption. We also evaluated mutual information metric on the design and results show that unrolled DES is less vulnerable. Further work involves testing this countermeasure with other algorithms like AES, etc. Also it could be interesting to partially unroll the algorithm like the rounds which are soft targets for an attacker.

Finally, we mention the potential advantage of algorithms unrolling against some fault attacks; for instance, it is impossible to inject faults via a setup time violation [22, 23, 24], produced by either under-powering or over-clocking the unrolled module. The resistance of partially or completely unrolled architectures against other DFAs is thus an interesting research direction.

Acknowledgments

This work has been partly financed by the french national research agency (ANR), through the ANR-07-ARFU-010 grant "SeFPGA" (Secured Embedded FPGAs). We acknowledge interesting discussions and encouragements from Renaud Pacalet from the LabSoC laboratory of TELECOM ParisTech at Sophia-Antipolis.

References

1. Kocher, P.C., Jaffe, J., Jun, B.: Differential Power Analysis. In: Wiener, M. (ed.) CRYPTO 1999. LNCS, vol. 1666, pp. 388–397. Springer, Heidelberg (1999)
2. Biham, E., Shamir, A.: Differential Fault Analysis of Secret Key Cryptosystems. In: Kaliski Jr., B.S. (ed.) CRYPTO 1997. LNCS, vol. 1294, pp. 513–525. Springer, Heidelberg (1997)
3. TELECOM ParisTech SEN research group: DPA Contest (2008–2009), http://www.DPAcontest.org/
4. Akkar, M.L., Giraud, C.: An Implementation of DES and AES Secure against Some Attacks. In: Koç, Ç.K., Naccache, D., Paar, C. (eds.) CHES 2001. LNCS, vol. 2162, pp. 309–318. Springer, Heidelberg (2001)
5. Tiri, K., Verbauwhede, I.: A Logic Level Design Methodology for a Secure DPA Resistant ASIC or FPGA Implementation. In: DATE 2004, Paris, France, pp. 246–251. IEEE Computer Society, Los Alamitos (2004)
6. Standaert, F.X., Örs, S.B., Preneel, B.: Power Analysis of an FPGA: Implementation of Rijndael: Is Pipelining a DPA Countermeasure? In: Joye, M., Quisquater, J.-J. (eds.) CHES 2004. LNCS, vol. 3156, pp. 30–44. Springer, Heidelberg (2004)
7. Brier, É., Clavier, C., Olivier, F.: Correlation Power Analysis with a Leakage Model. In: Joye, M., Quisquater, J.-J. (eds.) CHES 2004. LNCS, vol. 3156, pp. 16–29. Springer, Heidelberg (2004)
8. Guilley, S., Chaudhuri, S., Sauvage, L., Danger, J.L., Beyrouthy, T., Fesquet, L.: Updates on the Potential of Clock-Less Logics to Strengthen Cryptographic Circuits against Side-Channel Attacks. In: ICECS, Medina, Yasmine Hammamet, Tunisia. IEEE, Los Alamitos (2009)
9. Guilley, S., Hoogvorst, P., Pacalet, R.: A Fast Pipelined Multi-Mode DES Architecture Operating in IP Representation. Integration, The VLSI Journal 40, 479–489 (2007)
10. Roche, T., Tavernier, C.: Multi-Linear cryptanalysis in Power Analysis Attacks: MLPA. In: WEWoRC 2009, Graz, Austria (2009)
11. Aabid, M.A.E., Guilley, S., Hoogvorst, P.: Template Attacks with a Power Model. Cryptology ePrint Archive, Report 2007/443 (2007), http://eprint.iacr.org/2007/443/
12. Kocher, P.C., Jaffe, J., Jun, B.: Differential Power Analysis. In: Wiener, M. (ed.) CRYPTO 1999. LNCS, vol. 1666, pp. 388–397. Springer, Heidelberg (1999)
13. Guilley, S., Hoogvorst, P., Pacalet, R., Schmidt, J.: Improving Side-Channel Attacks by Exploiting Substitution Boxes Properties. In Presse Universitaire de Rouen et du Havre, ed.:BFCA, Paris, France, May 02-04, pp. 1–25 (2007), http://www.liafa.jussieu.fr/bfca/books/BFCA07.pdf

14. Guilley, S., Sauvage, L., Danger, J.L., Selmane, N., Pacalet, R.: Silicon-level solutions to counteract passive and active attacks. In: FDTC, 5th Workshop on Fault Detection and Tolerance in Cryptography, pp. 3–17. IEEE-CS, Washington (2008), Up-to-date version on HAL: http://hal.archives-ouvertes.fr/hal-00311431/en/

15. Quisquater, J.J., Standaert, F.X.: Physically Secure Cryptographic Computations: From Micro to Nano Electronic Devices. In: DSN, Workshop on Dependable and Secure Nanocomputing (WDSN)., Edinburgh, UK, 2 pages. IEEE Computer Society, Los Alamitos (2007) (invited talk)

16. Guilley, S., Flament, F., Pacalet, R., Hoogvorst, P., Mathieu, Y.: Security Evaluation of a Balanced Quasi-Delay Insensitive Library. In: DCIS, Grenoble, France, 6 pages. IEEE, Los Alamitos (2008); Session 5D – Reliable and Secure Architectures, full text in HAL: http://hal.archives-ouvertes.fr/hal-00283405/en/

17. Gierlichs, B., Batina, L., Tuyls, P., Preneel, B.: Mutual information analysis. In: Oswald, E., Rohatgi, P. (eds.) CHES 2008. LNCS, vol. 5154, pp. 426–442. Springer, Heidelberg (2008)

18. Prouff, E., Rivain, M.: Theoretical and Practical Aspects of Mutual Information Based Side Channel Analysis. In: Abdalla, M., Pointcheval, D., Fouque, P.-A., Vergnaud, D. (eds.) ACNS 2009. LNCS, vol. 5536, pp. 499–518. Springer, Heidelberg (2009)

19. Veyrat-Charvillon, N., Standaert, F.X.: Mutual Information Analysis: How, When and Why? In: Clavier, C., Gaj, K. (eds.) CHES 2009. LNCS, vol. 5747, pp. 429–443. Springer, Heidelberg (2009)

20. Standaert, F.X., Peeters, É., Rouvroy, G., Quisquater, J.J.: An Overview of Power Analysis Attacks Against Field Programmable Gate Arrays. Proceedings of the IEEE 94, 383–394 (2006) (invited paper)

21. Chari, S., Rao, J.R., Rohatgi, P.: Template Attacks. In: Kaliski Jr., B.S., Koç, Ç.K., Paar, C. (eds.) CHES 2002. LNCS, vol. 2523, pp. 13–28. Springer, Heidelberg (2003)

22. Faurax, O., Tria, A., Freund, L., Bancel, F.: Robustness of circuits under delay-induced faults: test of AES with the PAFI tool. In: IOLTS, Heraklion, Crete, Greece, pp. 185–186. IEEE Computer Society, Los Alamitos (2007)

23. Selmane, N., Guilley, S., Danger, J.L.: Setup Time Violation Attacks on AES. In: EDCC, The seventh European Dependable Computing Conference, Kaunas, Lithuania, pp. 91–96 (2008) ISBN: 978-0-7695-3138-0, doi:10.1109/EDCC-7.2008.11

24. Khelil, F., Hamdi, M., Guilley, S., Danger, J.L., Selmane, N.: Fault Analysis Attack on an FPGA AES Implementation. In: NTMS, Tangier, Morocco, pp. 1–5. IEEE, Los Alamitos (2008)

Fault Attacks Against EMV Signatures

Jean-Sébastien Coron[1], David Naccache[2], and Mehdi Tibouchi[2]

[1] Université du Luxembourg
6, rue Richard Coudenhove-Kalergi
L-1359 Luxembourg, Luxembourg
jean-sebastien.coron@uni.lu
[2] École normale supérieure
Département d'informatique, Groupe de Cryptographie
45, rue d'Ulm, F-75230 Paris CEDEX 05, France
{david.naccache,mehdi.tibouchi}@ens.fr

Abstract. At CHES 2009, Coron, Joux, Kizhvatov, Naccache and Paillier (CJKNP) exhibited a fault attack against RSA signatures with partially known messages. This fault attack allows factoring the public modulus N. While the size of the unknown message part (UMP) increases with the number of faulty signatures available, the complexity of CJKNP's attack increases exponentially with the number of faulty signatures.

This paper describes a simpler attack, whose complexity remains polynomial in the number of faults; consequently, the new attack can handle much larger UMPs. The new technique can factor N in a fraction of a second using ten faulty EMV signatures – a target beyond CJKNP's reach. We also show how to apply the attack even when N is unknown, a frequent situation in real-life attacks.

Keywords: Fault Attacks, Digital Signatures, RSA, ISO/IEC 9796-2, EMV.

1 Introduction

RSA [20] is certainly the most widely used signature scheme. To sign a message m with RSA, the signer first applies an encoding function μ to m, and then computes the signature $\sigma = \mu(m)^d \bmod N$. To verify the signature, the receiver checks that

$$\sigma^e = \mu(m) \mod N.$$

The Chinese Remainder Theorem (CRT) is often used to reduce the signer's computational load. This is done by computing:

$$\sigma_p = \mu(m)^d \mod p \quad \text{and} \quad \sigma_q = \mu(m)^d \mod q$$

and the signature σ is computed from σ_p and σ_q by Chinese Remaindering.

In [2], Boneh, DeMillo and Lipton showed that RSA implementations can be vulnerable to fault attacks (see also [15]). Assuming that the attacker can induce a fault when σ_q is computed while keeping the computation of σ_p correct, one gets

$$\sigma_p = \mu(m)^d \mod p \quad \text{and} \quad \sigma_q \neq \mu(m)^d \mod q$$

J. Pieprzyk (Ed.): CT-RSA 2010, LNCS 5985, pp. 208–220, 2010.
© Springer-Verlag Berlin Heidelberg 2010

and the resulting (faulty) signature σ satisfies

$$\sigma^e = \mu(m) \mod p \quad \text{and} \quad \sigma^e \neq \mu(m) \mod q .$$

Whereby the attacker can then factor N by

$$\gcd(\sigma^e - \mu(m) \bmod N, N) = p . \tag{1}$$

It is easy to see that Boneh et al.'s fault attack applies to any deterministic RSA encoding, e.g. the Full Domain Hash (FDH) [1] encoding where $\sigma = H(m)^d \mod N$ and $H : \{0,1\}^* \mapsto \mathbb{Z}_N$ is a hash function. The attack is also applicable to probabilistic signature schemes where the randomizer used to generate the signature is sent along with the signature, as in the PFDH signature scheme [7].

However, if the randomizer is only recovered when verifying the signature, or if some part of the message is unknown, the attack is thwarted. For example, consider a signature $\sigma = (m\|r)^d \mod N$. The random r is only recovered when verifying a *correct* signature. Given a faulty signature, the attacker cannot retrieve r nor infer $(m\|r)$ which would be necessary to compute $\gcd(\sigma^e - (m\|r) \bmod N, N) = p$.

At the CHES 2009 conference, Coron, Joux, Kizhvatov, Naccache and Paillier (CJKNP) showed how to extend Boneh et al.'s attack to RSA signatures with partially unknown messages (or unknown nonces) [4]. CJKNP's attack was illustrated with a probabilistic variant of the ISO/IEC 9796-2 standard [13], as used in the EMV specifications [10]. In ISO/IEC 9796-2 the encoded message has the form:

$$\mu(m) = 6\mathsf{A}_{16} \, \| \, m[1] \, \| \, H(m) \, \| \, \mathsf{BC}_{16}$$

where $m = m[1] \, \| \, m[2]$ is split into two parts. CJKNP show that if the unknown part of $m[1]$ is not too large (e.g. less than 160 bits for a 2048-bit RSA modulus), then a single faulty signature allows to factor N as in Boneh et al.'s attack. CJKNP's attack is based on a technique due to Herrmann and May [9] for finding small roots of linear equations modulo an unknown factor p of N; [9] is itself based on Coppersmith's technique [3] for finding small roots of polynomial equations using the LLL algorithm [18]. In addition, [4] introduced a multi-fault attack using an extension of Coppersmith's attack. Multiple faults make it possible to attack larger unknown message parts (UMPs). However, this comes at the cost of a complexity *exponential* in the number of faults.

This paper describes a simpler multiple fault attack. The new attack's complexity is polynomial in the number of faulty signatures. This allows us to tackle larger UMPs which were beyond [4]'s reach. For example, in a typical use case of EMV, ten faulty signatures are enough to factor N in a fraction of a second with our new attack, whereas the attack in [4] was completely impractical in such a situation.

Finally, we show that a similar technique can even recover N from a collection of valid signatures, so that the attack can be applied to protocols in which *public* RSA parameters are not available to outsiders, which do arise in practice (e.g. proprietary banking cards or e-passports).

2   Coron-Joux-Kizhvatov-Naccache-Paillier's Attack

2.1   ISO/IEC 9796-2 Standard with Partially Unknown Message

[4] considers a randomized version of the ISO/IEC 9796-2 standard. ISO/IEC 9796-2 is an encoding standard allowing partial or total message recovery [13,14]. The encoding can be used with hash functions $H \colon \{0,1\}^* \to \{0,1\}^{k_h}$ of various digest sizes k_h. When k_h, the size of m and the size of N (denoted k) are all multiples of 8, the ISO/IEC 9796-2 encoding of a message $m = m[1] \,\|\, m[2]$ is

$$\mu(m) = 6\mathsf{A}_{16} \,\|\, m[1] \,\|\, H(m) \,\|\, \mathsf{BC}_{16}$$

where $m[1]$ consists of the $k - k_h - 16$ leftmost bits of m and $m[2]$ represents the remaining bits of m. In [14] it is required that $k_h \geq 160$. This is also the case in the EMV specifications [10]. We note that a practical forgery attack (without faults) against ISO/IEC 9796-2 was recently described in [6], extending the attack in [5]. However the attack is only practical when $m[1]$ can be fully chosen by the adversary, which is not the case in EMV and in the randomized version of the ISO/IEC 9796-2 standard considered in this paper.

More precisely, [4] considers a message $m = m[1] \,\|\, m[2]$ of the form

$$m[1] = \alpha \,\|\, r \,\|\, \alpha', \quad m[2] = \text{DATA}$$

where r is a message part unknown to the adversary (UMP), α and α' are strings known to the adversary and DATA is some known or unknown string. The size of r is denoted by k_r and the size of $m[1]$ is $k - k_h - 16$ as required in ISO/IEC 9796-2. The encoded message is then

$$\mu(m) = 6\mathsf{A}_{16} \,\|\, \alpha \,\|\, r \,\|\, \alpha' \,\|\, H(\alpha \,\|\, r \,\|\, \alpha' \,\|\, \text{DATA}) \,\|\, \mathsf{BC}_{16} \qquad (2)$$

Therefore both r and $H(\alpha \,\|\, r \,\|\, \alpha' \,\|\, \text{DATA})$ are unknown; the total number of unknown bits inside $\mu(m)$ is then $k_r + k_h$.

2.2   Single Fault Attack

[4] describes a fault attack against the previous signature scheme. More precisely, one assumes that after injecting a fault the opponent has a faulty signature σ such that:

$$\sigma^e = \mu(m) \mod p, \quad \sigma^e \neq \mu(m) \mod q. \qquad (3)$$

From (2) one can write

$$\mu(m) = t + r \cdot 2^{n_r} + H(m) \cdot 2^8 \qquad (4)$$

where t is a known value. From (3) one gets:

$$\sigma^e = t + r \cdot 2^{n_r} + H(m) \cdot 2^8 \mod p.$$

This shows that $(r, H(m))$ must be a solution of the equation

$$a + b \cdot x + c \cdot y = 0 \mod p \tag{5}$$

where $a := t - \sigma^e \mod N$, $b := 2^{n_r}$ and $c := 2^8$ are known. This bivariate equation in x, y has a small root $(x_0, y_0) = (r, H(m))$. To solve this equation, one can use a result by Herrmann and May [9] based on Coppersmith's technique for finding small roots of polynomial equations [3].

Coppersmith's technique uses LLL to obtain two polynomials $h_1(x, y)$ and $h_2(x, y)$ such that

$$h_1(x_0, y_0) = h_2(x_0, y_0) = 0$$

holds over the integers. Then one computes the resultant between h_1 and h_2 to recover the common root (x_0, y_0). To that end, one must assume that h_1 and h_2 are algebraically independent. This ad hoc assumption makes the method heuristic; nonetheless it turns out to work quite well in practice. Then, given the root (x_0, y_0) one recovers the randomized encoded message $\mu(m)$ and factors N by GCD.

Assuming that $r < N^\gamma$ and $H(m) \leq N^\delta$, for a balanced RSA modulus one gets the condition:

$$\gamma + \delta \leq \frac{\sqrt{2} - 1}{2} \cong 0.207 \tag{6}$$

This means that for a 1024-bit modulus N, the total size of the unknowns x_0 and y_0 can be at most 212 bits. For ISO/IEC 9796-2 signatures with $k_h = 160$, the unknown r can thus be at most 52 bits long.

2.3 Extension to Several Faults Modulo the Same Factor

[4] shows how to extend the attack to multiple faults, in order to improve the bound on the UMP's size. More precisely, given ℓ faults, one gets a system of equations:

$$a_i + b \cdot x_i + c \cdot y_i = 0 \mod p$$

for $1 \leq i \leq \ell$, where a_i, b and c are known and x_i and y_i are unknown and small. The goal being still to recover p. Note that we can assume that $b = 1$ by multiplying the equations by $b^{-1} \mod N$.

[4] considers a more general system where instead of known constants b and c, one considers known b_i and c_i. More precisely, we are given ℓ different polynomials

$$f_u(x_u, y_u) = a_u + x_u + c_u y_u \tag{7}$$

where each polynomial f_u has a small root (ξ_u, ν_u) modulo p with $|\xi_u| \leq X$ and $|\nu_u| \leq Y$. Note that, as in the basic case, we re-normalized each polynomial f_u to equate the coefficient of x_u in f_u to one.

[4] shows how to extend Coppersmith's attack to these multiple polynomial equations, thereby obtaining a better bound on the UMP size. Theoretically, given a sufficiently large number of faults, the extended attack could tackle

cases where $\gamma + \delta$ is asymptotically close to $\frac{1}{2}$. However, the attack's complexity grows exponentially with the number of faults ℓ; hence aiming at $\gamma + \delta$ values significantly higher than the single fault maximum of 0.207 is totally impractical. We refer the reader to Table 2 illustrating how intractable the problem gets as $\gamma + \delta$ approaches $\frac{1}{2}$.

3   A New Multiple Faults Attack

The previous attack is only applicable for a small number of faults because the lattice dimension grows exponentially with the number of faults. This section describes a different attack that can take advantage of a large number of faults and thus handle much longer UMPs. Indeed, in the new attack, the lattice dimension remains equal to the number of faults, plus one.

As previously, we consider an encoding function given by equation (4)

$$\mu(m) = t + r \cdot 2^{n_r} + H(m) \cdot 2^8$$

Given a set of faulty signatures σ_i such that:

$$\sigma_i^e = \mu(m_i) = t + r_i \cdot 2^{n_r} + H(m_i) \cdot 2^8 \quad \mod p$$

we get a set of equations of the form

$$A_i + B \cdot x_i + D \cdot y_i = 0 \quad \mod p$$

where $A_i := t - \sigma_i^e \mod N$, $B := 2^{n_r}$ and $D := 2^8$ are known. As in previous section, we can assume that $B = 1$ by multiplying the equations by $B^{-1} \mod N$. This results in the following equations

$$a_i + x_i + c \cdot y_i = 0 \quad \mod p \tag{8}$$

for $1 \le i \le \ell$, where ℓ is the number of faulty signatures. Note that as opposed to equations (7) in the previous section, here we have a constant coefficient c, as in our fault attack.[1]

The new attack is similar to the one in [16]. Applying LLL [18] to the lattice spanned by the columns of the following matrix

$$\begin{pmatrix} \kappa a_1 & \kappa a_2 & \cdots & \kappa a_\ell & \kappa N \\ 1 & 0 & \cdots & 0 & 0 \\ & 1 & \ddots & \vdots & \vdots \\ & & \ddots & 0 & 0 \\ & & & 1 & 0 \end{pmatrix} \tag{9}$$

for a sufficiently large constant κ (as described in [17]), the attacker computes a short vector $(u_1, \ldots, u_\ell)$ such that

$$\sum_{i=1}^{\ell} u_i \cdot a_i = 0 \quad \mod N$$

---

[1] The attack in this section would not work with different c_i's.

This implies from (8)

$$\sum_{i=1}^{\ell} u_i \cdot x_i + c \cdot \left(\sum_{i=1}^{\ell} u_i \cdot y_i \right) = 0 \mod p$$

Letting

$$\alpha_0 := \sum_{i=1}^{\ell} u_i \cdot x_i \text{ and } \beta_0 := \sum_{i=1}^{\ell} u_i \cdot y_i \qquad (10)$$

this gives:

$$\alpha_0 + c \cdot \beta_0 = 0 \mod p$$

Therefore the vector (α_0, β_0) belongs to the lattice

$$L(c,p) = \{ (\alpha, \beta) \in \mathbb{Z}^2 \mid \alpha + c \cdot \beta = 0 \mod p \} \qquad (11)$$

Since the x_i's and y_i's are small, we see that if the u_i's are small, then (α_0, β_0) is a short vector in the lattice $L(c,p)$. More precisely, let v be a shortest non-zero vector of $L(c,p)$. If the u_i's are sufficiently small such that $\|(\alpha_0, \beta_0)\| < \|v\|$, then by definition of v we must have $\alpha_0 = \beta_0 = 0$. In this case we get:

$$\sum_{i=1}^{\ell} u_i \cdot x_i = \sum_{i=1}^{\ell} u_i \cdot y_i = 0$$

which means that the known vector $(u_1, \ldots, u_\ell)$ is orthogonal (in $\mathbb{Z}$) to the two unknown vectors $(x_1, \ldots, x_\ell)$ and $(y_1, \ldots, y_\ell)$.

Actually, the LLL reduction of lattice (9) yields many other vectors (u_i) which are orthogonal in $\mathbb{Z}$ to both (x_i) and (y_i). Assuming that we can generate $\ell - 2$ such vectors, we can then obtain a bi-dimensional lattice containing both vectors $\boldsymbol{x} = (x_i)$ and $\boldsymbol{y} = (y_i)$. Let $\boldsymbol{x'}$ and $\boldsymbol{y'}$ be a basis of this lattice. Such basis can be obtained by applying LLL a second time to the lattice spanned by the columns of:

$$\begin{pmatrix} \kappa' u_{1,1} & \cdots & \kappa' u_{1,\ell} \\ \vdots & & \vdots \\ \kappa' u_{\ell-2,1} & \cdots & \kappa' u_{\ell-2,\ell} \\ 1 & & \\ & \ddots & \\ & & 1 \end{pmatrix}$$

for a sufficiently large constant κ'.

Consider now a vector $\boldsymbol{v} = (v_i)$ that is orthogonal modulo N to both $\boldsymbol{x'}$ and $\boldsymbol{y'}$, that is:

$$\sum_{i=1}^{\ell} v_i \cdot x_i' = 0 \mod N, \qquad \sum_{i=1}^{\ell} v_i \cdot y_i' = 0 \mod N$$

Then since x and y belong to the lattice spanned by x' and y', we must have

$$\begin{cases} x = \alpha \cdot x' + \beta \cdot y' \\ y = \alpha' \cdot x' + \beta' \cdot y' \end{cases}$$

for some $\alpha, \alpha', \beta, \beta' \in \mathbb{Z}$. This implies:

$$\sum_{i=1}^{\ell} v_i \cdot x_i = 0 \quad \mod N, \qquad \sum_{i=1}^{\ell} v_i \cdot y_i = 0 \quad \mod N$$

Then from equation (8) this gives:

$$\sum_{i=1}^{\ell} v_i \cdot a_i = 0 \quad \mod p \tag{12}$$

Therefore $\gcd(\sum_i v_i \cdot a_i, N) = p$. Note that the previous computation can be simplified by restricting ourselves to the three first components of x' and y'; in that case, one obtains a unique (up to a multiplicative constant) tri-dimensional vector v orthogonal modulo N to both the first three components of x' and the first three components of y'. Then equation (12) still holds and as previously $\gcd(\sum_i v_i \cdot a_i, N) = p$.

It remains to justify why we can have $\|(\alpha_0, \beta_0)\| < \|v\|$, where v is a shortest non-zero vector of $L(c, p)$. We provide an argument similar to [19] (see also [8]) for higher lattice dimensions. We define a lattice $L(c, p)$ to be B-good if any non-zero vector has a norm $> B$; we say that the lattice is B-bad otherwise. Consider a fixed prime p. By definition of lattice $L(c, p)$ in (11), the value of c modulo p is determined by any non-zero vector in $L(c, p)$. Since there are at most $4B^2$ vectors in the disc of radius B, there are at most $4B^2$ lattices $L(c, p)$ which are B-bad. Therefore for a random c modulo p, the probability that a lattice is B-bad is at most $4B^2/p$. Taking $B := \sqrt{p}/3$, the probability that a lattice is B-bad is then at most $\frac{1}{2}$.

Therefore a lattice is B-good with probability at least $\frac{1}{2}$. This implies that if $\|(\alpha_0, \beta_0)\| \leq \sqrt{p}/3$, then with probability at least $\frac{1}{2}$ the vector (α_0, β_0) is shorter than the shortest non-zero vector in $L(c, p)$, which implies that $\alpha_0 = \beta_0 = 0$ as required.

Here we have considered a fixed p and a random c modulo p; however in our attack c is a fixed integer and p is random; therefore the previous analysis is heuristic only. More generally, if integers α_0 and β_0 have different sizes, we obtain that $\alpha_0 = \beta_0 = 0$ under the condition:

$$|\alpha_0| \cdot |\beta_0| < \frac{p}{3} \tag{13}$$

Using LLL we expect to obtain vectors (u_i) of norm roughly $N^{1/\ell}$, where ℓ is the number of faulty signatures (see [16]). Let X and Y be upper bounds for the unknowns x_i and y_i. We thus obtain from (10)

$$|\alpha_0| \leq N^{1/\ell} \cdot X, \quad |\beta_0| \leq N^{1/\ell} \cdot Y$$

From (13) we obtain the following bound

$$N^{2/\ell} \cdot X \cdot Y < \frac{p}{3}$$

With $X = N^\gamma$ and $Y = N^\delta$ this yields approximately

$$\frac{2}{\ell} + \gamma + \delta < \frac{1}{2} \tag{14}$$

Therefore, for a large number of faults ℓ we obtain the same asymptotic bound $\gamma + \delta < \frac{1}{2}$ as in [4]; however the lattice dimension in (9) is only $\ell + 1$ instead of being exponential in ℓ as in [4]. In Section 5 we provide the result of practical simulations validating the attack. We then apply the new technique to the EMV specifications in Section 6.

4   Recovering Unknown Moduli

In many practical situations, the modulus N may not be available to the attacker. While contrary to a basic cryptographic assumption, this is frequently the case in proprietary applications. The technique described in the previous section can be adapted to recover *unknown* moduli N from *correct* signatures when the public exponent e is small. Once N has been recovered, one can then apply the fault attack described in the previous section.

As previously, we consider an encoding function given by equation (4)

$$\mu(m) = t + r \cdot 2^{n_r} + H(m) \cdot 2^8$$

Given a set of ℓ correct signatures σ_i such that

$$\sigma_i^e = \mu(m_i) = t + r_i \cdot 2^{n_r} + H(m_i) \cdot 2^8 \bmod N$$

we obtain a set of ℓ equations of the form

$$A_i + B \cdot x_i + D \cdot y_i = 0 \quad \bmod N \tag{15}$$

where $A_i := t - \sigma_i^e$, $B := 2^{n_r}$ and $D := 2^8$ are known, but x_i, y_i and N are unknown. Note that as opposed to the previous section, A_i is not reduced modulo N; therefore the bit-size of A_i is approximately $e \cdot \log_2 N$.

As in the previous section, using LLL we can find a short vector (u_i) such that

$$\sum_{i=1}^{\ell} u_i \cdot A_i = 0$$

in $\mathbb{Z}$. This implies from (15)

$$B \cdot \left(\sum_{i=1}^{\ell} u_i \cdot x_i \right) + D \cdot \left(\sum_{i=1}^{\ell} u_i \cdot y_i \right) = 0 \quad \bmod N.$$

As previously, if the u_i's are sufficiently small, then we will have $\sum_i u_i \cdot x_i = \sum_i u_i \cdot y_i = 0$ over $\mathbb{Z}$. Then again, from $\ell - 2$ linearly independent vectors (u_i) one can recover a 2-dimensional lattice containing the two vectors (x_i) and (y_i).

We proceed by computing two vectors v_1 and v_2 which are both orthogonal in $\mathbb{Z}$ to any vector in this bi-dimensional lattice. This implies that both vectors are orthogonal in $\mathbb{Z}$ to the two vectors (x_i) and (y_i). Equation (15) implies that v_1 and v_2 are both orthogonal modulo N to the vector (A_i); therefore to recover N we simply compute the GCD of their respective scalar products with the vector (A_i).

Since the norm of the vector (A_i) is roughly N^e, we can expect (see [16]) to find a vector (u_i) of norm $\cong N^{e/(\ell-1)}$. Moreover, letting $\alpha_0 = \sum_i u_i \cdot x_i$ and $\beta_0 = \sum_i u_i \cdot y_i$, as in the previous section we must have $|\alpha_0| \cdot |\beta_0| < \frac{N}{3}$ so that $\alpha_0 = \beta_0 = 0$ holds with good (heuristic) probability. This gives the following approximate bound

$$N^{2e/(\ell-1)} \cdot X \cdot Y < \frac{N}{3}$$

i.e. using the previous notations

$$\frac{2e}{\ell - 1} + \gamma + \delta < 1 \qquad (16)$$

and the required number of correct signatures:

$$\ell > \frac{2e}{1 - \gamma - \delta} + 1$$

In other words, the number of required signatures is proportional to the public exponent e; this means the modulus recovery technique is practical only for small public exponents. We show in Section 5.2 that it then works well in practice.

5   Simulation Results

5.1   Multiple Fault Attack

We have simulated the fault attack described in Section 3 as follows: we first generate a correct $\sigma_p = \mu(m)^d \mod p$ and a random $\sigma_q \in \mathbb{Z}_q$ and then compute the faulty signature σ using the CRT. This mimics the process described in [4] where concrete faults are injected into devices generating randomized ISO/IEC 9796-2 signatures.

Simulation results are summarized in Table 1. We compute the attack's success rate for $\gamma + \delta = \frac{1}{3}$ for $12, 13$ and 14 faults. Theory predicts success with good probability when $\ell > 12$. Table 1 confirms this prediction for both balanced and unbalanced γ and δ configurations

Table 2 provides a comparison with [4]'s multi-fault attack. For large ℓ, the new attack has the asymptotic condition $\gamma + \delta < \frac{1}{2}$, identical to the theoretical asymptotic bound of [4]'s multi-fault variant. It is however considerably easier

Table 1. Attack Simulation Results Using SAGE. Random 1024-bit moduli. Single 2.5 GHz Intel CPU core.

Number of faults ℓ	12	13	14
Success rate with $\gamma = \delta = \frac{1}{6}$	13%	100%	100%
Success rate with $\gamma = \frac{1}{4}$, $\delta = \frac{1}{12}$	0%	100%	100%
Average CPU time (seconds)	0.19	0.14	0.17

to deal with cases where $\gamma + \delta$ approaches $\frac{1}{2}$ with the new attack than it is in [4]. Namely, as illustrated in Table 2 when $\gamma + \delta$ approaches $\frac{1}{2}$ [4]'s lattice dimension makes the attack completely impractical. In particular, we show in Section 6 that the new attack allows to attack EMV signature formats that were beyond [4]'s reach.

However we note that for smaller $\gamma + \delta$ values, [4] can be more practical since it requires fewer faulty signatures; for example for $\gamma + \delta = 0.214$ only 2 faulty signatures are required instead of eight in the new attack. In other words, the two techniques nicely complement each other and provide the attacker with a toolbox allowing to adapt his technique to the target's γ and δ configuration.

Table 2. Comparison of the new attack with [4] for a random 1024-bit modulus

$\gamma + \delta$	ℓ_{new}	ω_{new}	CPU time (new)	ℓ_{old}	ω_{old}	CPU time (old)
0.204	7	8	0.03 s	3	84	49 s
0.214	8	9	0.04 s	2	126	22 min
0.230	8	9	0.04 s	2	462	—
0.280	10	11	0.07 s	6	6188	—
0.330	14	15	0.17 s	8	2^{21}	—
0.400	25	26	1.44 s	—	—	—
0.450	70	71	36.94 s	—	—	—

Explanatory notes regarding Table 2: Table 2 provides, for several values of $\gamma + \delta$, the following information: the number of faulty signatures ℓ_{new} used in our simulation, the corresponding lattice dimension ω_{new}, and the running time of the new attack. For the attack described in [4], the table lists the minimal lattice dimension ω_{old} required to tackle the $\gamma + \delta$ values we consider and the corresponding number of faulty signatures ℓ_{old}. We find ω_{old} by exhaustive search over parameters (ℓ, t, m) with $\ell < 50$, $m < 80$. For $\gamma + \delta = 0.214$ and $\gamma + \delta = 0.23$, one can actually get away with slightly smaller lattice dimensions than indicated in the table (120 and 378 instead of 126 and 462) at the price of more faults (7 and 13 respectively).

5.2   Recovering Unknown Moduli

We have also implemented the technique described in Section 4 to recover N from correct signatures (when N is unknown to the attacker). As shown in Table 3

Table 3. Modulus recovery simulation in SAGE. Random 1024-bit moduli and $e = 3$. Single core 2.5 GHz Intel CPU.

Number of signatures ℓ	10	11	12	13
Success rate with $\gamma = \delta = \frac{1}{6}$	2%	59%	61%	61%
Success rate with $\gamma = \frac{1}{4}$, $\delta = \frac{1}{12}$	2%	62%	64%	64%
Average CPU time (seconds)	0.20	0.21	0.25	0.31

the attack is quite practical for small public exponent (e) values. More precisely, we give the success rates for $\gamma + \delta = \frac{1}{3}$ with 10 to 13 valid signatures for $e = 3$. In this case, the theoretical bound (16) predicts that the technique should succeed with good probability when $\ell > 10$; this is well verified for both balanced and unbalanced γ and δ configurations.

6   Application to EMV Signatures

6.1   The EMV Specification

EMV is a collection of industry specifications for the inter-operation of payment cards, POS terminals and ATMs. The EMV specification [10] uses ISO/IEC 9796-2 signatures to certify public-keys and to authenticate data. For instance, to authenticate itself, the payment card must issue a signature on data provided by the terminal. The signature algorithm is RSA with ISO/IEC 9796-2 using $e = 3$ or $e = 2^{16} + 1$. The bit length of all moduli is always a multiple of 8. EMV uses special message formats; 7 different formats are used, depending on the message type.

In the following, for clarity's sake, we analyze one of these formats only: the *Offline Dynamic Data Authentication, Dynamic Application Data* format, described in Book 2, Section 6.5, Table 15, page 67 of the EMV specifications [10]. The signing entity is the Card. The message m to be signed has the format $m = m[1]\|m[2]$ with :

$$m[1] = 0501_{16} \parallel \mathsf{L_{DD}} \parallel \mathsf{ICCDD} \parallel \mathsf{BB}_{16} \cdots \mathsf{BB}_{16}$$
$$m[2] = \mathrm{DATA}$$

where $\mathsf{L_{DD}}$ is a byte identifying the length (in bytes) of the ICC Dynamic Data string ICCDD, and DATA is some data provided by the terminal. In general, the ICC Dynamic Data string has the following form:

$$\mathsf{ICCDD} = \mathsf{L_{ICCDN}} \parallel \mathsf{ICCDN} \parallel \mathsf{ADD}$$

where $\mathsf{L_{ICCDN}}$ is one byte identifying the length (in bytes) of the time-variant ICC Dynamic Number ICCDN, and ADD consists of $\mathsf{L_{DD}} - \mathsf{L_{ICCDN}} - 1$ bytes of Additional Dynamic Data to be signed. It is specified that one must have $2 \le \mathsf{L_{ICCDN}} \le 8$.

As mentioned in the EMV specifications, the ICC Dynamic Number can be an unpredictable number or a counter incremented for every new signature. In a typical use case (as described, for example, as part of EMV Test 2CC.086.1 Case 07 [11]), ICCDN is a random 8-byte string generated by the card, and ADD is a variable 8-byte string, encoded according to [12]. In this case, we have:

$$m[1] = 0501_{16} \parallel 11_{16} \parallel 08_{16} \parallel \text{ICCDN} \parallel \text{ADD} \parallel \text{BB}_{16} \cdots \text{BB}_{16}$$

which can be rewritten as:

$$m[1] = \text{X} \parallel r \parallel \text{BB}_{16} \cdots \text{BB}_{16}$$

where X is a known value and r is a variable byte string of bit-size $k_r = 128$. This gives:

$$\mu(m) = 6\text{A}_{16} \parallel \text{X} \parallel r \parallel \text{BB}_{16} \cdots \text{BB}_{16} \parallel H(m) \parallel \text{BC}_{16} \tag{17}$$

where $H(m)$ is a 160-bit digest of the encoded message m. Note that the no-fault forgery attack from [6] does not apply because here $m[1]$ cannot be controlled by the adversary.

6.2   Fault Attack

The EMV format for $\mu(m)$ given in (17) is the same as the one considered in [4] and recalled in Section 2, and the same as the one considered in our new attack in Section 3. In the particular use case described above, the string X is known but the variables r and $H(m)$ are unknown to the attacker. Therefore the total number of unknown bits is:

$$k_r + k_h = 128 + 160 = 288$$

Hence, for a 1024-bit modulus N, we get:

$$\gamma + \delta = \frac{288}{1024} \approx 0.28$$

which is well beyond the range of practical applicability of [4], as shown in Table 2. However, as shown in Table 2 our new attack will factor N in a fraction of a second using about ten faulty signatures.

References

1. Bellare, M., Rogaway, P.: The Exact security of digital signatures: How to sign with RSA and Rabin. In: Maurer, U.M. (ed.) EUROCRYPT 1996. LNCS, vol. 1070, pp. 399–416. Springer, Heidelberg (1996)
2. Boneh, D., DeMillo, R.A., Lipton, R.J.: On the importance of checking cryptographic protocols for faults. Journal of Cryptology 14(2), 101–119 (2001)

3. Coppersmith, D.: Small solutions to polynomial equations, and low exponent vulnerabilities. Journal of Cryptology 10(4), 233–260 (1997)
4. Coron, J.-S., Joux, A., Kizhvatov, I., Naccache, D., Paillier, P.: Fault attacks on RSA signatures with partially unknown messages. In: Clavier, C., Gaj, K. (eds.) CHES 2009. LNCS, vol. 5747, pp. 444–456. Springer, Heidelberg (2009), eprint.iacr.org/2009/309
5. Coron, J.-S., Naccache, D., Stern, J.P.: On the security of RSA padding. In: Wiener, M. (ed.) CRYPTO 1999. LNCS, vol. 1666, pp. 1–18. Springer, Heidelberg (1999)
6. Coron, J.-S., Naccache, D., Tibouchi, M., Weinmann, R.P.: Practical cryptanalysis of ISO/IEC 9796-2 and EMV signatures. In: Halevi, S. (ed.) Advances in Cryptology - CRYPTO 2009. LNCS, vol. 5677, pp. 428–444. Springer, Heidelberg (2009), eprint.iacr.org/2009/203
7. Coron, J.-S.: Optimal security proofs for PSS and other signature schemes. In: Knudsen, L.R. (ed.) EUROCRYPT 2002. LNCS, vol. 2332, pp. 272–287. Springer, Heidelberg (2002)
8. Coron, J.-S., Joye, M., Naccache, D., Paillier, P.: Universal padding schemes for RSA. In: Yung, M. (ed.) CRYPTO 2002. LNCS, vol. 2442, pp. 226–241. Springer, Heidelberg (2002)
9. Herrmann, M., May, A.: Solving linear equations modulo divisors: On factoring given any bits. In: Pieprzyk, J. (ed.) ASIACRYPT 2008. LNCS, vol. 5350, pp. 406–424. Springer, Heidelberg (2008)
10. EMV, Integrated circuit card specifications for payment systems, Book 2. Security and Key Management. Version 4.2 (June 2008), http://www.emvco.com
11. EMV, EMVCo type approval terminal level 2 test cases. Version 4.2a (April 2009), http://www.emvco.com
12. ISO/IEC 8825-1:2002, Information technology – ASN.1 encoding rules: Specification of Basic Encoding Rules (BER), Canonical Encoding Rules (CER) and Distinguished Encoding Rules (DER) (2002)
13. ISO/IEC 9796-2, Information technology – Security techniques – Digital signature schemes giving message recovery – Part 2: Mechanisms using a hash-funcion (1997)
14. ISO/IEC 9796-2:2002 Information technology – Security techniques – Digital signature schemes giving message recovery– Part 2: Integer factorization based mechanisms (2002)
15. Joye, M., Lenstra, A., Quisquater, J.-J.: Chinese remaindering cryptosystems in the presence of faults. Journal of Cryptology 21(1), 27–51 (1999)
16. Nguyen, P., Stern, J.: Cryptanalysis of a fast public key cryptosystem presented at SAC 1997. In: Tavares, S., Meijer, H. (eds.) SAC 1998. LNCS, vol. 1556, pp. 213–218. Springer, Heidelberg (1999)
17. Nguyen, P., Stern, J.: Merkle-Hellman revisited: a cryptanalysis of the Qu-Vanstone cryptosystem based on group factorization. In: Kaliski Jr., B.S. (ed.) CRYPTO 1997. LNCS, vol. 1294, pp. 198–212. Springer, Heidelberg (1997)
18. Lenstra, A., Lenstra Jr., H., Lovász, L.: Factoring polynomials with rational coefficients. In: Mathematische Annalen, vol. 261, pp. 513–534. Springer, Heidelberg (1982)
19. Fujisaki, E., Okamoto, T., Pointcheval, D., Stern, J.: RSA-OAEP is secure under the RSA assumption. Journal of Cryptology 17(2), 81–104 (2004)
20. Rivest, R., Shamir, A., Adleman, L.: A method for obtaining digital signatures and public key cryptosystems. Communications of the ACM, 120–126 (1978)

Revisiting Higher-Order DPA Attacks:
Multivariate Mutual Information Analysis

Benedikt Gierlichs[1], Lejla Batina[1,2], Bart Preneel[1], and Ingrid Verbauwhede[1]

[1] K.U. Leuven, ESAT/SCD-COSIC and IBBT
Kasteelpark Arenberg 10, B-3001 Leuven-Heverlee, Belgium
firstname.lastname@esat.kuleuven.be
[2] Radboud University Nijmegen, CS Dept./Digital Security group
Heyendaalseweg 135, 6525 AJ Nijmegen, The Netherlands
lejla@cs.ru.nl

Abstract. Security devices are vulnerable to side-channel attacks that perform statistical analysis on data leaked from cryptographic computations. Higher-order (HO) attacks are a powerful approach to break protected implementations. They inherently demand multivariate statistics because multiple aspects of signals have to be analyzed jointly. However, most works on HO attacks follow the approach to first apply a pre-processing function to map the multivariate problem to a univariate problem and then to apply established 1st order techniques. We propose a novel and different approach to HO attacks, Multivariate Mutual Information Analysis (MMIA), that allows to directly evaluate joint statistics without pre-processing. While this approach can benefit from a good power model, it also works without an assumption. We present the first experimental results for 2nd and 3rd order MMIA as well as state-of-the-art HO attacks based on real measurements. A thorough empirical evaluation confirms the advantage of the new approach: 3rd order MMIA attacks require about 800 measurements to achieve 100% success while state-of-the-art HODPA requires 1000 measurements to achieve about 40% success. As a consequence, the security provided by the masking countermeasure needs to be reconsidered as 3rd and possibly higher order attacks become more practical.

1 Introduction

Embedded devices such as smart cards, mobile phones, and RFID tags are becoming increasingly pervasive. In order to secure the applications, these devices execute cryptographic algorithms and protocols to authenticate data and entities and to protect the confidentiality of sensitive data. An embedded device is often physically accessible and it is very likely that such a device falls into the hands of a malicious user. The physical accessibility has led to a number of very powerful attacks that include physical tampering and side-channel attacks. A typical example is Differential Power Analysis (DPA) [11]. The technique explores weaknesses of implementations rather than algorithms, allowing an attacker to extract the secret of a device by monitoring its power dissipation, if

J. Pieprzyk (Ed.): CT-RSA 2010, LNCS 5985, pp. 221–234, 2010.

no special countermeasures are taken. A successful DPA attack is subject to two conditions: i) there exists an intermediate variable in the implementation that is correlated with the power consumption and ii) this variable can be predicted with knowledge of the plaintext (or ciphertext) and by guessing a small part of the key and possibly other constants.

In order to protect devices against DPA, one can get rid of the second condition by data randomization or masking [4,8]. The idea is to conceal intermediate values through addition or multiplication with random values, which makes it impossible to correctly predict the intermediate variable.

However, this so-called 1st order masking succumbs to higher-order DPA attacks (HODPA) as originally proposed by Messerges [13] and Chari et al. [2]. These HODPA attacks are based on the joint statistical properties of multiple aspects of the signal, typically joint analysis of the power consumption at two (or more) points in time. In this case one would think of multivariate analysis but all established techniques [2,10,13,15,17,21] rely on a pre-processing step to map the multivariate problem to a univariate problem before attacking the result with a standard DPA attack. The sole exception is a recent and independent work by Prouff and Rivain [16], which we will address in Sect. 4.1.

HODPA attacks imply higher costs in terms of number of samples and computational complexity. In addition, the identification of points in time at which to take the signals is a hard problem. Another common issue is the pre-processing step: while this problem has been studied by many authors, finding the optimal transformation is still an open problem. Furthermore, it is evident that none of the solutions is generic as each pre-processing is tightly linked to a leakage model, that is not always met in practice [18]. Eventually, it is unclear how the pre-processing functions can be generalized for attacks of order higher than two without accepting enormous drawbacks.

Our contribution solves all but one of the aforementioned problems. At CHES 2008 we introduced a side-channel distinguisher called Mutual Information Analysis (MIA) [7]. This 1st order attack is effective without any knowledge or restrictive assumption about the particular dependencies between processed data and observable power consumption. Our proposal, Multivariate MIA (MMIA), inherits this important property. Also the issue of pre-processing unravels because MMIA is explicitly multivariate, which furthermore allows to easily adapt it to attacks of order higher than two. What remains is the problem of identifying the points in time at which to take the signals. Our theory is confirmed by an extensive empirical evaluation of MMIA, the approach of [16], and state-of-the-art HODPA attacks against 1st and 2nd order masked software implementations of a DES like mini-cipher.

This paper is organized as follows. In Sect. 2 we introduce our notation and summarize related work on MIA, the masking countermeasure, and in particular HODPA attacks. In Sect. 3 we discuss our motivation and formulate a generic 2nd order attack problem, for which we present a sound solution in Sect. 4 and compare it to the approach of [16]. Section 5 deals with the empirical evaluation of our proposal and state-of-the-art HODPA attacks. We conclude our work in Sect. 6.

2   Preliminaries

A device performs a cryptographic computation $E_k(x)$ under some key k from a keyspace $\mathcal{K} = \{0,1\}^m$. Since the key is unknown it is modeled as a random variable (RV) $\mathbf{K}$ on $\mathcal{K}$ with a priori uniform probability mass function (PMF). The information leakage of the device, due to its physical properties, is called side-channel leakage. Side-channel key-recovery attacks exploit this side-channel leakage to recover the key k. Typically, a differential attack consists of statistically analyzing the side-channel leakage related to a well chosen, key and input dependent intermediate value of the computation $E_k(x)$. We model this intermediate value as a RV $\mathbf{W}_k$ on a space $\mathcal{W} = \{0,1\}^n$ where n is the word length of the device. The corresponding side-channel leakage is denoted $\mathbf{L}_k$. Non-profiled side-channel attacks usually ignore the presence of noise and measurement inaccuracy, and assume that the measurement of a physical observable (here power consumption) provides direct access to the leakage $\mathbf{L}_k$. In this idealized case $\mathbf{L}_k$ exists on a space with at most 2^n elements.

The central problem in differential attacks is to decide whether two RVs are statistically (in-)dependent. The RV $\mathbf{L}_k$ is the measured side-channel leakage. The other RV $\mathbf{P}_{k'}$ is the predicted side-channel leakage of the same computation that depends on a key guess k', the input, and a leakage model. DPA attacks apply a statistical test $T(\mathbf{L}_k, \mathbf{P}_{k'})$ for all key guesses k' that measures whether the RVs are statistically dependent or not. The value of k' that leads to the highest dependence is an adversary's best guess. In this context, statistical tests are frequently called distinguishers in the literature [19].

MIA's core is the mutual information based distinguisher that implements the test T by computing the mutual information values

$$\mathbf{I}(\mathbf{L}_k; \mathbf{P}_{k'}) = \mathbf{H}(\mathbf{L}_k) - \mathbf{H}(\mathbf{L}_k | \mathbf{P}_{k'}), \tag{1}$$

between $\mathbf{L}_k$ and $\mathbf{P}_{k'}$ for all key hypotheses k'. The value of k' that leads to the highest value of mutual information is an adversary's best guess. In the above formula $\mathbf{H}(\cdot)$ denotes Shannon entropy. Throughout the paper we assume that the sampling of probability distributions is sufficient such that entropy estimations $\mathbf{H}(\cdot)$ are good. In our experiments we apply our histogram method from [7] to estimate said distributions.

2.1   Masking

A typical way to protect an implementation of a cryptographic algorithm against such 1st order attacks is to mask the occurring intermediate values. Masking is usually implemented by combining the intermediate values with random data and by adapting the algorithm accordingly. The effect of masking is that each intermediate value that is predictable by an attacker is pairwise uncorrelated to the masked intermediate values that are actually processed.

In the following we formalize these notions. Let $\mathbf{M}$ denote a RV on $\mathcal{W}$ with uniform PMF. Masking is implemented by replacing the intermediate value $\mathbf{W}_k$

by $\mathbf{W}_{k,\mathbf{M}} = \mathbf{W}_k \circ \mathbf{M}$ where $\circ$ is a suitable operation. In masked block cipher implementations, for example, one often uses the exclusive-or operation and replaces $\mathbf{W}_k$ by $\mathbf{W}_{k,\mathbf{M}} = \mathbf{W}_k \oplus \mathbf{M}$. Non-linear operations that are typically implemented as S-box tables S-box($\mathbf{W}_k^{in}$) can be implemented with a recomputed S-box such that $\mathbf{W}_{k,\mathbf{M}}^{out} = \text{S-box}'(\mathbf{W}_{k,\mathbf{M}}^{in}) = \text{S-box}(\mathbf{W}_k^{in}) \oplus \mathbf{M}$, see Fig. 1 (left).

If $\mathbf{M}$ is a RV with uniform PMF, intermediate results $\mathbf{W}_k$ predictable by an adversary (i.e. $k' = k$) are not correlated to computed intermediate results $\mathbf{W}_k \oplus \mathbf{M}$. It follows that the actual leakage $\mathbf{L}_{k,\mathbf{M}}$ of a masked value is independent of any predictable leakage $\mathbf{P}_k$ and therefore $\mathbf{I}(\mathbf{L}_{k,\mathbf{M}}; \mathbf{P}_k) = 0$.

2.2 Higher-Order Attacks

The mounting point for 2nd order attacks is the fact that the side-channel leakage $\mathbf{L}_{k,\mathbf{M}}$ of a masked value depends on a predictable value $\mathbf{W}_k$ and an unpredictable value $\mathbf{M}$. The core idea is to jointly analyze the leakage of the masked value and the mask (or a second value masked with the same mask) to establish a relation to the predictable $\mathbf{W}_k$ or its predictable leakage $\mathbf{P}_k$. In the example in Fig. 1 (left) one could for example combine the leakage $\mathbf{L}_M$ of the mask at time τ_1 with the leakage $\mathbf{L}_{k,\mathbf{M}}$ of the masked S-box output at time τ_4.

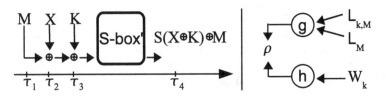

Fig. 1. Left: Masked S-box lookup with recomputed S-box. Right: Schematic of 2nd order DPA with pre-processing functions g and h that output correlated values.

2nd order DPA requires a suitable pre-processing that combines the leakage of two masked RVs to construct a signal that is correlated to the unmasked intermediate result (or a function thereof) and can be attacked with 1st order DPA. More formally that is: one looks for functions $g(\mathbf{L}_\mathbf{M}, \mathbf{L}_{k,\mathbf{M}})$ and $h(\mathbf{W}_k)$ that yield highly correlated values [14], see Fig. 1 (right). These values can be attacked with 1st order DPA.

Most proposals for the pre-processing do not consider the function h but focus on a function g whose outputs are correlated with $\mathbf{P}_k$. Such proposals typically assume that the leakage at all relevant time instants approximately follows the Hamming weight model and use Pearson's correlation coefficient (ρ) as distinguisher. More precisely and sticking to the above example, the adversary evaluates

$$\rho(g(\mathbf{L}_\mathbf{M}, \mathbf{L}_{k,\mathbf{M}}), \text{HW}(\mathbf{W}_{k'})) \qquad (2)$$

for all hypotheses k'. We provide background information on HODPA attacks in Appendix 6. The state-of-the-art combining function is the normalized product [17]

$$g(\mathbf{L_M}, \mathbf{L}_{k,\mathbf{M}}) = (\mathbf{L_M} - E(\mathbf{L_M})) \cdot (\mathbf{L}_{k,\mathbf{M}} - E(\mathbf{L}_{k,\mathbf{M}})) \tag{3}$$

where $E(\cdot)$ denotes expectation or the empirical mean. This combining function was first hinted by Chari et al. [2] and more thoroughly analyzed by Prouff, Rivain, and Bevan [17]. They showed that this combining function maximizes the Pearson coefficient in Eq. (2) for the correct key guess k in the Hamming weight model. We note, however, that this combining function is not necessarily optimal, since it was not shown that wrong key guesses can not lead to similarly high Pearson coefficients.

3 Motivation

An adversary faces three essential problems when mounting a HODPA:

1. How to identify the points of interest τ_j when the interesting intermediate values leak?
2. How to model the power consumption at these points in time? This question is particularly interesting as the power consumption model need not be the same at several instants, e.g. while a random number generator is active vs. during a table lookup.
3. How to choose the functions g and h for the pre-processing? This question is particularly interesting because the answer is tightly linked to the previous question.

In this paper we will not deal with the first problem but assume that the instants τ_j are known.[1] The same assumption is made in all related literature except for [15].

In our view most previous work discusses 2nd and HODPA in specific contexts and under restrictive assumptions, that are not always met in practice [15,18] but the drawbacks have been accepted. In particular we point out that most contributions assume i) the linear Hamming weight or distance model and implicitly require them for the attack to work, due to the choice of g (and h where applicable); and ii) that the leakage functions associated to the computation of different intermediate results, where different parts of the device may be active, are the same or very similar.

It is surprising that, while DPA and its higher-order variants have been published more than 10 years ago, the problem of finding an optimal pre-processing (except for very specific contexts) remains unsolved.

A weakness of earlier work is that tight link between leakage model(s) and pre-processing. It is evident that a pre-processing tailored to specific leakage functions looses all meaning if the leakage models are not met.

---

[1] Some advice can be found in Appendix 6.

Two works that relax the above mentioned assumptions or that could be accordingly adapted [1,14] deal with variants of template attacks [3], which consider an adversary who is able to characterize the leakage function(s) of the target device and the implementation as well as the electrical properties of the measurement setup. The adversarial context of a profiled attack is, however, beyond the scope of this paper. The third work relaxing above assumptions is a recent and independent paper by Prouff and Rivain [16], which we will address in Sect. 4.1.

3.1 Problem Statement

We formulate a generic 2nd order DPA problem that relaxes all (but one) of the above mentioned assumptions and requirements. Informally speaking, we ask "what is possible" if i) the power models at the two (known) instants are unknown and possibly substantially different, which implies ii) that the best choice for g and h in the pre-processing is *a priori* unknown. Further, the sought method should naturally extend to attacks of order greater than two. A sound solution would be a powerful tool that allows successful HO attacks in a range of scenarios otherwise resistant or inaccessible to standard attacks due to intrinsic errors introduced in the pre-processing [2,10,13,15,18,21].

Formally, let $\mathbf{L}_{\tau_1}$ denote the leakage of a masked value at time τ_1 and $\mathbf{L}_{\tau_2}$ denote the leakage of the corresponding mask (or a value masked with the same mask) at instant τ_2. Given leakage $(\mathbf{L}_{\tau_1}, \mathbf{L}_{\tau_2})$ determine k with non-negligible advantage over a random guess. Note that solving the problem does not necessarily require a transformation step (i.e. a pre-processing).

4 Extending MIA to Multivariate Analysis

It is natural to look for methods that can solve the above stated problems without transforming them to other problems while relying on possibly wrong assumptions. In [7] we showed that MIA is a 1st order attack that works without restrictive assumptions about the leakage function. We can thus use MIA to solve the problem of unknown leakage functions at instants τ_1 and τ_2. Since MIA is well suited to exploit dependencies between RVs without making a restrictive assumption about *how* the RVs are related, it appears natural to also use this technique to solve the second problem, i.e. combining the information contained in $\mathbf{L}_{\tau_1}$ and $\mathbf{L}_{\tau_2}$.

Our extension of MIA to a multivariate scenario is straight forward: one merely computes the multivariate mutual information of three RVs

$$\mathbf{I}(\mathbf{P}_{k'}; \mathbf{L}_{k,\mathrm{M}}; \mathbf{L}_{\mathrm{M}}) . \tag{4}$$

In [5,6,12] it is shown that

$$\mathbf{I}(\mathbf{X}; \mathbf{Y}; \mathbf{Z}) = \mathbf{I}(\mathbf{X}; \mathbf{Y}) - \mathbf{I}(\mathbf{X}; \mathbf{Y}|\mathbf{Z}) \tag{5}$$

where both terms on the right hand side of the equation can be computed using Eq. (1).

Depending on the source, Eq. (5) is either called multivariate mutual information or mutual interaction. It is clear that Eq. (5) can have positive and negative values depending on the relation between the arguments. For example, if $\mathbf{X}$ and $\mathbf{Y}$ are independent but possibly related through $\mathbf{Z}$ as in our context, then

$$\mathbf{I}(\mathbf{X};\mathbf{Y};\mathbf{Z}) = \mathbf{I}(\mathbf{X};\mathbf{Y}) - \mathbf{I}(\mathbf{X};\mathbf{Y}|\mathbf{Z}) = 0 - \mathbf{I}(\mathbf{X};\mathbf{Y}|\mathbf{Z}) \leq 0$$

and one says that $\mathbf{Z}$ explains the (in-)dependence of $\mathbf{X}$ and $\mathbf{Y}$. Note that the choice of how to substitute the arguments is arbitrary, any combination works. The MMIA key recovery attack decides for the key hypothesis k' that minimizes expression (4). For the more general case of nth order MMIA attacks one computes

$$\mathbf{I}(\mathbf{X}_1;\ldots;\mathbf{X}_{N+1}) = \mathbf{I}(\mathbf{X}_1;\ldots;\mathbf{X}_N) - \mathbf{I}(\mathbf{X}_1;\ldots;\mathbf{X}_N|\mathbf{X}_{N+1}).$$

We want to emphasize that our proposal has one clear advantage: there is no need to assume leakage functions neither to choose the functions g and h for the pre-processing. This makes it generic and applicable in virtually any scenario.

4.1   Generalized MIA

Simultaneous but independent of this work, Prouff and Rivain proposed a different extension of MIA to multivariate analysis [16]. In the case of a 2nd order attack they suggest to evaluate the classical formula for mutual information

$$\mathbf{I}(\mathbf{P}_{k'};(\mathbf{L}_{k,\mathbf{M}},\mathbf{L}_{\mathbf{M}})) \tag{6}$$

between the predicted leakage and the pair of leakage observations. For n^{th} order attacks they suggest to compute $\mathbf{I}(\mathbf{X}_1;(\mathbf{X}_2,\ldots,\mathbf{X}_{N+1}))$. We note that this approach shares all the advantages over HODPA we pointed out for our approach. In the next section we investigate how both approaches perform in practice and how they compare to state-of-the-art HODPA.

5   Reality Check

We study our approach and confront its performance with HODPA using the normalized product preprocessing [2,17] and the generalized MIA variant of [16] in two scenarios.

We consider a Boolean masking scheme as for example described in [8] for DES or triple-DES. For simplicity we focus on a representative mini-cipher consisting of a masked S-box lookup of the first DES S-box (S_1). In practice, we precompute the values $\mathbf{M} = m_i$ and $\mathbf{W}_k \circ \mathbf{M} = S_1(x_i \oplus k) \oplus m_i$ for each encryption i on a desktop computer and send them to the card, which successively moves the values over its data-bus generating leakage $\mathbf{L}_M$ and $\mathbf{L}_{k,\mathbf{M}}$. The data bus is reset to $0x00$ before and after each memory access. All attacks are provided with the physical measurements of $\mathbf{L}_M$ and $\mathbf{L}_{k,\mathbf{M}}$. Note that unmasked values are never processed by the card and that $\mathbf{M}$ as well as $\mathbf{W}_{k,\mathbf{M}}$ exist on $\{0,1\}^4$.

For our experiments we use an 8-bit RISC microcontroller based smartcard. The power measurements represent the voltage drop over a 10Ω resistor inserted in the smart-card's GND. We implemented two scenarios: A) 1st order masking exactly as described above; B) 2nd order masking, i.e. the S-Box output is concealed by two masks $\mathbf{M}_1$ and $\mathbf{M}_2$ and the card generates leakage $\mathbf{L}_{M_1}$, $\mathbf{L}_{M_2}$, and $\mathbf{L}_{k,M_1,M_2}$. In both scenarios the leakage of the device is very close to the linear Hamming weight model and affected by little noise. Given the mask values, we obtain Pearson coefficients $\rho > 0.99$.

We apply the MIA variants and HODPA assuming the Hamming weight leakage model. In this configuration the scenarios are very well suited to study the impact of the pre-processing in HODPA, i.e. a potential loss of information, versus the multivariate approaches without pre-processing. We also apply the attacks without assuming the Hamming weight leakage model but instead assume a generic leakage model, namely the identity function. In this configuration our experiments allow to study how much an attack depends on a good leakage model.

More precisely, all attacks target the same unmasked intermediate result $\mathbf{W}_k = S_1(\mathbf{X} \oplus k)$ which does not give rise to 1st order leakage. When assuming the Hamming weight leakage model we make leakage predictions of the form $\mathbf{P}_{k'} = \mathrm{HW}(\mathbf{W}_{k'})$. Without this assumption predictions are of the form $\mathbf{P}_{k'} = \mathbf{W}_{k'}$.

For the MIA variants the assumed leakage model further affects the number of bins for the histograms which we use to estimate densities [7]. The densities are required to compute entropy and mutual information values. When making the Hamming weight assumption we use five bins, since the Hamming weight of a 4-bit variable can take five distinct values, and otherwise we use sixteen bins, since $2^4 = 16$.

For HODPA we use the normalized product combining function (see Eq. (3)) and Pearson's correlation coefficient.

Following the framework for the comparison of side-channel distinguishers [20] we use the first-order success rate to assess the performance of the attacks. The first-order success rate expresses the probability that, given n measurements, the attack's best guess is the correct key. For each scenario, we acquired a set of 100 000 power curves using random masks and plaintexts. To evaluate the attacks under the Hamming weight assumption in the 2nd order attack scenario, for example, we split the set into 1000 subsets v_i ($i = 1, \ldots, 1000$) of 100 curves and do the following:

```
for n := 10 to 100
counter ← 0
    for i := 1 to 1000
        i. select the first n curves from set v_i
        ii. run the attack for k' ∈ {0,1}^6
        iii. increase counter if attack successful
        compute success rate for n curves as counter/1000.
```

In other configurations the attacks require more measurements and we use less subsets each containing more curves instead. In both scenarios a HODPA attack is considered successful if the correct key guess leads to the highest correlation coefficient in absolute terms, a MMIA attack (this paper) is considered successful if the correct key guess minimizes the multivariate mutual information, and a generalized MIA attack [16] is considered successful if the correct key guess maximizes the mutual information.

Figure 2 shows the results of the 2nd order attacks in scenario A.

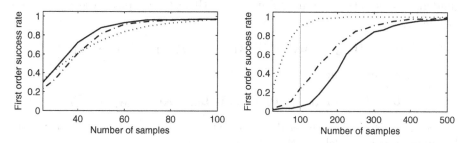

Fig. 2. First-order success rates of 2nd order attacks: MMIA (solid), generalized MIA (dash-dotted), HODPA (dotted), left: assuming a HW leakage model, right: assuming an identity leakage model; the vertical line shall remind the reader that the X-axis' scale is not the same

On the left hand side of the figure we show how the attacks perform under the Hamming weight assumption. We can see that all three attacks perform very similar. They require about 40 measurements to achieve a success rate of 50% and about 100 measurements to reach a success rate close to 100%. The results for both MIA variants using the (correct) Hamming weight assumption are particularly interesting as these attacks take uncertainty about the leakage functions out of the equation and show the impact of sound joint statistics. The fact that the HODPA performs similarly well in this experiment supports the optimality claim for normalized product 2nd order DPA in the Hamming weight model of [17].

On the right hand side of the figure we show how the attacks perform under the assumption of an identity leakage model. We can see that all attacks are affected by this change, but that the performance of both MIA variants decreases more than the performance of HODPA. HODPA requires about 200 measurements to reach a success rate close to 100% while both MIA variants require about 500 measurements. This observation can be explained by two facts. First, the identity function and the Hamming weight of a 4-bit variable have a strong linear correlation [16]. This explains why the performance of the HODPA decreases not that much. Second, the MIA variants now use sixteen instead of five bins for the histograms. Therefore more samples are required to obtain good density estimations. This explains why the performance of both MIA variants decreases similarly and quite drastically.

We conclude that each 2nd order attack that uses the correct power model and sound joint statistics works very efficient. However, we remind that scenario A is a somewhat easy target because device's leakage almost perfectly follows the Hamming weight model and because the noise level in the measurements is low. In [16] it was shown that generalized MIA is less affected by an increase of noise than HODPA. It is reasonable to assume that MMIA behaves similarly.

In scenario B we want to study how the MIA variants and HODPA scale with respect to attacks of order greater than two, and extended scenario A to 2nd order masking. The S-box output values are now concealed by two independent random masks. The combining function of HODPA computes the normalized product

$$(\mathbf{L_{M_1}} - E(\mathbf{L_{M_1}})) \cdot (\mathbf{L_{M_2}} - E(\mathbf{L_{M_2}})) \cdot (\mathbf{L}_{k,\mathbf{M_1},\mathbf{M_2}} - E(\mathbf{L}_{k,\mathbf{M_1},\mathbf{M_2}})) .$$

The MIA variants compute

$$\mathbf{I}(\mathbf{L_{M_1}}; \mathbf{L_{M_2}}; \mathbf{L}_{k,\mathbf{M_1},\mathbf{M_2}}; \mathbf{P}_{k'}) \text{ and } \mathbf{I}((\mathbf{L_{M_1}}, \mathbf{L_{M_2}}, \mathbf{L}_{k,\mathbf{M_1},\mathbf{M_2}}); \mathbf{P}_{k'})$$

respectively. Note that the leakage behavior of the card is still close to the Hamming weight model and that the noise level is low.

Figure 3 shows our experimental results for the 3rd order attacks in scenario B. On the left hand side of the figure we show how the attacks perform under

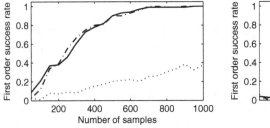 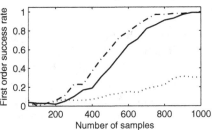

Fig. 3. First-order success rates of 3rd order attacks: MMIA (solid), generalized MIA (dash-dotted), HODPA (dotted), left: assuming a HW leakage model, right: assuming an identity leakage model

the Hamming weight assumption. We can see that both MIA variants perform again very similar. They require about 800 measurements to achieve a success rate close to 100%. HODPA performs significantly worse. Even using 1000 measurements the attack merely achieves a success rate of about 40%. As before the results for both MIA variants using the (correct) Hamming weight assumption are particularly interesting because they allow us to focus on the impact of sound joint statistics. The fact that HODPA performs significantly worse can not be assigned to a wrong assumption in the leakage model and can therefore be best explained by information loss during the pre-processing. Note that the optimality claim of [17] only holds for 2nd order attacks.

On the right hand side of the figure we show how the attacks perform under the assumption of an identity leakage model. We can see that the performance of HODPA is only slightly decreased by this change. Using 1000 measurements the attack achieves a success rate of about 30%. This observation supports our theory that the efficiency of HODPA in a 3rd order attack scenario is mostly limited by sub-optimal pre-processing. The impact of the change in the assumed leakage model on the MIA variants is more visible but much weaker than in scenario A. Both attacks require about 900 measurements to achieve close to 100% success. This rather minor increase (recall that both attacks now generate histograms using sixteen instead of five bins) indicates that the large number of samples is not required to estimate each separate density sufficiently well. Instead, it is required to extract the complex interrelation from all of them at once.

We conclude that an information theoretic approach using multivariate statistics is clearly preferable over HODPA in attacks of order greater than two. While the prevailing opinion is that the measurement cost of HO attacks grows exponentially with the order of the attack [2,15,18], we demonstrate that attacks of order up to three are realistic and practical.

6 Conclusion

Confronted with a new problem, one typically first tries to transform it into another problem for which one knows the solution. HODPA attacks seem to be such a problem. They inherently demand multivariate statistics because multiple aspects of signals have to be analyzed jointly. However, most publications on HO-attacks follow the approach to first apply a pre-processing function to map the multivariate problem to a univariate problem and then to apply established 1st order techniques. All proposed pre-processing functions have drawbacks that are accepted at the price of an exponential growth of the measurement cost with the attack order.

We propose a novel and different approach for HO attacks that does not suffer from intrinsic errors but solves the initial problem directly. Further, we present the first experimental results for the considered 2nd and 3rd order attacks based on real measurements. The empirical evidence confirms the advantage of MMIA over established HODPA in particular in 3rd order attack scenarios. Our results also show that MMIA and generalized MIA [16] perform similar in practice. As a consequence, the security provided by the masking countermeasure needs to be reconsidered as 3rd and possibly higher order attacks become more practical. The typically implemented combination of masking and temporal randomization [9] should render attacks using either MIA variant more difficult.

Directions for further research are: a detailed comparison of MMIA and generalized MIA, combining functions for HODPA of order greater than two, and methods for the identification of the interesting points in time without device profiling step.

Acknowledgments and Disclaimer

This work was supported in part by the IAP Programme P6/26 BCRYPT of the Belgian State, by FWO project G.0300.07, by the European Commission under contract number ICT-2007-216676 ECRYPT NoE phase II, and by the K.U. Leuven-BOF (OT/06/40).

The information in this document reflects only the author's views, is provided as is and no guarantee or warranty is given that the information is fit for any particular purpose. The user therefore uses the information at its sole risk and liability.

References

1. Agrawal, D., Rao, J.R., Rohatgi, P., Schramm, K.: Templates as Master Keys. In: Rao, J.R., Sunar, B. (eds.) CHES 2005. LNCS, vol. 3659, pp. 15–29. Springer, Heidelberg (2005)
2. Chari, S., Jutla, C.S., Rao, J.R., Rohatgi, P.: Towards Sound Approaches to Counteract Power-Analysis Attacks. In: Wiener, M. (ed.) CRYPTO 1999. LNCS, vol. 1666, pp. 398–412. Springer, Heidelberg (1999)
3. Chari, S., Rao, J.R., Rohatgi, P.: Template Attacks. In: Kaliski Jr., B.S., Koç, Ç.K., Paar, C. (eds.) CHES 2002. LNCS, vol. 2523, pp. 172–186. Springer, Heidelberg (2003)
4. Coron, J.-S., Goubin, L.: On Boolean and Arithmetic Masking against Differential Power Analysis. In: Paar, C., Koç, Ç.K. (eds.) CHES 2000. LNCS, vol. 1965, pp. 231–237. Springer, Heidelberg (2000)
5. Cover, T.M., Thomas, J.A.: Elements of Information Theory. John Wiley & Sons, Chichester (2006)
6. Fano, R.M.: Transmission of Information: A Statistical Theory of Communications. MIT Press, Cambridge (1961)
7. Gierlichs, B., Batina, L., Tuyls, P., Preneel, B.: Mutual Information Analysis - A Generic Side-Channel Distinguisher. In: Oswald, E., Rohatgi, P. (eds.) CHES 2008. LNCS, vol. 5154, pp. 426–442. Springer, Heidelberg (2008)
8. Goubin, L., Patarin, J.: DES and Differential Power Analysis (The "Duplication" Method). In: Koç, Ç.K., Paar, C. (eds.) CHES 1999. LNCS, vol. 1717, pp. 158–172. Springer, Heidelberg (1999)
9. Herbst, C., Oswald, E., Mangard, S.: An AES Smart Card Implementation Resistant to Power Analysis Attacks. In: Zhou, J., Yung, M., Bao, F. (eds.) ACNS 2006. LNCS, vol. 3989, pp. 239–252. Springer, Heidelberg (2006)
10. Joye, M., Paillier, P., Schoenmakers, B.: On Second-Order Differential Power Analysis. In: Rao, J.R., Sunar, B. (eds.) CHES 2005. LNCS, vol. 3659, pp. 293–308. Springer, Heidelberg (2005)
11. Kocher, P., Jaffe, J., Jun, B.: Differential Power Analysis. In: Wiener, M. (ed.) CRYPTO 1999. LNCS, vol. 1666, pp. 388–397. Springer, Heidelberg (1999)
12. McGill, W.J.: Multivariate Information Transmission. Psychometrika (19), 97–116 (1954)
13. Messerges, T.S.: Using Second-Order Power Analysis to Attack DPA Resistant Software. In: Paar, C., Koç, Ç.K. (eds.) CHES 2000. LNCS, vol. 1965, pp. 238–251. Springer, Heidelberg (2000)

14. Oswald, E., Mangard, S.: Template Attacks on Masking - Resistance Is Futile. In: Abe, M. (ed.) CT-RSA 2007. LNCS, vol. 4377, pp. 243–256. Springer, Heidelberg (2006)
15. Oswald, E., Mangard, S., Herbst, C., Tillich, S.: Practical Second-Order DPA Attacks for Masked Smart Card Implementations of Block Ciphers. In: Pointcheval, D. (ed.) CT-RSA 2006. LNCS, vol. 3860, pp. 192–207. Springer, Heidelberg (2006)
16. Prouff, E., Rivain, M.: Theoretical and Practical Aspects of Mutual Information Based Side Channel Analysis. In: Abdalla, M., Pointcheval, D., Fouque, P.-A., Vergnaud, D. (eds.) ACNS 2009. LNCS, vol. 5536, pp. 499–518. Springer, Heidelberg (2009)
17. Prouff, E., Rivain, M., Bevan, R.: Statistical Analysis of Second Order Differential Power Analysis. IEEE Transactions on Computers (58-6), 799–811 (2009)
18. Schramm, K., Paar, C.: Higher Order Masking of the AES. In: Pointcheval, D. (ed.) CT-RSA 2006. LNCS, vol. 3860, pp. 208–225. Springer, Heidelberg (2006)
19. Standaert, F.-X., Gierlichs, B., Verbauwhede, I.: Partition vs. Comparison Side-Channel Distinguishers. In: Lee, P.J., Cheon, J.H. (eds.) ICISC 2008. LNCS, vol. 5461, pp. 253–267. Springer, Heidelberg (2009)
20. Standaert, F.-X., Malkin, T., Yung, M.: A Unified Framework for the Analysis of Side-Channel Key Recovery Attacks. In: Joux, A. (ed.) EUROCRYPT 2009. LNCS, vol. 5479, pp. 443–461. Springer, Heidelberg (2009)
21. Waddle, J., Wagner, D.: Towards Efficient Second-Order Power Analysis. In: Joye, M., Quisquater, J.-J. (eds.) CHES 2004. LNCS, vol. 3156, pp. 1–15. Springer, Heidelberg (2004)

Appendix

A   More Background on HODPA

Early works mentioned two essential options for the pre-processing: the product of the two leaked values and the absolute value of their difference. The first work showing a practical higher-order attack to defeat the masking countermeasure came from Messerges [13]. He assumed that the device leaks the Hamming weight of intermediate values (i.e. $\mathbf{L}_k = \mathrm{HW}(\mathbf{W}_k)$) and proposed to compute the absolute difference $|\mathbf{L}_{k,\mathbf{M}} - \mathbf{L}_\mathbf{M}|$ in the pre-processing (abs-diff-DPA). Messerges showed that, when focusing on a single bit,

$$|\mathrm{HW}(\mathbf{W}_k \oplus \mathbf{M}) - \mathrm{HW}(\mathbf{M})| = \mathrm{HW}(\mathbf{W}_k \oplus \mathbf{M} \oplus \mathbf{M}) = \mathrm{HW}(\mathbf{W}_k). \qquad (7)$$

Thus, the pre-processing reveals the unmasked $\mathbf{L}_k = \mathrm{HW}(\mathbf{W}_k)$ which can be attacked with 1st order DPA. If one wants to attack more than a single bit simultaneously Eq. (7) changes to

$$\mathrm{HW}(\mathbf{W}_k) = \mathrm{HW}(\mathbf{W}_k \oplus \mathbf{M}) + \mathrm{HW}(\mathbf{M}) - 2 \cdot \mathrm{HW}((\mathbf{W}_k \oplus \mathbf{M}) \wedge \mathbf{M}) \qquad (8)$$

where $\wedge$ denotes bitwise AND. However, an adversary cannot evaluate this function because $\mathbf{W}_k \oplus \mathbf{M}$ and $\mathbf{M}$ are unknown.

The Hamming weight assumption was also used by Oswald et al. in [15]. They showed that the idea of Eq. (7) can still be used to attack multiple bits although the equality no longer holds. For 8-bit variables, the Pearson correlation coefficient (ρ) between the predictable $HW(\mathbf{W}_k)$ and the output of the abs-diff pre-processing decreases to 0.24.

Chari et al. [2] suggested to use the product $HW(\mathbf{W}_k \oplus \mathbf{M}) \cdot HW(\mathbf{M})$ in the pre-processing (product-DPA). Their technique does not require the ideal Hamming weight model but still makes some restrictive assumptions about the leakage and power consumption behavior and is in practice more vulnerable to deviations from the model. Waddle and Wagner [21] were the first to clearly split higher-order attacks into pre-processing and attack step as we present them here. They proposed a few variants of product-DPA that differ in complexity. The work of Joye et al. [10] introduces a more theoretical approach to 2nd order DPA. The authors analyzed single bit 2nd order abs-diff-DPA in the Hamming weight model, as introduced by Messerges, and in the Hamming distance model. They suggest to use a power of the absolute difference in the pre-processing, which yields a slightly higher coefficient ρ [15].

B   Identifying the Points of Interest

One approach towards identifying these instants may be to examine the empirical variance of several power traces obtained during processing of constant input data. In this case, the variance in the power traces is mostly caused by the masking and thus reveals the points in time when masked values are processed. Another approach is to select a small time window based on an educated guess and to perform an exhaustive search over all pairs of time instants [15].

Differential Cache-Collision Timing Attacks on AES with Applications to Embedded CPUs

Andrey Bogdanov[1], Thomas Eisenbarth[2], Christof Paar[2], and Malte Wienecke[2]

[1] Dept. ESAT/SCD-COSIC, Katholieke Universiteit Leuven, Belgium
andrey.bogdanov@esat.kuleuven.be
[2] Horst Görtz Institute for IT Security
Ruhr University Bochum, Germany
{thomas.eisenbarth,christof.paar,malte.wienecke}@rub.de

Abstract. This paper proposes a new type of cache-collision timing attacks on software implementations of AES. Our major technique is of differential nature and is based on the internal cryptographic properties of AES, namely, on the MDS property of the linear code providing the diffusion matrix used in the MixColumns transform. It is a chosen-plaintext attack where pairs of AES executions are treated differentially. The method can be easily converted into a chosen-ciphertext attack. We also thoroughly study the physical behavior of cache memory enabling this attack.

On the practical side, we demonstrate that our theoretical findings lead to efficient real-world attacks on embedded systems implementing AES at the example of ARM9. As this is one of the most wide-spread embedded platforms today [7], our experimental results might make a revision of the practical security of many embedded applications with security functionality necessary. To our best knowledge, this is the first paper to study cache timing attacks on embedded systems.

1 Introduction

Side-channel attacks and cache timing leakage. Though side-channel leakage seems to have been extensively used by state security agencies for decades to obtain secret information [9], the idea of applying side-channel attacks to implementations of cryptographic algorithms appeared in the scientific literature rather recently. The first side-channel attack published was timing analysis proposed by Kocher [9] in 1996 where he observes the execution time of keyed cryptographic algorithms to recover the key and points out the usefulness of timing analysis applied to software implementations of block ciphers.

Probably the most widely known timing attacks on symmetric algorithms belong to the class of cache timing attacks on block ciphers with S-boxes. This is not least due to the literally ubiquitous usage of block ciphers, and first of all, of the U.S. encryption standard AES [2] in the overwhelming majority of security applications, both in PCs and embedded systems.

J. Pieprzyk (Ed.): CT-RSA 2010, LNCS 5985, pp. 235–251, 2010.

As the name of cache timing attacks suggests, they utilize the particularities of microcontrollers and microprocessors with cache memory which frequently exhibit key-dependent timing. Cache timing attacks on many block ciphers with S-boxes become possible since S-box invocations in software are often implemented as indexed table look-up operations that can require different execution times for different inputs due to RAM cache hits and misses. When the inputs to S-boxes are key-dependent, this timing information frequently turns out sufficient to recover the entire key. Today, several variants of cache timing attacks on AES are known [4], [12], [5], [1].

Generally speaking, side-channel analysis methods strongly depend on a concrete implementation of the attacked cryptographic algorithm. There is also no exception for cache timing attacks on AES where the choice of the optimal attack method can greatly vary depending on the implementation at hand. For AES, one can basically distinguish between first round and final round approaches: While first round attacks [1], [5] tend to be applicable to large classes of implementations at the cost of more encryption samples required, final round attacks [5] target only the T-box based 32-bit implementation [6] of AES having the advantage of being more efficient in this particular case. In this paper, we pursue and expand the more generic approach of first round attacks.

Main idea of our attack. The major idea behind our new cache timing attack is to choose pairs of plaintexts in a specific way, so that five AES S-boxes (one in round 2 and four in round 3) process either pairwisely equal or pairwisely distinct values in two adjacent AES executions. If for a plaintext pair the five S-boxes process pairwisely equal values, it is called a *wide collision*.

In our attack, we measure the average time of every second AES execution from each pair. We are interested in the average number c of S-box collisions (S-box pairs processing equal values) between the two AES runs in a pair. If a wide collision has occurred for a pair of plaintexts, c will be by 5 higher compared to c when there is no wide collision. We hope to detect enough wide collisions against the background if there are enough samples available. After this, we construct four systems of nonlinear equations with respect to parts of the key which are then resolved by brute force for the key recovery.

This technique becomes possible due to the fact that AES uses a maximum distance separable (MDS) code to construct its diffusion matrix for MixColumns operation. MDS codes are known to provide linear transforms with the maximum possible branch number [6], which is 5 for the parameter choice of AES. Interestingly enough, it is precisely the excellent cryptographic properties of AES, due to the optimal selection of the diffusion matrix making it resistant to differential and linear cryptanalysis, that enable our cache timing attack techniques.

Cache timing attacks and embedded security. Security in embedded systems is constantly and quickly becoming more crucial with the spread of embedded devices in one's everyday life. It is getting even more important in the age of pervasive computing.

Side-channel attacks have been known to impose a serious threat to embedded systems such as smart cards or other embedded microcontrollers for the last decade. However, as applied to symmetric key algorithms, the toolbox of the attacker was mainly limited to techniques based on information leakage via power consumption and electromagnetic radiation of the devices [11], [10]. At the same time, timing analysis have been only very rarely utilized to analyze embedded implementations of block ciphers, being mainly applied in the domain of desktop and server PCs. This is partially due to the fact that many lightweight and low-cost embedded systems have been providing hardware implementations of symmetric key algorithms. Besides that, many lightweight platforms based on 8-bit or 16-bit CPUs run at such low frequencies that microarchitectural performance optimization such as caches are not necessary.

This apparent disregard of and disbalance against cache timing attacks in the context of embedded security does not seem justified anymore though, since the embedded landscape is rapidly changing nowadays. As the computing world goes pervasive, a steadily growing number of embedded applications require more computing power. 32-bit RISC ARM-type CPUs have become a standard choice in many embedded applications such as banking and payment terminals, mobile communications, JavaCard applications, mobile TV, multimedia, toll collect systems, smart phones, electronic tachographs, PDAs etc. More and more security-related functionality is being put into software instead of hardware. Even some smart card microcontrollers are migrating towards powerful and universal computing architectures based on an ARM core [16]. At the same time, ARM microprocessors do have cache memory with nontrivial behavior and are as a rule operated under multi-process operating systems such as Linux or Windows Embedded/Mobile. As opposed to almost all lightweight 8-bit microcontrollers, these two facts make many embedded systems vulnerable to cache timing attacks, first of all those based on ARM-type CPUs.

Aiming to close this gap, we tackle the problem of applying cache timing attacks to embedded devices at the example of ARM9 microprocessors. Our findings presented in Table 2 show that our cache timing based techniques of new type apply well to the OpenSSL software implementation of AES on ARM9. This indicates that numerous real-world embedded applications relying on AES and using ARM-type CPUs can turn out vulnerable to cache-timing attacks. This might force us to reconsider the practical security of many embedded systems currently in use. Furthermore, based on our results, we recommend to take the threat of cache timing attacks into account when designing and evaluating new embedded systems with security functionality.

Organization of this paper. The remainder of the paper is organized as follows. In Section 2, the most relevant previous work is briefly outlined including the advanced methods of expanded second-round attacks. Section 3 presents our new differential attack technique based the diffusion properties of AES. We deal with the physical cache behavior enabling our attack in Section 4. The attacked embedded platform, its impact as well as our experiments and practical results are provided in Section 5. We conclude in Section 6.

2   Previous Work on Cache-Collision Timing Attacks

In 1998 J. Kelsey *et al.* [8] analyzed the cache behavior of modern processors as a side channel against ciphers with large lookup tables like S-boxes. This proposal was established by D. Page [14] in the year 2002, who described and simulated a theoretical attack on DES. The first real-world implementation of such an attack was developed by Y. Tsunoo *et al.* [17] against DES and Triple-DES.

In general, cache attacks can be divided into three basic classes: trace driven, access driven, and time driven attacks. In trace driven attacks, the adversary is allowed to observe every single memory and cache access. Therefore, he knows when and where a collision occurs [14]. The access driven attacks provide the information which set of the cache is accessed by the cryptographic progress. For this, the cache is filled with data of the attacker. After the encryption the attacker checks which data is still present in the cache [13].

Attacks presented in this paper belong, however, to the class of the time driven attacks. Here, information is obtained by observing the execution time which is influenced by cache hits and cache misses. In this case, the attacker can only capture the total execution time of the encryption and then make a statistical evaluation to extract key-related information. The basic idea of timing attacks was introduced by Kocher [9]. A considerably higher number of encryption samples is needed compared to trace driven attacks. However, time-driven attacks correspond to an attacker with most restricted attack potential and are typically much more realistic, thus, being valid for numerous real-world applications, especially on embedded systems.

The cache based timing techniques developed in this paper target implementations of the Advanced Encryption Standard (AES)[6]. AES is a symmetric block cipher standardized by the National Institute of Standards and Technology (NIST). Nowadays, it is the most used cipher. The algorithm behind AES, Rijndael [6], was designed by J. Daemen and V. Rijmen. The most common ways to implement the cipher are the straightforward implementation, which is used on 8-bit microprocessors, and the 32-bit transformation table implementation. The latter combines different round functions to five transformation tables, or T-tables. During an encryption one table is used for the last round, the remaining rounds are processed with the other four lookup tables. Since cache attacks exploit the cache hits of lookup tables, the 32-bit T-box implementation is a well suited target, because it offers five large lookup tables.

2.1   First-Round Attack

The first round attack is a basic attack which takes advantage of cache line collisions evoked in the first round of the encryption. A cache line collision appears if two entries of the same cache line are accessed. Since the first round of the 32-bit implementation is realized with four tables, only four input values of the round p_i' access the same table. For example, the values p_0', p_4', p_8', and p_{12}' are processed by the first transformation table $\mathbf{T}_0$. The first round input p'_i itself is computed by an XOR combination of the plaintext p_i and the corresponding key value k_i:

$$p'_i = p_i \oplus k_i \qquad \text{for } 0 \leq i < 16. \qquad (1)$$

With a cache line collision the adversary can create a relation between two different key bytes. Such a collision is evoked if, for example, $\langle p'_i \rangle = \langle p'_j \rangle$, for $i, j \in \{0, 4, 8, 12\}$ and $i \neq j$, i.e., the most significant bits of the values are equal, ignoring the $(\log_2 \gamma)$ least significant bits, where γ indicates the number of table entries in one cache line. The resulting relation is $\Delta_{i,j} = \langle k_i \oplus k_j \rangle = \langle p_i \oplus p_j \rangle$. Using such relation the size of the key space is reduced from 2^{128} to 2^{68} possible keys, if $\gamma = 8$.

Since the execution time is influenced by cache hits, the cache line collisions can be detected using statistic methods, like calculating the the average encryption time of a sample with the same relation $\Delta_{i,j}$.

2.2 Second-Round Attack

The second round attack is based on the first round attack, but also considers collisions between the first and the second round of the encryption. To do so, the input values of the second round p'' with access to the same transfomation table are analyzed. For the first table $\mathbf{T}_0$ the following equations describe how the input values are computed:

$$p''_0 = 2 \bullet \mathbf{S}[p'_0] \oplus 3 \bullet \mathbf{S}[p'_5] \oplus \mathbf{S}[p'_{10}] \oplus \mathbf{S}[p'_{15}] \oplus k_{16} \qquad (2)$$

$$p''_4 = 2 \bullet \mathbf{S}[p'_4] \oplus 3 \bullet \mathbf{S}[p'_9] \oplus \mathbf{S}[p'_{14}] \oplus \mathbf{S}[p'_3] \oplus k_{20} \qquad (3)$$

$$p''_8 = 2 \bullet \mathbf{S}[p'_8] \oplus 3 \bullet \mathbf{S}[p'_{13}] \oplus \mathbf{S}[p'_2] \oplus \mathbf{S}[p'_7] \oplus k_{24} \qquad (4)$$

$$p''_{12} = 2 \bullet \mathbf{S}[p'_{12}] \oplus 3 \bullet \mathbf{S}[p'_1] \oplus \mathbf{S}[p'_6] \oplus \mathbf{S}[p'_{11}] \oplus k_{28} \qquad (5)$$

where $\mathbf{S}[x]$ and $\bullet$ stand for the AES S-box lookup for the value x and the finite field multiplication in $GF(2^8)$ as used in the AES. The key values k_i are the key bytes generated by the key scheduling algorithm. These values depend on the initial key. For instance, the value k_{24} is equivalent to $(\mathbf{S}[k_{13}] \oplus k_0 \oplus 01_{(16)} \oplus k_4 \oplus k_8)$ for AES-128. If a first round look up collides with a second round lookup, e.g., if $\langle p'_0 \rangle = \langle p''_8 \rangle$, we gain the following equation:

$$\langle p_0 \oplus k_0 \rangle = \langle 2 \bullet \mathbf{S}[p_8 \oplus k_8] \oplus 3 \bullet \mathbf{S}[p_{13} \oplus k_{13}] \oplus \mathbf{S}[p_2 \oplus k_2]$$
$$\oplus \mathbf{S}[p_7 \oplus k_7] \oplus \mathbf{S}[k_{13}] \oplus k_0 \oplus 01_{(16)} \oplus k_4 \oplus k_8 \rangle \qquad (6)$$

which leads to

$$\langle p_0 \rangle = \langle 2 \bullet \mathbf{S}[p_8 \oplus k_8] \oplus 3 \bullet \mathbf{S}[p_{13} \oplus k_{13}] \oplus \mathbf{S}[p_2 \oplus k_2]$$
$$\oplus \mathbf{S}[p_7 \oplus k_7] \oplus \mathbf{S}[k_{13}] \oplus 01_{(16)} \oplus \Delta_{4,8} \rangle. \qquad (7)$$

The adversary can now divide a large sample of plaintexts and encryption times into 2^{32} sets considering every combination for the key values k_2, k_7, k_8 and k_{13}, so that (7) is solved. The set with the correct key values should have the lowest encryption time. In a similar way, the complete key can be extracted.

2.3   Expanded Second-Round Attack

O. Acıiçmez et al.[1] improved the idea of the second round attack and created a chosen plaintext attack. For the expanded second round attack collisions between the first and the second round are considered, e.g., as described in (6). The difference in this attack is that the plaintext values p_0, p_2, p_7, and p_{13} are fixed for the entire sample. By combining all invariable parameters into one constant c, (6) is simplified to:

$$\langle p_0 \rangle = \langle 2 \bullet \mathbf{S}[p_8 \oplus k_8] \oplus c \rangle. \tag{8}$$

Since p_0 is a fixed value as well, the appearance of a cache line collision depends only on the value of $\langle p_8 \rangle$, i.e., γ values of p_8 evoke a cache collision. The adversary takes advantage of this fact by collecting a sample of encryption time and plaintext, where the fixed values remain the same. This sample is divided into 2^8 sets according to the value of p_8 of each plaintext. The sets which evoke the cache line collision have a lower average encryption time. To confirm the results a reference sample can be taken, where one fixed plaintext byte has another value. Using a similar procedure each key value can be reconstructed separately.

3   Differential Cache-Collision Attack Using Diffusion

Our cache timing technique is based on the notion of a wide collision where five pairs of AES S-boxes process pairwisely equal values. Though it can be made applicable to all AES versions, we will introduce it here at the example of AES-128. In order to be able to recover the key, the adversary needs to detect such wide collisions. Correspondingly, the attack flow consists of an online stage, a collision detection stage and a key recovery stage (the latter two being offline stages):

- In the *online stage*, pairs of chosen 16-byte plaintexts (P_1, P_2) are sent to the AES encryption routine. The adversary measures the time t required by the CPU to encrypt P_2, that is, the second plaintext in each pair.
  The output of the online stage to the next stages consists of the set of plaintext pairs (P_1, P_2) and the corresponding execution time values t.
- In the *collision detection stage*, the time values t are used to tell which plaintext pairs (P_1, P_2) lead to a wide collision. It is expected that if a wide collision occurs, t will be lower (results of five table lookups already in the cache memory). Otherwise, we expect t to be higher. Thus, the collision detection stage accepts sets of plaintext pairs and times output by the online stage and returns the set of plaintext pairs (P_1, P_2) that most probably lead to wide collisions.

- In the key recovery stage, one reconstructs AES key candidates from the list of plaintext pairs (P_1, P_2) which most probably result in wide collisions. These key candidates are then checked using a known plaintext-ciphertext pair.

Now, having realized the importance of wide collisions for our attack, we will first introduce this notion more formally. Then we will return to the online as well as key recovery stages afterwards. Collision detection is dealt with in Section 4.

3.1   Wide Collisions

In the attack, we always consider plaintexts pairwisely. More precisely, the pairs of plaintexts (P_1, P_2) are divided into pairs of main diagonals of the 4×4-byte AES state. A diagonal of P_1 is paired with the corresponding diagonal of P_2. In this way, four pairs are formed, marked with the same coloring:

$$
P_1 = \begin{bmatrix} a_0 & d_1 & c_2 & b_3 \\ b_0 & a_1 & d_2 & c_3 \\ c_0 & b_1 & a_2 & d_3 \\ d_0 & c_1 & b_2 & a_3 \end{bmatrix}
\qquad
P_2 = \begin{bmatrix} e_0 & h_1 & g_2 & f_3 \\ f_0 & e_1 & h_2 & g_3 \\ g_0 & f_1 & e_2 & h_3 \\ h_0 & g_1 & f_2 & e_3 \end{bmatrix}
\tag{9}
$$

One of these pairs is the pair (A, E), where $A = \{a_i\}$ and $E = \{e_i\}$, for $0 \le i < 4$. In the following description, we show how to extract a subset of key bytes at the example of the diagonal pair (A, E). The remaining key bytes can be extracted in a similar way using the other three diagonal pairs.
Let us form the plaintexts P_1 and P_2 in the following way:

- Byte values on the main diagonals A and E are chosen randomly and independently of each other with the only restriction that $A \ne E$ (four byte positions should not collide simultaneously).
- The remaining bytes of P_1 and P_2 are pairwisely equal but randomly chosen as well.

Then one obtains[1]:

$$
P_1 = \begin{bmatrix} a_0 & x_4 & x_8 & x_{12} \\ x_1 & a_1 & x_9 & x_{13} \\ x_2 & x_6 & a_2 & x_{14} \\ x_3 & x_7 & x_{11} & a_3 \end{bmatrix}
\qquad
P_2 = \begin{bmatrix} e_0 & x_4 & x_8 & x_{12} \\ x_1 & e_1 & x_9 & x_{13} \\ x_2 & x_6 & e_2 & x_{14} \\ x_3 & x_7 & x_{11} & e_3 \end{bmatrix}
\tag{10}
$$

---

[1] In equations (10), (11), (13), and (14), the grey values mark the differing values of both states.

Now we will follow the propagation of this difference on the main diagonal up to the S-box layer of round 3. So, after the first round of the encryption, the plaintexts P_1 and P_2 are transformed into:

$$P_1' = \begin{bmatrix} a_0' & x_4' & x_8' & x_{12}' \\ a_1' & x_9' & x_{13}' & x_1' \\ a_2' & x_{14}' & x_2' & x_6' \\ a_3' & x_3' & x_7' & x_{11}' \end{bmatrix} \qquad P_2' = \begin{bmatrix} e_0' & x_4' & x_8' & x_{12}' \\ e_1' & x_9' & x_{13}' & x_1' \\ e_2' & x_{14}' & x_2' & x_6' \\ e_3' & x_3' & x_7' & x_{11}' \end{bmatrix} \qquad (11)$$

Note that P_1' and P_2' differ only in the first column.

The MixColumns transform of the first column in the first round can provide collisions in up to three byte positions (four collisions would imply the non-bijectivety of AES which is not the case):

$$a_i' = e_i' \qquad \text{for some } i\text{'s in } 0 \leq i < 4. \qquad (12)$$

Consider the first byte position with $i = 0$ as an example. Here we have two possibilities: either $a_0' = e_0'$ or $a_0' \neq e_0'$.

If the byte values collide $a_0' = e_0'$, which occurs with probability $1/256$, one obtains 4 more byte collisions in the second round, as P_1' and P_2' are transformed by SubBytes and ShiftRows into P_1'' and P_2'':

$$P_1'' = \begin{bmatrix} a_0'' & x_4'' & x_8'' & x_{12}'' \\ x_9'' & x_{13}'' & x_1'' & a_1'' \\ x_2'' & x_6'' & a_2'' & x_{14}'' \\ x_{11}'' & a_3'' & x_3'' & x_7'' \end{bmatrix} \qquad P_2'' = \begin{bmatrix} e_0'' & x_4'' & x_8'' & x_{12}'' \\ x_9'' & x_{13}'' & x_1'' & e_1'' \\ x_2'' & x_6'' & e_2'' & x_{14}'' \\ x_{11}'' & e_3'' & x_3'' & x_7'' \end{bmatrix} \qquad (13)$$

and the MixColumns operation of the second round outputs two equal columns:

$$P_1''' = \begin{bmatrix} x_0''' & y_4''' & y_8''' & y_{12}''' \\ x_1''' & y_5''' & y_9''' & y_{13}''' \\ x_2''' & y_6''' & y_{10}''' & y_{14}''' \\ x_3''' & y_7''' & y_{11}''' & y_{15}''' \end{bmatrix} \qquad P_2''' = \begin{bmatrix} x_0''' & z_4''' & z_8''' & z_{12}''' \\ x_1''' & z_5''' & z_9''' & z_{13}''' \\ x_2''' & z_6''' & z_{10}''' & z_{14}''' \\ x_3''' & z_7''' & z_{11}''' & z_{15}''' \end{bmatrix} \qquad (14)$$

Only the values of the first columns are pairwisely equal, which leads to 4 S-boxes in the SubBytes layer of the third round to process pairwisely equal byte values and 5 S-box collisions in total. This is called a *wide collision*.

However, if $a_0' \neq e_0'$, which occurs with probability $255/256$, one has $a_0'' \neq e_0''$ in (13). That is, only one byte position differs in the first columns of P_1'' and P_2''. Due to the MDS property of the MixColumns matrix acting on 4 byte values, all elements in the first column of P_1''' and P_2''' will be pairwisely different, since the matrix has branch number 5. This leads to 5 S-box non-collisions in total and is called a *wide non-collision*.

The average difference between the numbers of colliding and non-colliding S-boxes for wide collision and wide non-collision is 5. This discrepancy in the number of S-box collisions makes wide collisions much easier to detect against the background of wide non-collisions.

The intuition behind our cache timing collision attack is then that the average AES encryption time is detectably lower in the presence of a wide collision. We will deal with this kind of statistics in Section 4. In this section, we will discuss how the online stage is arranged and describe the procedure of key recovery based on a set of detected wide collisions for each diagonal.

3.2   Online Phase

In the online phase, the plaintexts P_1 and P_2 are generated in the way described above. A pair of diagonals[2] (A, E) is randomly chosen with the property $A \neq E$. For each of the 4 diagonals, this random choice of 8 byte values (4 for P_1 and 4 for P_2) is performed n times. That is, altogether, the online procedure described below is performed $4 \cdot n$ times.

For a fixed choice of A and E, the remaining state values of P_1 and P_2 are randomly chosen as well, but they are pairwisely equal for both plaintexts P_1 and P_2. That is, P_1 and P_2 are equal up to the main diagonals. For a fixed (A, E), this random choice is performed I times: For each of the $4 \cdot n$ choices, we run I such iterations. Each of these I iterations is repeated r times to ensure the stability of time measurements: We say that each iteration has r rounds.

In each of the r rounds, both plaintexts P_1 and P_2 are sent to the consecutive encryption. The time t required by the encryption of P_2 is captured. To improve the resolution of the time measurements, one can clear the cache memory before encrypting P_1. Moreover, the time interval between the two encryptions should be possibly short to avoid numerous cache accesses by other processes that might clear parts of the cache memories. With the cache cleared, the first encryption behaves like a random encryption in terms of cache usage and fills the cache memory with some lookup entries. Physically, to detect a wide collision, we want to observe how many entries on average are added to the cache by the encryption of P_2 after encrypting P_1. Therefore, t contains information about wide collisions and is stored together with the corresponding diagonal values (A, E) for the offline analysis.

Note that it is also possible to work with the complete encryption time for both plaintexts, but then more measurements are necessary to cancel out added by the encryption of P_1.

Thus, the major parameters influencing the complexity of the online stage (also referred to as *online complexity*) in our attack are n, I and r. The total number of AES encryptions required by the attack in the online phase will be $8 \cdot n \cdot I \cdot r$.

---

[2] Again, we explain the online phase at the example of the main diagonals A and E without loss of generality. All the other diagonals are attacked in a similar way.

3.3   Key Recovery

Algebraically, if a wide collision is detected, at least one byte at the end of the first AES round collides. So we have $a'_i = e'_i$ for some $0 \leq i < 4$ (see also (11) and (12)). Every such expression binds the four key bytes on the same main diagonal. For instance, for $a'_0 = e'_0$ we will have the following equation:

$$02 \cdot S(k_0 \oplus a_0) \oplus 03 \cdot S(k_5 \oplus a_1) \oplus 01 \cdot S(k_{10} \oplus a_2) \oplus 01 \cdot S(k_{15} \oplus a_3)$$
$$= \tag{15}$$
$$02 \cdot S(k_0 \oplus e_0) \oplus 03 \cdot S(k_5 \oplus e_1) \oplus 01 \cdot S(k_{10} \oplus e_2) \oplus 01 \cdot S(k_{15} \oplus e_3),$$

where k_0, k_5, k_{10} and k_{15} form the 4-byte subkey of the main diagonal. One obtains similar nonlinear equations with respect to 4-byte key chunks for all other possible byte collisions after the first round.

To recover each 4-byte subkey corresponding to each diagonal, we need at least four equations of type (15) and, thus, at least 16 in total for the full key recovery. For each diagonal, parameter n is chosen in a way that more than four collision candidates will be normally proposed by the collision detection stage. Assume that the collision detection stage proposes $4 + m$ collisions, $m \in \{0, 1, 2, \dots\}$. As a rule, the detection error probability will be nonzero, so that some of the proposed $4 + m$ collision candidates will be non-collisions. Therefore, the key recovery procedure has to deal with this type of errors. We propose to do that in two steps as follows.

In the first step, we consider all possible $\binom{4+m}{4}$ choices of 4 collisions out of suggested $4+m$ collisions. We also consider all possible 2^{32} subkey candidates for this diagonal. For each choice of 4 pairs (A_i, B_i), $0 \leq i < 4$, and for each subkey candidate, we perform AddRoundKey, SubBytes, ShiftRows and MixColumns transforms as applies to the target diagonal. If the current choice of pairs (A_i, B_i) leads to collisions between all four pairs in some position[3] in the output column, the 4-byte subkey candidate survives and is added to the short list of subkey candidates. This process is visualized in Figure 1. On average, for each subkey test, one has to perform an amount of operations roughly comparable to 25% of an AES round, as a key candidate will only rarely survive the check with the first pair of diagonals. The complexity of the first step is approximately $4 \cdot \frac{1}{10} \cdot \frac{1}{4} \cdot \binom{4+m}{4} \cdot 2^{32} \approx \binom{4+m}{4} \cdot 2^{28.7}$ AES encryptions and can be optimized by taking into account that the values of subkey candidates are adjacent.

For the second step, consider how many subkey candidates survive for each of the four diagonals after the first step of key recovery. Since the positions of collisions after the first round are unknown, for each of the four diagonals there will be $4^4 = 256$ subkey candidates if $m = 0$. If $m > 0$, we expect to have $256 \cdot \binom{4+m}{4}$ subkey candidates.

In the second step of key recovery, all partial keys from the four short lists (one list for each diagonal) are concatenated to perform a final key test by computing a full AES encryption using a known plaintext-ciphertext pair. This final key test is executed for each key candidate. Having the estimated number

---

[3] Note that this position does not have to be the same for all four pairs of diagonals.

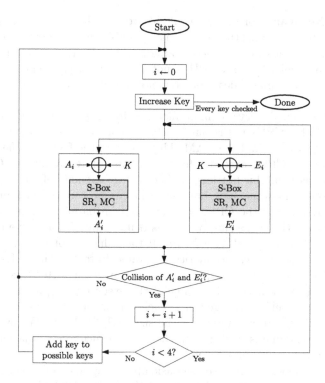

Fig. 1. Finding candidate subkeys from four diagonal pairs (A_i, E_i) for $0 \le i < 4$. K is a 4-byte subkey corresponding to some diagonal to be tested

of surviving subkey candidates after the first step in mind, the complexity of the second step of key recovery can be computed as $2^{32} \cdot \binom{4+m}{4}^4$ as one has to inspect each combination of subkeys.

Thus, the offline complexity of our attack is dominated by the second step of key recovery and can be estimated as $2^{32} \cdot \binom{4+m}{4}^4$ AES encryptions. See Section 5 for our experimental results.

4   Physical Behavior of Cache Hits on Embedded Platforms

Classical timing attacks [9,21] are applicable in cases where the implemented algorithm features a data-dependent runtime. Hence, resistance is easily achievable by building constant run-time code.

The microarchitecture of modern CPUs contains several measures to speed up the execution of programs by methods such as instruction level parallelism, several caches and branch prediction units. The behavior of these microarchitectural measures is usually not considered by implementers, since most code written for modern embedded systems is supposed to be portable to different platforms. Even if an implementer would like to take the behavior of the underlying platform into account, this is often not possible since the processor interacts with different threads in an unpredictable manner. Furthermore, in some cases timing relevant behavior of a CPU is not documented [4]. Hence, code with a constant execution time is desirable, but not always possible.

Cache Behavior. A very common microarchitectural feature found on almost all modern 32-bit CPUs is the data cache. Cache is a small, but fast memory between the processor and the RAM. This is due to the fact that the processing speed of modern CPUs exceeds the access time of RAM by far. Caches are intended to overcome this bottleneck. The storage capacity of a cache is smaller compared to the main memory, but the cache can be accessed at a much higher speed than RAM. For the CPU, the cache is transparent. When a value from RAM is queried, the cache simply returns it if a copy of that value is currently in the cache. This is called a *cache hit*. If the value is not in the cache (a so-called *cache miss*), the cache queries the value from the larger RAM, passes it to the CPU and stores it in the cache. Of course the latter takes additional time, resulting in an increased runtime of the executed program for each cache miss. The cache itself is arranged in 2^l cache lines. Each of these lines can hold 2^b bytes. This leads to a complete cache size of $2^{(b+l)}$ bytes. For every queried value, a full cache line is loaded from the RAM, hence a few adjacent values to the queried one are also prefetched.

Several techniques to improve the basic operating mode of a cache have been proposed in order to improve the ratio of the cache hits and cache misses. In direct-mapped cache every data from the main memory can only be stored in one specific cache line. This allows a very simple and fast verifying method to check if the data is cached at the cost of a rather high number of cache misses. In a fully associative cache the data can be stored in every cache line. To determine, if the needed data is cached, all entries must be checked. This takes a long time compared to the direct-mapped cache, but has the advantage that the amount of cache misses is very low. A combination of the advantages of both models is the n-way set associative cache. An entry can be stored in n possible cache lines. These cache lines are combined into one cache set. The n-way set associative cache is quite common in practice, as it provides a good tradeoff between cache hit time and cache miss ratio.

Target Platform. Since our goal is to evaluate the threat of cache timing attacks to modern embedded platforms, we chose a rather powerful ARM9 processor as target. Modern ARM 16/32 bit processors are used for many embedded applications where the demand for computing power is high and the power consumption is restricted. ARM cores can be found in most modern smart phones, portable game consoles and PDA's, but also in various other embedded electronics [7].

The hardware used as target is the Embest SBC2440-II single board computer. The board hosts a Samsung S3C2440A [15] microprocessor featuring an ARM9 core, namely an ARM920T [3]. Besides the ARM, the CPU provides functionality to handle the boards interfaces, such as an LCD controller. It is a typical chip to be found in modern PDAs, such as Nokia N810, palmOne Treo 600, etc. The CPU can be clocked at up to 400 MHz. We operated the Embest SBC2440II board with an open source ARM-Linux with a 2.6.13 linux kernel. The board alternatively features WindowsCE.

Cache Architecture of the ARM. The ARM920T core features a Harvard memory architecture with separate data and instruction cache. The caches have a size of 16 KB each and are divided into 512 cache lines of 8 four-byte words. Both caches are arranged in 64-way set-associative caches with each having eight sets of 64 cache lines. Each entry can be located in just one set, but in this set it can be stored in any of the set's cache lines. The bits 7 to 5 of the address define the set where the entry is located. The cache line itself can be determined by comparing the tag of the address, bits 31 to 8, with the tags stored in the cache. The bits 4 to 2 specify the word in the cache line and the bytes in a word can be addressed with the bits 1 to 0.

Attack Conditions. As a target for the attack we used the T-box implementation of the AES provided by the openSSL package [20]. Although our attack is not limited to the T-box implementation, we analyzed the cache behavior for this case, as the T-box implementation is the most common in practice. We assume that the attacker can encrypt two consecutive chosen plaintexts (or decrypt two consecutive chosen ciphertexts). The attacker can also measure the execution time of the second encryption only. Hence our attack is applicable in cases where the AES output can not be accessed. Other attacks such as all final round attacks are not possible in this case. The attack can be easily performed in cases where the attacker has full access to the system; a realistic assumption for many embedded applications. On many PDAs, smart phones etc., the user is able to execute own code directly or after jailbreaking the device. Hence, depending on the specific application, the adversary is able to query a commercial application protected by an AES and to overcome the security by measuring its execution time. Cache timing attacks can also be a viable measure to circumvent security protection mechanisms on the operating system layer such as sandboxing.

Cache Behavior of the ARM. As described earlier, wide collisions have a much stronger influence on the execution time than normal collisions. Figure 2 presents a histogram over the encryption time of the encryption of the second plaintext. The encryption time is visibly decreased in the case a wide collision occurs (shown in light gray). If no wide collision occurs, the execution time is slightly higher (dark gray bars). The difference between the two sets can be exploited by the attacker to perform the previously described attack.

Them measurement setup should try to minimize the risk of non-collisions being detected as collisions. One big source of noise is structure of the cache, namely the existence of cache lines. In case of the T-box implementation up to eight table entries are loaded into a cache line for each cache miss. Consequently, even though no wide cache collision occurs, several unexpected collisions may

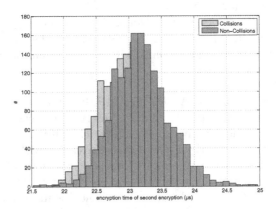

Fig. 2. Histogram comparing the execution time of AES encryptions with and without wide collisions on the target platform

occur. Please keep in mind that difference in Figure 2 is taken from real measurements. Effects like the non-perfect description of the cache behavior mentioned above are already included.

5   Experimental Results

All measurements were performed on the ARM platform described in Section 4. The ARM Board is set up as a server running the AES implementation of openSSL 0.9.8K and queried via the Ethernet interface of the board. Every challenge queries two consecutive AES encryptions.

As described earlier, four wide collisions per column are sufficient to extract the corresponding four bytes of the AES key with a remaining uncertainty of 2^8 key candidates. Hence, after finding four wide collisions for each of the four columns, a full AES-128 key can be recovered with a remaining computational complexity of 2^{32} AES computations. The amount of computations has to be increased if we do not assume a perfect collision detection, i.e., in the case we accept a certain number of false positives.

Our test setup has been optimized to minimize the number of false positives in the collision detection process. We evaluated different test settings by trying different test parameters. Besides the number different diagonals n, we tested I different plaintexts per diagonal. The tested parameters are $n = 256$, $n = 512$, and $n = 1024$ diagonals with $I = 200$, $I = 400$ and $I = 800$ different plaintexts each. To increase the reliability of the timing measurements, we repeat each measurement r times, where $r = 20$ or $r = 40$. Since measurement noise mostly increases the measured computation times, we calculated the average encryption time for one plaintext by using only the j fastest out of the r timing samples. Chosen parameters for j were the 2, 5, 10 and $j = r$ fastest measurements.

Table 1. Number of false positives m for one column with a success probability of 65%

Test	Parameters			False positives			
	n	I	r	$j = 2$	$j = 5$	$j = 10$	$j = r$
1	800	600	40	1.78	1.56	1.33	1.56
2	1024	400	20	2.11	1.22	1.22	1.67
3	1024	400	40	0.89	0.89	0.67	0.89
4	1024	600	15	1.67	1.89	2	1.78
5	1024	800	40	0.13	0.13	0.38	0.38

Table 1 shows the expected number of false positives for one column (hence, four collisions and four revealed key bytes). All non-collisions that have a timing lower than or within the group of the four fastest real collisions are considered false positives. If we accept one false positive per column (i.e. a total of four false positives), the remaining key space is increased to about 2^{41} key candidates for the entire AES key. If two non-collisions are accepted per column, the complexity of the key search rises to approximately $2^{47.6}$, for three to $2^{52.5}$, and for four false positives to $2^{56.5}$ possibilities. With a highly optimized AES implementation such as the one of Hamburg [19], up to 2^{38} AES encryptions can be performed per hour on a modern core2duo desktop processor, resulting in a feasible attack, even in the case of one false positive per column.

Table 2. Number of measurements N and the size of the remaining key space K for our differential wide collision attack and the expanded second round attack by Aciiçmez et. al. [1]

Attack	Test	N	Remaining key space K $[2^x]$			
			$j = 2$	$j = 5$	$j = 10$	$j = r$
Our attack	2	65,536,000	49.7	43.8	43.8	47.4
	3	131,072,000	41.6	42.2	41.2	42.2
	5	262,144,000	36.9	36.9	38.7	37.5
Aciiçmez et. al. [1]		128,000,000				32

Table 2 summarizes the complexity of the key recovery step as number of remaining key candidates and complexity of the online phase as the needed number of traces for the different test cases described in Table 1. The results of our differential wide collision attack are compared to the expanded second round attack of Aciiçmez et. al. [1] on the same target platform, revealing that our attack has an increased complexity for the key recovery, but can successfully be performed with a lower number of measurements. All parameters of both attacks are for an expected success rate of 90%. Depending on the chosen attack parameters, we can approximately halve the number of needed measurements when compared to the expanded second round attack. In test setup 2, 66 million measurements suffice, with a remaining key space of $2^{43.8}$, resulting in a more realistic attack.

6 Conclusion

We presented a novel differential collision attack making use of the MDS properties of the AES algorithm. The attack outperforms previous attacks in the same adversarial scenario in terms of needed measurements and decreases the remaining key space far enough to be easily computable on a modern desktop PC.

We furthermore presented the first evaluation of the vulnerability of embedded platforms to cache timing attacks and showed that cache attacks are feasible in practical setups. We want to stress that cache attacks pose a serious threat, especially on embedded platforms. Developers of embedded software solutions relying security functionality should consider the threat of cache timing attacks when designing their systems.

Acknowledgements. During the work on this paper, Andrey Bogdanov was supported partially by the Fund for Scientific Research - Flanders (FWO) and partially by the Chair for Embedded Security at the Ruhr-University of Bochum.

References

1. Aciiçmez, O., Schindler, W., Koç, Ç.K.: Cache based remote timing attack on the AES. In: Abe, M. (ed.) CT-RSA 2007. LNCS, vol. 4377, pp. 271–286. Springer, Heidelberg (2006)
2. Advanced Encryption Standard. FIPS. Publication 197. National Bureau of Standards, U.S. Department of Commerce (2001)
3. ARM Limited. ARM920T Technical Reference Manual, 1 edn.
4. Bernstein, D.J.: Cache-timing attacks on AES. Technical report, Department of Mathematics, Statistics and Computer Science, The University of Illinois at Chicago, 2005, cr.yp.to/antiforgery/cachetiming-20050414.pdf
5. Bonneau, J., Mironov, I.: Cache-collision timing attacks against AES. Technical report, Computer Science Department, Stanford University and Microsoft Research, Mountain View, CA (2006)
6. Daemen, J., Rijmen, V.: The Design of Rijndael. Springer, Heidelberg (2002)
7. ARM INC. ARM Powered Products, http://www.arm.com/markets/mobile_solutions/app.html
8. Kelsey, J., Schneier, B., Wagner, D., Hall, C.: Side channel cryptanalysis of product ciphers. Journal of Computer Security, 97–110 (1998)
9. Kocher, P.C.: Timing Attacks on Implementations of Diffie-Hellman, RSA, DSS and Other Systems. In: Koblitz, N. (ed.) CRYPTO 1996. LNCS, vol. 1109, pp. 104–113. Springer, Heidelberg (1996)
10. Kocher, P.C., Jaffe, J., Jun, B.: Differential Power Analysis. In: Wiener, M. (ed.) CRYPTO 1999. LNCS, vol. 1666, p. 388. Springer, Heidelberg (1999)

11. Mangard, S., Oswald, E., Popp, T.: Power Analysis Attacks and Countermeasures for Cryptographic Smart Cards: Revealing the Secrets of Smart Cards. Springer, Heidelberg (2007)
12. Neve, M., Seifert, J., Wang, Z.: Cache time-behavior analysis on AES (2006), http://www.cryptologie.be/document/Publications/AsiaCSSfull06.pdf
13. Osvik, D.A., Shamir, A., Tromer, E.: Cache attacks and countermeasures: The case of AES. In: Pointcheval, D. (ed.) CT-RSA 2006. LNCS, vol. 3860, pp. 1–20. Springer, Heidelberg (2006)
14. Page, D.: Theoretical use of cache memory as a cryptanalytic side-channel. Technical report (2002)
15. Samsung Electronics. S3C2440A 32-Bit CMOS Microcontroller User's Manual, 1 edn.
16. ST33F1M. Smartcard MCU with 32-bit ARM, http://www.st.com/stonline/books/pdf/docs/15066.pdf
17. Tsunoo, Y., Saito, T., Suzaki, T., Shigeri, M., Miyauchi, H.: Cryptanalysis of DES implemented on computers with cache. In: Walter, C.D., Koç, Ç.K., Paar, C. (eds.) CHES 2003. LNCS, vol. 2779, pp. 62–76. Springer, Heidelberg (2003)
18. Tsunoo, Y., Tsujihara, E., Shigeri, M., Kubo, H., Minematsu, K.: Improving cache attacks by considering cipher structure. Int. J. Inf. Secur. 5(3), 166–176 (2006)
19. Hamburg, M.: Accelerating AES with Vector Permute Instructions. In: Clavier, C., Gaj, K. (eds.) CHES 2009. LNCS, vol. 5747, pp. 18–32. Springer, Heidelberg (2009)
20. OpenSSL 0.9.8.K. Openssl: The open source toolkit for ssl/tls, http://www.openssl.org/ (accessed June 18, 2009)
21. Dhem, J.-F., Koeune, F., Leroux, P.-A., Mestré, P., Quisquater, J.-J., Willems, J.-L.: A Practical Implementation of the Timing Attack. In: Schneier, B., Quisquater, J.-J. (eds.) CARDIS 1998. LNCS, vol. 1820, pp. 167–182. Springer, Heidelberg (2000)

Usable Optimistic Fair Exchange

Alptekin Küpçü and Anna Lysyanskaya

Brown University, Providence, RI, USA
{kupcu,anna}@cs.brown.edu

Abstract. Fairly exchanging digital content is an everyday problem. It has been shown that fair exchange cannot be done without a trusted third party (called the *Arbiter*). Yet, even with a trusted party, it is still non-trivial to come up with an efficient solution, especially one that can be used in a p2p file sharing system with a high volume of data exchanged.

We provide an efficient optimistic fair exchange mechanism for bartering digital files, where receiving a payment in return to a file (buying) is also considered fair. The exchange is optimistic, removing the need for the Arbiter's involvement unless a dispute occurs. While the previous solutions employ costly cryptographic primitives for every file or block exchanged, our protocol employs them only once per peer, therefore achieving $O(n)$ efficiency improvement when n blocks are exchanged between two peers. The rest of our protocol uses very efficient cryptography, making it perfectly suitable for a p2p file sharing system where *tens* of peers exchange *thousands* of blocks and they do not know beforehand which ones they will end up exchanging. Therefore, our system yields to one-two orders of magnitude improvement in terms of both computation and communication (80 seconds vs. 84 minutes, 1.6MB vs. 100MB). Thus, for the first time, a provably secure (and privacy respecting when payments are made using e-cash) fair exchange protocol is being used in real bartering applications (e.g., BitTorrent) [14] without sacrificing performance.

1 Introduction

Fairly exchanging digital content is an everyday problem. A fair exchange scenario commonly involves Alice and Bob. Alice has something that Bob wants, and Bob has something that Alice wants. A fair exchange protocol guarantees that at the end either each of them obtains what (s)he wants, or neither of them does (see [40] for more details and examples).

In this paper, we consider a general file exchange (bartering) scenario, inspired by the BitTorrent [22] peer-to-peer file sharing protocol. Alice has several files (BitTorrent blocks) of interest to Bob, and Bob has several files (blocks) of interest to Alice. They do not know ahead of time how many or which blocks they will end up exchanging. They want to perform a fair exchange: Alice should get Bob's file (block) if and only if Bob gets Alice's file (block). In a signature fair exchange [4,3,2], there is a verification mechanism (i.e., the public key) that enables the sender to verifiably encrypt the signature so that the receiver can check that the encrypted signature verifies. No such efficient verifiable encryption method is currently known for exchanging files. Therefore, a compensation is required after the fact if one of the parties cheat. In our scenario, we

J. Pieprzyk (Ed.): CT-RSA 2010, LNCS 5985, pp. 252–267, 2010.

are assuming that Alice/Bob will be equally happy to get a payment in return to her/his file. Thus, exchanging a file with a payment (buying) is also considered fair, as in some previous works [4,8,18,37,36].

One of the hardest points in creating a usable optimistic fair exchange protocol suitable for p2p file sharing applications is that the peers to contact and the content to exchange are not pre-defined. BitTorrent clients keep connecting to different peers to obtain different blocks. Fault-tolerance issues, connectivity problems, and availability of data blocks are all factors affecting from whom which block should be obtained. Our protocol uniquely addresses these issues by removing the need to know what content to exchange with whom beforehand.

In a nutshell, in our protocol, Alice sends a verifiable escrow of a payment (e.g., e-coin) to Bob first. Then, they exchange encrypted files. Afterward, Alice sends Bob an escrow of her key with her signature on the escrow. Then, Bob sends Alice the key to his file. Finally, Alice sends Bob the key to her file. Since Bob has a verifiable escrow of an e-coin and an escrow of a key before he sends his key to Alice, he is protected. In the worst case, if Alice does not provide the correct key and the key escrow contains garbage, Bob can go to the Arbiter and obtain Alice's payment. The escrow of the payment cannot contain garbage, because it was formed using a *verifiable* escrow. After the exchange of the verifiable escrow, the rest of our protocol can be repeated as many times as necessary to exchange multiple files (even if the number and content of the files were not known in advance), unless there is a dispute.

We provide two versions of the protocol: In the first one (the one described briefly above) only one party provides a verifiable escrow. This version requires the use of timeouts for dispute resolution purposes. We provide another version that needs both parties to provide verifiable escrows but requires no timeouts. Both versions are very efficient since they use only *one* (resp. *two*) expensive primitives (verifiable escrow and payment) regardless of the number of files exchanged. We stress the fact that our timeouts can be very large (e.g., one day or week) to allow for unexpected situations in which the participants act honestly (e.g., network failure), and thus require very loose synchronization (e.g., one hour difference), and users can freely participate in other exchanges without waiting for the timeout.

Previous Work: It is well-known that a fair exchange protocol is impossible without a trusted third party (TTP) [43] (called the *Arbiter*) that ensures that Alice cannot take advantage of Bob, and vice versa. Without loss of generality, Alice will have to send the last message of the protocol, and we want to protect Bob in case she chooses not to do so. Without an arbiter, gradual release type of protocols where parties send pieces to each other in rounds can provide only weaker forms of fairness, and are much less efficient [11,13].

Luckily, the impossibility result [43] does not require that the Arbiter be involved in each transaction, but simply that the Arbiter exists. If Alice and Bob are both well-behaved, there is no need for the Arbiter to do anything (or even know an exchange took place). Micali [39], Asokan, Schunter and Waidner [2], and Asokan, Shoup and Waidner [4,3] investigated this *optimistic* fair exchange scenario in which the Arbiter gets involved only in case of a dispute. Two such protocols [4,30] were analyzed in [46] (see also [7]).

Asokan, Shoup and Waidner (ASW) [4] gave the first provably secure and completely fair optimistic exchange protocol for exchanging digital signatures. Later on, Belenkiy et al. [8] gave a protocol for buying digital content in exchange for e-cash, building on top of the ASW protocol. They provided an optimization for the Arbiter so that, unlike in the ASW protocol, the amount of work that the Arbiter is required to do depends only logarithmically on the size of the file. They also assume there is an additional TTP (which we call the *Tracker*) that provides a means of verification that the file actually contains the right content (e.g., using hashes). Such entities certifying hashes already exist in current BitTorrent systems [22].

Belenkiy et al. [8] used e-cash (introduced by Chaum [20]), in particular, endorsed e-cash [18] in their constructions. The reason is that other forms of payments (signatures or electronic checks used in [4,37]) do not provide any privacy. In our protocols, any form of payment can be employed, but we will also use endorsed e-cash in our sample instantiation since it is efficient and anonymous.

Contributions: We present the most efficient fair exchange known to us, where *the efficiency is comparable to a simple unfair exchange if performed multiple times between the same pair of users, even when peers do not know beforehand which blocks they will end up exchanging*. Using the best previous work (Belenkiy et al. barter protocol [8]), n pairs of blocks can be exchanged using n transactions, each of which requires a costly step involving expensive cryptographic primitives (a verifiable escrow and an e-coin). Our contribution is a very efficient fair exchange protocol using which this can be done with only *one* (or *two* if we do not want to employ timeouts) step in total that involves the same expensive primitives (verifiable escrow and payment). This is a property that is unique to our protocol: Instead of employing the costly primitives for every file or block that is exchanged, we employ them once per peer, even when peers do not know beforehand which blocks they will end up exchanging. Then, exchanging multiple files/blocks between peers involves only very efficient cryptography (i.e., symmetric- and public-key encryption, and digital signatures). In a real setting where BitTorrent peers exchange thousands of blocks with only tens of peers, there is *one or two orders of magnitude improvement in terms of both computation and communication* (80 seconds vs. 84 minutes computational overhead and 1.6MB vs. 100MB communication overhead for a 2.8GB file —for detailed numbers, see Section 3.2). This means that, with no (i.e., neglectable) efficiency loss, our fair exchange protocol can be used to exchange files instead of the unfair protocol currently used by BitTorrent or similar file sharing protocols.

We stress the fact that the timeouts used for dispute resolution purposes in one of our protocols can be very large (e.g., one day or week) to allow for unexpected situations in which the participants act honestly (e.g., network failure), and thus require very loose synchronization (e.g., one hour difference), and *users can freely participate in other exchanges without waiting for the timeout.*

We take the idea of using verifiable escrow from ASW [4], and the subprotocols of Belenkiy et al. [8] that increase the efficiency of the Arbiter (proving and disproving keys). The Arbiter does absolutely no work in our protocols, as long as no dispute occurs. *Our protocols can make use of any type of payments*, but we will show an instantiation using e-cash since it also provides privacy. Our performance evaluation numbers will use endorsed e-cash [18] as the payment mechanism. Note that other (non-anonymous) forms of payments (e.g., electronic checks [21]) will be more efficient.

Our additional contribution is definitional. We give a general definition of fair exchange of digital content (not just digital signatures) provided that it can be verified using some verification algorithm (defined in Section 2.2). Furthermore, our fairness definition covers polynomially many exchanges between an honest party and an adversary controlling polynomially-many other participants (see [27] for an example fair exchange protocol that is fair for a single exchange but stops being fair in a multi-user setting). We then prove our protocol's security based on this definition. We sum up the most important properties of our protocols below.

Security of our protocol: Our protocols provably satisfy the following condition (waiting for at most one timeout period if timeouts are used, or without waiting at all if no timeouts are used), as long as at least one of the trading parties (Alice and Bob) is honest:

- Either Alice and Bob both get their corresponding files,
- Or Alice gets Bob's file and Bob gets Alice's payment (turns into a buy protocol in effect),
- Or neither of them gets anything.

Efficiency of our protocol: We have the following properties regarding efficiency:

- An honest user can reuse her e-coin for other exchanges without waiting for the completion of the protocol.
- The overhead of our costly step – verifiable escrow and e-cash – is constant $O(1)$, instead of linear $O(n)$ as in previous best results, when n files or blocks are exchanged.

Already, the Brownie Project [14] is using our protocols in their BitTorrent deployment.

2   Definitions

Barter is an exchange of two items, which are digital files in our case. We assume that the reader is familiar with encryption and signature schemes, and hash functions.

2.1   Notation

An escrow is a ciphertext under the public key of some trusted third party (TTP). A *verifiable* escrow [4,19,15] means that the recipient can verify that the contents of the ciphertext satisfy some relation (therefore stating that the ciphertext contains the expected content). A contract (a.k.a. label, condition, or tag) attached to such a ciphertext defines the conditions under which the TTP should decrypt and give away the encrypted secret [47]. The label is public and it is integrated with the ciphertext in a such way that it cannot be modified. We will use $E_{Arb}(a;b)$ to denote an escrow of the secret a under the Arbiter's public key, with the contract b. Similarly, $VE_{Arb}(a;b)$ will denote a verifiable escrow.

Any payment protocol that can efficiently be verifiably escrowed and is secure can be used in our protocols. Furthermore, if privacy is desired, the payments should be anonymous as in e-cash [20]. We provide an instantiation using endorsed e-cash [18] (which is an extension of compact e-cash [17]), since it satisfies all these requirements. Endorsed e-cash splits a coin into an unendorsed coin (denoted $coin'$) and endorsement (denoted end). One can think of $coin'$ as an encrypted coin and end as the key. One can check if the endorsement end in a given verifiable escrow [19] matches the given unendorsed coin $coin'$ (without learning the endorsement end). Furthermore, given only the unendorsed part $coin'$, no other party (except the owner) can come up with a valid endorsement end. Endorsed e-cash moreover has the ability to catch double-spenders. Hence, if one uses two different $coin', end$ pairs trying to spend the same coin twice, (s)he will be caught (and, since her identity is revealed, can be punished). Note that if a party tries to deposit the same coin twice (using the same $coin', end$ pair), the operation can easily be denied by checking against a list of past transactions. Lastly, only matching $coin', end$ pairs can be linked, unendorsed coins and endorsements prepared for different exchanges remain unlinkable.

Wherever used, K_P will denote a symmetric key of a party P, generated through an encryption scheme's key generation algorithm. We let $c = Enc_K(f)$ denote that the ciphertext c is an encryption of the plaintext f under the symmetric key K. Similarly, $f = Dec_K(c)$ will denote that the plaintext f is the decryption of the ciphertext c under the symmetric key K. Our protocol can make use of any secure symmetric encryption scheme (see the book by Katz and Lindell [33] for definitions and constructions).

Let pk_P and sk_P denote public and secret keys for a party P. Then $sign_{sk}(x)$ will denote a signature on x under the secret key sk which can be verified using the corresponding public key pk. Our protocol can make use of any secure public-key encryption scheme [24,28] and any secure signature scheme [31].

Furthermore, let H_k be a family of (universal one-way) hash functions [41], where k is the security parameter, and let $hash$ be a hash function uniformly choosen from the family H_k of hash functions. Then, $h_x = hash(x)$ will denote that h_x is the hash of x under the hash function $hash$. We now introduce a definition we frequently use in the paper.

Definition 1. *We say that a key K **decrypts correctly**, or is the **correct key** with respect to a plaintext hash h_f and a ciphertext c, if the plaintext $f' = Dec_K(c)$ has the property $hash(f') = h_f$.*

Finally, a negligible probability denotes a probability that is a negligible function of the security parameter (e.g., the key-length of an encryption scheme). A negligible function of n is a function which is smaller than any inverse polynomial over n with $n > N$ for sufficiently large N (e.g., $neg(n) = 2^{-n}$).

2.2 (Optimistic) Fair Exchange

In this section we will give a general definition of fair exchange. Unlike in ASW, our definitions will not be specific to signature exchange, and we will consider polynomially-many exchanges between an honest user and an adversary controlling polynomially-many other users. Furthermore, we separate and clearly define the roles of all trusted parties. While providing models and definitions for a general framework of (optimistic) fair exchange applicable to a broad range of protocols, we will also show its extensions to our case.

MODEL: The model is adapted from the ASW definition [4], with clarifications and generalizations. There are three players; *Alice* and *Bob* exchanging two digital items, and the *Arbiter*[1] for conflict resolution. All players are assumed to be polynomial time interactive Turing machines. We make no assumption about the underlying network capability.[2] Any message that does not confirm with the protocol specification will be discarded by the honest parties. Any input which does not verify according to the protocol will be resolved as stated by the protocol or the protocol will be aborted if no resolution is applicable. It is important that the Arbiter resolves conflicts on the same exchange *atomically*.[3] Thus, it will only interact with either Alice or Bob at any given time instance, until that interaction ends as specified by the protocol.[4] Sensitive communication (e.g., exchange of decryption keys for files or endorsement of an e-coin) will be carried out over a secure (and possibly authenticated) channel (e.g., SSL can be used to connect to the Arbiter, a secure key exchange with no public key infrastructure can be used for the communication between Alice and Bob).

For protocols using a *timeout*[5], we assume that the adversary cannot prevent the honest party from reaching the Arbiter before the timeout. If no timeouts are defined, we assume the adversary cannot prevent the honest party from reaching the Arbiter eventually. Hence, the honest party is assumed to be able to reach the Arbiter as defined by the protocol. Even with timeouts, this is not an unrealistic assumption since our timeouts can be large (e.g., one day or week).

In our model, we have two additional players, namely the *Tracker* (also in [4,8,22])[6] providing verification algorithms, and the *Bank* dealing with monetary parts of the system.

SETUP PHASE: Before the fair exchange protocol is run, we assume there is a setup phase. In this one-time pre-exchange phase, the Arbiter generates his public-private key pair (for the (verifiable) escrow schemes) and publishes his public key(s) so that both Alice and Bob obtain it. Optionally, the Arbiter may learn public keys of Alice and Bob in the setup phase, but our focus is on the case where the Arbiter does not need to know anything (and learns almost nothing) about Alice or Bob. *The adversary*

---

[1] One of the TTPs in ASW.

[2] Clients will have a local *message timeout* mechanism like the TCP timeout, which is small (e.g., one minute). The receiver deals with a *message timeout* exactly as it would deal with a non-verifying input.

[3] We present a trade-off between non-atomicity and performance of the Arbiter later on.

[4] For ease of the Arbiter to find the correct exchange, a random exchange ID can be incorporated into the messages. Since this is only a minor implementation efficiency issue, we do not want to complicate our definitions with that.

[5] This is not the *message timeout*, it is the *timeout* specified by the protocol, which is generally much longer (e.g., one day or week).

[6] ASW has the corresponding TTP in their file exchange scheme. In their signature exchange protocol, the public key infrastructure providing the public keys can be seen as the Tracker.

cannot interfere with the setup phase.[7] In the setup phase, the Bank and the Tracker also generate their public-private key pairs and publish their public keys.

Definition 2. *Let* SP *denote the security parameters of the system (e.g., key lengths of the primitives used). Let* PP *denote all the public values in the system, including* SP, *public keys of the trusted parties, and possibly some public parameters. Let* PPGen(SP) *be the randomized procedure which generates the public values given the security parameters. Then, define our* $PP = (pk_{arb}, pk_{bank}, pk_{tracker}, timeout, SP,$ *and additional parameters for primitives used).*

From now on, we need to talk about multiple exchanges taking place. Alice has files $f_A^{(1)}, .., f_A^{(n)}$ to be exchanged with Bob, and Bob has $f_B^{(1)}, .., f_B^{(n)}$ to be exchanged with Alice (n is a polynomial in SP).[8] In general, we can consider these files as some strings in $\{0,1\}^*$, therefore consider fair exchange of anything that is verifiable. Without loss of generality, the Tracker gives Alice a *verification algorithm* $V_{f_B^{(i)}}$ for each file $f_B^{(i)}$, and Bob a verification algorithm $V_{f_A^{(i)}}$ for each file $f_A^{(i)}$ before the exchange takes place.

Assume that the content to be exchanged and associated verification algorithms are output by a generation algorithm Gen(SP) that takes the security parameters as input and outputs some content to be exchanged, with associated verification algorithms, and possibly some public information about the content. This procedure involves a trusted party H and the Tracker. The parties trust the Tracker in that any input accepted by that verification algorithm will be the content they want. In other words, they are going to be happy with any content that verifies under that verification algorithm. In particular, the content generation process is trusted. The adversary cannot generate "junk" files and ask the Tracker to create verification algorithms for them. BitTorrent forum sites and ratings provide a level of defense against this in practice.

Definition 3. *Content and verification algorithms are secure if* $\forall$ PPT *adversaries* $\mathcal{A}$ *and* $\forall$ *auxiliary inputs* $z \in \{0,1\}^{poly(SP)}$ *we have (over the randomness of the generation algorithms, the adversary, and possibly the verification algorithms)*

$$Pr[PP \leftarrow PPGen(SP); (f_H^{(1)}, V_{f_H^{(1)}}, pub_{f_H^{(1)}}, .., f_H^{(n)}, V_{f_H^{(n)}}, pub_{f_H^{(n)}}) \leftarrow Gen(SP);$$
$$(f_{\mathcal{A}}^{(1)}, .., f_{\mathcal{A}}^{(n)}) \leftarrow \mathcal{A}(V_{f_H^{(1)}}, pub_{f_H^{(1)}}, .., V_{f_H^{(n)}}, pub_{f_H^{(n)}}, PP, z):$$
$$\exists i \in [1..n] \mid (V_{f_H^{(i)}}(f_H^{(i)}) \neq accept \vee V_{f_H^{(i)}}(f_{\mathcal{A}}^{(i)}) = accept)] = neg(SP)$$

The definition above models the case in which the files to be exchanged cannot be found by the adversary by some other means[9] (and hence exchanging files makes

---

[7] This is the standard trusted setup assumption that says Alice and Bob have the correct public key of the Arbiter.

[8] Note that Alice or Bob can represent multiple entities controlled by the adversary.

[9] We assume that the adversary cannot just "guess" an honest participant's file, in which case the exchange is trivially unfair.

sense for the adversary), even with the help of associated verification algorithms and public information[10].

To provide evidence on the generality and applicability of our definition, we present several example verification algorithms for various tasks. For example, a file verification can be performed using hashes. So, each verification algorithm $V_{f_A^{(i)}}$ for Alice's file $f_A^{(i)}$ contains the definition of hash function used $-hash-$[11], and the hash value $h_{f_A^{(i)}} = hash(f_A^{(i)})$. The ith verification algorithm computes the hash of the given input according to the description of the hash function, and accepts it if and only if the computed hash matches $h_{f_A^{(i)}}$. As another example, consider the ASW signature exchange protocol, in which each verification algorithm contains the signature scheme's description[11], the signature public key of Alice pk_A[11], and the message m_i to be signed. When it receives a signature as input, the ith verification accepts the signature if and only if it is a valid signature on message m_i under the public key pk_A using the signature scheme. As yet another example, an e-coin verification algorithm can take a coin to verify, and use the Bank's public key while verifying the non-interactive proofs given. Such an algorithm is a part of the specification of every e-cash scheme (e.g., see [18,17]). Verifiable encryption schemes (e.g., [19]) and, in general, proof systems also specify a verification algorithm in their definitions. Such algorithms can be used directly in a fair exchange protocol, satisfying our definition as long as they are secure according to Definition 3.

To summarize, in the setup phase, public values are generated using PPGen(SP). The files and the verification algorithms are generated jointly by the Tracker and some trusted content generator (e.g., movie distributor) using the Gen(SP) procedure. In the context of BitTorrent, this means that we trust the content generator about the content, and the Tracker about the verification algorithms. A "highly rated" BitTorrent user will be trusted about the content, or alternatively, comments on the forum sites will warn against bogus content. From now on, we assume the content and the verification algorithms used are secure and trusted.

Definition 4. *Fair Exchange Protocol: A fair exchange protocol is composed of three interactive algorithms: Alice running algorithm A, Bob running algorithm B, and the Arbiter running the trusted algorithm T. The content and verification algorithms used need to be secure according to Definition 3. The security of the exchange is then defined in terms of completeness (when Alice and Bob are both honest) and fairness (when either Alice or Bob is malicious).*

COMPLETENESS for a (non-optimistic) fair exchange states that the interactive run of A, B and T by *honest parties* results in A getting B's files and B getting A's files (assuming an ideal network):

---

[10] For example, if movies are being exchanged, a lot of information is publicly available about such a movie file, such as actors, length, and release date. But these do not enable people to come up those movie files.

[11] Possibly different for each verification algorithm.

$$Pr[(f_B^{(1)}, .., f_B^{(n)}) \leftarrow A(f_A^{(1)}, .., f_A^{(n)}, V_{f_B^{(1)}}, .., V_{f_B^{(n)}}, PP) \overset{T(sk_{arb})}{\longleftrightarrow}$$
$$B(f_B^{(1)}, .., f_B^{(n)}, V_{f_A^{(1)}}, .., V_{f_A^{(n)}}, PP) \rightarrow (f_A^{(1)}, .., f_A^{(n)})] = 1$$

where the notation describes that A, B and T can all communicate (in a three-way interaction) following the protocol, and at the end A outputs $f_B^{(i)}$ and B outputs $f_A^{(i)}$ for all $i : 1..n$.

OPTIMISTIC COMPLETENESS for an optimistic fair exchange states that the interactive run of A and B by *honest parties* results in A getting $f_B^{(i)}$ and B getting $f_A^{(i)}$ for all $i : 1..n$ (the Arbiter's algorithm T is not involved, assuming an ideal network). A protocol satisfying optimistic completeness also satisfies completeness. Our *optimistic completeness* definition is:

$$Pr[(f_B^{(1)}, .., f_B^{(n)}) \leftarrow A(f_A^{(1)}, .., f_A^{(n)}, V_{f_B^{(1)}}, .., V_{f_B^{(n)}}, PP) \leftrightarrow$$
$$B(f_B^{(1)}, .., f_B^{(n)}, V_{f_A^{(1)}}, .., V_{f_A^{(n)}}, PP) \rightarrow (f_A^{(1)}, .., f_A^{(n)})] = 1$$

Fairness states that at the end of the protocol, either Alice and Bob both get content that passes the verification algorithms given to them, or neither Alice nor Bob gets anything that passes the verification, in each of the n exchanges, even when one of them is malicious.[12] This definition is easy to satisfy using a (non-optimistic) fair exchange protocol since Alice and Bob can both hand their files to the Arbiter, and then the Arbiter can send Bob's files to Alice and Alice's files to Bob, if they pass respective verifications. Thus, below, we will define the more interesting case; fairness for an *optimistic* fair exchange. It is important to note that the ASW definition of fairness applies only to a single exchange, whereas our definition covers polynomially-many exchanges between an honest party and other players all controlled by the adversary.

FAIRNESS: We have an honest player H, and an adversarial player $\mathcal{A}$. The honest player runs algorithm A in exchanges where he plays the role of Alice, algorithm B in exchanges where he plays the role of Bob, and the Arbiter runs the algorithm T, all as defined by the protocol. H has files $f_H^{(1)}, .., f_H^{(n)}$ to be exchanged with the adversary, and $\mathcal{A}$ has $f_{\mathcal{A}}^{(1)}, .., f_{\mathcal{A}}^{(n)}$ to be exchanged with H. The adversary is assumed to control all other players, and hence all interactions of the honest player are with parties controlled by the adversary, which is the worst possible scenario covering multiple exchanges.

First there is the trusted setup phase as explained above, getting the security parameters as input, generating secure content and verification algorithms, along with some associated public information, and giving the appropriate values to each party. Since the setup phase is trusted, $\forall i : 1..n V_{f_H^{(i)}}, V_{f_{\mathcal{A}}^{(i)}}, PP$ are trusted. Then parties proceed with the fairness game explained below, the honest party outputting X and the adversary outputting Y. At the end of the game, we require the fairness condition holds on X, Y, the verification algorithms $V_{f_H^{(1)}}, V_{f_{\mathcal{A}}^{(1)}}, .., V_{f_H^{(n)}}, V_{f_{\mathcal{A}}^{(n)}}$, and the public values PP with high probability against all PPT adversaries $\mathcal{A}$, and all polynomially-long auxiliary inputs.

Pr [Setup; FairnessGame: FairnessCondition] $= 1 - neg(SP)$

---

[12] On the contrary, completeness definition only deals with honest participants.

FAIRNESS GAME: There are three types of interaction in our fairness game. Type 1 interactions are between H and $\mathcal{A}$. Type 2 interactions are between H and T. Type 3 interactions are between $\mathcal{A}$ and T.[13] The adversary can arbitrarily interleave type $1,2,3$ interactions, but cannot prevent type 2 interactions from happening until the timeout if timeouts are used, or eventually otherwise. The game ends when the honest party H produces its final output (including aborts and resolutions) in all the started protocols. Without loss of generality, in the fairness game we assume both parties want to exchange different content in different exchanges ($\forall i \neq j \quad f_H^{(i)} \neq f_H^{(j)}$ and $f_{\mathcal{A}}^{(i)} \neq f_{\mathcal{A}}^{(j)}$ and $\forall i,j \quad f_H^{(i)} \neq f_{\mathcal{A}}^{(j)}$).[14]

FAIRNESS CONDITION: Recall that the honest party's output was X and the adversary's output was Y at the end of the fairness game. A general fairness condition would be $\forall i : 1..n \quad [\exists x \in X : V_{f_{\mathcal{A}}^{(i)}}(x) = accept \Leftrightarrow \exists y \in Y : V_{f_H^{(i)}}(y) = accept]$ meaning that either H and $\mathcal{A}$ both get what they want or both don't, in each exchange.

Our protocol with payments has a very straightforward generalization of the fairness property. Our fairness condition states that either they both parties get each other's file, or one of them gets the other's file whereas the other gets his payment, or they both get nothing at each exchange. We believe that a broad range of optimistic fair exchange protocols can adapt the definition above using straightforward extensions whenever necessary.

TIMELY RESOLUTION: Lastly, as pointed out by ASW [4], an optimistic fair exchange protocol must provide timely resolution: Alice and Bob must be able to have disputes resolved within a finite and limited time. In our protocol without timeouts, resolution is immediate. In our protocol with timeouts, we guarantee resolution at the timeout (which is finite and fixed). We furthermore show that timeouts do not render our system less usable (Alice and Bob can freely participate in other exchanges without waiting for the timeout), and so in general we can use our more efficient protocol with timeouts.

3   Efficient Optimistic Barter Protocol

3.1   Barter with Timeouts

We will show a particular instantiation of our protocol, using endorsed e-cash [18] as the payment and hashes as the file verification algorithms. Full version of our

---

[13] In the implementation, T may need to have a way to differentiate which one of Alice and Bob he is talking to, which can easily be done in our protocols without learning who Alice and Bob are. When necessary, using one-way function values whose pre-image is known by only one of the parties will suffice.

[14] If the honest party already has the adversary's file, the exchange will be trivially fair due to the completeness property. If the adversary already has the honest party's file, then there is no hope for fairness since the adversary can just abort the protocol but he already has the file. Similar arguments hold for exchanging the same file multiple times.

paper discusses generalizations to our protocols [35]. Before the protocol begins, we assume Alice has withdrawn an e-coin from the Bank. Every time Alice and Bob wants to exchange two files (every time before step 2 of the protocol below), Alice generates her fresh key K_A and Bob generates his fresh key K_B for a symmetric encryption scheme. Alice and Bob both have their files (f_A, f_B), have the encrypted versions of their files $(c_A = Enc_{K_A}(f_A), c_B = Enc_{K_B}(f_B))$, have the hashes of their files and encryptions (Alice has $h_{f_A} = hash(f_A), h_{c_A} = hash(c_A)$, and Bob has $h_{f_B} = hash(f_B), h_{c_B} = hash(c_B)$). Besides, the Tracker provides them with the respective verification algorithms: Alice gets h_{f_B}, Bob gets h_{f_A}.[15] Everyone uses the same time zone (e.g., GMT), and the *timeout* is a globally known parameter[16]. If anything goes wrong prior to step 5 (no resolution protocol is applicable), the protocol will be aborted. The protocol proceeds as follows (summarized in Figure 1):

1. Alice creates a fresh public-secret key pair pk_A, sk_A for a signature scheme. Alice sends a fresh unendorsed e-coin *coin'* to Bob, along with a verifiable escrow $v = VE_{Arb}(end; pk_A)$ of the endorsement *end*, labeled with the signature scheme's public key.
2. Alice sends Bob ciphertext c_A of her file.[17] Bob calculates $h_{c_A} = hash(c_A)$.[18]
3. Bob sends Alice ciphertext c_B of his file. Alice calculates $h_{c_B} = hash(c_B)$.
4. Alice sends Bob an escrow $e = E_{Arb}(K_A; h_{f_A}, h_{f_B}, h_{c_A}, h_{c_B}, time)$ and her signature $s = sign_{sk_A}(e)$ on that escrow. The escrow e should encrypt a key and should be labeled with four hash values $h_{f_A}, h_{f_B}, h_{c_A}, h_{c_B}$, and a *time* value. If any of the hash values do not match Bob's knowledge of those values, or if the *time* value is deviated too much from Bob's knowledge of the time (e.g., almost one timeout difference), then Bob aborts.[19] Moreover, if the signature s on the escrow e does not verify with the public key pk_A sent in step 1 as part of the verifiable escrow v, Bob aborts the protocol.
5. Bob sends Alice his key K_B. Alice checks if the key K_B decrypts the ciphertext c_B correctly. If not, Alice does not proceed with the next step, and runs *AliceResolve*, although she might have to run it again just after the timeout to be able to resolve.
6. Alice sends Bob her key K_A. Bob checks if the key K_A decrypts the ciphertext c_A correctly. If not, he runs *BobResolve*; he must do so before the timeout.[20]

---

[15] We are abusing the notation by using hash values as verification algorithms provided by the Tracker hoping that the actual verification procedure of hashing the files and comparing the result with values given by the Tracker is obvious.

[16] It can easily be a per-exchange parameter known to (or agreed by) both parties.

[17] Alice and Bob can use their choice of (symmetric) encryption schemes (not necessarily the same). This only requires us to add the definition of the encryption scheme used to the messages exchanged.

[18] These will be Merkle hashes [38] for efficiency reasons.

[19] We do not require tight synchronization. So, for example, the *time* value can just contain hours, and not minutes and seconds.

[20] Bob can run BobResolve immediately after a *message timeout*. He need not wait for a long time for Alice.

Once step 1 is completed, cheap steps 2-6 can be repeated to exchange more files, as long as no dispute occurs. Alice and Bob need not know beforehand how many or which files/blocks to exchange. When-

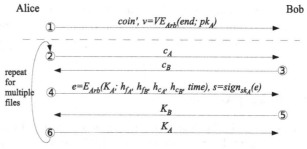

Fig. 1. Our Barter Protocol with Timeouts

ever they decide to exchange blocks (before every step 2), it is enough for them to just obtain their hashes from the Tracker. Actually, in BitTorrent, once you ask for hash of a file, the Tracker provides you with the hashes of all the blocks in that file already. Thus, connecting the Tracker for each block is not necessary in real life.

Below we present the resolution protocols in case of a dispute between Alice and Bob. The Arbiter never gets involved in a transaction unless there is a dispute.

BobResolve. Bob needs to contact the Arbiter before the timeout for resolution (current time $<$ *time* in escrow e + *timeout*), since otherwise the Arbiter is not going to honor his request. Assuming Bob resolves before the timeout, he provides the Arbiter with the escrow e and signature s that he received in step 4, and also the verifiable escrow v he received in step 1 from Alice. The escrow e should be labeled with four hash values $h_{f_A}, h_{f_B}, h_{c_A}, h_{c_B}$, and a *time* value. The verifiable escrow v should be labeled with a public key pk_A for a signature scheme. If the labels of the escrows are ill-formed, the Arbiter will not honor the request. The Arbiter checks the signature s using the public key in the verifiable escrow v, and if it verifies, he asks Bob to present his correct key K_B that verifies using the VerifyKey protocol in [8] (i.e., it decrypts a ciphertext with hash h_{c_B} to a plaintext with hash h_{f_B}). If Bob succeeds in giving the correct key, the Arbiter stores the key K_B, decrypts the escrow e and hands in the key K_A from the escrow to Bob. Bob checks if K_A decrypts Alice's file f_A correctly. If not, he proves this to the Arbiter using the technique in [8] and gets the endorsement *end* in the verifiable escrow v from the Arbiter.[21] Notice that only Bob may succeed in the BobResolve protocol with the Arbiter because any other party will fail to provide the correct key matching hashes of Bob's files). The subprotocols from [8] can be found in the full version of our paper [35].

AliceResolve. When Alice contacts the Arbiter for resolution, she asks for Bob's key K_B. If such a key exists, then the Arbiter sends K_B to her.[22] K_B has already been verified,

---

[21] The Arbiter can abort this trade forgetting the K_B in such a case. This is not necessary according to our definition (and can even be considered unfair), but it can be used as a way to punish cheating Alice even more. In the worst case, if non-atomicity of the Arbiter is allowed for efficiency reasons, Alice can obtain K_B before Bob proves K_A to be incorrect, effectively turning our protocol into a buy protocol.

[22] If the Arbiter is allowed to be non-atomical for efficiency reasons, then he needs to ask Alice for her key K_A, verifying it using the VerifyKey protocol in [8] before giving her K_B. This represents a tradeoff between the atomicity and efficiency of the Arbiter, which can be resolved arbitrarily, although it can also be used as a tougher punishment for cheaters.

so Alice does not need to perform any further action. If such a key does not exist yet, Alice should come back after the timeout. If, even after the timeout K_B does not exist, then Alice is assured that it will never exist, and can consider that particular trade as aborted.

3.2 Efficiency Analysis

The efficiency of Alice's and Bob's parts in the protocol can be further improved, as we show in the full version of our paper [35], although this would require the Arbiter to perform more work. Since such trusted third parties can become the bottlenecks of the system, we prefer having the least amount of work to be done by the Arbiter, and let users perform slightly more work instead.

We consider a concrete instantiation of our protocol using endorsed e-cash [18], Camenisch-Shoup verifiable escrow [19], AES encryption [25], DSS signatures [42], and RSA-OAEP public key encryption for (non-verifiable) escrow [10]. Our protocol has only neglectable overhead over just doing an unfair exchange. Sending the ciphertexts in steps 2 and 3 just corresponds to sending the files in any (even unfair) exchange.[23] The keys sent in steps 5 and 6 are extremely short messages (16 bytes each for 128-bit AES keys). For a fair exchange, step 4 is still very cheap since the only primitives used are an ordinary (non-verifiable) escrow (just a public key encryption), and a signature (A DSS signature created using a 1024-bit key is about 40 bytes, while an RSA-OAEP encryption with a 1024-bit key is about 128 bytes).

Assuming IO and CPU can be overlapped, encryption of files will not add any time. Furthermore, signatures and escrows take only a few milliseconds. The most time consuming step is sending the blocks themselves, which has to be done in any case (and encryption does not increase size). The only real overhead is the first step, where the verifiable escrow (and endorsed e-cash, if used) is costly (see below).

Our protocol, in addition to guaranteeing fair barter efficiently, is optimized for multi-barter situations. One such situation is a file sharing scenario as in BitTorrent [22,8]. The peers Alice and Bob are expected to have a long-term barter relationship. Hence, **step 1 needs to be carried out only once per peer, and remaining cheap steps 2-6 would be repeated for each block, whereas previous protocols required a costly step like step 1 to be performed for each block**. This greatly amortizes the costly step 1 in our protocol, when multiple blocks (or files) are exchanged, **even when the files/blocks to be exchanged are not pre-defined** (they need to be defined only before each execution of step 2).

To give some numbers, consider an average BitTorrent file of size $2.8GB$ made up of about 2,500 blocks [32]. Using previous optimistic fair exchange protocols, this requires 2,500 costly steps (one per block). Our C++ implementation using endorsed e-cash [18] and Camenisch-Shoup verifiable escrow [19] takes about 2 seconds of computation for step 1 (most of which is the verifiable escrow) on an average computer ($2GHz$). This corresponds to $2500 \times 2 seconds = $ **84 minutes** of computation overhead. Considering a BitTorrent client that connects to about 40 peers, using our protocol,

---

[23] We can in general assume that the I/O and CPU can be pipelined so that the encryption will not add more time to uploading the files.

this overhead becomes just **80 seconds**. Our network overhead is similarly neglectable (around $40KB$ per peer, almost all of which is the one-time cost of step 1, about half of it being endorsed e-cash). This corresponds to about $2500 \times 40KB = $ **100 MB** total overhead using previous schemes, and only **1.6 MB** total overhead using our scheme (for a $2.8GB$ file).

As for the Arbiter, he checks a signature, sometimes decrypts a (verifiable) escrow, and performs the VerifyKey protocol of Belenkiy et al. [8]. The signature check and ordinary escrow decryption takes only milliseconds, the verifiable escrow decryption, when necessary, can take a few hundred milliseconds. The bottleneck is the data that the Arbiter needs to download for the VerifyKey protocol, which is about $22chunks \times 16KB = 352KB$ [8]. An important point to note is that *the amount of data the Arbiter's needs to download is independent of the size of the file that is being exchanged.*[24]

Without considering distributed denial of service (DDoS) attacks, let us provide some numbers for evaluation. To have an idea, consider a p2p system of $1,700,000$ users, exchanging $2.8GB$ files on the average [32]. Exchanging two such files means exchanging $5.6GB$ of data. If 1% of all users are malicious, this can correspond to $17,000$ exchanges requiring an arbiter at a given time (where one user is honest and the other is malicious. If both of them are malicious, this number reduces to half of it). We said, in case of a dispute, a peer should upload $352KB$ of data to the Arbiter. Assume that the same upload speed is used when trading files and contacting the Arbiter. If we assume the worst case scenario where the Arbiter can handle only one user at a time and every user is active at all times, this requires having 2 arbiters; with 10% malicious user ratio, we need 11 arbiters. Under the very realistic assumption that an arbiter can handle 25 users at a time (e.g., assuming 25 times as fast download speed of the Arbiter as the upload speed of the users [23]), we will need 1 arbiter in this system (even with 10% malicious user ratio). When we use our protocol without timeouts, these numbers will double (but if our arbiter can handle 25 users at a time, we still need only 1 arbiter). Some more efficiency evaluation, limitations and possible solutions, a generalized version of our protocols, security proofs and privacy discussion can be found in the full version of this paper [35]. The full version also includes the version of our protocol that does not require timeouts.

4   Conclusion

There already are many scenarios where peers trade content [22,32]. These systems unfortunately rely on the honesty of the peers for providing fairness, partly because of the high cost incurred by the previous fair exchange protocols [2,3,4,5,8,18,40]. Our protocols uniquely limit the use of the costly primitives (verifiable escrow and e-cash) to once (or twice) per peer, as opposed to per file/block. We have shown in Section 3.2 that there are one or two orders of magnitude efficiency gains over previous protocols. Besides, most of the existing systems already rely on similar trusted parties [2,3,4,5,8,17,18,20,22,32,40,43]. Therefore, for the first time, by using our protocols, such bartering systems will experience almost no performance loss, while the benefit

---

[24] Merkle proofs are logarithmic in number of the blocks in the file, but are much smaller in size than the data blocks themselves in practice.

of providing fairness guarantees will be very noticeable indeed (e.g., see [8] for how the use of fair exchange can solve the free-riding problem of BitTorrent). Already, the Brownie Project [14] is adopting our protocols in their BitTorrent deployment.

References

1. Asokan, N., Janson, P.A., Steiner, M., Waidner, M.: The state of the art in electronic payment systems. IEEE Computer 30, 28–35 (1997)
2. Asokan, N., Schunter, M., Waidner, M.: Optimistic Protocols for Fair Exchange. In: CCS (1997)
3. Asokan, N., Shoup, V., Waidner, M.: Asynchronous protocols for optimistic fair exchange. In: IEEE Security and Privacy (1998)
4. Asokan, N., Shoup, V., Waidner, M.: Optimistic fair exchange of digital signatures. IEEE Journal on Selected Areas in Communications 18(4), 591–610 (2000)
5. Ateniese, G.: Efficient verifiable encryption (and fair exchange) of digital signatures. In: CCS (1999)
6. Avoine, G., Vaudenay, S.: Optimistic Fair Exchange Based on Publicly Verifiable Secret Sharing. In: Wang, H., Pieprzyk, J., Varadharajan, V. (eds.) ACISP 2004. LNCS, vol. 3108, pp. 74–85. Springer, Heidelberg (2004)
7. Backes, M., Datta, A., Derek, A., Mitchell, J.C., Turuani, M.: Compositional analysis of contract-signing protocols. Theoretical Computer Science 367(1-2), 33–56 (2006)
8. Belenkiy, M., Chase, M., Erway, C.C., Jannotti, J., Küpçü, A., Lysyanskaya, A., Rachlin, E.: Making P2P Accountable without Losing Privacy. In: WPES (2007)
9. Belenkiy, M., Chase, M., Erway, C.C., Jannotti, J., Küpçü, A., Lysyanskaya, A.: Incentivizing Outsourced Computation. In: NetEcon (2008)
10. Bellare, M., Rogaway, P.: Optimal Asymmetric Encryption. In: De Santis, A. (ed.) EURO-CRYPT 1994. LNCS, vol. 950, pp. 92–111. Springer, Heidelberg (1995)
11. Ben-Or, M., Goldreich, O., Micali, S., Rivest, R.L.: A fair protocol for signing contracts. IEEE Transactions on Information Theory 36(1), 40–46 (1990)
12. Blakley, G.R.: Safeguarding cryptographic keys. In: National Computer Conference (1979)
13. Boneh, D., Naor, M.: Timed commitments. In: Bellare, M. (ed.) CRYPTO 2000. LNCS, vol. 1880, p. 236. Springer, Heidelberg (2000)
14. Brownie Project, http://cs.brown.edu/research/brownie
15. Camenisch, J., Damgård, I.: Verifiable Encryption, Group Encryption, and Their Applications to Group Signatures and Signature Sharing Schemes. In: Okamoto, T. (ed.) ASI-ACRYPT 2000. LNCS, vol. 1976, p. 331. Springer, Heidelberg (2000)
16. Camenisch, J., Hohenberger, S., Kohlweiss, M., Lysyanskaya, A., Meyerovich, M.: How to Win the Clonewars: Efficient Periodic N-times Anonymous Authentication. In: CCS (2006)
17. Camenisch, J.L., Hohenberger, S., Lysyanskaya, A.: Compact e-cash. In: Cramer, R. (ed.) EUROCRYPT 2005. LNCS, vol. 3494, pp. 302–321. Springer, Heidelberg (2005)
18. Camenisch, J., Lysyanskaya, A., Meyerovich, M.: Endorsed e-cash. IEEE Security and Privacy (2007)
19. Camenisch, J., Shoup, V.: Practical verifiable encryption and decryption of discrete logarithms. In: Boneh, D. (ed.) CRYPTO 2003. LNCS, vol. 2729, pp. 126–144. Springer, Heidelberg (2003)
20. Chaum, D.: Bling signatures for untraceable payments. In: CRYPTO (1982)
21. Chaum, D., den Boer, B., van Heyst, E., Mjolsnes, S., Steenbeek, A.: Efficient offline electronic checks. In: EUROCRYPT (1990)

22. Cohen, B.: Incentives build robustness in bittorrent. In: Kaashoek, M.F., Stoica, I. (eds.) IPTPS 2003. LNCS, vol. 2735, Springer, Heidelberg (2003)
23. Cohen, L.: Testimony of Larry Cohen, President of Communications Workers of America (May 2007)
24. Cramer, R., Shoup, V.: A Practical Public Key Cryptosystem Provably Secure Against Adaptive Chosen Ciphertext Attack. In: Krawczyk, H. (ed.) CRYPTO 1998. LNCS, vol. 1462, p. 13. Springer, Heidelberg (1998)
25. Daemen, J., Rijmen, V.: The Design of Rijndael: AES–the Advanced Encryption Standard. Springer books (2002)
26. Dingledine, R., Mathewson, N., Syverson, P.: Tor: The second-generation onion router. In: USENIX Security (2004)
27. Dodis, Y., Lee, P.J., Yum, D.H.: Optimistic Fair Exchange in a Multi-user Setting. In: Okamoto, T., Wang, X. (eds.) PKC 2007. LNCS, vol. 4450, pp. 118–133. Springer, Heidelberg (2007)
28. Dolev, D., Dwork, C., Naor, M.: Nonmalleable cryptography. SIAM Journal on Computing (2000)
29. Fujisaki, E., Okamoto, T., Pointcheval, D., Stern, J.: RSA-OAEP Is Secure under the RSA Assumption. Journal of Cryptology 17(2), 81–104 (2004)
30. Garay, J., Jakobsson, M., MacKenzie, P.: Abuse-free optimistic contract signing. In: Wiener, M. (ed.) CRYPTO 1999. LNCS, vol. 1666, p. 449. Springer, Heidelberg (1999)
31. Goldwasser, S., Micali, S., Rivest, R.: A Digital Signature Scheme Secure Against Adaptive Chosen Message Attack. SIAM Journal on Computing (1988)
32. Iosup, A., Garbacki, P., Pouwelse, J., Epema, D.H.J.: Correlating Topology and Path Characteristics of Overlay Networks and the Internet. In: GP2PC (2006)
33. Katz, J., Lindell, Y.: Introduction to Modern Cryptography. Chapman and Hall/CRC Press, Boca Raton (2007)
34. Küpçü, A., Lysyanskaya, A.: Optimistic Fair Exchange with Multiple Arbiters. Cryptology ePrint Archive, Report 2009/069 (2009), http://eprint.iacr.org/2009/069
35. Küpçü, A., Lysyanskaya, A.: Usable Optimistic Fair Exchange. Cryptology ePrint Archive, Report 2008/431 (2008), http://eprint.iacr.org/2008/431
36. Lindell, Y.: Legally Enforceable Fairness in Secure Two-Party Computation. In: Malkin, T.G. (ed.) CT-RSA 2008. LNCS, vol. 4964, pp. 121–137. Springer, Heidelberg (2008)
37. Markowitch, O., Saeednia, S.: Optimistic fair exchange with transparent signature recovery. In: Syverson, P.F. (ed.) FC 2001. LNCS, vol. 2339, p. 329. Springer, Heidelberg (2002)
38. Merkle, R.: A digital signature based on a conventional encryption function. In: Pomerance, C. (ed.) CRYPTO 1987. LNCS, vol. 293, pp. 369–378. Springer, Heidelberg (1988)
39. Micali, S.: Simultaneous Electronic Transactions. U.S. Patent, No. 5,666,420 (1997)
40. Micali, S.: Simple and fast optimistic protocols for fair electronic exchange. In: PODC (2003)
41. Naor, M., Yung, M.: Universal one-way hash functions and their cryptographic applications. In: STOC (1989)
42. NIST. Digital Signature Standard (DSS). FIPS, PUB 186-2 (2000)
43. Pagnia, H., Gärtner, F.C.: On the impossibility of fair exchange without a trusted third party. Technical Report, TUD-BS-1999-02 (1999)
44. Paillier, P.: Public-key cryptosystems based on composite degree residuosity classes. In: Stern, J. (ed.) EUROCRYPT 1999. LNCS, vol. 1592, p. 223. Springer, Heidelberg (1999)
45. Shamir, A.: How to Share a Secret. ACM Communications (1979)
46. Shmatikov, V., Mitchell, J.C.: Finite-state analysis of two contract signing protocols. Theoretical Computer Science 283(2), 419–450 (2002)
47. Shoup, V., Gennaro, R.: Securing threshold cryptosystems against chosen ciphertext attack. In: Nyberg, K. (ed.) EUROCRYPT 1998. LNCS, vol. 1403, pp. 1–16. Springer, Heidelberg (1998)

Hash Function Combiners in TLS and SSL

Marc Fischlin, Anja Lehmann, and Daniel Wagner

Darmstadt University of Technology, Germany
www.minicrypt.de

Abstract. The TLS and SSL protocols are widely used to ensure secure communication over an untrusted network. Therein, a client and server first engage in the so-called handshake protocol to establish shared keys that are subsequently used to encrypt and authenticate the data transfer. To ensure that the obtained keys are as secure as possible, TLS and SSL deploy *hash function combiners* for key derivation and the authentication step in the handshake protocol. A robust combiner for hash functions takes two candidate implementations and constructs a hash function which is secure as long as at least one of the candidates is secure. In this work, we analyze the security of the proposed TLS/SSL combiner constructions for pseudorandom functions resp. message authentication codes.

1 Introduction

Hash functions are an important primitive for cryptographic protocols and are currently used for various tasks that require, among others, collision resistance or, in keyed settings, behavior of a pseudorandom function or a MAC. However, recent attacks [27,26,5,22] against the most widely deployed hash functions MD5 and SHA1 caused a decrease of confidence, especially concerning long-term security. Hence, approaches like robust combiners [13] which allow to obtain less vulnerable hash functions are of great interest and have triggered a series of research [1,18,6,8,19,11].

In general, a hash combiner takes two hash functions H_0, H_1 and combines them into a failure-tolerant function such that this function remains secure as long as at least one of the two functions H_0 or H_1 is secure. For example, the classical combiner for collision-resistance simply concatenates the outputs of both hash functions $\mathsf{Comb}(M) = H_0(M) \| H_1(M)$. If a hash function is supposed to be used as a pseudorandom function (PRF), then the exclusive-or of the outputs $\mathsf{Comb}(k_0 \| k_1, M) = H_0(k_0, M) \oplus H_1(k_1, M)$ with independent keys k_0, k_1 yields a robust design. Combiners that preserve even multiple properties in a robust manner where proposed in [9,11].

Interestingly, the fact that combiners give better security assurances has been acknowledged by the designers of TLS and its predecessor SSL, long before they have been investigated more thoroughly by theoreticians. Both TLS and SSL use various combinations of MD5 and SHA1 instead of relying only on a single hash

J. Pieprzyk (Ed.): CT-RSA 2010, LNCS 5985, pp. 268–283, 2010.

function. The specification of TLS even explicitly states: "In order to make the PRF as secure as possible, it uses two hash algorithms in a way which *should* guarantee its security if either algorithm remains secure" [24] .

The SSL protocol [23] was published in 1994 by Netscape to provide secure communication between two parties over an untrusted network, and subsequently formed the basis for the TLS protocol [24,25]. Nowadays, both protocols are ubiquitously present in various applications such as electronic banking, online shopping or secure data transfer. However, neither TLS nor SSL were accompanied with rigorous security proofs. An important step was recently done by Morrissey et al. [17] and Gajek et al. [12] who gave the first security analysis of the handshake protocol of TLS. The handshake protocol is the essential part of TLS/SSL as it allows a client and server to negotiate security parameters, such as shared symmetric keys or trusted ciphers, without having any common secrets yet. The established keys and cryptographic algorithms are subsequently used to protect the data transfer, i.e., the confidentiality and authenticity of the entire communication relies on the security of the key agreement. Thus, it is of crucial importance that the handshake protocol provides reliable parameters. Ideally, this statement should be fortified by comprehensive security proofs.

OUR RESULTS. In this work, we scrutinize the design of the (non-standard) hash combiners, deployed in the TLS and SSL handshake protocols, regarding the suitability for the respective purposes. As already mentioned, secure key derivation is one of the main tasks of the handshake phase. Both TLS and SSL use hash combiners to compute the master secret out of the pre-master secret, which is assumed to be a shared random string. To achieve secure key-derivation, robustness with respect to pseudorandomness is required.

While TLS (mainly) reverts to the standard design for PRF combiners, i.e., it xors the outputs of the two hashes, SSL applies the cascade $H_0(k, (H_1(k, M)))$ as the pseudorandom function for key derivation. For SSL we prove that the combiner is not robust and not even preserving, i.e., even two secure PRFs may yield an insecure combiner. This stems from the fact that both hash functions are invoked with the same master key. By using individual keys for each underlying function, we show that the security of the SSL combiner is somewhat between robustness and property-preservation. In the case of TLS, we prove that the combiner is a secure PRF if either of H_0, H_1 is a pseudorandom function. Interestingly, the TLS construction is neither optimal in terms of security nor efficiency. We therefore also discuss possible tweaks to obtain better security bounds while saving on computation.

TLS and SSL also use hash combiners for the finished message in the handshake protocol, which is basically a message authentication code generated for the shared master secret and all previous handshake messages. This concludes the key exchange phase in TLS/SSL and authenticates the previous communication. Ideally, the combiners used for this purpose should be robust for MACs, i.e., rely only on unforgeability instead of pseudorandomness of the hash function.[1] We show that in TLS the combiner for authentication requires the additional assumption of at least one hash function being collision resistant. The combiner used in SSL is again neither robust nor preserving, due to the same problem of using the master secret as key for both functions. We discuss that a modified version which splits the key into independent halves, is a secure MAC when at least one hash function is simultaneously unforgeable and collision resistant.

In summary, we give the first formal treatment of the hash combiners deployed in the TLS and SSL protocols. Our results essentially show that the choices in TLS are sound as they follow common design criteria for such combiners (but still leave space for improvements), whereas the SSL design for combiners requires much stronger assumptions. Our result, together with other steps like the security proofs in [17,12], strengthen the confidence in the important protocols TLS and SSL.

2   Preliminaries

In this section we present the preliminaries for our investigation of the combiners in SSL/TLS.

2.1   Hash Functions and Their Properties

Since we give all results in terms of concrete security we adopt Rogaway's approach [21] of defining hash functions as single instances (instead of families) and considering constructive reductions between security properties. For security notions without secret keys like collision-resistance the adversary is implicitly based on (the description of) H, whereas for security properties involving secret keys like pseudorandomness or the MAC property, the adversary also gets black-box access to the hash function $H(k, \cdot)$ with secret key k (we often write $H(k||\cdot)$ if the key is simply prepended to the message). In this case we call H a keyed hash function and usually denote the key space by K.

Most recent hash functions such as MD5, SHA1 apply the Merkle-Damgård construction [16,7] to obtain a variable-input length function out of a fixed-input length compression function $h : \{0,1\}^n \times \{0,1\}^\ell \to \{0,1\}^n$ and an initial vector IV. To compute a digest one divides (and possibly pads) the message $M = m_0 m_1 \ldots m_{k-1}$ into blocks m_i of ℓ bits and computes the digest $H(M) = \mathrm{iv}_k$ as

$$\mathrm{iv}_0 = \mathrm{IV}, \qquad \mathrm{iv}_{i+1} = h(\mathrm{iv}_i, m_i) \qquad \text{for } i = 0, 1, \ldots, k-1.$$

---

[1] The devil's advocate may claim that we can already start from the assumption that one of the hash function is a PRF, as we require this for the key derivation step anyway. However, it is a common principle to revert to the minimal requirements for such sub protocols and their designated purpose. Suppose, for example, that both hash functions turn out to be *not* pseudorandom, that key derivation becomes insecure and confidentiality of the subsequently transmitted data is breached. Then, if one of the function is nonetheless still a good MAC, a secure authentication step in the finished message via the robust MAC-combiner would still guarantee authenticity of the designated partner.

COLLISION-RESISTANCE. Let $H : \{0,1\}^* \rightarrow \{0,1\}^n$ be a hash function. The *collision-finding advantage* of an adversary $\mathcal{A}$ is

$$\mathbf{Adv}_H^{cr}(\mathcal{A}) := \text{Prob}\left[(M, M') \leftarrow \mathcal{A}() : M \neq M' \wedge H(M) = H(M')\right].$$

We again note that, formally, for any hash function there is a very efficient algorithm $\mathcal{A}$ with advantage 1, namely, the one which has a collision hardwired into it and simply outputs this collision. However, based on current knowledge it is usually infeasible to specify this algorithm constructively (cf. [21]).

PSEUDORANDOMNESS. Let $H : K \times \{0,1\}^* \rightarrow \{0,1\}^n$ be a keyed hash function with key space K. We define the advantage of a distinguisher $\mathcal{A}$ as

$$\mathbf{Adv}_H^{prf}(\mathcal{A}) = \left| \text{Prob}\left[\mathcal{A}^{H(k,\cdot)} = 1\right] - \text{Prob}\left[\mathcal{A}^{f(\cdot)} = 1\right]\right|$$

where the probability in the first case is over $\mathcal{A}$'s coin tosses and the choice of $k \overset{\$}{\leftarrow} K$, and in the second case over $\mathcal{A}$'s coin tosses and the choice of the random function $f : \{0,1\}^* \rightarrow \{0,1\}^n$.

MESSAGE AUTHENTICATION (UNFORGEABILITY). Let $H : K \times \{0,1\}^* \rightarrow \{0,1\}^n$ be a keyed (deterministic) hash function with key space K. We define the *forgeability advantage* of an adversary $\mathcal{A}$ as

$$\mathbf{Adv}_H^{mac}(\mathcal{A}) = \text{Prob}\left[k \overset{\$}{\leftarrow} K, (M, \sigma) \leftarrow \mathcal{A}^{H(k,\cdot)} : H(k, M) = \sigma \wedge M \text{ not queried}\right]$$

HASH FUNCTION COMBINERS. A hash function combiner Comb for hash functions H_0, H_1 "merges" the two functions H_0, H_1 into a single hash function. The combiner is called *preserving* [13] for some property like collision-resistance if Comb^{H_0,H_1} has this property given that *both* hash functions have this property. In a sense, this ensures a minimalistic security guarantee. The combiner is called *robust* [13] if it obeys the property if at least *one* of the two functions H_0, H_1 has the corresponding property. Note that, in terms of our concrete security statements, collision-resistance robustness for example is formulated by demanding that the probability of finding collisions in a combiner is bounded from above by the minimum of finding collisions for the individual hash functions.

2.2 HMAC

Each hash function can be used as a pseudorandom function or MAC by replacing the initial value IV with a randomly chosen key k of the same size. A more convenient technique was proposed by Bellare et al. [2] with the HMAC/NMAC algorithms, which are message authentication codes built from iterated hash functions. Recall that a MAC takes a secret key k, message m and outputs a tag σ. The HMAC algorithm takes, in its more general version, two keys k_{in}, k_{out} and applies an iterated hash function H like MD5 and SHA1 in a nested manner:

$$\text{HMAC}(k_{in}, k_{out})(M) = H(IV, k_{out} || H(IV, k_{in} || M)) \tag{1}$$

In practice, HMAC typically uses only a single key k from which it derives dependent keys $k_{in} = k \oplus \mathsf{ipad}$ and $k_{out} = k \oplus \mathsf{opad}$ for fixed constants $\mathsf{ipad} = 0x3636\ldots36, \mathsf{opad} = 0x5c5c\ldots5c$.

Originally, Bellare et al. [2] proved HMAC – resp. its theoretical counterpart NMAC – to be pseudorandom functions when the underlying compression function h is pseudorandom and collision-resistant. Subsequently, the proof was restated on the sole assumption that the compression function is pseudorandom [4]. As the security claims are given for NMAC, Bellare [4] introduced the notion of a "dual" pseudorandom function function $\bar{h} : \{0,1\}^n \times K \to \{0,1\}^n$ with $\bar{h}(m, k) = h(k, m)$. If both $\bar{h}$ and h are pseudorandom, the security of NMAC carries over to HMAC. For the single-keyed HMAC-version, the security of $\bar{h}$ must hold for related-key attacks as well. That is, the adversary is granted access to two oracles $\bar{h}(k \oplus \mathsf{opad}, \cdot), \bar{h}(k \oplus \mathsf{ipad}, \cdot)$ with dependent keys.

2.3   The SSL/TLS Handshake Protocol

The SSL and TLS protocols consist of two layers: the record layer and the handshake protocol. The record layer encrypts all data with a cipher and session key that have been negotiated by the handshake protocol. Thus the handshake protocol is a key-exchange protocol layered above the record layer and initializes and synchronizes a cryptographic state between a server and a client. Both versions of the handshake protocol, for TLS and for SSL, vary mainly in the implementation of the exchanged messages, i.e., the overall structure of the handshake part is the same and can be summarized as the sequence of the following steps [20]:

(1) The client conveys its willingness to engage in the protocol by sending a list of supported cipher algorithms and a random number, that is subsequently used for key-derivation.

(2) The server responds by choosing one of the proposed ciphers, and sending its certified public key as well as a random nonce.

(3) The client verifies the validity of the received certificate and sends a randomly chosen *pre-master secret* encrypted under the server's public key. (An alternative to having the client choose the pre-master secret is to engage in a key exchange protocol like signed Diffie-Hellman. Since our analysis below only assumes that the pre-master secret is random we omit the details about its generation.)

(4) Both client and server individually compute a *master secret* from the exchanged random nonces and the pre-master secret. Once the master key is computed, it can be used to obtain further application keys.

(5) Finally, the master secret is confirmed by the *finished message*, where each party sends a MAC over the transcript of the conversation using the new master key. This is also the first transmission which uses the secure channel for the derived keys.

3   Derivation of the Master Secret

In this section we analyze the functions that are deployed by TLS and SSL to derive a secret master key from a shared pre-master key. The basic requirement of key derivation is that the obtained key should be indistinguishable from a randomly chosen one. In particular, the key-derivation function must be pseudo-random. For more discussion see [15]. We will show that the combiner proposed by TLS is PRF-robust, i.e., security of one of the underlying hash function suffices, whereas the SSL combiner requires assumptions on both hash functions in order to produce random looking output.

3.1   The PRF-Combiner Used in TLS

The TLS key derivation obtains the master secret (ms) from the pre-master secret (pms) by invoking the following hash combiner:

$$ms =$$
$$\mathsf{Comb}_{\mathsf{TLS-prf}}^{\mathsf{MD5,SHA1}}(pms, \text{``master secret''}, \mathrm{ClientRandom}\|\mathrm{ServerRandom})[0..47]$$

The pre-master secret is assumed to be a random value both parties have agreed upon, and *ClientRandom* and *ServerRandom* are public random nonces exchanged in the handshake protocol. By introducing a specific label (here "master secret") to the input, the combiner can subsequently be used for further (key-derivation) computations, while guaranteeing distinct inputs for each application. The appendix $[0..47]$ indicates that the master secret consists of the first 48 bytes of the combiners output.

Basically, the combiner $\mathsf{Comb}_{\mathsf{TLS-prf}}^{H_0,H_1}$ xors the output of a function P which gets called twice based on two distinct hash functions H_0 and H_1. To this end, the combiner also splits the key $K = k_0\|k_1$ with $|k_1| = |k_0|$ into independent halves:

$$\mathsf{Comb}_{\mathsf{TLS-prf}}^{H_0,H_1}(k_0\|k_1, M) = \mathsf{P}_{H_0}(k_0, M) \oplus \mathsf{P}_{H_1}(k_1, M) \tag{2}$$

The underlying function P_{H_b} makes several queries to HMAC_{H_b} and produces byte strings of (arbitrary) length that is a positive multiple of n.

$$\mathsf{P}_{H_b}(k, M) = \mathsf{HMAC}_{H_b}(k, A(1)\|M) \| \mathsf{HMAC}_{H_b}(k, A(2)\|M)\|\ldots \tag{3}$$

with $A(0) = M$ and $A(i) = \mathsf{HMAC}_{H_b}(k, (A(i-1))$.

ANALYSIS OF $\mathsf{Comb}_{\mathsf{TLS-prf}}^{H_0,H_1}$. We show that the TLS-combiner for key derivation is a pseudorandom function if at least one of the two hash functions H_0, H_1 is based on a pseudorandom compression function. To this end, we first show that P_{H_b} inherits the pseudorandomness of the underlying hash function.

Note that the P_{H_b} construction uses the HMAC transform to obtain a PRF, which gets keyed via the input data, out of a standard hash function H_b with fixed IV. It was shown in [4] that HMAC is a pseudorandom function, when the underlying compression-function is a *dual PRF*, i.e., it has to be a secure

PRF when keyed by either the data input or the chaining value. Thus, while functional-wise HMAC uses the cryptographic hash function only as a black-box, the security guarantee is still based on the underlying compression function h_b. We therefore consider each hash function $H_b : \{0,1\}^* \to \{0,1\}^n$ as the Merkle-Damgård iteration of a compression function $h_b : \{0,1\}^n \times \{0,1\}^\ell \to \{0,1\}^n$. By applying the results of [4] we can conclude that HMAC_{H_b} is a pseudorandom function, when h_b is a dual PRF.

Next, we show that the design of the P_H construction preserves the pseudorandomness of HMAC_H. For a modular analysis – and for the sake of readability – we simplify the description of P_H by replacing HMAC and the hash function H by the same function H, and prove that the modified function P'_H is a pseudorandom function if H is. Furthermore, we make a rather syntactical change of P to obtain a function that is efficiently computable on its own: According to the TLS specification, the P construction produces output of arbitrary length from which the combiner takes as much bytes as required, e.g., the first 48 bytes in case of the derivation of the master secret. In the following we slightly deviate from that notation and assume that P gets also parametrized by an integer c which indicates that an output of length $c \cdot n$ is requested. Overall, we analyze the following function P':

$$\mathsf{P}'_H(k, M, c) = \tag{4}$$
$$\mathsf{H}(k, A(0) \parallel M) \parallel \mathsf{H}(k, A(1) \parallel M) \parallel \cdots \parallel \mathsf{H}(k, A(c-1) \parallel M)$$

where $A(0) = M$, $A(i) = \mathsf{H}(k, (A(i-1)))$.

Lemma 1. *Let* $\mathsf{H} : \{0,1\}^n \times \{0,1\}^* \to \{0,1\}^n$ *be a pseudorandom function with key space* $\{0,1\}^n$, *and let* $\mathsf{P}'_H : \{0,1\}^n \times \{0,1\}^* \to \{0,1\}^{c \cdot n}$ *be defined by (4) above. For all adversaries* $\mathcal{A}$ *running in time* t, *making* q *queries of length at most* l *and with* $c \le c_{\max}$, *there exist an adversary* $\mathcal{B}$ *such that*

$$\mathbf{Adv}^{prf}_{\mathsf{P}'}(\mathcal{A}) \le \mathbf{Adv}^{prf}_{\mathsf{H}}(\mathcal{B}) + q \cdot \binom{c_{max}}{2} \cdot 2^{-n}$$

where $\mathcal{B}$ *makes at most* $2c_{max} \cdot q$ *queries, each of length at most* $l + n$ *and runs in time at most* $t + \mathcal{O}(c_{max})$.

Proof. Assume that there is an adversary $\mathcal{A}$ that can distinguish the function $\mathsf{P}'_H(k, \cdot)$ from a random function $F : \{0,1\}^* \to \{0,1\}^n$ with advantage $\mathbf{Adv}^{prf}_{\mathsf{P}'}(\mathcal{A})$. Given $\mathcal{A}$ we show how to obtain an adversary $\mathcal{B}$ against the underlying hash function $\mathsf{H}(k, \cdot)$. Recall that $\mathcal{A}$ has black-box access to an oracle that is either the keyed construction $\mathsf{P}'_H(k, \cdot, \cdot)$ or a random function $F : \{0,1\}^* \to \{0,1\}^*$ (where, formally, F also takes the parameter c as additional input and outputs strings of length cn). The distinguisher $\mathcal{B}$ has to simulate this oracle with the help of its own oracle, which is either the keyed hash-function $\mathsf{H}(k, \cdot)$ or a random function $f : \{0,1\}^* \to \{0,1\}^n$. To this end, for any query (M, c) of $\mathcal{A}$, the adversary $\mathcal{B}$ mimics the construction P' but replaces each evaluation of the underlying hash function H by the response of its oracle on the corresponding query. If $\mathcal{A}$ stops outputting its guess d, algorithm $\mathcal{B}$ stops with output d too.

If the oracle of $\mathcal{B}$ was the hash function H, then $\mathcal{B}$ perfectly simulates the construction P'. Thus, the output distribution of $\mathcal{B}$ equals the one of $\mathcal{A}$ with access to P'.

In the case that the oracle of $\mathcal{B}$ was the truly random function f, we have to show that processing its random answers in the P' construction yields random values again. Recall that for each query (M, c) the adversary $\mathcal{B}$ now computes the sequence $f(A(0)\|M) \;\|\; f(A(1)\|M) \;\|\ldots\| \; f((A(c-1)\|M)$ where $A(i) = f(A(i-1))$ starting with $A(0) = M$. As long as $A(i) \neq A(j)$ for all $i \neq j \in \{0, 1, \ldots c-1\}$ holds for each query, the function f gets evaluated in the outer iterations on distinct and unique values, such that the corresponding outputs from f are independently and uniformly distributed. Thus, it remains to show that the probability for collisions on the $A(i)$ values, which are derived using f in a cascade, is small. Assume that for a query (M, c) a collision occurred, i.e., there exist (unique) indices $i^* \in \{0, \ldots, c-1\}$ and $j^* \in \{0, \ldots, i^*-1\}$ such that $f(A(i^*-1)) = A(j^*)$ but $A(i^*-1) \neq A(j)$ for all $j = 0, 1, \ldots, i^*-2$. That is, f has never been invoked on the value $A(i^*-1)$ but maps to an value $A(j^*)$ which is an previous answer of (the cascade of) f. Since f is a truly random function, such a collision can only occur with probability $q \cdot \binom{c_{\max}}{2} \cdot 2^{-n}$ where q denotes the number of $\mathcal{A}$ queries and $c_{\max}$ is the largest value for c that appeared in the simulation. Overall, $\mathcal{B}$ distinguishes H from f with probability:

$$\mathrm{Prob}\left[\mathcal{B}^{\mathsf{H}(k,\cdot)} = 1\right] - \mathrm{Prob}\left[\mathcal{B}^f = 1\right]$$

$$\geq \mathrm{Prob}\left[\mathcal{A}^{\mathsf{P}_{\mathsf{H}}(k,\cdot,\cdot)} = 1\right] - \mathrm{Prob}\left[\mathcal{A}^F = 1\right] - q \cdot \binom{c_{\max}}{2} \cdot 2^{-n}.$$

This proves the claim.   □

Putting the results of [4] and Lemma 1 together, we now obtain that the pseudorandomness of h_b is preserved by the corresponding construction HMAC_{H_b} and, in turn, by $\mathsf{P}'_{\mathsf{HMAC}_{H_b}}$ which equals P_{H_b}. Furthermore, XOR is a robust combiner for pseudorandom functions, and thus, if least one of $\mathsf{P}_{H_0}, \mathsf{P}_{H_1}$ is a PRF, also $\mathsf{Comb}_{\mathsf{TLS-prf}}^{H_0,H_1}$ provides outputs that are indistinguishable from random. This, together with the fact that the key is divided into independent halves, implies the following theorem:

Theorem 1. *Let $H_b : \{0,1\}^n \times \{0,1\}^* \to \{0,1\}^n$ for $b \in \{0,1\}$ be a hash function with underlying compression function $h_b : \{0,1\}^n \times \{0,1\}^\ell \to \{0,1\}^n$. Let $\mathsf{Comb}_{\mathsf{TLS-prf}}^{H_0,H_1}$ be defined as in (2). For all adversaries $\mathcal{A}$ running in time t, making q queries of length at most l and such that $c \leq c_{\max}$, there exist adversaries $\mathcal{A}_0, \mathcal{A}_1$ such that*

$$\mathbf{Adv}_{\mathsf{Comb}_{\mathsf{TLS-prf}}}^{prf}(\mathcal{A})$$

$$\leq \min\left\{\mathbf{Adv}_{\mathsf{HMAC}_{h_0}}^{prf}(\mathcal{A}_0), \mathbf{Adv}_{\mathsf{HMAC}_{h_1}}^{prf}(\mathcal{A}_1)\right\} + q \cdot \binom{c_{max}}{2} \cdot 2^{-n}$$

where each of $\mathcal{A}_0, \mathcal{A}_1$ makes at most $2c_{max} \cdot q$ queries of length at most $l + n$ and runs in time at most $t + \mathcal{O}(c_{max}(1 + 2q \cdot T_{\bar{b}}))$ where $T_{\bar{b}}$ denotes the time required for one evaluation of $\mathsf{P}_{\bar{b}}$ (as defined in (3)).

IMPROVEMENTS. When the combiner $\mathsf{Comb}_{\mathsf{TLS-prf}}^{H_0, H_1}$ is used for key derivation, the underlying construction P ensures that sufficiently many output bytes are produced. However for the purpose of range extension of a PRF, the construction P is neither optimal in terms of efficiency nor security. Namely, if one assumes HMAC_H to be a secure PRF, one could simply augment the input M by a fixed-length encoded counter $\langle i \rangle$, which ensures distinct inputs for each PRF evaluation:

$$\mathsf{P}_{H_b}^*(k, M) = \mathsf{HMAC}_{H_b}(k, M || \langle 1 \rangle) \,||\, \mathsf{HMAC}_{H_b}(k, M || \langle 2 \rangle) || \cdots$$

Replacing P with P* would result in better security bounds, as one gets rid of the probability $q \cdot \binom{c_{\max}}{2} \cdot 2^{-n}$ of a collision on the $A(i)$ values. In terms of efficiency, the above construction only requires half of the PRF evaluations as needed in the original P function.

Another solution is to use solely the outputs of $A(\cdot)$, i.e., without feeding them into HMAC again:

$$\mathsf{P}_{H_b}^*(k, M) = A(1) \,||\, A(2) \,||\, A(3) \,||\, \cdots$$

with $A(i)$ being the i-th cascade of $\mathsf{HMAC}(k, M)$ as defined in (3). With this construction one inherits the same security bound as in the original solution, but invokes HMAC after the first evaluation only one shorter inputs, e.g., 128 bits in the case of MD5 and 160 bits for SHA1, which decreases the computational costs.

3.2   The PRF-Combiner Used in SSL

In the SSL protocol the following construction gets repeated until sufficient key material for the master secret is generated:

$$ms = \mathsf{MD5}(pms||(\mathsf{SHA1}(\text{``A''}||pms||\mathrm{ClientRandom}||\mathrm{ServerRandom})))||$$
$$\mathsf{MD5}(pms||(\mathsf{SHA1}(\text{``BB''}||pms||\mathrm{ClientRandom}||\mathrm{ServerRandom})))||$$
$$\mathsf{MD5}(pms||(\mathsf{SHA1}(\text{``CCC''}||pms||\mathrm{ClientRandom}||\mathrm{ServerRandom})))|| \cdots$$

Both functions get keyed by the input data, where in the case of the outer hash function the key is prepended to the message, and for the inner hash the key is somewhat embedded in the message. Due to length-extension attacks, key-prepending approaches must be accompanied by prefix-free encoding, otherwise the hash function can not serve as a pseudorandom function, as shown in [3]. For the analysis we assume that the hash function takes care of that issue, and thus that a hash function $H_b : \{0,1\}^* \rightarrow \{0,1\}^n$ is a secure PRF when keyed via the first n bits of the data input.

On a more abstract level, each repetition of the SSL-combiner above for prefixes "A", "BB", "CCC" etc. can be represented as the following construction:

$$\mathsf{Comb}_{\mathsf{SSL-prf}}^{H_0, H_1}(k, M) = H_0(k \,||\, H_1(k||M)), \tag{5}$$

e.g., where $H_1(k\|M)$ implements $\mathsf{SHA1}(\text{"CCC"}\|k\|M)$ for the fixed value "CCC". To be a robust combiner for pseudorandom functions, the SSL-combiner needs to be robust for H_0 and each such function H_1. From now on we fix an arbitrary H_1.

ANALYSIS OF $\mathsf{Comb}_{\mathsf{SSL-prf}}^{H_0,H_1}$. The cascade $\mathsf{Comb}_{\mathsf{SSL-prf}}^{H_0,H_1}$ of two hash functions is not a robust design for pseudorandomness, because as soon as the outer function becomes insecure the combiner, too, can be easily distinguished from a random function: Consider as an example the constant function $H_0(x) = 0^n$ that maps any input to zeros, which is obviously distinguishable from random. Then, also the combiner $\mathsf{Comb}_{\mathsf{SSL-prf}}^{H_0,H_1}(k, M) = H_0(k\|H_1(k\|M))$ becomes a constant function, independently of the strength of the inner hash function H_1. Hence,

Proposition 1. *The combiner* $\mathsf{Comb}_{\mathsf{SSL-prf}}^{H_0,H_1}$ *is not PRF-robust.*

Actually, $\mathsf{Comb}_{\mathsf{SSL-prf}}^{H_0,H_1}$ is not even PRF-preserving, i.e., there exist two functions H_0, H_1 that are both secure pseudorandom functions, but become easily distinguishable when used in the SSL-combiner. The problem arises from the fact that the same secret key is used for both functions, which contradicts the general design paradigm of provably robust combiners.

For the counter example let $H_1 : K \times \{0,1\}^* \to \{0,1\}^n$ be a pseudorandom function. Define $H_0(k, x)$ now as follows: if $x = H_1(k, 0^n)$ then return 0^n, else output $H_1(k\|1\|x)$. Then H_0 basically inherits the pseudorandomness of H_1 because any distinguisher with access to $H_0(k, \cdot)$ only retrieves replies $H_1(k\|1\|x)$ to queries $x \in \{0,1\}^*$, unless it is able to predict the value $H_1(k\|0^n)$. The latter would contradict the pseudorandomness of H_1, though. But when both functions are combined into $H_0(k \| H_1(k\|M))$, the combiner returns 0^n for input 0^n and is obviously therefore not a pseudorandom function.

In order to allow any reasonable statement about the security of the construction $\mathsf{Comb}_{\mathsf{SSL-prf}}^{H_0,H_1}$, we assume in the following that the combiner splits the key into two independent halves, and invokes the hash functions on distinct shares:

$$\mathsf{Comb}_{\mathsf{SSL-prf*}}^{H_0,H_1}(k_0\|k_1, M) = H_0(k_0 \| H_1(k_1\|M))$$

Note that the first discussed counter example is still valid, as it did not require any dependencies of the individual keys. Thus, even $\mathsf{Comb}_{\mathsf{SSL-prf*}}^{H_0,H_1}$ is not a robust combiner in general. However, the security can be considered to be somewhat above property-preservation, since we can relax the assumption on one hash function while the combiner still preserves the pseudorandomness property of the stronger function.

In the case that the outer hash function H_0 is a secure pseudorandom function, the inner hash function only needs to ensure that for distinct queries $M \neq M'$ of an adversary to the combiner, the function H_0 gets evaluated on different values too. Thus, it suffices for H_1 to be *weakly collision-resistant*, which is defined similarly to collision-resistance, except that here the function is keyed with a secret key and the adversary only gets black-box access to the function.

If the inner hash function H_1 is a pseudorandom function, an adversary that queries the combiner gets to see images of H_0 only for random domain points.

Thus, it is not necessary that the outer function is a full-fledged PRF as well. In this case, already the assumption of H_0 being a *weak pseudorandom function* is sufficient. We discuss both cases as well as the issue of combining two insecure functions in more detail in the full paper.

3.3   Application Key Derivation in TLS and SSL

Both combiners $\mathsf{Comb}_{\mathsf{TLS-prf}}^{H_0,H_1}$ and $\mathsf{Comb}_{\mathsf{SSL-prf}}^{H_0,H_1}$ are used to obtain a shared master secret from a pre-shared key. However, subsequently, the same functions are deployed to derive further keys, e.g., for encryption or message authentication. To this end, the freshly computed master secret is used instead of the pre-master secret that was assumed to be a random value. For TLS we have shown that the combiner $\mathsf{Comb}_{\mathsf{TLS-prf}}^{H_0,H_1}$ provides a master secret that is indistinguishable from random when at least one hash function is a PRF. Thus, our result carries over to the application key derivation, that uses the combiner with the derived master secret. The same holds for SSL, but under stronger assumptions on the underlying hash functions.

4   Finished-Message

In this section we investigate the TLS/SSL combiners that are used to compute the so-called *finished*-message of the handshake protocols. The finished message is the last part of the key exchange and is realized by a message authentication code which is computed over the transcript of the previous communication. Thus, the combiners that are used for this application should optimally be robust for MAC, i.e., only rely on the unforgeability property instead of the stronger PRF-assumption.

We note that the finished message itself is already secured through the negotiated application keys. This complicates the holistic security analysis of this step. But since we are at foremost interested in the design of the combiners and their designated purpose, we only touch this issue briefly at the end of Section 4.1 (where we address this issue for TLS; the same discussion holds for SSL).

4.1   The MAC-Combiner Used in TLS

To compute the finished MAC, the TLS protocol applies the same combiner as for the derivation of the master secret, but already uses the new master key. As the key is known only at the very end of the protocol, the MAC cannot be computed iteratively during the communication. To circumvent the need of storing the entire transcript until the master secret is available, TLS hashes the transcript iteratively and then computes the MAC over the short hash value only:

$$\sigma_{\mathsf{finished}} =$$
$$\mathsf{Comb}_{\mathsf{TLS-prf}}^{\mathsf{MD5,SHA1}}\left(ms,\ \mathsf{FinishedLabel},\ \mathsf{MD5(transcript)}\|\mathsf{SHA1(transcript)}\right)[0..11]$$

A further input to the combiner is the FinishedLabel which is either the ASCII string "client" or "server", which ensures that the MAC values of both parties are different, otherwise an adversary could simply return a finished tag back to its sender. The appendix $[0..11]$ indicates again that the first 12 bytes of the combiner output are used as the MAC.

Recall that the combiner $\mathsf{Comb}_{\mathsf{TLS-prf}}^{H_0,H_1}$ is based on the construction P which produces arbitrary length output by invoking the underlying hash function in an iterative and nested manner. However, this range extension is only necessary when the combiner is used for key derivation. To compute the finished message, only the first 12 byte of the combiners output are used, which is shorter than the digests of both applied hash functions (16 bytes for MD5 and 20 bytes for SHA1). Thus, we can omit the P part from the construction and simplify the combiner as follows:

$$\mathsf{Comb}_{\mathsf{TLS-mac}}^{H_0,H_1}(k_0\|k_1, M) = \tag{6}$$
$$\mathsf{HMAC}_{H_0}(k_0, H_0(M)\|H_1(M)) \;\oplus\; \mathsf{HMAC}_{H_1}(k_1, H_0(M)\|H_1(M))$$

Verification for the above MAC-combiner is done by recomputing the tag and comparing it to the given tag.

ANALYSIS OF $\mathsf{Comb}_{\mathsf{TLS-mac}}^{H_0,H_1}$. We have already shown that the combiner construction $\mathsf{Comb}_{\mathsf{TLS-prf}}^{H_0,H_1}$, which can be seen as the more complex version of $\mathsf{Comb}_{\mathsf{TLS-mac}}^{H_0,H_1}$, is robust for pseudorandom functions. Thus, if one is willing to assume that at least one hash function behaves like a random function, the combiner can be used directly as a MAC, as well.

However, ideally, the combiner $\mathsf{Comb}_{\mathsf{TLS-mac}}^{H_0,H_1}$ should be a secure MAC on the sole assumption that at least one of the underlying hash functions H_0, H_1 is unforgeable rather than being a pseudorandom function. Unfortunately, hashing the transcript before the MAC gets computed, imposes another assumption on the hash functions (even when starting from the PRF assumption), namely at least one hash function needs to be collision-resistant. Otherwise an adversary could try to induce a collision on the input to the HMAC functions, which immediately gives a valid forgery for the entire MAC function. Under the assumption that such a collision is unlikely, we show that the combiner $\mathsf{Comb}_{\mathsf{TLS-mac}}^{H_0,H_1}$ is MAC-robust.

To this end, we first prove that the xor of two deterministic MACs (like HMAC_{H_b}) invoked directly with the message yields a robust combiner:

$$\mathsf{Comb}_{\oplus}^{H_0,H_1}(k_0\|k_1, M) = H_0(k_0, M) \oplus H_1(k_1, M) \tag{7}$$

In the context of aggregate authentication, Katz and Lindell [14] gave a similar result by showing that multiple MAC tags, computed by (possibly) different senders on multiple (possibly different) messages, can be securely aggregated into a shorter tag by simply xoring them.

Lemma 2. *Let* $H_0, H_1 : \{0,1\}^n \times \{0,1\}^* \to \{0,1\}^n$ *be deterministic message authenticated codes, and let* $\mathsf{Comb}_{\oplus}^{H_0,H_1}$ *be defined by (7). For any adversary* $\mathcal{A}$

against $\mathsf{Comb}_{\oplus}^{H_0,H_1}$ *making at most q queries and running in time at most t, there exist adversaries $\mathcal{A}_0, \mathcal{A}_1$ such that*

$$\mathbf{Adv}^{mac}_{\mathsf{Comb}_{\oplus}}(\mathcal{A}) \leq \min\left\{\mathbf{Adv}^{mac}_{H_0}(\mathcal{A}_0), \mathbf{Adv}^{mac}_{H_1}(\mathcal{A}_1)\right\}$$

where $\mathcal{A}_b$ for $b = 0, 1$ makes at most q queries and runs in time at most $t + \mathcal{O}(qT_{\bar{b}})$ where $T_{\bar{b}}$ denotes the time for one evaluation of $H_{\bar{b}}$.

Due to space constraints the prove is delegated to the full version of the paper.

Complementing the above Lemma 2 with the probability of finding collisions on the concatenated combiner $H_0(M) \| H_1(M)$ yields Theorem 2.

Theorem 2. *Let $H_0, H_1 : \{0,1\}^n \times \{0,1\}^* \to \{0,1\}^n$ be hash functions, and let $\mathsf{Comb}_{\mathrm{TLS-mac}}^{H_0,H_1}$ be defined by (6). For any adversary $\mathcal{A}$ against $\mathsf{Comb}_{\mathrm{TLS-mac}}^{H_0,H_1}$ making at most q queries and running in time at most t, there exist adversaries $\mathcal{A}_0, \mathcal{A}_1, \mathcal{B}_0, \mathcal{B}_1$ such that*

$$\mathbf{Adv}^{mac}_{\mathsf{Comb}_{\mathrm{TLS-mac}}}(\mathcal{A}) \leq \min\left\{\mathbf{Adv}^{mac}_{\mathsf{HMAC}_{H_0}}(\mathcal{A}_0), \mathbf{Adv}^{mac}_{\mathsf{HMAC}_{H_1}}(\mathcal{A}_1)\right\}$$
$$+ \min\left\{\mathbf{Adv}^{cr}_{H_0}(\mathcal{B}_0), \mathbf{Adv}^{cr}_{H_1}(\mathcal{B}_1)\right\}$$

where $\mathcal{A}_b$ for $b = 0, 1$ makes at most q queries and runs in time at most $t + \mathcal{O}(qT_{\bar{b}})$ where $T_{\bar{b}}$ denotes the time for one evaluation of $\mathsf{HMAC}_{H_{\bar{b}}}$, and $\mathcal{B}_b$ runs in time $t + \mathcal{O}(qT_b)$.

Note that for both properties, unforgeability and collision-resistance, it suffices that either one of the hash functions has this property (instead of one hash function with obeying both property simultaneously). This is similar to the difference between weak and strong combiners in [9].

So far, we have reduced the security of the combiner $\mathsf{Comb}_{\mathrm{TLS-mac}}^{H_0,H_1}$ of H_0, H_1 to the collision-resistance of the hash functions and the unforgeability of the HMAC transforms HMAC_{H_0} and HMAC_{H_1}. Preferably, the security of HMAC_{H_b} should in turn only rely on the unforgeability of the underlying hash resp. compression function. However, such a reduction for the plain HMAC transform is still unknown. The previous results for this issue either require stronger assumptions than MAC (yet, weaker than PRF), or additional keying-techniques for the compression function. In the full version we briefly recall the two most relevant approaches for our scenario.

THE PROBLEM OF CHOPPING. Theorem 2 states that the TLS-combiner for the finished message is robust for message authentication codes even when starting from the unforgeability assumption which is significantly weaker than assuming a PRF. However, according to the TLS specification, not the entire output of the combiner is used as tag, but only the first 12 bytes. Since the unforgeability notion is not closed under chopping transformations, a shortened output of a MAC loses any security guarantees. To allow usage of a chopped fraction of the combiners output, on either has to assume that one of the underlying MACs is secure for truncation, or one needs to make the stronger assumption that at least one of the two hash functions is a secure PRF.

Is UNFORGEABILITY ENOUGH? When using MACs in a stand-alone fashion, unforgeability clearly gives sufficient security guarantees. However, in TLS (and SSL) the tag for the finished message is computed under the master secret, from which further application keys for encryption and authentication are derived. The tag itself is now encrypted and authenticated with these derived keys. On one hand, this may help to prevent the tag in the finished message from leaking some information about the master secret. On the other hand, this causes critical circular dependencies between these values, possibly even enabling leakage of entire keys. This problem has already been noticed in other works (e.g., in [17] where the analysis of the handshake protocol assumes that the tag is sent without securing it with the application keys; or more explicitly in the context of delayed-key authentication in [10]). It is beyond the scope of this work about combiners, though.

4.2   The MAC-Combiner Used in SSL

The SSL-construction for the finished message resembles the HMAC construction, but appends the inner key to the message instead of prepending it. This stems from the same problem as in TLS, namely that the MAC should be computed iteratively as soon as the communication starts, although the necessary key is negotiated only at the end. To obtain a robust design, SSL uses the concatenation of the HMAC-like construction based on the MD5 and SHA1 functions:

$$\sigma_{\text{finished}} = \text{HMAC}^*_{\text{MD5}}(ms, \text{Label}||\text{transcript}) \; || \; \text{HMAC}^*_{\text{SHA1}}(ms, \text{Label}||\text{transcript})$$

where HMAC^*_H is defined as:

$$\text{HMAC}^*_H(k, M) = H(k||\text{opad}|| \; H(M||k||\text{ipad})) \tag{8}$$

with opad, ipad being the same fixed patterns as in HMAC. The structure of HMAC^* then allows to accomplish the bulk of the computation without knowing the key k.

Overall, the MAC combiner of SSL can be described as follows:

$$\text{Comb}^{H_0,H_1}_{\text{SSL-mac}}(k, M) = \text{HMAC}^*_{H_0}(k, M) \; || \; \text{HMAC}^*_{H_1}(k, M) \tag{9}$$

ANALYSIS OF $\text{Comb}^{H_0,H_1}_{\text{SSL-mac}}$. In contrast to the TLS-combiner, SSL uses the entire master secret as key for both hash functions. This approach results in a construction $\text{Comb}^{H_0,H_1}_{\text{SSL-mac}}$ that is not even MAC-preserving, although concatenation is MAC-robust when used with distinct keys for each hash function [13].

Proposition 2. *The combiner* $\text{Comb}^{H_0,H_1}_{\text{SSL-mac}}$ *is not MAC-preserving (and thus not MAC-robust either).*

Consider two secure MACs H_0, H_1, that on input of a secret key k and a message M outputs a tag σ_b. Assume furthermore that both MACs ignore parts of their key, i.e., H_0 ignores the left half of its input key and H_1 ignores the right part. We now derive functions H_b^* that can still be unforgeable when used alone, but become totally insecure when being plugged into the combiner. The first MAC H_0^* behaves like H_0 but also leaks the left half k_l of the secret key, i.e., $H_0^*(k, M) = k_l || H_0(k, M)$. The second function is defined analogously, but outputs the right half of the key: $H_1^*(k, M) = k_r || H_1(k, M)$. Even though each tag is now accompanied with a part of the key, it remains hard to create a forgery. When we use now both functions H_0^*, H_1^* as in the SSL-combiner[2] we obtain: $H_0^*(k, M) || H_1^*(k, M) = k_l || \sigma_0 || k_r || \sigma_1$ which allows to easily reconstruct the entire secret key and subsequently forge tags for any message.

In the full paper we also discuss an improved version of the SSL-Combiner, that is a robust MAC if at least one compression function is *simultaneously* collision-resistant and unforgeable. Note that this is a stronger assumption than for the TLS combiner, where both properties can be possessed by possibly different functions.

Acknowledgments

We thank the anonymous reviewers for valuable comments. The first two authors are supported by the Emmy Noether Program Fi 940/2-1 of the German Research Foundation (DFG).

References

1. Boneh, D., Boyen, X.: On the Impossibility of Efficiently Combining Collision Resistant Hash Functions. In: Dwork, C. (ed.) CRYPTO 2006. LNCS, vol. 4117, pp. 570–583. Springer, Heidelberg (2006)
2. Bellare, M., Canetti, R., Krawczyk, H.: Keying hash functions for message authentication. In: Koblitz, N. (ed.) CRYPTO 1996. LNCS, vol. 1109, pp. 1–15. Springer, Heidelberg (1996)
3. Bellare, B.M., Canetti, R., Krawczyk, H.: Pseudorandom Functions Revisited: The Cascade Construction and Its Concrete Security. In: FOCS 1996, pp. 514–523. IEEE Computer Society Press, Los Alamitos (1996)
4. Bellare, M.: New proofs for NMAC and HMAC: Security without collision-resistance. In: Dwork, C. (ed.) CRYPTO 2006. LNCS, vol. 4117, pp. 602–619. Springer, Heidelberg (2006)
5. De Cannière, C., Rechberger, C.: Preimages for reduced SHA-0 and SHA-1. In: Wagner, D. (ed.) CRYPTO 2008. LNCS, vol. 5157, pp. 179–202. Springer, Heidelberg (2008)
6. Canetti, R., Rivest, R., Sudan, M., Trevisan, L., Vadhan, S.P., Wee, H.M.: Amplifying collision resistance: A complexity-theoretic treatment. In: Menezes, A. (ed.) CRYPTO 2007. LNCS, vol. 4622, pp. 264–283. Springer, Heidelberg (2007)
7. Damgård, I.B.: A Design Principle for Hash Functions. In: Brassard, G. (ed.) CRYPTO 1989. LNCS, vol. 435, pp. 416–427. Springer, Heidelberg (1990)

---

[2] Invoking the combiner directly on H_b^* instead of $HMAC_{H_b^*}^*$ still proves our statement as the HMAC transform can inherit the behavior H^*. We omit the additional level for the sake of simplicity.

8. Fischlin, M., Lehmann, A.: Security-Amplifying Combiners for Hash Functions. In: Menezes, A. (ed.) CRYPTO 2007. LNCS, vol. 4622, pp. 224–243. Springer, Heidelberg (2007)
9. Fischlin, M., Lehmann, A.: Robust Multi-Property Combiners for Hash Functions. In: Canetti, R. (ed.) TCC 2008. LNCS, vol. 4948, pp. 375–392. Springer, Heidelberg (2008)
10. Fischlin, M., Lehmann, A.: Delayed-Key Message Authentication for Streams. In: Micciancio, D. (ed.) TCC 2010. LNCS, vol. 5978, pp. 288–305. Springer, Heidelberg (2010)
11. Fischlin, M., Lehmann, A., Pietrzak, K.: Robust Multi-Property Combiners for Hash Functions Revisited. In: Aceto, L., Damgård, I., Goldberg, L.A., Halldórsson, M.M., Ingólfsdóttir, A., Walukiewicz, I. (eds.) ICALP 2008, Part II. LNCS, vol. 5126, pp. 655–666. Springer, Heidelberg (2008)
12. Gajek, S., Manulis, M., Pereira, O., Sadeghi, A.-R., Schwenk, J.: Universally Composable Security Analysis of TLS. In: Baek, J., Bao, F., Chen, K., Lai, X. (eds.) ProvSec 2008. LNCS, vol. 5324, pp. 313–327. Springer, Heidelberg (2008)
13. Herzberg, A.: On Tolerant Cryptographic Constructions. In: Menezes, A. (ed.) CT-RSA 2005. LNCS, vol. 3376, pp. 172–190. Springer, Heidelberg (2005)
14. Katz, J., Lindell, A.Y.: Aggregate Message Authentication Codes. In: Malkin, T.G. (ed.) CT-RSA 2008. LNCS, vol. 4964, pp. 155–169. Springer, Heidelberg (2008)
15. Krawczyk, H.: On Extract-then-Expand Key Derivation Functions and an HMAC-based KDF (2008), http://webee.technion.ac.il/~hugo/kdf/kdf.pdf
16. Merkle, R.: One Way Hash Functions and DES. In: Brassard, G. (ed.) CRYPTO 1989. LNCS, vol. 435, pp. 428–446. Springer, Heidelberg (1990)
17. Morrissey, P., Smart, N., Warinschi, B.: A Modular Security Analysis of the TLS Handshake Protocol. In: Pieprzyk, J. (ed.) ASIACRYPT 2008. LNCS, vol. 5350, pp. 55–73. Springer, Heidelberg (2008)
18. Pietrzak, K.: Non-Trivial Black-Box Combiners for Collision-Resistant Hash-Functions don't Exist. In: Naor, M. (ed.) EUROCRYPT 2007. LNCS, vol. 4515, pp. 23–33. Springer, Heidelberg (2007)
19. Pietrzak, K.: Compression from Collisions, or why CRHF Combiners have a Long Output. In: Wagner, D. (ed.) CRYPTO 2008. LNCS, vol. 5157, pp. 413–432. Springer, Heidelberg (2008)
20. Rescorla, E.: SSL and TLS - Designing and Building Secure Systems. Addison Wesley, Reading (2001)
21. Rogaway, P.: Formalizing Human Ignorance. In: Nguyên, P.Q. (ed.) VIETCRYPT 2006. LNCS, vol. 4341, pp. 211–228. Springer, Heidelberg (2006)
22. Stevens, M., Sotirov, A., Appelbaum, J., Lenstra, A., Molnar, D., Osvik, D.A., de Weger, B.: Short Chosen-Prefix Collisions for MD5 and the Creation of a Rogue CA Certificate. In: Halevi, S. (ed.) Advances in Cryptology - CRYPTO 2009. LNCS, vol. 5677, pp. 55–69. Springer, Heidelberg (2009)
23. Hickman, K.E.B.: The SSL Protocol (Internet Draft). Technical report (1994)
24. Dierks, T., Allen, C.: The TLS Protocol Version 1.0. Technical Report RFC 2246 (1999)
25. Dierks, T., Allen, C.: The TLS Protocol Version 1.2. Technical Report (TLS 1.2) RFC 4346 (2006)
26. Wang, X., Yu, H.: How to break MD5 and other hash functions. In: Cramer, R. (ed.) EUROCRYPT 2005. LNCS, vol. 3494, pp. 19–35. Springer, Heidelberg (2005)
27. Wang, X., Yin, Y.L., Yu, H.: Finding collisions in the full SHA-1. In: Shoup, V. (ed.) CRYPTO 2005. LNCS, vol. 3621, pp. 17–36. Springer, Heidelberg (2005)

Improving Efficiency of an 'On the Fly' Identification Scheme by Perfecting Zero-Knowledgeness

Bagus Santoso[1], Kazuo Ohta[2], Kazuo Sakiyama[2], and Goichiro Hanaoka[1]

[1] National Institute of Advanced Industrial Science and Technology (AIST),
Akihabara Daibiru 1003, 1-18-13 Sotokanda, Chiyoda-ku, Tokyo 101-0021, Japan
[2] The University of Electro-Communications,
1-5-1 Chofugaoka Chofu-shi, Tokyo 182-8585, Japan

Abstract. We present a new methodology for constructing an efficient identification scheme, and based on it, we propose a lightweight identification scheme whose computational and storage costs are sufficiently low even for cheap devices such as RFID tags. First, we point out that the efficiency of a scheme with *statistical* zero-knowledgeness can be significantly improved by enhancing its zero-knowledgeness to *perfect* zero-knowledge. Then, we apply this technique to the Girault-Poupard-Stern (GPS) scheme which has been standardized by ISO/IEC.

The resulting scheme shows a perfect balance between communication cost, storage cost, and circuit size (computational cost), which are crucial factors for implementation on RFID tags. Compared to GPS, the communication and storage costs are reduced, while the computational cost is kept sufficiently low so that it is implementable on a circuit nearly as small as GPS. Under standard parameters, the prover's response is shortened 80 bits from 275 bits to 195 bits and in application using coupons, storage for one coupon is also reduced 80 bits, whereas the circuit size is estimated to be larger by only 328 gates. Hence, we believe that the new scheme is a perfect solution for *fast* authentication of RFID tags.

Keywords: Identification scheme, RFID, zero-knowledge, impersonation.

1 Introduction

1.1 Background

Recently, there has been a tremendous increase of uses of resource-constrained electronic devices such as IC cards and RFID tags in daily life. In particular, there has been a huge demand for *very fast* online authentication using such devices in several applications such as mass-transit, toll-collection, and mass authentication of goods as anti-counterfeiting strategies. To develop a secure authentication scheme which meets this demand is a hard task, especially when we consider implementation on RFID tags which have fierce constraints regarding silicon area and power consumption. Although symmetric cryptography can be one possible solution [8], it is only applicable in closed systems

J. Pieprzyk (Ed.): CT-RSA 2010, LNCS 5985, pp. 284–301, 2010.

where a trustable key distribution is available, since all parties must know the secret key. One promising solution is offered by Girault-Poupard-Stern (GPS),[1] a three-pass public-key identification scheme proposed by Girault, Poupard, and Stern [9,10,24,1], which is derived from Schnorr identification scheme [27]. The feasibility of implementing GPS in very small size circuits has been shown recently by McLoone and Robshaw [19,18]. GPS can provide fast online authentication using lightweight devices, because in practice, the most complex calculation of prover in GPS can be done offline without interaction with the verifier and only a very simple calculation is necessary on online phase when interacting with the verifier. GPS has been accepted by European NESSIE project [21] and has also been standardized by ISO/IEC [16]. GPS has variants [25,23,22], which derive similar feature, i.e., simple online calculation on the prover.

Motivation. Although GPS and its variants look promising for fast online authentication by small circuits, they require a certain amount of memory and certain length of response on the prover in order to guarantee the statistical zero-knowledgeness in practice. On the other hand, in fiercely resource-constrained provers such as RFID tags, the data rate and the memory is very limited.

As an illustration of the data rate limitation, the usual data rate of an ordinary passive (batteryless) RFID tag satisfying ISO 18000-6C (EPC C1G2) ranges from 27 to only 128 kbps [7]. Since in GPS, the length of prover's response with standard parameters is 275 bits (see Sect. 5.1), under the modest 40 kbps data rate [28], the total time for sending the whole response is about 7 ms. The longer the prover's response is, the longer the time for sending the whole response will be. And the longer sending time is, the longer a stable power supply is necessary. Although a higher data rate can shorten the sending time, a higher data rate also needs more power. On the other hand, the power supply of the provers (tags) is not large or stable in general, especially for a *passive* RFID tag whose power is supplied only by the reader through electromagnetic wave. Thus, a lower communication cost (shorter prover's response) is highly preferred.

As an illustration of the storage limitation, a very recently developed RFID tag [13] still provides only 1536 bits of user memory. This is not sufficient for even the smallest assumed number of coupons for GPS (5 coupons) in [19], even when the coupon is optimized using technique in [11]. Hence, it is a very important issue to investigate whether we can reduce cost of communication and memory without sacrificing either the security or the computational cost.

1.2   Our Results

Our approach (see Sect. 2) results in a new scheme which is not only has *perfect* zero-knowledgeness (compared to GPS which is only *statistical* zero-knowledge), but also successfully reduces the size of prover's randomness required to mask

---

[1] Unless noted otherwise, GPS referred in this paper is the one described in GPS' final paper [10] where the underlying group is defined as a generic group, not as a group constructed from a composite integer like the one in [16].

the secret value in prover's responses. This induces smaller memory and communication costs compared to GPS, while maintaining the small computational cost of online response which is the most important feature of GPS.

Illustration of Practical Advantage. Under recommended standard parameters [10], our estimation shows that the time cost of calculating online prover's response in our new scheme on CMOS technology under 100 kHz clock rate, compared to that of GPS [19], differs by only approximately 0.8 ms (with the same size of circuit). Also, under the condition that the calculation of prover's response is as fast as GPS in [19], we estimate that our new scheme can be implemented within 1970 gates, which is an acceptable circuit size for RFID tags (200-4000 gates [6]). Just for comparison, applying the same estimation to Schnorr scheme results in 5294 gates (see Table 4 in Sect. 5.2).

And in the implementation using *use & throw coupons* [20, 19] under the same standard parameters, while GPS needs 435 bits of memory per coupon, the new scheme only needs 355 bits (80 bits reduced). A further optimization using the technique in [11] results in 349 bits and 269 bits per coupon for GPS and the new scheme respectively. Thus, at least the new scheme can fit 5 coupons (the assumed smallest number coupons in [19]) into a recently developed RFID tag [13]. In applications with 20 coupons (the largest number of coupons assumed in [19]), our new scheme reduces totally 1600 bits of memory from GPS.

In term of communication cost, the prover's response in our new scheme is at most 195 bits, while that in GPS is at most 275 bits (=80 bits reduced). Thus, the total time for sending the whole response of our new scheme under the modest 40 kbps data rate is about 5 ms (2 ms faster than that of GPS). Remind that a lower communication cost is highly preferred for RFID tags, since the current data rate of tags are not that high and the power supply of the provers (tags) is not so large or stable in general, especially in the case of an RFID tag whose power is supplied only by the reader through electromagnetic wave. Particularly, if the RFID tag is moving with considerable speed as in typical applications of a very fast authentication (e.g., toll-gate, mass transit system [10]), it is difficult to maintain the stability of the power generated for the tag within sufficient time.

In [19], it is stated that one may try to use PRNG with seed to regenerate prover's randomness and thus avoid the storage problem with coupons. However, this implies additional resources such as gates or time for regenerating the randomness, as well as power supply proportional to the size of the randomness. Since the size of prover randomness of GPS is larger than that of our scheme, GPS with PRNG might need larger additional gate circuits or more time for regenerating the randomness than ours, which means a larger power supply. Remind that this is generally not desirable for RFID tags.[2]

As shown above, the new scheme reduces both the communication and storage costs of GPS without sacrificing either the computational cost of online response or the circuit size for fast authentication. As a comparison, although Schnorr scheme indeed reduces both the communication and storage costs of GPS much more than the new scheme, as shown above, it requires either too

---

[2] The detailed evaluation will be in the full version of this paper.

large computational cost or too large circuit for fast authentication on current RFID tags[3]. Hence, since the new scheme shows a perfect balance between communication cost, storage cost, and circuit size (computational cost) which are the most crucial factors for implementation on RFID tags, we believe that the new scheme is a perfect solution for *fast* authentication of RFID tags.

Additionally, since the time cost in practice for producing one coupon in our scheme is only 2/3 of that in GPS (according to our machine experiment described in Sect. 5.2), the new scheme offers an improvement of robustness against a certain *denial-of-service* attack in the framework proposed by Hofferek et al. [14], where the recalculation of coupons in the idle time of prover has been proposed as the countermeasure (see Sect. 5.1).

On Security Assumption. The security of our scheme against impersonation under *serial* active attack is proven under a slightly stronger computational assumption than GPS, i.e., the discrete logarithm with short exponent in our scheme is assumed to be hard with *publicly known order* [26] (as opposed to unknown order in GPS). However, as noted in [19], for implementation in devices with very limited resources such as RFID tags, the group from elliptic curve (with known order) is more suitable than the group from RSA-like modulus (with unknown order), by which GPS is mostly described [16, 21]. Hence, in practice, there is no difference on computational assumption between GPS and our scheme. Furthermore, by using a computational assumption resembling one-more discrete logarithm problem [2], we are able to prove the security of our new scheme against impersonation under *concurrent* active attack, while it is still unknown for the case of GPS.

Generalization. A further generalization of our approach results in a general scheme which can be seen as a loose generalization of previous works such as [27], [5], [3] (see Sect. 5.4 for the detailed explanation).

1.3 Related Work

The original scheme of GPS was proposed by Girault in 1991 [9], derived from the work of Schnorr [27]. However, it took several years for its security proof to be established. The first security proof was provided by Poupard and Stern in 1998 [24] for the case of group from RSA-like modulus. Since then, numerous variants of GPS such as [25, 22, 23] are proposed. The final security proof for the case of generic group was completed recently [10]. Several efforts to implement GPS on RFID tags have been proposed by McLoone and Robshaw [19, 18].

A more general scheme which also covers our scheme has been previously proposed by Burmester et al. [4]. However, the possible significant reduce of complexity on the calculation of prover's response using the setting as the one in our new scheme has not been noticed. And also, on the theoretical side, the

---

[3] One may see the new scheme as a trade-off of Schnorr scheme with less computation and more communication (maintaining *perfect* ZK).

procedure of simulator for proving the zero-knowledgeness has not been formally described, and security against impersonation under either serial attacks or concurrent attacks has not been discussed.

Roadmap. Sect. 2 explains the essence of our approach. In Sect. 2.2, we explain diagrammatically the intuition of construction of our proposed scheme. In Sect. 3, we show the description of our proposed scheme. In Sect. 4, we state the theorems on the security of our proposed scheme against impersonation under concurrent active attacks. Several issues on practical advantages are discussed in Sect. 5. The detailed proofs of lemmas and theorems are in the full version.

2   Our Technique

In this section, we explain our methodology for improving the efficiency of identification schemes.

2.1   Illustration of Our Technique

General Problem. First, we give an observation which explains a general problem on efficiency in identification schemes with statistical zero-knowledge. In such a scheme, we notice that there exists a parameter ψ which determines statistical distance between the distribution of the distinguisher's view in the real world and that in the simulation. Certain values in the protocol (e.g. size of the prover's randomness) are required to be sufficiently large according to parameter ψ. Hence, in general, existence of this parameter significantly affects the efficiency of identification schemes (with statistical zero-knowledge) in practice.

Our Approach. Based on the above observation, we propose a strategy for improving the efficiency of identification schemes with statistical zero-knowledge in practice. Roughly speaking, our strategy is as follows. Assume that we have an identification scheme with statistical zero-knowledge. As mentioned above, according to a parameter ψ, certain values in the protocol have to be sufficiently large. For improving this scheme, one promising approach is *to (somehow) enhance its zero-knowledgeness to be perfect without changing its essential mechanisms*. Namely, if such modification is possible, then we can remove parameter ψ. This implies that certain values which are related to ψ become reduced. For example, in GPS scheme (which is statistically zero-knowledge), the prover's randomness is required to be larger than $195 + \psi$ bits where $\psi \geq 80$ for 80-bit security. However, if we can modify the GPS scheme to be perfectly zero-knowledge (without changing essential mechanisms), then it becomes only ≈ 195 bits.

We illustrate our technique as follows.

Preliminaries. Let $\mathcal{S}$ denote the set of all secret keys. Let $\mathcal{X}_s$ and $X(s)$ be the set of all generatable view in real world and the random variable induced by the view respectively, with the prover's secret key $s \in \mathcal{S}$. Also, define $\mathcal{X}$ as the union of all $\mathcal{X}_s$ for $s \in \mathcal{S}$, i.e., $\bigcup_{s \in \mathcal{S}} \mathcal{X}_s$.

Observation. Consider a protocol scheme $\mathcal{C}$ which has been proven to be statistical zero-knowledge. Assume that $\mathcal{C}$ has a non-empty set $\mathcal{D} \subset \mathcal{X}$ such that for any $x \in \mathcal{D}$, $\exists s_i, s_j (s_i \neq s_j) : \Pr[X(s_i) = x] \neq \Pr[X(s_j) = x]$ holds. Notice that as long as there exists such non-empty set $\mathcal{D}$, it is impossible for $\mathcal{C}$ to reach perfect zero-knowledgeness, since any *perfect* simulation of view itself will reveal some knowledge about the secret key. This contradicts the perfect zero-knowledgeness.

Our technique. We construct a transform f such that: (1) for any $x \in \mathcal{D}$, $\forall s_i, s_j (s_i \neq s_j) : \Pr[f(X(s_i)) = x] = \Pr[f(X(s_j)) = x]$ holds, and (2) f can be embedded into $\mathcal{C}$ without changing $\mathcal{C}$'s essential mechanism or feature. Although we still have to concretely construct an appropriate simulator to prove the zero-knowledgeness of $\mathcal{C}$ combined with transform f, at least, such transform f opens the possibility for reaching the perfect zero-knowledge.

In the next subsection, we show the application of our technique on GPS.

2.2 Minimum Modification to Gain Perfect Zero-Knowledgeness

Here we explain diagrammatically the intuition of how we construct our new scheme from GPS based on the approach described above. The most important point is how to achieve perfect zero-knowledge without sacrificing the simplicity of online prover's response which is the essential mechanism in GPS for providing online fast authentication. We start by a brief description of GPS.

Notations. In a multiplicative group $\mathcal{G}$, for any element $g \in \mathcal{G}$, we let $\mathrm{ord}(g)$ denote the order of g in $\mathcal{G}$. For any integer a, $|a|$ means the bit-length of a. For any two integers a and b, $a|b$ means that a divides b. Let $k \in \mathbb{N}$ denote the general security parameter for the rest of this paper. Unless noted otherwise, in this paper, all polynomials are positive integer polynomials in k and all algorithms are randomized polynomial time algorithms. We say that a value x *is polynomial* if x is related to k such that x can be represented by some polynomial in k.

GPS Scheme. Let P and V be a prover and a verifier in GPS and $\langle g \rangle$ be a multiplicative group. P holds a secret key $s \in [0, S)$ and publishes $I = g^{-s}$. In each elementary round: (1) P sends to V $x = g^r$ where r is chosen randomly from $[0, A)$, (2) V sends to P a random $c \in [0, B)$, and (3) P sends $y = r + cs$ (without modulus calculation) to V to be verified. V accepts P for this round if and only if $x = g^y I^c$ holds. A complete identification requires ℓ repetitions of elementary round where P is accepted if and only if V accepts P for all ℓ rounds. To ensure security in GPS, A has to be set much larger than $B \times S$.

Remark 1. Unless noted otherwise, we assume that $s \neq 0$ and $S \leqq \mathrm{ord}(g) \leqq BS$ hold. The theorem on zero-knowledgeness (Theorem 2) later still holds for $s = 0$.

The Diagram. As described above, in GPS, any response y from an honest prover is determined by the value of $y' = r + cs$ where s is the secret key and $(x = g^r, c, y = y')$ is the round's communication transcript. Thus, in the case,

(c, y') is sufficient to represent the valid (x, c, y), since $x = g^{y'} I^c$ holds and we know that $y = y'$. Namely, any (x, c, y) can be represented by one coordinate point in two-dimensional diagram of c and y'. Fig. 1(a) shows the distribution of (c, y') of GPS when $A \geq BS$. The horizontal axis represents the value of c and the vertical axis represents the value of y'. A pair (c, y') from one round is plotted as a circle ($\bullet$), representing a tuple (x, c, y') where $x = g^{y'} I^c$ holds.

Let $\mathcal{A}_0$ be the area containing all (c, y')'s such that $y' \in [cs, A - 1]$, and $\mathcal{A}_1$ be the area containing all (c, y')'s where $y' \in [\max\{A, cs\}, A - 1 + cs]$, and $\mathcal{A}_2$ be the area containing all (c, y') where $y' \in [0, \min\{cs - 1, A - 1\}]$.[4] Since in the case of an honest prover, $r \geq 0$ holds and there is no case where $y' < cs$, no concrete tuple is represented by a member of $\mathcal{A}_2$ in real GPS interactions. Also, let a set $\mathcal{S}_{(r)}$ be a collection of all circles which represent (c, y')'s such that $y' - cs = r$. It is easy to see that the number of circles belong to a set $\mathcal{S}_{(r)}$ is B and the total number of distinct sets of $\mathcal{S}_{(r)}$ is A.

Problem for Perfect Simulation. Since there is one-to-one correspondence between each (x, c, y) and (c, y'), it is clear that we can simulate the distribution of (x, c, y) by simulating the distribution of (c, y'). And intuitively, simulating *perfectly* the distribution of (c, y') means that we are drawing *exactly* the same diagram as Fig. 1(a) *without* any knowledge about secret key s more than the public key. However, in this case, simulating (or drawing) any point (c, y') in $\mathcal{A}_1$ means that we reveal some knowledge about secret key s, i.e., $s \in [(y' - A)/c, S - 1]$. Especially, it is easy to see that an exact drawing of all circles in $\mathcal{S}_{(A-1)} \subset \mathcal{A}_1$ as shown in Fig. 1(a) implies the secret key s being revealed. Therefore, it seems that unless we know the value of the secret key s, we will never be able to correctly draw all circles in Fig. 1(a). One can see $\mathcal{A}_1$ as a subset of the set $\mathcal{D}$ described in "Our Approach" above (Sect. 2.1).[5]

The Solution. The idea for a solution comes from our observation that area $\mathcal{A}_1$ in Fig. 1(a) which covers all pairs (c, y')'s where $y' \geq A$ has the same size as area $\mathcal{A}_2$ which represents all (c, y') points where $y' < cs$. If we can find a transform f which is bijective such that it maps uniquely each tuple (x, c, y') represented by a (c, y') in $\mathcal{A}_1$ into another tuple (x, c, y'') represented by a (c, y'') in $\mathcal{A}_2$, then f can be used to transform the response of GPS such that we get a new scheme in which all members of $\mathcal{A}_1$ are moved to the area $\mathcal{A}_2$ as shown in Fig. 1(b). By such f, we do not have to worry about revealing the secret s when drawing the distribution of (c, y'), since we do not need to draw $\mathcal{A}_1$ or $\mathcal{S}_{(A-1)}$ anymore.

We construct f as follows. W pick a bijective map f which maps (x, c, y') represented by (c, y') in $\mathcal{A}_1$ into $(x, c, y' - A)$ represented by $(c, y' - A)$ in $\mathcal{A}_2$. Note that each (c, y') in area $\mathcal{A}_1$ represents a tuple of communication transcript (x, c, y') where $x = g^{y'} I^c$ holds. For such f, it is sufficient to find the condition where $g^{y'-A} I^c = x = g^{y'} I^c$ holds. Namely, it is sufficient to have $g^A = 1$ hold, i.e., A being a multiple of $\operatorname{ord}(g)$. Therefore, we conclude that for the case of

---

[4] For the case of $A \geq BS$ (Fig. 1(a)), $y' \in [A, A - 1 + cs]$ and $y' \in [0, cs - 1]$ are sufficient to describe $\mathcal{A}_1$ and $\mathcal{A}_2$ respectively.

[5] Similar argument can also be explained using $\mathcal{A}_2$ and $\mathcal{S}_{(0)}$.

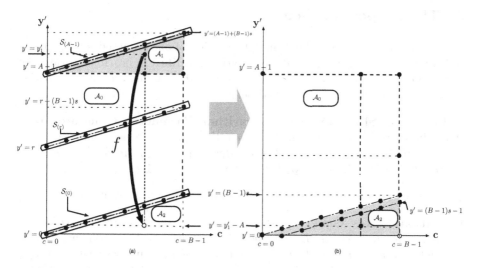

Fig. 1. Distribution of tuples of communication transcripts and transformation when $\operatorname{ord}(g)|A$ and $A - 1 \geq (B - 1)s$

$A \geq BS$, a bijective map f defined as $f(y') \overset{def}{=} y' - A$ with condition A being a multiple of $\operatorname{ord}(g)$ is sufficient for our purpose.

Minimum Modification for Perfect Zero-Knowledge. Based on above f, we can modify GPS as follows. First, we require A to be a multiple of $\operatorname{ord}(g)$, i.e., $\operatorname{ord}(g)|A$, and such that $A \geq BS$ holds. Then, we modify the calculation of y (prover's response) of GPS as follows.

(Step 1) $y' = r + cs$,
(Step 2) if $y' < A$ then set $y = y'$, otherwise $y = y' - A$.

This modification maintains the simplicity of calculation of the prover's response, and as proven by Theorem 2 later, this modification is proven sufficient for us to achieve *perfect* zero-knowledge (with dishonest verifier).

2.3   General Modification to Gain Perfect Zero-Knowledgeness

We discover that above modification is an instance of a more general modification as follows. First, we require A to be such that $\operatorname{ord}(g)|A$ holds. Then, we modify the calculation of y (prover's response) of GPS as follows: $y = r + cs \bmod A$. Notice that our minimum modification in previous subsection is an instance of this general modification when $A \geq BS$. The basic idea is to solve a similar "problem for perfect simulation" in previous subsection for the case $BS < A$. The detailed explanation will be in the full version of this paper. For simplicity, in this paper, we describe our proposed scheme based on this general modification.

3   The Proposed Scheme

In this section, we provide the formal description of our proposed scheme in the style of [2], based on the general modification described in previous section. An instance of our scheme will be always associated to a *short discrete logarithm parameter generator* $\mathcal{K}_{\text{dlse}}$ described below.

Definition 1 (Short Discrete Logarithm Parameter Generator[6]). *A short discrete logarithm parameter generator $\mathcal{K}_{\text{dlse}}$ is an algorithm such that on input the security parameter k generates a tuple $(\mathcal{G}, g, \text{ord}(g), S, s)$ where $\mathcal{G}$ is a multiplicative group, $g \in \mathcal{G}$, $S \in \mathbb{N}$, and $s \in [0, S-1]$.*

The next is the algorithm of the key generation of our proposed scheme.

Definition 2 (Key Generation). *Let $\mathcal{K}_{\text{dlse}}$ be a short discrete logarithm parameter generator. And let B and ℓ be two functions which are polynomial in k. Also let κ be a positive real number. Our proposed scheme associated to $\mathcal{K}_{\text{dlse}}$ and setup parameters (κ, B, ℓ) has its key generation algorithm $\mathcal{K}$ as follows. On input the security parameter k, $\mathcal{K}$ executes $\mathcal{K}_{\text{dlse}}$ with input k and obtains a tuple $(\mathcal{G}, g, \text{ord}(g), S, s)$. $\mathcal{K}$ calculates $I = g^{-s}$. Then, $\mathcal{K}$ chooses the smallest integer A such that $\kappa BS \leq A$ and $\text{ord}(g)|A$ hold. Finally, $\mathcal{K}$ outputs the public key $pk = (\mathcal{G}, g, S, B, \ell, I, A)$ and the secret key $sk = s$.*

The Elementary Round. An elementary round of identification between prover P and verifier V in our scheme proceeds as follows.

(Step 1) P picks randomly r from $[0, A-1]$, computes the *commitment* $x = g^r$ and sends x to V.

(Step 2) Receiving x from P, V picks randomly a *challenge* c from $[0, B-1]$ and sends c to P.

(Step 3) Upon receiving c from V, P checks whether $c \in [0, B-1]$ and calculates $y = r + c \times s \bmod A$. P sends the *response* y to V and V checks whether $y \in [0, A-1]$. If $x = g^y I^c$ holds, V accepts this round.

A *complete identification* consists of repetition of above elementary round for ℓ times. Here the verifier V *accepts* the prover P in a complete identification if and only if V accepts all ℓ consecutive elementary rounds.

Simple Calculation Version. As shown in previous section, if we set the setup parameter $\kappa = 1$ so that $A \geq BS$ holds, the "response" phase can be transformed into very simple two steps as follows.
(Step 1) $y' = r + cs$, (Step 2) if $y' < A$ then set $y = y'$, otherwise $y = y' - A$. These are sufficient for realizing mod A, since when $A \geq BS$, cs never exceeds the value of A, and thus we only need to check whether $r + cs$ will exceed A.

The next theorem guarantees that every honest prover is always accepted.

---

[6] Similar to GPS in [10], no specific generator is fixed. The difference with GPS is that we make the order of group public here (see also Sect. 1.2 and Sect. 4.1 for more detail).

Theorem 1. *If* ord$(g)|A$ *holds, an honest prover* P *is always accepted by an honest verifier* V *in a complete identification of our proposed scheme.*

The next theorem says that any interaction between an honest prover and any verifier (including dishonest one) reveals no knowledge about secret key.

Theorem 2. *Our proposed scheme is perfect zero-knowledge under the following conditions: (1)* ord$(g)|A$*, (2) B and ℓ are polynomial.*

Proof Sketch. Here is the construction of an expected polynomial algorithm M to simulate one round of interaction between an honest prover P and an arbitrary verifier $\widehat{V}$. The input to M is public key $pk = (\mathcal{G}, g, S, B, \ell, I, A)$.

(Step 1) Pick $\bar{c}$ randomly from $[0, B - 1]$. Then pick $\bar{y}$ randomly from $[0, A - 1]$.

(Step 2) Calculate $\bar{x} = g^{\bar{y}} I^{\bar{c}}$. Send $\bar{x}$ to $\widehat{V}$.

(Step 3) Upon receiving c from $\widehat{V}$, if $c \neq \bar{c}$ then go back to **Step 1** with another $(\bar{c}, \bar{y})$, otherwise return $(\bar{x}, \bar{c}, \bar{y})$.

The expected number of execution for simulating one round is $B = poly(k)$. Thus, the total expected running time for simulating ℓ rounds is $O(\ell B) = poly(k)$.

It is sufficient to prove that for any (α, β, γ) such that $\alpha \in \langle g \rangle$, $\beta = \mathsf{chal}_{\widetilde{V}}(\alpha)$, $\gamma \in [0, A - 1]$, the probability that $(\bar{x} = \alpha, \bar{c} = \beta, \bar{y} = \gamma)$ holds taken over random tape of M and the probability that $(x = \alpha, c = \beta, y = \gamma)$ holds taken over random tape of real prover P are same, where $\mathsf{chal}_{\widetilde{V}}(x)$ denotes the internal function of $\widehat{V}$ which determines the challenge upon receiving the commitment x.

Since ord$(g)|A$ holds, there is exactly $A/\text{ord}(g)$ values of r satisfying $g^r = \alpha$ among $r \in [0, A - 1]$ for P and there is exactly $A/\text{ord}(g)$ values of $\bar{y}$ satisfying $g^{\bar{y}} I^{\beta} = \alpha$ among $\bar{y} \in [0, A - 1]$. And for P, it is easy to see that there is exactly one value of y satisfying $y = \gamma$ among $A/\text{ord}(g)$ values of $y \in [0, A - 1]$ such that $g^y I^{\beta} = \alpha$ holds. This also holds for M with $\bar{y}$. Thus, for both P and M, the overall probability is $(A/\text{ord}(g))/A \times 1/(A/\text{ord}(g)) = 1/A$. ∎

4   Security against Impersonation

Here we discuss the security of our scheme against an adversary whose goal is impersonation: posing itself as a prover and successfully making an honest verifier to accept without having valid secret key. The adversary can launch various attacks on a number of honest provers to gain some knowledge before the final impersonation attempt. To formalize this scenario, we recall the definition of impersonation under serial active attacks (IMP-SA) and concurrent active attacks (IMP-CA) [2]. An adversary $\mathcal{A}$ on an identification scheme $\mathsf{ID} = (\mathcal{K}, P, V)$ where $\mathcal{K}$ is the key-generation algorithm, P is the prover, and V is the verifier, is a pair of algorithms $(\widehat{V}, \widehat{P})$, the *cheating verifier* and *cheating prover*, respectively. $\mathcal{A}$ performs the following game IMP.

Stage 1: $\mathcal{K}$ runs on input k and produces (pk, sk). $\widehat{V}$ is initialized with pk and a random tape $\omega_{\widehat{V}}$. Cheating verifier $\widehat{V}$ interacts with a $poly(k)$ different clones of prover P, all having independent random tapes and being initialized with pk, sk. Eventually, $\widehat{V}$ outputs some information $\widehat{hist}$ and stops.

Stage 2: Cheating prover $\widehat{P}$ is initialized with input $\widehat{hist}$, whereas verifier V is initialized with pk and freshly chosen random tape ω_V. $\widehat{P}$ and V interact. $\mathcal{A}$ wins if and only if V accepts in this interaction.

The difference between serial active and concurrent active attacks is that in *Stage 1*, the former allows $\mathcal{A}$ to only interact with the clones of prover *sequentially*, i.e., only a single prover at a time, whereas the later allows $\mathcal{A}$ to interact *arbitrarily* with many clones of prover at the same time. An adversary is called an *IMP-SA adversary* (resp. *IMP-CA adversary*) if it executes above game in serial (resp. concurrent) active attacks. The comparison of security between our proposed scheme and GPS is summarized in the following table.

Table 1. Security comparison between Schnorr, GPS, and our proposed scheme

	Schnorr [27]	GPS [10, 19]	Proposed Scheme
Zero-Knowledgeness	perfect	statistical	**perfect**
Security against IMP-SA (assumption)	DL	DLSE *)	DLSE [26]
Security against IMP-CA (assumption)	OMDL	unknown	OMDL***)

IMP-SA=impersonation under serial attack, IMP-CA=impersonation under concurrent attack, DL=discrete logarithm problem, DLSE=DL with short exponent, OMDL=one more discrete logarithm problem *)In GPS, DLSE is allowed with unknown order (slightly weaker than DLSE in [26]). ***)See Definition 3 for the detailed description of OMDL in our scheme.

4.1 Against Impersonation under Serial Active Attacks

The security proof of our proposed scheme against impersonation against serial active attacks can be easily proven using the same idea for the case of GPS in [10], i.e., using zero-knowledge simulator to simulate prover in *Stage 1* and trying to get two pairs of challenge and response with different challenges in *Stage 2* in order to solve the discrete logarithm problem associated to the key generator. The only difference from GPS is that we assume that the discrete logarithm problem (with short exponent) is still hard although the group order of g is public (same as the one defined in [26]). We omit the detailed security proof in this paper. The full proof will be in the full version of this paper.

4.2 Against Impersonation under Concurrent Active Attacks

In order to prove the security against impersonation under concurrent active attacks, we need to assume the intractability of *one more discrete logarithm problem with one short exponent (omdl-ose)*, which is a slight modification of *one-more discrete logarithm problem* described in [2].

Definition 3 (One More Discrete Logarithm Problem with One Short Exponent (OMDL-OSE)). *Let $\mathcal{I}$ be a randomized algorithm and let $\mathcal{K}_{\mathsf{dlse}}$ be a short discrete logarithm parameter generator. The game of "$\mathcal{I}$ attacking $\mathcal{K}_{\mathsf{dlse}}$ in omdl-ose" is as follows. First, $\mathcal{K}_{\mathsf{dlse}}$ runs on input k and outputs $(\mathcal{G}, g, \mathrm{ord}(g), S, s_0)$. Let $W_0 = g^{-s_0}$. $\mathcal{I}$ runs on inputs k and $(\mathcal{G}, g, \mathrm{ord}(g), S, W_0)$. $\mathcal{I}$ also has access to two oracles: (1)$DLOG_{g,\mathrm{ord}(g)}(\cdot)$ which on input $Y \in \mathcal{G}$, returns $y \in \mathbb{Z}_{\mathrm{ord}(g)}$ such that $g^y = Y$ holds, (2)challenge oracle O_{chal} which each time it is invoked (without any input), returns a randomly chosen challenge point $W \in \mathcal{G}$. Let $W_1, \ldots, W_n$ denote the challenges returned by O_{chal}. $\mathcal{I}$ can query its $DLOG_{g,\mathrm{ord}(g)}$ at most n times. $\mathcal{I}$ wins if and only if it outputs $s_0, \ldots, s_n \in \mathbb{Z}_{\mathrm{ord}(g)}$ where $W_i = g^{s_i}$ holds. The probability of algorithm $\mathcal{I}$ winning the game is denoted by $\mathbf{Adv}_{\mathcal{K}_{\mathsf{dlse}},\mathcal{I}}^{\mathsf{omdl-ose}}(k)$, taken over the coin tosses of $\mathcal{K}_{\mathsf{dlse}}$, $\mathcal{I}$ and its oracles. $\mathcal{I}$ is said to (t, ε)-expectedly-omdl-ose-break $\mathcal{K}_{\mathsf{dlse}}$ if its expected running time is at most t and satisfies $\mathbf{Adv}_{\mathcal{K}_{\mathsf{dlse}},\mathcal{I}}^{\mathsf{omdl-ose}}(k) \geq \varepsilon$ where t and ε are polynomials in k. $\mathcal{K}_{\mathsf{dlse}}$ is (t, ε)-expectedly-omdl-ose-secure if there is no algorithm (t, ε)-expectedly-omdl-ose-breaks it.*

Let $\mathcal{A}$ be an IMP-CA adversary. Let $\mathbf{Adv}_{\mathsf{ID},\mathcal{A}}^{\mathsf{imp-ca}}(k)$ denote the probability $\mathcal{A}$ wins, taken over the coin tosses of $\mathcal{K}$, the random tape $\omega_{\mathcal{A}}$ of $\mathcal{A}$, the random tapes of prover clones, and the random tape ω_V of V. An adversary $\mathcal{A}$ is said to (t, ε)-*breaks* identification scheme ID if it runs in time at most t and satisfies $\mathbf{Adv}_{\mathsf{ID},\mathcal{A}}^{\mathsf{imp-ca}}(k) \geq \varepsilon$, where t and ε are polynomials in k. And we say that an identification scheme ID is (t, ε)-*secure against impersonation under concurrent attack* if there is no algorithm (t, ε)-breaks it.

The following theorem states the upper bound of $\mathcal{A}$'s success probability.

Theorem 3. *Let $\mathsf{ID} = (\mathcal{K}, P, V)$ be our scheme associated to short discrete logarithm parameter generator $\mathcal{K}_{\mathsf{dlse}}$ and setup parameters (κ, B, ℓ), where B, ℓ are polynomial and $\kappa > 0$. Let $\mathcal{A} = (\widehat{V}, \widehat{P})$ be an IMP-CA adversary of running time at most t_κ attacking ID, where t_κ is polynomial. Then there is an algorithm $\mathcal{I}$ attacking $\mathcal{K}_{\mathsf{dlse}}$ in omdl-ose such that for every k, the followings holds.*

$$\mathbf{Adv}_{\mathsf{ID},\mathcal{A}}^{\mathsf{imp-ca}}(k) \leq 1/B^\ell + \mathbf{Adv}_{\mathcal{K}_{\mathsf{dlse}},\mathcal{I}}^{\mathsf{omdl-ose}}(k).$$

Furthermore, the expected running time of $\mathcal{I}$ is at most $4t_\kappa + O((n\ell + 1)B)$ and $n = poly(k)$ is the number of clones of prover with which $\widehat{V}$ interacts.

Proof Sketch. Mainly, the proof is similar to the one for Schnorr ID scheme in [2]. We construct $\mathcal{I}$ as follows. Note that $\mathcal{I}$ is given an access to the IMP-CA adversary $\mathcal{A}(\widehat{V}, \widehat{P})$ attacking our scheme. As described in the game of $\mathcal{I}$ attacking $\mathcal{K}_{\mathsf{dlse}}$ in omdl-ose (Def. 3), $\mathcal{I}$ gets inputs $(\mathcal{G}, g, \mathrm{ord}(g), S, W_0)$. To simulate provers in *Stage 1* against concurrent active attacks of cheating verifier $\widehat{V}$, $\mathcal{I}$ uses the given challenge oracle O_{chal} and discrete logarithm oracle $DLOG_{g,\mathrm{ord}(g)}$ as follows. First, $\mathcal{I}$ sets the public key $I = W_0$ for all provers. The procedure to simulate the i-th prover for $\widehat{V}$ is as follows (w.l.o.g., here we assume that $\ell = 1$).

- Upon request of commitment from $\widehat{V}$, $\mathcal{I}$ queries the challenge oracle O_{chal} to retrieve randomly selected $W_i \in \mathcal{G}$, and sends W_i to $\widehat{V}$ as commitment.

Receiving the challenge c_i, $\mathcal{I}$ sends (W_i/I^{c_i}) to $DLOG_{\mathrm{ord}(g),g}$ and retrieves $z_i \in \mathbb{Z}_{ord(g)}$ such that $g^{z_i} = W_i/I^{c_i}$ holds. Then $\mathcal{I}$ randomly selects $j_i \in [0, A/\mathrm{ord}(g) - 1]$ and sends $\overline{y_i} = j_i \times \mathrm{ord}(g) + z_i$ to $\widehat{V}$ as the response.

Since $\mathrm{ord}(g)|A$ holds, the selected W_i from O_{chal} and the commitment of real prover have the same distribution, and so do $\overline{y_i}$ and the response of real prover.

In *Stage 2*, using the well-known rewinding technique, $\mathcal{I}$ gets two pairs of challenge and response for the same commitment with different challenges from $\widehat{P}$, and and obtains (σ, τ) such that $g^\sigma = I^\tau$ holds, where $\sigma \in (-A, A)$ and $\tau \in [1, B)$. If τ divides σ and $\sigma/\tau \in [0, S-1]$, $\mathcal{I}$ sets $s_0' = -\sigma/\tau$. Otherwise, $\mathcal{I}$ sets $\widetilde{s_0} = (-\sigma/d)(\tau/d)^{-1} \pmod{\mathrm{ord}(g)/d}$ where $d = \gcd(\tau, \mathrm{ord}(g))$ and finds $j \in [0, d-1]$ such that $\widetilde{s_0} + j \times \frac{\mathrm{ord}(g)}{d} \in [0, S-1]$ and $I = g^{\widetilde{s_0} + j \times \frac{\mathrm{ord}(g)}{d}}$ hold. Then $\mathcal{I}$ outputs $s_0' = \widetilde{s_0} + j \times \frac{\mathrm{ord}(g)}{d}$, and $s_i' = z_i - c_i s_0' \bmod \mathrm{ord}(g)$. One can easily verify that $W_i = g^{s_i'}$ for any $i \in [0, n]$. ∎

Corollary 1 (Security of Our Proposed Scheme against IMP-CA). *If short discrete logarithm parameter generator with public order of group $\mathcal{K}_{\mathrm{dlse}}$ is (t', ε')-expectedly-omdl-ose-secure, then our proposed scheme associated to $\mathcal{K}_{\mathrm{dlse}}$ and (κ, B, ℓ) is (t_κ, ε)-secure against impersonation under concurrent active attack, where $\varepsilon = 1/B^\ell + \varepsilon'$, $t_\kappa = t'/4 - O((n\ell+1)B)/4$ and n is the number of clones of prover with which $\widehat{V}$ interacts.*

5   Discussion

5.1   Practical Advantages of Our Scheme

The most important improvement of our scheme is that we can have the prover's randomness much smaller than GPS without worrying that the secret key may leak, since the perfect zero-knowledgeness is guaranteed regardless the size of prover's randomness as long as $\mathrm{ord}(g)|A$ holds. Especially, for the case of $A \geq BS$, we offer improvement of efficiency on memory cost of prover's randomness and communication cost of prover's response, while maintaining the fast online authentication of GPS.[7] We illustrate these advantages of our scheme as follows.

Lower Communication Cost. Smaller size of prover's randomness (=A) makes the prover's response in our proposed scheme shorter than that in GPS. As shown in Table 2, under standard parameters, the prover's response in our scheme is at most 195 bits, while that in GPS is at most 275 bits (=80 bits reduce≈ 28% reduction from GPS). Thus, under the modest 40 kbps data rate of an RFID tag, the total time for sending the whole response in GPS is $275/40 \approx 7$ ms, while that in our scheme is $195/40 \approx 5$ ms (2 ms faster).

Lower Storage Cost. Let assume an application using coupons with recommended parameters as the ones in Table 2. A coupon consists of a pair $(r, \mathrm{Hash}(g^r))$ where $\mathrm{Hash}(\cdot)$ is a standard hash-function, e.g., SHA-1. First, we assume the output

---

[7] See the discussion on the time cost of online response in Sect. 5.2.

Table 2. Comparison on prover's efficiency among Schnorr, GPS, our proposed scheme using standard parameters [10, 17], i.e, $\ell=1$, $|B| = 35$, $|S| = 160$, $|\mathrm{ord}(g)| = 180$

		Previous Works		Proposed Scheme	
		Schnorr [27]	GPS [10, 19]	$A < BS$	$\mathbf{A \geq BS}$
Size of A (in bits)		160	275	193	**195**
Storage requirement	one coupon (in bits)	320	435	353	**355**
	20 coupons (in bits)	6400	8700	7060	**7100**
	in % of Schnorr's	100	136	110	111
Size of prover's response y (in bits)		160	276	193	**195**

size of Hash($\cdot$) to be 160 bits. For this case, in order to guarantee zero-knowledge of GPS in practice, [10] recommended the size for A to be $|A| = |B| + |S| + 80 = 35 + 160 + 80 = 275$ bits. Thus, every coupon in GPS needs $|A| + 160 = 435$. On the other hand, in our scheme, since the condition $\mathrm{ord}(g)|A$ is sufficient to ensure perfect zero knowledge, A does not have to be so large, i.e., one coupon needs only 355 bits. For applications with 20 coupons (the largest number of coupons assumed in [19]), we reduce $8700 - 7100 = 1600$ bits (> 1 kbits). One may use the technique of [11] to reduce the output size of hash function. Note that this technique needs a longer challenge to keep the same security level, which means a longer prover's randomness in GPS and our scheme. Based on calculation in [11], assuming 9-collision free hash function, the output size of hash function can be reduced to 71 bits with 3-bits longer prover's randomness. Thus, using this technique, the coupon size for GPS and our scheme become $35 + 160 + 80 + 3 + 71 = 349$ bits and 269 bits respectively.

5.2   Performance Comparison Using Machine Experiment

$\mathcal{G}$ is generated using a Quad-core Xeon 3.0 GHz processor. The key-generation and elementary rounds are performed by a Pentium IV 3.0 GHz processor. The schemes are implemented using C with GMP library version 4.1.4 [12] on Linux 2.6. For precise measurement of the clock cycles, we use RDTSC (ReaD Time Stamp Counter) internal command of Pentium processor [15]. Table 3 shows the average results of 10^6 trials. Here we used the standard parameters for guaranteeing 80 bit security against key only attack and 35 bit security against impersonation as recommended in [10,17], i.e., $\ell=1$, $|B| = 35$, $|S| = 160$, $|\mathrm{ord}(g)| = 180$.

Table 3. Results of experiment on PC

		Previous Works		Proposed Scheme	
		Schnorr [27]	GPS [10, 19]	$A < BS$	$\mathbf{A \geq BS}$
Size of A (in bits)		160	275	193	**195**
Cost of online response	(in cycles)	2628.00	815.14	1937.15	977.65
	in % of Schnorr's	100	31.01	73.71	**37.20**
Cost of computing g^r	(in $\times 1000$ cycles)	6537	9850	6537 (=Schnorr's)	
	in % of Schnorr's	100	150.67	**100 (=Schnorr's)**	

Table 4. Estimation based on the result of experiment on PC

	Previous Works		Proposed Scheme	
	Schnorr [27]	GPS [10, 19]	$A < BS$	**A $\geq$ BS**
Time response on a circuit with the same size as GPS' [19] (in ms)	12.93	4.01	9.53	**4.81**
Circuit size with time response as fast as GPS' [19] (in gates)	5294	1642	3903	**1970**

Here we evaluate/estimate the practical performance of our scheme and compare to that of GPS based on the result of machine experiment in Table 3.

Low Cost Online Response is Maintained. As shown in Table 3, in implementation using PC, the time cost of online response in our scheme is almost the same as that in GPS. In [19], an implementation on UMC 180 nm CMOS technology using 16-bit adder with similar parameters performs the online response calculation in 401 clock cycles under 100 kHz clock rate, i.e., $401 \times 10^{-6} = 4.01$ ms. By assuming that a similar architecture is constructible, we roughly estimate the time cost for our scheme if it is implemented on a circuit with the same size as GPS'. The estimation results are shown in Table 4. The estimation method is as follows. Since the implementation of GPS on PC performs online response within ≈ 815 clock cycles according to Table 3, by assuming rough linear comparison, an implementation of our scheme with $A \geq BS$ on the same CMOS technology may perform online response within $\approx 978/815 \times 401 \approx 481$ cycles. Thus, under 100 kHz clock rate, the online response of our scheme with $A \geq BS$ might be performed within $481 \times 10^{-6} = 4.81$ ms (differs by only 0.8 ms from that of GPS). Even for the case of $A < BS$ with $|A| = 193$, according to Table 3, for the case of $A < BS$ with $|A| = 193$, we get $\approx 1937/815 \times 401 \times 10^{-6} \approx 9.53$ ms. This is still lower than the implementation of AES proposed by Feldhofer et al. [8], i.e., 10.16 ms, while Schnorr scheme is not, i.e., 12.93 ms (see Table 4).

Circuit Size with Response as Fast as GPS is Small Enough for RFID Tags. Here, we estimate the size of the circuit which can provide the calculation of prover's response as fast as GPS, for our scheme and Schnorr scheme. Our estimation uses a rough linear comparison based on the size of GPS' circuit on UMC 180 nm CMOS technology with 16-bit adder shown in [19] (=1642 gates) and the cost of online response of each scheme from the result of experiment on PC. For our scheme with $A \geq BS$, we estimate the size of circuit which calculates prover's response in 4.01 ms (same as GPS in [19]) to be $\approx 977.65/815.14 \times 1642 \approx 1970$ gates. For Schnorr scheme, we estimate the size of circuit to be $\approx 2628/815.14 \times 1642 \approx 5294$ gates. Similarly, for our scheme with $A < BS$, we get $\approx 2628/815.14 \times 1642 \approx 3903$ gates. Thus, under the condition that the response calculation of the circuit is as fast as GPS' circuit in [19], based on our estimation and the current technology that an RFID tag only consist of 200-4000 gates [6], we conclude that our scheme is feasible for RFID tags of current technology, while Schnorr scheme is not.

More Robustness against Denial of Service Attack. Hofferek et al. [14] noted that in memory-constrained devices such as RFID tags where the number of identifications are limited by the number of stored coupons, i.e., pairs of pre-selected r and pre-computed $x = g^r$, a malice verifier can launch a *denial-of-service* attack by keeping requesting authentication from a tag, and few requests are sufficient to exhaust all coupons. In [14], recalculation of new pairs of r and g^r in the idle time of the RFID tags is proposed in order to prevent the immediate exhaustion of coupons. Since $\mathrm{ord}(g)|A$ holds and $\mathrm{ord}(g)$ is known in our scheme, we can generate r in two steps: (1)pick a random $r' \in [0, \mathrm{ord}(g) - 1]$, (2)pick a random $r_0 \in [0, A/\mathrm{ord}(g)-1]$ and set $r = r_0 \times \mathrm{ord}(g)+r'$. Then we can substitute g^r by $g^{r'}$, whose cost is only about the order of $|\mathrm{ord}(g)|$. As shown in Table 3, within a fixed period of time, under standard parameters, our scheme produces coupons approximately $3/2$ times faster than GPS. Thus, in this scenario, our scheme offers more robustness against the denial of service attack.

Trade-off between Memory and Time Costs. As shown in Table 3, our scheme offers a trade-off between the memory and the time costs. For fixed BS, the larger A is, the larger the storage for prover's randomness r is, but the smaller the time cost for calculating response $y = r + cs \bmod A$ is, as subtractions for reducing $r + cs$ into $(r + cs) \bmod A$ becomes less, and vice versa. Schnorr scheme can be seen as an instance of our scheme with $A < BS$ and $A=\mathrm{ord}(g)=160$ bits.

5.3   Parameter Settings

The choices of the multiplicative group $\mathcal{G}$ and the base g of $\mathcal{G}$ can be considered similar to GPS except one must be careful that in our proposed scheme, $\mathcal{G}$ and g must be chosen such that discrete logarithm in subgroup $\langle g \rangle$ is still averagely hard even though the order of g is revealed. The choices of ℓ, S and B can be considered as same as in GPS. For further detailed analysis, see [10, 17].

5.4   Previous Works as Instances of Our Proposed Scheme

Our scheme (the general modification) can be seen as a loose generalization of previous discrete-logarithm based identification schemes. The Schnorr scheme [27] can be seen as the case of $A = \mathrm{ord}(g)$ and $S = \mathrm{ord}(g)$. The basic discrete-log scheme of Chaum-Evertse-van de Graff with group from prime modulus N can be seen as the case of $\mathcal{G} = \mathbb{Z}_N$, $A = \mathrm{ord}(g)$. The Brickell-McCurley scheme [3] can also be seen as the case of $\mathcal{G} = \mathbb{Z}_p$ where p is a prime and $S = \mathrm{ord}(g)$, if we drop the requirement to guarantee the hardness of factoring $p - 1$.

References

1. Baudron, O., Boudot, F., Bourel, P., Bresson, E., Corbel, J., Frisch, L., Gilbert, H., Girault, M., Goubin, L., cois Misarsky, J.F., Nguyen, P., Patarin, J., Pointcheval, D., Poupard, G., Stern, J., Traoré, J.: GPS - an Asymmetric identification scheme for on the fly authentication of low cost smart cards Ver 2.0 (October 2001)

2. Bellare, M., Palacio, A.: GQ and Schnorr identification schemes: Proofs of security against impersonation under active and concurrent attacks. In: Yung, M. (ed.) CRYPTO 2002. LNCS, vol. 2442, pp. 162–177. Springer, Heidelberg (2002)
3. Brickell, E.F., McCurley, K.S.: An interactive identification scheme based on discrete logarithms and factoring. J. Cryptology 5(1), 29–39 (1992)
4. Burmester, M., Desmedt, Y., Beth, T.: Efficient zero-knowledge identification schemes for smart cards. Comput. J. 35(1), 21–29 (1992)
5. Chaum, D., Evertse, J.H., van de Graaf, J., Peralta, R.: Demonstrating possession of a discrete logarithm without revealing it. In: Odlyzko, A.M. (ed.) CRYPTO 1986. LNCS, vol. 263, pp. 200–212. Springer, Heidelberg (1987)
6. Addressing Insecurities and Violations of Privacy. In: Cole, P.H., Ranasinghe, D.C. (eds.) Networked RFID Systems and Lightweight Cryptography. Springer, Heidelberg (2008)
7. Dobkin, D.M.: The RF in RFID:physical layer operation of passive UHF tags and readers: 4. UHF RFID Protocols (July 2009),
http://www.enigmatic-consulting.com/Communications_articles/RFID/RFID_protocols.html
8. Feldhofer, M., Dominikus, S., Wolkerstorfer, J.: Strong authentication for RFID systems using the AES algorithm. In: Joye, M., Quisquater, J.-J. (eds.) CHES 2004. LNCS, vol. 3156, pp. 357–370. Springer, Heidelberg (2004)
9. Girault, M.: Self-certified public keys. In: Davies, D.W. (ed.) EUROCRYPT 1991. LNCS, vol. 547, pp. 490–497. Springer, Heidelberg (1991)
10. Girault, M., Poupard, G., Stern, J.: On the fly authentication and signature schemes based on groups of unknown order. Journal of Cryptology 19(4), 463–487 (2006)
11. Girault, M., Stern, J.: On the length of cryptographic hash-values used in identification schemes. In: Desmedt, Y.G. (ed.) CRYPTO 1994. LNCS, vol. 839, pp. 202–215. Springer, Heidelberg (1994)
12. GNU Multiple Precision Arithmetic Library (2004), http://www.swox.com/gmp
13. Hitachi, Ltd. Secure RFID μ-Chip Hibiki (UHF) (March 2009),
http://www.hitachi.co.jp/Prod/mu-chip/mu-chip_hibiki_secure.pdf
14. Hofferek, G., Wolkerstorfer, J.: Coupon recalculation for the GPS authentication scheme. In: Grimaud, G., Standaert, F.-X. (eds.) CARDIS 2008. LNCS, vol. 5189, pp. 162–175. Springer, Heidelberg (2008)
15. Intel Corporation. RDTSC–Read Time-Stamp Counter,
http://www.intel.com/software/products/documentation/vlin/mergedprojects/analyzer_ec/mergedprojects/reference_olh/mergedprojects/instructions/instruct32_hh/vc275.htm
16. ISO/IEC. International Standard ISO/IEC 9798 Part 5: Mechanisms Using Zero Knowledge Techniques (December 2004)
17. Lenstra, A.K., Verheul, E.R.: Selecting cryptographic key sizes. Journal of Cryptology 14(4), 255–293 (2001)
18. McLoone, M., Robshaw, M.J.B.: New architectures for low-cost public key cryptography on RFID tags. In: ISCAS, pp. 1827–1830. IEEE, Los Alamitos (2007)
19. McLoone, M., Robshaw, M.J.B.: Public key cryptography and RFID tags. In: Abe, M. (ed.) CT-RSA 2007. LNCS, vol. 4377, pp. 372–384. Springer, Heidelberg (2006)
20. Naccache, D., M'Raïhi, D., Vaudenay, S., Raphaeli, D.: Can D.S.A. be improved? In: De Santis, A. (ed.) EUROCRYPT 1994. LNCS, vol. 950, pp. 77–85. Springer, Heidelberg (1995)

21. NESSIE. Final report of European project IST-1999-12324: New European Schemes for Signatures Integrity and Encryption, GPS - Public Report No. NES/DOC/RHU/WP3/004/b (February 2004)
22. Okamoto, T., Katsuno, H., Okamoto, E.: A fast signature scheme based on new on-line computation. In: Boyd, C., Mao, W. (eds.) ISC 2003. LNCS, vol. 2851, pp. 111–121. Springer, Heidelberg (2003)
23. Pointcheval, D.: The composite discrete logarithm and secure authentication. In: Imai, H., Zheng, Y. (eds.) PKC 2000. LNCS, vol. 1751, pp. 113–128. Springer, Heidelberg (2000)
24. Poupard, G., Stern, J.: Security analysis of a practical "on the fly" authentication and signature generation. In: Nyberg, K. (ed.) EUROCRYPT 1998. LNCS, vol. 1403, pp. 422–436. Springer, Heidelberg (1998)
25. Poupard, G., Stern, J.: On the fly signatures based on factoring. In: Proc. of the 6th CCS, pp. 48–57. ACM Press, New York (1999)
26. Santoso, B., Ohta, K.: A new 'on the fly' identification scheme: an asymptoticity trade-off between ZK and correctness. IEICE Transactions on Fundamentals of Electronics, Communications and Computer Sciences E92.A (1), 122–136 (2009)
27. Schnorr, C.P.: Efficient identification and signatures for smart cards. In: Brassard, G. (ed.) CRYPTO 1989. LNCS, vol. 435, pp. 239–252. Springer, Heidelberg (1990)
28. SkyeTek, Inc. SkyeModule M7: compact 900 MHz UHF RFID reader/writer, http://www.skyetek.com/Portals/0/Documents/Products/SkyeModule_M7_DataSheet.pdf

Linear Cryptanalysis of Reduced-Round PRESENT

Joo Yeon Cho

[1] Helsinki University of Technology, Finland
[2] Nokia A/S, Denmark
joo.cho@tkk.fi

Abstract. PRESENT is a hardware-oriented block cipher suitable for resource constrained environment. In this paper we analyze PRESENT by the multidimensional linear cryptanalysis method. We claim that our attack can recover the 80-bit secret key of PRESENT up to 25 rounds out of 31 rounds with around $2^{62.4}$ data complexity. Furthermore, we showed that the 26-round version of PRESENT can be attacked faster than key exhaustive search with the 2^{64} data complexity by an advanced key search technique. Our results are superior to all the previous attacks. We demonstrate our result by performing the linear attacks on reduced variants of PRESENT. Our results exemplify that the performance of the multidimensional linear attack is superior compared to the classical linear attack.

Keywords: Block Ciphers, Lightweight Cryptography, PRESENT, Multidimensional Linear Cryptanalysis.

1 Introduction

PRESENT [3] is a lightweight SPN block cipher proposed by Bogdanov et al. at CHES 2007. PRESENT is designed for resource restricted applications such as RFID and sensor networks. Due to the impressive hardware performance and the strong security, PRESENT has drawn a lot of attention from the lightweight cryptographic community.

On the other hand, the cryptanalysis on PRESENT has been also actively performed so far. In [15], Wang presented a differential cryptanalysis that could attack the 16-round variant with 2^{64} chosen texts and 2^{65} memory accesses. In [1], Albrecht et al. presented a differential attack using algebraic techniques that can recover an 80-bit key of the 16-round variant with similar complexity to [15] and a 128-bit key of the 19-round variant by 2^{113} computations. In [4], Collard et al. presented a statistical saturation attack that can recover the key of the 24 round variant with 2^{57} chosen texts and 2^{57} time complexity under the condition that the parts of plaintexts are fixed to a constant value. More recently, Ohkuma presented a linear attack on 24-round variant with $2^{63.5}$ known texts [13]. He claimed that due to the linear hull effect, the linear approximation of PRESENT using weak keys could have much stronger correlation than the one expected by designers.

J. Pieprzyk (Ed.): CT-RSA 2010, LNCS 5985, pp. 302–317, 2010.

In this paper, we analyze PRESENT by a multidimensional linear attack method. We observe that PRESENT has a large number of linear approximations that hold with the same order of magnitude of correlations due to the simple structure of the round function. As shown in [7], a multidimensional linear attack can be efficiently applied to such cipher. Our attack is different from Ohkuma's attack [13] since Ohkuma presented the linear attack using a single linear approximation which can have the strongest correlation if weak keys are used. According to our analysis, the 25-round variant of PRESENT using the 80-bit key can be attacked faster than key exhaustive search with around $2^{62.4}$ data complexity. Furthermore, an advanced key search technique enables us to attack the 26-round version of PRESENT with 2^{64} data complexity. Our results are superior to all the previous attacks presented in the open literature. We demonstrate our claim by performing the multidimensional linear attacks on reduced variants of PRESENT.

This paper is organized as follows. In Section 2, the structure of PRESENT is briefly described and the framework of multidimensional linear attack is presented. In Section 3, linear characteristics are derived and their capacities are computed. In Section 4, the attack algorithm using linear characteristics is described. In Section 5, our attacks are applied to reduced variants of PRESENT and the experimental results are presented. Section 6 concludes this paper.

2   Preliminaries

2.1   Brief Description of PRESENT

PRESENT is a SPN block cipher that consists of 31 rounds. The encryption block length is 64 bits and the key lengths is 80 bits or 128 bits. Each of the 31 rounds consists of three layers: addRoundKey, SboxLayer and pLayer. The AddRoundKey is a 64-bit eXclusiveOR operation with a round key. The SboxLayer is a 64-bit nonlinear transform using a single S-box 16 times in parallel. The S-box is a nonlinear bijective mapping $S : \mathbb{F}_2^4 \mapsto \mathbb{F}_2^4$ given in Table 4. The pLayer is a bit-by-bit permutation $P : \mathbb{F}_2^{64} \mapsto \mathbb{F}_2^{64}$ given in Table 5. The design idea of SboxLayer and pLayer is adapted from Serpent [2] and DES block cipher [10], respectively. The structure of PRESENT is illustrated in Figure 1.

The key scheduling algorithm has two versions depending on whether the key size is 80 bits or 128 bits. Since the key schedule is not directly relevant to out attack, we do not describe the key schedule algorithm here. For complete description of PRESENT we refer to the paper [3].

2.2   Multidimensional Linear Cryptanalysis Using χ^2 Method

Multidimensional linear cryptanalysis is an extension of Matsui's classical linear cryptanalysis [9] in which multiple linear approximations are optimally exploited. The general framework of the multidimensional linear cryptanalysis adapting Matsui's algorithm 2 was presented by Hermelin et al. in [8]. In their paper,

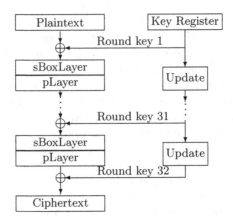

Fig. 1. Overview of PRESENT

Hermelin et al. studied two statistical methods: the log-likelihood ratio (LLR) and the χ^2. We apply the χ^2 statistic method to PRESENT since the LLR method is not proper to PRESENT-like structure. The detailed explanation will be given in Section 4.4.

The brief framework of the χ^2 method is as follows. Let V_n denote the space of n-dimensional binary vectors. A function $f : V_n \rightarrow V_m$ with $f = (f_1, \cdots, f_m)$ where f_i is a linear approximation is called a vectorial linear approximation of the dimension m. The correlation of f_i is defined as $c(f_i) = 2^{-n} [\#(f_i(a) = 0) - \#(f_i(a) = 1)]$ where $a \in V_n$.

Let p be the probability distribution of m-dimensional linear approximations. The capacity of $p = (p_0, \ldots, p_{2^m-1})$ is defined by $C_p = \sum_{i=0}^{2^m-1} \frac{(p_i - u_i)^2}{u_i}$ where $u = (u_0, \ldots, u_{2^m-1})$ is the uniform distribution. It is well known that the C_p is equal to the sum of the square of correlations of all $2^m - 1$ linear approximations.

Suppose l is the length of the target key. For all values of $k \in [0, 2^l - 1]$, one obtains the empirical probability distributions $Q_k = (q_{k,0}, \ldots, q_{k,2^m-1})$ by measuring the frequency of m-dimensional vectors which are Boolean values of m linear independent approximations. Then the candidate keys are sorted according to their χ^2-statistics defined as

$$\mathcal{D}(k) = 2^m \sum_{i=0}^{M} (q_{k,i} - 2^{-m})^2, \quad M = 2^m - 1 \tag{1}$$

which represents the l_2-distance of the Q_k from the uniform distribution.

If the right key is ranked in the position of d from the top out of 2^l key candidates, we say that the attack has the advantage of $(l - \log_2 d)$ [14]. The advantage of the χ^2-method using statistic (1) is derived in Theorem 1 in [8] by

$$advantage = \frac{(NC_p - 4\Phi^{-2}(2P_s - 1))^2}{8M}, \quad \Phi(x) = \int_{-\infty}^{x} \frac{1}{\sqrt{2\pi}} e^{-t^2/2} dt \tag{2}$$

where P_s is the success probability, N is the amount of data and C is the capacity.

2.3   Notations

Let S_i denote the i-th S-box in the SboxLayer and P denote the permutation in the pLayer. Let K_r denote the r-th round key and $K_r^{[i]}$ denote the i-th bit of the K_r. The $K_r^{[i..j]}$ denote the bit string from $K_r^{[i]}$ to $K_r^{[j]}$. We use $E_K(X)$ for representing the average value of X over all possible values of K. In our notation of the bit masks, we identify $\mathbb{F}_2^4$ with $\mathbb{Z}_{16}$. We use the little endian for bit notation through the paper, that is, the least significant bit is counted at the rightmost.

3   Linear Characteristics of PRESENT

We define a *linear trail* as a single path of linear approximations concatenated over multiple rounds. It is a common belief that the linear characteristic with multiple linear trails has a larger correlation than one with a single linear trail due to the linear hull effect [11]. In this section, we derive a linear characteristic of PRESENT that has multiple linear trails. Each linear trail exploits the linear approximations of S-boxes which have a *single active bit* in the input and output masks. The linear masks having more than one active bit affect at least two S-boxes in the consecutive round due to the permutation layer, which yield much less correlations in the multiple rounds of PRESENT.

Definition 1. *A single-bit linear trail is a linear trail where the input and output masks of linear approximations of all intermediate S-boxes are of Hamming weight one.*

We call a single-bit linear trail as just a linear trail unless specified otherwise.

3.1   Single Bit Linear Trails

Let $\pi(\alpha, \beta)$ denote a linear approximation of S-box S where $\alpha, \beta \in \mathbb{F}_2^4$ are an input and output mask of S, respectively. The correlation of $\pi(\alpha, \beta)$ is denoted by $\rho(\alpha, \beta)$. We observe that the S-box has the following properties:

S1. For $\alpha, \beta \in \{2, 4, 8\}$, $\rho(\alpha, \beta) = \pm 2^{-2}$ except that $\rho(8, 4) = 0$;
S2. For $\alpha \in \{1, 2, 4, 8\}$, $\rho(\alpha, 1) = \rho(1, \alpha) = 0$.

According to Property S1 and S2, the S-box holds eight linear approximations which has a single active bit in both the input and output linear masks.

Let us define $\mathcal{S} = \{S_5, S_6, S_7, S_9, S_{10}, S_{11}, S_{13}, S_{14}, S_{15}\}$ and $\mathcal{B} = \{4i + 1, 4i + 2, 4i + 3 | 0 \leq i \leq 15, S_i \in \mathcal{S}\}$. Then, the permutation P of the pLayer has the following properties:

P1. If $x \in \mathcal{B}$, then $P(x) \in \mathcal{B}$;
P2. All the outputs of S_0, S_4, S_8 and S_{12} turn into the least significant bits of the inputs of S-boxs next round by the permutation. Also, the outputs of S_1, S_2 and S_3 turn into the input of S_0, S_4, S_8 and S_{12} next round.

Due to Property S2 and P2, the linear trails passing any bit position that does not included in $\mathcal{B}$ do not have correlations. Hence, by Property S1 and P1, any r-round linear trail with an input mask α and an output mask β takes the following path:

$$\pi(\alpha, 2^{v_1}) \to \pi(2^{u_2}, 2^{v_2}) \to \cdots \to \pi(2^{u_{r-1}}, 2^{v_{r-1}}) \to \pi(2^{u_r}, \beta)$$

where $u_i, v_i \in \{1, 2, 3\}$ and $(u_i, v_i) \neq (3, 2)$ for $1 \leq i \leq r$.

3.2 n-Round Linear Characteristic

Let $\Omega^{(1)}$ denote the 1-round linear characteristic which has all the single bit linear trails of nine S-boxes of $\mathcal{S}$, as shown in Figure 2. Due to Property S1, the $\Omega^{(1)}$ contains $9 \times 8 = 72$ linear trails, each of which has $\pm 2^{-2}$ correlation. Since $x \mapsto P(x)$ is a one-to-one mapping, Property P1 implies that $\{P(x)|x \in \mathcal{B}\} = \mathcal{B}$. Hence, we can construct the n-round linear characteristic, which is denoted by $\Omega^{(n)}$, by concatenating $\Omega^{(1)}$ iteratively n times as follows:

$$\Omega^{(n)} = \underbrace{\Omega^{(1)} \circ \cdots \circ \Omega^{(1)}}_{n \text{ times}}.$$

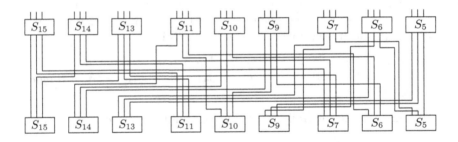

Fig. 2. Linear trails in the 1-round linear characteristic

We can expect that the number of linear trails grows exponentially according to the increment of the number of rounds. Let $\zeta^{(r)}(i, j)$ denote a bundle of linear trails which start from the i-th bit of input and end up at the j-th bit of output over $\Omega^{(r)}$. Each $\zeta^{(r)}(i, j)$ is extended to $\zeta^{(r+1)}(i, k)$ for some $k \in \mathcal{B}$ via two or three single-bit linear approximations of the S-box.

Let $\theta^{(r)}(i, j)$ denote the correlation of $\zeta^{(r)}(i, j)$. If the $\theta_j^{(r)}$ is defined as a summation of the correlations of all linear trails that reach the j-th bit of output over $\Omega^{(r)}$, then $\theta_j^{(r)} = \sum_{i \in \mathcal{B}} \theta^{(r)}(i, j)$. The actual value of $\theta_j^{(r)}$ depends on the round keys involved in each linear trail. Suppose K is a user-supplied key. For any $i, j \in \mathcal{B}$, the $\theta_j^{(r)}(K)$ is recursively expressed as

$$\theta_j^{(r)}(K) = \sum_{i=1}^{3} (-1)^{K_r^{[\nu]}} \rho(2^i, 2^j \bmod 4) \theta_\nu^{(r-1)}(K), \quad \nu = P^{-1}(4\lfloor j/4 \rfloor + i) \qquad (3)$$

where P^{-1} is an inverse mapping of P.

The average value of $\theta_j^{(r)}$ over all possible values of K is recursively computed by the following algorithm:

1. Initialize $\theta_j^{(0)} = 1$ for all $j \in B$. Set $r = 1$.
2. For each $j \in B$,
   (a) compute $\theta_j^{(r)}(K)$ using (3) for all possible values of $K \in \mathbb{F}_2^{27}$;
   (b) assign $\theta_j^{(r)} = E_K(|\theta_j^{(r)}(K)|) = 2^{-27} \sum_K |\theta_j^{(r)}(K)|$.
3. Repeat Step 2 for $r = 2, 3, \ldots, n$.

Above the algorithm can be much simplified by the following theorem: (In this theorem, the correlation potential means the square of the correlation.)

Theorem 1. *(Theorem 7.9.1 [6], Theorem 1 [11]) The average correlation potential between an input and an output selection pattern is the sum of the correlation potentials of all linear trails between the input and output selection patterns.*

By Theorem 1, the average value of $|\theta_j^{(n)}|$ is obtained by summing the absolute values of correlations of all linear trails in the $\zeta^{(r)}(i, j)$ for all $i \in B$. Hence, the average value of $|\theta_j^{(n)}|$ can be computed simply by the following algorithm:

1. Initialize $\theta_j^{(0)} = 1$ for all $j \in B$. Set $r = 1$.
2. For each $j \in B$, compute

$$\theta_j^{(r)} = \sum_{i=1}^{3} |\rho(2^i, 2^j \bmod 4)| \, \theta_\nu^{(r)}, \quad \nu = P^{-1}(4\lfloor j/4 \rfloor + i).$$

3. Repeat Step 2 for $r = 2, 3, \ldots, n$.

Theorem 1 concerns a single linear approximation that has multiple linear trails. For the multidimensional linear cryptanalysis, Theorem 1 can be extended as follows:

Proposition 1. *The expected capacity of an m-dimensional linear approxima-tion is the sum of the square of the expected correlations of all the linear trails that all the $2^m - 1$ one-dimensional linear approximations have.*

3.3 $(n + 4)$-Round Linear Characteristic

Let us define U as the 2-round characteristic which starts from S_5, S_9 and S_{13} and ends with nine S-boxes of S. Each input S-box of U takes arbitrary value from 1 to 15 as the input mask and each output S-box takes a single-bit output mask only. We also define V as the 2-round characteristic which starts from nine S-boxes of S and ends up at S_5, S_6 and S_7. Each input S-box of V takes a single-bit linear mask and each output S-box takes arbitrary value from 1 to 15 as the output mask. For a positive integer n, the $n + 4$ round linear characteristic is constructed by adding U and V to $\Omega^{(n)}$ at the top and the bottom respectively as shown in Figure 3.

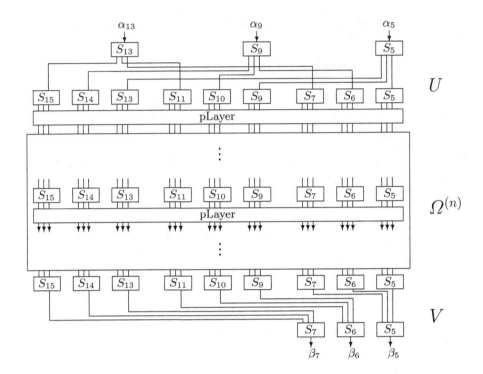

Fig. 3. $(n + 4)$ rounds linear characteristic

Let $C_p^{(n+4)}$ denote the capacity of $U \circ \Omega^{(n)} \circ V$. By the definition of the capacity and due to Theorem 1, the average value of $C_p^{(n+4)}$ is the sum of the square of correlations of all linear trails over the $U \circ \Omega^{(n)} \circ V$, which is calculated by the following theorem:

Theorem 2. *Let us assume that the round keys of PRESENT are statistically independent. For a positive integer n, the expected capacity of $U \circ \Omega^{(n)} \circ V$ over the secret key K is*

$$2^{-8} \sum_{i \in B} \left(\theta_i^{(n)} \right)^2.$$

Proof. Let α_i be an input mask of $S_i \in \{S_5, S_9, S_{13}\}$ of the U and β_j be an output mask of $S_j \in \{S_5, S_6, S_7\}$ of the V. For a fixed α_i, the U has nine linear trails holding with correlations of $\rho(\alpha_i, 2^u) \cdot 2^{-2}$ for some $u \in \{1, 2, 3\}$. Similarly, for a fixed β_j, the V has nine linear trails with the correlations of $\rho(2^v, \beta_j) \cdot 2^{-2}$ for some $v \in \{1, 2, 3\}$. We define $B[\alpha_i]$ and $B[\beta_j]$ as the sets of input and output bit positions where $\Omega^{(n)}$ is linked with U and V for fixed α_i and β_j, respectively. Obviously, $\#B[\alpha_i] = \#B[\beta_j] = 9$ for any α_i and β_j.

Let $c^{(n+4)}(\alpha_i, \beta_j)$ denote the correlation of the linear approximation with the input mask α_i and the output mask β_j over $U \circ \Omega^{(n)} \circ V$. Then, for a fixed key

K, we can write

$$c^{(n+4)}(\alpha_i, \beta_j; K) = \sum_{x \in B_1} \sum_{y \in B_2} (-1)^k \cdot \rho(\alpha_i, 2^u) \cdot 2^{-2} \cdot \theta^{(n)}(x, y) \cdot 2^{-2} \cdot \rho(2^v, \beta_j)$$

where k denotes a parity of the relevant round key bits. Since

$$E_K \left[(-1)^{k_s} (-1)^{k_t} \right] = \begin{cases} 1 & \text{if } s = t, \\ 0 & \text{if } s \neq t \end{cases}$$

under the assumption that the round key bits are statistically independent,[1] we get

$$E_K \left[\left(c^{(n+4)}(\alpha_i, \beta_j; K) \right)^2 \right] = E_K \left[2^{-8} \sum_{x \in B_1} \sum_{y \in B_2} \rho(\alpha_i, 2^u)^2 \cdot \left(\theta^{(n)}(x, y) \right)^2 \cdot \rho(2^v, \beta_j)^2 \right].$$

Parseval's theorem says that $\sum_{\alpha_i=0}^{15} \rho(\alpha_i, 2^u)^2 = \sum_{\beta_j=0}^{15} \rho(2^v, \beta_j)^2 = 1$ for any $u, v \in \{1, 2, 3\}$. Hence, the average value of $C_p^{(n+4)}(\alpha_i, \beta_j; K)$ is obtained by computing

$$E_K[C_p^{(n+4)}(\alpha_i, \beta_j; K)] = \sum_{\alpha_i=0}^{15} \sum_{\beta_j=0}^{15} E_K \left[\left(c^{(n+4)}(\alpha_i, \beta_j; K) \right)^2 \right]$$

$$= 2^{-8} \cdot E_K \left[\sum_{x \in B_1} \sum_{y \in B_2} \left(\theta^{(n)}(x, y; K) \right)^2 \right].$$

The $C_p^{(n+4)}(K)$ is the sum of $C_p^{(n+4)}(\alpha_i, \beta_j; K)$ for all pairwise combinations of $\{\alpha_i, \beta_j\}$ where $\alpha_i \in \{\alpha_5, \alpha_9, \alpha_{13}\}$ and $\beta_j \in \{\beta_5, \beta_6, \beta_7\}$. Since $B[\alpha_5] \cup B[\alpha_9] \cup B[\alpha_{13}] = B$ and $B[\beta_5] \cup B[\beta_6] \cup B[\beta_7] = B$, we conclude that

$$E_K[C_p^{(n+4)}(K)] = \sum_{i \in \{5,9,13\}} \sum_{j \in \{5,6,7\}} E_K[C_p^{(n+4)}(\alpha_i, \beta_j; K)]$$

$$= 2^{-8} \cdot E_K \left[\sum_{x \in B} \sum_{y \in B} \left(\theta^{(n)}(x, y; K) \right)^2 \right]$$

$$= 2^{-8} \cdot E_K \left[\sum_{y \in B} \left(\theta_y^{(n)}(K) \right)^2 \right] = 2^{-8} \cdot \sum_{i \in B} \left(\theta_i^{(n)} \right)^2. \qquad \Box$$

Theorem 2 implies that the expected capacity of $U \circ \Omega^{(n+4)} \circ V$ is the sum of the square of correlations of all linear trails starting from the second round and ending to the second last round. We calculated the average capacities of

---

[1] The statistical behaviour of $k_s \oplus k_t$ was experimentally verified to follow unbiased binomial distribution by a summer school student.

Table 1. Evaluation of capacities of $n + 4$ round characteristics

round	capacity	round	capacity
6	$2^{-8.42}$	18	$2^{-39.71}$
7	$2^{-11.00}$	19	$2^{-42.32}$
8	$2^{-13.61}$	20	$2^{-44.94}$
9	$2^{-16.22}$	21	$2^{-47.55}$
10	$2^{-18.82}$	22	$2^{-50.16}$
11	$2^{-21.43}$	23	$2^{-52.77}$
12	$2^{-24.04}$	24	$2^{-55.38}$
13	$2^{-26.66}$	25	$2^{-57.99}$
14	$2^{-29.27}$	26	$2^{-60.61}$
15	$2^{-31.88}$	27	$2^{-63.22}$
16	$2^{-34.49}$	28	$2^{-65.83}$
17	$2^{-37.10}$	29	$2^{-68.44}$

$(n + 4)$-round linear characteristics for $2 \leq n \leq 25$ by Theorem 2. The results are displayed in Table 1.

In the next section, we present the multidimensional linear attacks using the $(n + 4)$-round linear characteristics.

4   Multidimensional Linear Attacks on PRESENT

4.1   Selection of Linear Independent Approximations

Suppose n is a positive integer. The dimension of input and output masks of the $U \circ \Omega^{(n)} \circ V$ is $4 \times 3 = 12$ each. As mentioned before, the linear trails passing more than one S-box at each round have much less correlations than single-bit linear trails. Thus, it is sufficient to take nine linear characteristics individually, each of which has an 8-dimensional linear characteristic with 4-bit input and 4-bit output. Then, the number of linear approximations spanned for our attack is $9 \times (2^8 - 1)$ in total.

We use eight unit vectors as the linear independent approximations. Even though each unit vector does not have any correlation, all linear approximations can be obtained by spanning these unit vectors. The merit of this approach is that the evaluation of Boolean values of linear approximations is not needed; The probability distribution of the linear approximations can be obtained by just measuring the frequencies of the concatenated value of input and output of the linear characteristic. Hence, the time complexity of the attack can be reduced by at least a factor of m where m is the dimension of the linear approximations.

4.2   Attack Algorithm

We target to attack the n-round version of PRESENT. Our attack uses the $(n - 2)$-round linear characteristic $U \circ \Omega^{(n-6)} \circ V$ from the second round to

$(n-1)$-th round and recovers the 32 bits of the round key in the first round and the last round. The inputs of the U are connected to S_4, S_5, S_6 and S_7 of the first round and the outputs of the V are connected to S_1, S_5, S_9 and S_{13} of the n-th round. Thus, we target to recover the 16 bits of the K_1, which are $K_1^{[16..19]}||K_1^{[20..23]}||K_1^{[24..27]}||K_1^{[28..31]}$, and the 16 bits of the K_n, which are $P(K_n^{[4..7]})||P(K_n^{[20..23]})||P(K_n^{[36..39]})||P(K_n^{[52..55]})$.

Let k_e and k_d be the targeted 16 bits of K_1 and K_n, respectively. Then, we recover the k_e and k_d in the following way:

1. Prepare $9 \cdot 2^{32} \cdot 2^8$ counters and initialize them by zero.
2. Collect N plaintext-ciphertext pairs.
3. For $K = 0, \ldots, 2^{32-1}$,
   (a) Partially encrypt each plaintext one round by the 16 bits of k_e and decrypt the corresponding ciphertext one round by the 16 bits of k_d where $K = k_d||k_e$.
   (b) Extract three input values $\alpha_5, \alpha_9, \alpha_{13}$ of U and three output values $\beta_5, \beta_6, \beta_7$ of V.
   (c) Obtain nine 8-bit values by pairwise concatenating α_i and β_j for $i = 5, 9, 13$ and $j = 5, 6, 7$.
   (d) Increment nine counters indicated by K and $(\alpha_i||\beta_j)$.
4. Repeat Step 3 for all N text pairs.
5. Compute the l_2 distance using (1) between the probability distribution for each K and uniform distribution.
6. Sort out the candidate keys according to their l_2 distances.
7. Search the right key from the top of the sorted keys.

4.3 Attack Complexity

The amount of data required for χ^2 statistic method is obtained from (2) as follows:

$$N = \left(\sqrt{advantage \cdot 8 \cdot M} + 4\Phi^{-2}(2P_s - 1) \right) / C^{(r)}$$

where P_s is the success probability and $C^{(r)}$ is the capacity. Since the number of linear approximations available for the attack is $9 \times (2^8 - 1)$, the full advantage (32 bits) of the attack with the success probability 0.95 is achieved by the data complexity of $N = \left(\sqrt{32 \cdot 8 \cdot 9 \cdot (2^8 - 1)} + 4\Phi^{-2}(2 \cdot 0.95 - 1) \right) / C^{(r)} \approx 2^{9.6}/C^{(r)}$.

According to the Step 3 and 4 of the attack algorithm, we needs to perform both 1-round encryption and decryption for each plaintext-ciphertext pair and each guessed key. A naive implementation of these steps requires $N \cdot 2^{32}$ operations. We can reduce the computational complexity greatly by removing the repeated computations.

Let x and y be a 16-bit plaintext and a 16-bit ciphertext used for our attack. The Step 3-(a) of the attack algorithm is to compute $z_k = (P(S(x \oplus k_e))||(S^{-1}(P^{-1}(y \oplus k_d)))$ where k_e and k_d denote the guessed 16-bit keys of the first round and the last round, respectively. Thus, the probability distribution Q_k of z_k is obtained by mapping $(x \oplus k_e)||(y \oplus k_d) \mapsto z_k$ for all $k = (k_e||k_d) \in \mathbb{F}_2^{32}$

with N pairs of data. This step can be divided into two sub-steps for efficient computations: First, the table Q^* is obtained by measuring the frequency of $(x\|y) \in \mathbb{F}_2^{32}$. Next, the mapping $(Q^*, k) \mapsto Q_k$ can be done by 2^{32} times access of Q^* for each candidate k. Hence, the Step 3 can be done by $2^{32} \cdot 2^{32}$ operations in total.[2] Since computing the l_2 distance requires $9 \cdot 2^8$ operations for each candidate key, the total time complexity of the attack is $2^{64} + 9 \cdot 2^8 \cdot 2^{32} \approx 2^{64}$. For the memory complexity, the Q^* needs $2^{32} \cdot 4 = 2^{34}$ bytes of memory and some additional memory is required for storing temporary values of computations.

Without increasing the amount of data complexity, we can recover another 32 bits of the round key by changing the input S-boxes of U and the output S-boxes of V over the $U \circ \Omega^{(n)} \circ V$; if the attack uses the linear characteristic starting with S_7, S_{11}, S_{15} and ending with S_{13}, S_{14}, S_{15}, we can recover the $K_1^{[48..63]}$ in the first round and the $K_n^{[12..15]}, K_n^{[28..31]}, K_n^{[44..47]}$ and $K_n^{[60..63]}$ in the last round. In this manner, we can recover $32 \cdot 2$ bits of the round keys in the first and the last round key in total. The remaining $80 - 64 = 16$ bits of key can be obtained by exhaustive key search. Hence, the time complexity of the attack is around $2 \cdot 2^{64} + 2^{16} \approx 2^{65}$ in total.

Attack on 26-round PRESENT. Our attack can be extended to 26-round version of PRESENT with the 24-round characteristic holding with the capacity of $2^{-55.38}$. If the attack uses the full range of text pairs, which is 2^{64}, the theoretical advantage of attack is expected to be 8 by (2). This means that the right key is possibly ranked within the position of $2^{32-8} = 2^{24}$ out of 2^{32} candidates with the probability of 0.95. Hence, we apply the following attack scenario to the 26-round PRESENT: First, the multidimensional linear attack targeting 32 bits of the round key is performed with 2^{64} text pairs. As a result, the 2^{24} candidate keys are obtained. Second, the remained secret key bits $(80 - 32 = 48)$ are combined with the 2^{24} candidate keys from the top in order and the key exhaustive search is performed.

From this scenario, the secret key can be found within the time complexity of $2^{64} + 2^{48} \cdot 2^{24} \approx 2^{72}$. Note that the theoretical estimation is always the lower bound since we use the correlations of only single linear trails. We compare our attacks with previous attacks against various rounds versions of PRESENT in Table 2.

4.4   Discussion

Weakness of bit permutation. Our attack is mainly based on the observation that PRESENT has a large number of linear approximations with the same magnitude of correlations. It seems that this weakness is caused by the lack of diffusion property of the bit permutation. Even though the bit permutation is desirable for efficient hardware implementation, it has a potential weakness that input bits and output bits have one-to-one correspondence. Hence, a single-bit linear approximations of an S-box of any round can be connected to another

---

[2] The computational complexity may be further reduced by applying Fast Fourier Transform at the cost of the increased memory complexity [5].

Table 2. Comparison of data and time complexity of the attacks against PRESENT (CP: Chosen Plaintext, KP: Known Plaintext)

round	data	time	source
16	2^{64}CP	2^{65}	Differential [15]
19	-	2^{113}	Differential + Algebraic [1]
24	2^{57}CP	2^{57}	Saturation [4]
24	$2^{63.5}$KP	-	Linear [13]
25	$2^{62.4}$KP	2^{65}	Linear (this paper)
26	2^{64}KP	2^{72}	Linear (this paper)

single-bit linear approximation of next round through the permutation layer. Since the S-box of PRESENT has multiple linear approximations of which linear masks have a single active bit, one can construct multiple single-bit linear trails over arbitrary number of rounds. Note that this weakness does not appear in the linear transformation functions of Serpent [2] or AES [6] since any single output bit of the linear transformation is expressed as a boolean function of at least two input bits.

A simple remedy to prevent our attack is to revise the S-box in such a way that a single-bit linear approximations of S-box does not have significant correlations. However, we did not investigate how this remedy affects the other aspects of the security of PRESENT.

Correlation and Piling Up Lemma. The designers of PRESENT proved in Theorem 2 of [3] that the maximum correlation of a linear approximation of four rounds of PRESENT is 2^{-6}. As a result, the maximal correlation of a 28-round linear approximation was estimated to be $(2^{-6})^7 = 2^{-42}$ by Piling Up lemma [9]. Thus, the linear attack using the 28-round linear approximation would require more than 2^{84} data. On the other hand, according to our analysis, the capacity of the 28 round PRESENT is estimated to be around $2^{-65.8}$ by the correlation theorem [12] so that 30 round of PRESENT can be (theoretically) attacked by around 2^{75} data.

The difference between the designers' estimation and our result is originated from the fact that the designers considered a single linear approximation using a single linear trail holding with the strongest correlation, whereas our attack takes into account multiple linear approximations, each of which has multiple linear trails holding with strong correlations. Due to the existence of large amount of linear trails in PRESENT, the data complexity of the attack is reduced significantly compared to the estimate by a correlation of a single linear approximation.

The χ^2 and LLR method. Finally, we justify the reason why the LLR method is not used for the attack on PRESENT even if the LLR method showed a better performance than the χ^2 method in the attacks on SERPENT [7]. As described in [7], the LLR method is more advantageous compared to the χ^2 method if the pre-computed profile of the probability distribution is accurate. However, the distribution of linear approximations in PRESENT heavily depends on the key values so that the space for the profile of the probability distribution becomes too large. On the other hand, the χ^2 method does not need to know the distribution accurately; We only need to detect a large deviation of the probability distribution from the uniform distribution. It is an open problem whether there is a way to apply the LLR method efficiently for the attacks against PRESENT.

5   Experiments

We performed the multidimensional linear attacks on the reduced-round PRESENT by experiments in order to verify our theoretical analysis. We chose $r = 6, 7, 8, 9$ rounds version of PRESENT and applied our attack algorithm. We targeted to recover the 16 bits of the last round key by using $(r - 1)$-round linear characteristics. The plaintexts were randomly generated and encrypted by r-round PRESENT. The experiments were repeated with the randomly chosen 200 secret keys and the average values of advantage were calculated. Figure 4 illustrates the relationship between the advantage of the attacks and the required amount of plaintexts for each experiments. The dashed lines represent theoretically estimations drawn by (2) and the solid lines are empirical results. We can see that the estimation of the full advantage of the attack is well matched with empirical results up to 9 rounds PRESENT. Due to the restriction of computational resources, we could not perform the attack algorithm which recovers 32 bits of the round key of the r-round version by the $(r - 2)$-round characteristic. However, based on our experimental results, we can conclude that our estimates of attack complexity against further rounds PRESENT are reasonable.

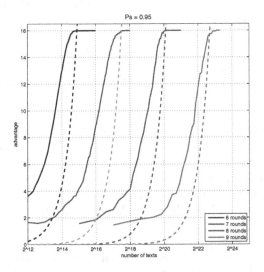

Fig. 4. Empirical evaluation of linear attacks on reduced variants of PRESENT

6   Conclusion

Modern block ciphers often prove the resistance of linear cryptanalysis by counting the minimum number of the active S-boxes involved in the best linear approximation. Even though PRESENT provides a provable security against linear cryptanalysis according to this rule, our attack shows that the resistance of the classical linear cryptanalysis does not always thwart the multidimensional linear attacks. Even though a simple, iterative structure of the cipher is desirable for the hardware-oriented block ciphers, such ciphers may have possibility to retain a large number of linear approximations by which a multidimensional linear attack can be applied efficiently. It is interesting to see that our attack can be applied to some other ciphers that have simple structures, like AES.

Acknowledgment

I wish to thank Kaisa Nyberg for very useful comments that helped to improve the paper. I am grateful to anonymous reviewers of Asiacrypt 2009 and CT-RSA 2010 for their valuable comments.

References

1. Albrecht, M., Cid, C.: Algebraic techniques in differential cryptanalysis. In: Dunkelman, O. (ed.) FSE 2009. LNCS, vol. 5665, pp. 193–208. Springer, Heidelberg (2009)
2. Anderson, R., Biham, E., Knudsen, L.: Serpent: A proposal for the Advanced Encryption Standard. In: First Advanced Encryption Standard (AES) conference (1998)
3. Bogdanov, A., Knudsen, L.R., Leander, G., Paar, C., Poschmann, A., Robshaw, M.J.B., Seurin, Y., Vikkelsoe, C.: PRESENT: An ultra-lightweight block cipher. In: Paillier, P., Verbauwhede, I. (eds.) CHES 2007. LNCS, vol. 4727, pp. 450–466. Springer, Heidelberg (2007)
4. Collard, B., Standaert, F.: A statistical saturation attack against the block cipher PRESENT. In: Fischlin, M. (ed.) Topics in Cryptology – CT-RSA 2009. LNCS, vol. 5473, pp. 195–210. Springer, Heidelberg (2009)
5. Collard, B., Standaert, F., Quisquater, J.: Improving the time complexity of matsui's linear cryptanalysis. In: Nam, K.-H., Rhee, G. (eds.) ICISC 2007. LNCS, vol. 4817, pp. 77–88. Springer, Heidelberg (2007)
6. Daemen, J., Rijmen, V.: The Design of Rijndael- AES, the Advanced Encryption Standard. Springer, Heidelberg (2002)
7. Hermelin, M., Cho, J., Nyberg, K.: Multidimensional linear cryptanalysis of reduced round Serpent. In: Mu, Y., Susilo, W., Seberry, J. (eds.) ACISP 2008. LNCS, vol. 5107, pp. 203–215. Springer, Heidelberg (2008)
8. Hermelin, M., Cho, J., Nyberg, K.: Multidimensional Extension of Matsui's Algorithm 2. In: Dunkelman, O. (ed.) FSE 2009. LNCS, vol. 5665, pp. 209–227. Springer, Heidelberg (2009)
9. Matsui, M.: Linear cryptoanalysis method for DES cipher. In: Helleseth, T. (ed.) EUROCRYPT 1993. LNCS, vol. 765, pp. 386–397. Springer, Heidelberg (1994)

10. National Bureau of Standards, FIPS PUB 46-3: Data Encryption Standard (DES), National Institute for Standards and Technology (January 1977)
11. Nyberg, K.: Linear approximation of block ciphers. In: De Santis, A. (ed.) EUROCRYPT 1994. LNCS, vol. 950, pp. 439–444. Springer, Heidelberg (1995)
12. Nyberg, K.: Correlation theorems in cryptanalysis. Discrete Applied Mathematics 111, 177–188 (2001)
13. Ohkuma, K.: Weak keys of reduced-round PRESENT for linear cryptanalysis. In: Preproceeding of SAC 2009 (2009)
14. Selçuk, A.: On probability of success in linear and differential cryptanalysis. Journal of Cryptology 21(1), 131–147 (2008)
15. Wang, M.: Differential cryptanalysis of reduced-round PRESENT. In: Vaudenay, S. (ed.) AFRICACRYPT 2008. LNCS, vol. 5023, pp. 40–49. Springer, Heidelberg (2008)

Appendix

A   Correlation Table of S-box of PRESENT

Given an input mask α and an output mask β where $\alpha, \beta \in \mathbb{F}_2^4$, the correlation of the linear approximation $\alpha \cdot x \oplus \beta \cdot S(x) = 0$ of the S-box is measured as follows:

$$c(\alpha, \beta) = 2^{-4}(\#(\alpha \cdot x \oplus \beta \cdot S(x) = 0) - \#(\alpha \cdot x \oplus \beta \cdot S(x) = 1))$$

where the $\cdot$ notation stands for the standard inner product. The correlation table of the S-box is given in Table 3.

Table 3. Correlation table of S-box of PRESENT: $c(\alpha, \beta)$

$\alpha\backslash\beta$	1	2	3	4	5	6	7	8	9	a	b	c	d	e	f
1	0	0	0	0	-2^{-1}	0	-2^{-1}	0	0	0	0	0	-2^{-1}	0	2^{-1}
2	0	2^{-2}	2^{-2}	-2^{-2}	-2^{-2}	0	0	2^{-2}	-2^{-2}	0	2^{-1}	0	2^{-1}	-2^{-2}	2^{-2}
3	0	2^{-2}	2^{-2}	2^{-2}	-2^{-2}	-2^{-1}	0	-2^{-2}	2^{-2}	-2^{-1}	0	0	0	-2^{-2}	-2^{-2}
4	0	-2^{-2}	2^{-2}	-2^{-2}	-2^{-2}	0	2^{-1}	-2^{-2}	-2^{-2}	0	-2^{-1}	0	0	-2^{-2}	2^{-2}
5	0	-2^{-2}	2^{-2}	-2^{-2}	2^{-2}	0	0	2^{-2}	2^{-2}	-2^{-1}	0	2^{-1}	0	2^{-2}	2^{-2}
6	0	0	-2^{-1}	0	0	-2^{-1}	0	0	-2^{-1}	0	0	2^{-1}	0	0	0
7	0	0	2^{-1}	2^{-1}	0	0	0	0	-2^{-1}	0	0	0	0	2^{-1}	0
8	0	2^{-2}	-2^{-2}	0	0	-2^{-2}	2^{-2}	-2^{-2}	2^{-2}	0	0	-2^{-2}	2^{-2}	2^{-1}	2^{-1}
9	2^{-1}	-2^{-2}	-2^{-2}	0	0	2^{-2}	-2^{-2}	-2^{-2}	-2^{-2}	-2^{-1}	0	-2^{-2}	2^{-2}	-2^{-2}	2^{-2}
a	0	2^{-1}	0	2^{-2}	2^{-2}	2^{-2}	-2^{-2}	0	0	0	-2^{-1}	2^{-2}	2^{-2}	-2^{-2}	2^{-2}
b	-2^{-1}	0	0	-2^{-2}	-2^{-2}	2^{-2}	-2^{-2}	-2^{-1}	0	0	0	2^{-2}	2^{-2}	2^{-2}	-2^{-2}
c	0	0	0	-2^{-2}	-2^{-2}	-2^{-2}	-2^{-2}	2^{-1}	0	0	-2^{-1}	-2^{-2}	2^{-2}	2^{-2}	-2^{-2}
d	2^{-1}	2^{-1}	0	-2^{-2}	-2^{-2}	2^{-2}	2^{-2}	0	0	0	0	2^{-2}	-2^{-2}	2^{-2}	-2^{-2}
e	0	2^{-2}	2^{-2}	-2^{-1}	2^{-1}	-2^{-2}	-2^{-2}	-2^{-2}	-2^{-2}	0	0	-2^{-2}	-2^{-2}	0	0
f	2^{-1}	-2^{-2}	2^{-2}	0	0	-2^{-2}	-2^{-2}	-2^{-2}	2^{-2}	2^{-1}	0	2^{-2}	2^{-2}	0	0

B   The S-Box and Permutation Tables of PRESENT

The S-box and the permutation tables of PRESENT are given in Table 4 and Table 5, respectively.

Table 4. S-box table of PRESENT in hexadecimal notation

x	0	1	2	3	4	5	6	7	8	9	A	B	C	D	E	F
$S(x)$	C	5	6	B	9	0	A	D	3	E	F	8	4	7	1	2

Table 5. Permutation table of PRESENT

i	0	1	2	3	4	5	6	7	8	9	10	11	12	13	14	15
$P(i)$	0	16	32	48	1	17	33	49	2	18	34	50	3	19	35	51
i	16	17	18	19	20	21	22	23	24	25	26	27	28	29	30	31
$P(i)$	4	20	36	52	5	21	37	53	6	22	38	54	7	23	39	55
i	32	33	34	35	36	37	38	39	40	41	42	43	44	45	46	47
$P(i)$	8	24	40	56	9	25	41	57	10	26	42	58	11	27	43	59
i	48	49	50	51	52	53	54	55	56	57	58	59	60	61	62	63
$P(i)$	12	28	44	60	13	29	45	61	14	30	46	62	15	31	47	63

Dependent Linear Approximations: The Algorithm of Biryukov and Others Revisited

Miia Hermelin[1] and Kaisa Nyberg[1,2]

[1] Aalto University, School of Science and Technology
[2] Nokia, Finland

Abstract. Biryukov, et al., showed how it is possible to extend Matsui's Algorithm 1 to find several bits of information about the secret key of a block cipher. Instead of just one linear approximation, they used several linearly independent approximations that were assumed to be statistically independent. Biryukov, et al., also suggested a heuristic enhancement to their method by adding more linearly and statistically dependent approximations.

We study this enhancement and show that if all linearly dependent approximations with non-negligible correlations are used, the method of Biryukov, et al., is the same as the convolution method presented in this paper. The data complexity of the convolution method can be derived without the assumption of statistical independence. Moreover, we compare the convolution method with the optimal ranking statistic log-likelihood ratio, and show that their data complexities have the same order of magnitude in practice. On the other hand, we show that the time complexity of the convolution method is smaller than for the other two methods.

Keywords: Matsui's Algorithm 1, linear cryptanalysis, multidimensional cryptanalysis, method of Biryukov, convolution method.

1 Introduction

Linear cryptanalysis of block ciphers makes use of probabilistic relations between the plaintext and ciphertext data and the secret key. Such a relation is called a linear approximation of the block cipher. Given a sufficient amount of data derived from the cipher, Matsui's Algorithm 1 [1] can be used in recovering one bit of information about the secret key.

First, Kaliski and Robshaw [2] showed that by using multiple linear approximations, the data complexity can be reduced and later, Biryukov, et al., [3] that multiple bits of information about the secret key can be obtained. However, these methods rely on the assumption that the linear approximations used in the attack are statistically independent. Murphy noted that this is not true in general [4]. Hermelin, et al., investigated this problem in practice using a reduced round Serpent and showed that strong linear approximations are not usually statistically independent [5].

It was observed already in [3] that including more strong linear approximations seemed only to improve the results even if the used approximations were neither linearly nor statistically independent. The practical experiments performed in [5] also showed that when using multiple linear approximations the larger the number of strong approximations was in the method of Biryukov, et al., the closer the observed data complexity became to the data complexities of the methods based on χ^2 and the Kullback-Leibler distance [5].

J. Pieprzyk (Ed.): CT-RSA 2010, LNCS 5985, pp. 318–333, 2010.

These observations suggest that the assumption about statistical independence of the linear approximations could and should be relaxed when applying in practice the method presented Biryukov, et al., which we will call the Biryukov method, for brevity. In this paper we give theoretical justification that this is really the case. For this purpose, we investigate the Biryukov method in the case, where the set of linear approximations is the full linear span of the given set of linear approximations. Completed in this manner the method can be shown to be equivalent to a new method, which we will call the convolution method. The convolution method is interesting, first because it does not rely on the assumption about statistical independence. Secondly, it has the same time complexity as the Biryukov method would have if only the linearly independent approximations are used. Thirdly, the data complexity of the convolution method is at most the same as the data complexity of the Biryukov method.

Previously, the log-likelihood ratio (LLR) was used in [6] for realising another Algorithm 1 type linear attack. In this work we also compare the convolution method and the LLR-method in theory by modelling the problem of finding the correct key information bit as a multiple hypothesis testing problem. While the LLR is the optimal solution with the smallest data complexity, the data complexity of the convolution method is of the same order of magnitude. The key ranking problem in the Algorithm 1 type attacks is also investigated and the existing approaches are compared.

The structure of this paper is as follows: In Sect. 2, some basic notation is given. The linear approximation of a block cipher and the basic Biryukov method is studied in Sect. 3. Section 3.3 studies the completed Biryukov method and presents the convolution method. Statistical analysis of the convolution method is done in Sect. 5. It is shown that the convolution method or the completed Biryukov method do not require the assumption about statistical independence. Section 6 studies the data, time and memory complexities for convolution method, the completed Biryukov method Biryukov and LLR-method.

2 Probability Distributions and Boolean Functions

The space of n-dimensional binary vectors is denoted by $\mathbb{Z}_2^n$. The sum modulo 2 is denoted by $\oplus$. The inner product for $a = (a^1, \ldots, a^n), b = (b^1, \ldots, b^n) \in \mathbb{Z}_2^n$ is defined as $a \cdot b = a^1 b^1 \oplus \cdots \oplus a^n b^n$. Then the vector a is called the (linear) mask of b. The Hamming weight w_H of a binary vector $a \in \mathbb{Z}_2^n$ is $w_H(a) = \#\{i = 1, \ldots, n : a^i = 1\}$, the number of non-zero components in a.

A function $f : \mathbb{Z}_2^n \mapsto \mathbb{Z}_2$ is called a Boolean function. A linear Boolean function is a mapping $x \mapsto u \cdot x$. A function $f : \mathbb{Z}_2^n \mapsto \mathbb{Z}_2^m$ with $f = (f_1, \ldots, f_m)$, where f_i are Boolean functions, is called a vector Boolean function of dimension m. A linear Boolean function from $\mathbb{Z}_2^n$ to $\mathbb{Z}_2^m$ is represented by an $m \times n$ binary matrix U. The m rows of U are denoted by $u_1, \ldots, u_m$, where each u_i is a linear mask.

The correlation between a Boolean function $f : \mathbb{Z}_2^n \mapsto \mathbb{Z}_2$ and zero is

$$c(f) = c(f, 0) = 2^{-n} \left(\#\{x \in \mathbb{Z}_2^n : f(x) = 0\} - \#\{x \in \mathbb{Z}_2^n : f(x) \neq 0\} \right)$$

and it is also called the correlation of f.

We denote random variables $\mathbf{X}, \mathbf{Y}, \ldots$ by capital boldface letters, their domains by $\mathcal{X}, \mathcal{Y}, \ldots$ and their realisations $x \in \mathcal{X}, y \in \mathcal{Y}, \ldots$ by small letters. Let $\mathbf{X}$ be a random variable taking on values in $\mathcal{X} = \{0, 1, \ldots, M\}$. The discrete probability distribution (p.d.) of $\mathbf{X}$ is vector a $p = (p_0, \ldots, p_M)$ if $\Pr(\mathbf{X} = \eta) = p_\eta$, for all $\eta \in \mathcal{X}$. Then we denote $\mathbf{X} \sim p$. We denote the uniform p.d. by θ.

Let $f : \mathbb{Z}_2^n \mapsto \mathbb{Z}_2^m$ and $\mathbf{X} \sim \theta$, where $\mathbf{X}$ takes on values in $\mathbb{Z}_2^n$. If $\mathbf{Y} = f(\mathbf{X})$, then the p.d. of $\mathbf{Y}$ is called the p.d. of f and we say that the random variable $\mathbf{Y}$ is associated with f. Let $f_1, \ldots, f_m : \mathbb{Z}_2^n \mapsto \mathbb{Z}_2^m$ be Boolean functions and for each f_i the associated random variable is $\mathbf{Y}_i$. Then we say that the Boolean functions $f_1, \ldots, f_m$, are statistically independent (s.i.), if the random variables $\mathbf{Y}_1, \ldots \mathbf{Y}_m$, are s.i.

3   Multidimensional Matsui's Algorithm 1

3.1   Linear Approximation of a Block Cipher

Let f be an encryption function of a block cipher with block size n. We denote by x the plaintext, by K the expanded key, that is, a vector consisting of all (fixed) round key bits and by $y = f(x, K)$ the ciphertext. Then an m-dimensional linear approximation of the block cipher is a vector Boolean function

$$\mathbb{Z}_2^n \times \mathbb{Z}_2^n \to \mathbb{Z}_2^m, \ (x, y) \mapsto Ux \oplus Wy \oplus VK, \tag{1}$$

where U and W are $m \times n$ binary matrices and the modulo 2 addition $\oplus$ is calculated component-wise for the vectors. The matrix V has also m rows and it divides the expanded keys, and therefore also the keys, to 2^m equivalence classes $z = VK \in \mathbb{Z}_2^m$. The task is to find the right inner key class, denoted by z_0.

The most complex task in linear cryptanalysis is to determine the p.d. p of the Boolean function (1). A method for determining an approximation p given the biases of $2^m - 1$ one-dimensional linear approximations related to (1) was presented in [5]. We will henceforth assume that a good approximation of the p.d. p of (1) is available.

We make the usual assumption that the plaintexts $x_1, \ldots, x_N$, are the realised values of N independent and identically distributed (i.i.d.) random variables, each following the uniform distribution. Then for all $t = 1, \ldots, N$, the observed values $Ux_t \oplus Wy_t \oplus z, z \in \mathbb{Z}_2^m$, are realisations of i.i.d. random variables following p. Hence, for each $z \in \mathbb{Z}_2^m$, the values $Ux_t \oplus Wy_t, t = 1, \ldots, N$, are the realisations of i.i.d. random variables following p^z, a fixed permutation of p determined by z. Then all the p.d.'s $p^z, z \in \mathbb{Z}_2^m$, are each other's permutations, and in particular,

$$p_{\eta \oplus a}^z = p_\eta^{z \oplus a}, \quad \text{for all} \quad z, \eta, a \in \mathbb{Z}_2^m. \tag{2}$$

The goal of Alg. 1. is to determine z_0 using the empirical data of N plaintext-ciphertext pairs $(x_t, y_t), t = 1, \ldots, N$. For each key $z \in \mathbb{Z}_2^m$ we give a mark defined by $F(z) = T((x_1, y_1), \ldots, (x_N, y_N); z)$, where T is a suitable ranking statistics with data as the variable [7] [8]. The key z is a parameter of T. Given the data, the keys are ordered in increasing or decreasing order according to their marks $F(z)$. The key z' with the highest mark is chosen to be the right key candidate. The error probability $\Pr(z' \neq z_0)$ should decrease if the amount of data N is increased. The best statistics gives the smallest error for a given N. The ranking statistic proposed by Biryukov, et al., is described in the next section.

3.2   Method of Biryukov, et al.

The basic version of the Biryukov method uses m linearly independent approximations $u_i \cdot x \oplus w_i \cdot y \oplus v_i \cdot K, i = 1, \ldots, m$, where the ith approximation has a non-negligible correlation c_i. Biryukov, et al., assumed that the approximations are s.i., that is, if $\mathbf{X}_i$ is a binary random variable associated with the ith approximation $u_i \cdot x \oplus w_i \cdot y \oplus z$, then the random variables $\mathbf{X}_1, \ldots, \mathbf{X}_m$, are s.i.

For each $i = 1, \ldots, m$, let ρ_i denote the empirical correlation of the ith approximation calculated using the data (x_t, y_t), $t = 1, \ldots, N$ as follows:

$$\rho_i = 2N^{-1}\{t = 1, \ldots, N : u_i \cdot x_t \oplus w_i \cdot y_t = 0\} - 1.$$

Denote $z = (z^1, \ldots, z^m)$ such that z^i is the ith bit of the key z. Denote the theoretical and empirical correlation vectors by $\mathbf{c}_z = ((-1)^{z^1} c_1, \ldots, (-1)^{z^m} c_m)$ and $\boldsymbol{\rho} = (\rho_1, \ldots, \rho_m)$, respectively. The mark for each $z \in \mathbb{Z}_2^m$ is given by the ℓ_2 distance between the two correlation vectors:

$$b(z) = ||\mathbf{c}_z - \boldsymbol{\rho}||_2^2.$$

The key z' minimising $b(z)$ is chosen to be the right key.

Later Murphy noted that the assumption about statistical independence of the linear approximations does not hold in general [4]. In particular, linearly dependent approximations are also statistically dependent. Murphy also suggested to use the traditional measure of covariance of two linear approximations in verifying the assumption about linear independence. This method has been subsequently used by other researchers, for example in [9]. The most natural way is to use the converse of the Piling Up lemma [1], which we give in the Appendix 7.

Biryukov, et al., proposed a heuristic enhancement to their method [3]. They added approximations that were linearly dependent of the m original approximations. Ultimately, they could use all $2^m - 1$ one-dimensional approximations in the span of the original approximations. We call this method the full Biryukov method and we will study it in the next section.

3.3   The Full Biryukov Method

In this method, the empirical correlation $\rho(a)$ for each $a \in \mathbb{Z}_2^m$ is calculated using the data (x_t, y_t), $t = 1, \ldots N$ as follows:

$$\rho(a) = 2N^{-1}\{t = 1, \ldots, N : Ux_t \oplus Wy_t \oplus = 0\} - 1$$

The ηth component of the theoretical correlation vector $\mathbf{c}_z$ is now $(-1)^{\eta \cdot z} c(\eta)$ and the vector of empirical correlations is $\boldsymbol{\rho} = (\rho(0), \ldots, \rho(2^m - 1))$. Similarly to the basic version, the mark is given by

$$B(z) = ||\mathbf{c}^z - \boldsymbol{\rho}||_2^2 = \sum_{a \in \mathbb{Z}_2^m} ((-1)^{a \cdot z} c(a) - \rho(a))^2$$

and the key z' that minimises $B(z)$ is chosen to be the right key.

Next we analyse this full method. Our analysis is based on the observation that there exists another statistic which is equivalent to the $B(z)$ statistic, in the sense that both will produce exactly the same key ranking. Moreover, this equivalent statistic gives a more efficient way of ranking the candidate keys, and in particular, to determine the most likely key candidate.

4   Convolution Method

We now show how to make the full Biryukov method more efficient in practice. We obtain the empirical distribution $q = (q_0, \ldots, q_{2^m-1})$ of the multidimensional approximation $Ux \oplus Wy \oplus VK$ by computing

$$q_\eta = N^{-1}\#\{t = 1, \ldots, N : Ux_t \oplus Wy_t = \eta\}, \quad \text{for all} \quad \eta \in \mathbb{Z}_2^m. \tag{3}$$

The mark $B(z)$ of the full Biryukov method can also be written as

$$B(z) = -2 \sum_{a \in \mathbb{Z}_2^m} (-1)^{a \cdot z} c(a)\rho(a) + \sum_{a \in \mathbb{Z}_2^m} (\rho(a)^2 + c(a)^2),$$

where the latter sum does not depend on z. On the other hand, by equation (3) in [10], we have

$$c(a) = \sum_{\eta \in \mathbb{Z}_2^m} (-1)^{a \cdot \eta} p_\eta \quad \text{and} \quad \rho(a) = \sum_{\eta \in \mathbb{Z}_2^m} (-1)^{a \cdot \eta} q_\eta.$$

Using the previous formulas for correlations we have

$$\sum_{a \in \mathbb{Z}_2^m} (-1)^{a \cdot z} c(a)\rho(a) = 2^m \sum_{\eta \in \mathbb{Z}_2^m} q_\eta p_{\eta \oplus z}. \tag{4}$$

But the sum is just the zth component of the convolution $q * p$ of the p.d.'s p and q. Hence, finding the minimum of $B(z)$ is equivalent to finding the maximum of the zth component of the convolution of q and p, that is, z is the mode of the p.d. $q * p$. We now propose the following mark

$$G(z) = (p * q)_z, \tag{5}$$

and the key z' that maximises $G(z)$ is chosen to be the right key. We call this new method based on $G(z)$ the convolution method. We have the following result.

Theorem 1. *The key z' minimises $B(z)$ if and only if it maximises $G(z)$. Hence, the full Biryukov method and the convolution method are equivalent.*

Both methods are also equivalent to the maximum likelihood decoding. The problem is to decode the code where the channel has error probability distribution p and the original message is $z \in \mathbb{Z}_2^m$. The message is sent N times over the channel with noise $Ux_t \oplus Wy_t \sim p^z$, at each time $t = 1, \ldots, N$. The receiver obtains sequence $z \oplus Ux_t \oplus Wy_t, t = 1, \ldots, N$, with observed empirical p.d. q that should approximate p. Then $q * p^z$ gives an empirical p.d. for $z = (Ux_t \oplus Wy_t) \oplus (Ux_t \oplus Wy_t \oplus z)$ and the key candidate z is given as the mode of the p.d. $q * p^z$.

While the two methods have the same data complexities, the convolution method has smaller time complexity. The basic and full Biryukov methods have time complexities $m2^m$ and 2^{2m}, respectively. This is because we have to compute the rank $b(z)$ or $B(z)$, respectively, for each $z \in \mathbb{Z}_2^m$. In the convolution method we do not have to consider each key or p.d. p^z separately. It suffices to compute only one convolution $p * q$ and determine its mode. The convolution is computed using FFT with time complexity $m2^m$. Hence, with the same data the convolution method outputs the same key class as the full Biryukov method, but the time complexity for the convolution method is the same as for the basic Biryukov method. In [6] Hermelin, et al., studied the optimal method based on the LLR-statistic. We prove in the next section that the data complexities of the convolution method and the LLR-method are approximately equal.

More accurate descriptions for the algorithms for the different methods are given in Section 6.2. In the next section, we study the statistical properties of the convolution method.

5   Statistical Analysis

Finding the right key z_0 is actually a multiple hypothesis testing problem. Section 5.2 studies the problem and how to solve it. The next section gives some necessary theory about discrete random variables and multinomial probability distributions needed in multiple hypothesis testing problems.

5.1   Multinomial Distribution

Let $\mathbf{X}_1, \ldots, \mathbf{X}_N$, be i.i.d. random variables drawn from space $\mathcal{X} = \{0, 1, \ldots, M\}$ by a discrete p.d. $s = (s_0, \ldots, s_M)$, where M is some positive integer. Let $\mathbf{Q} = (\mathbf{Q}_0, \ldots, \mathbf{Q}_M)$ be a vector of random variables where for each $\eta \in \mathcal{X}$,

$$\mathbf{Q}_\eta = N^{-1} \# \{i = 1, \ldots, N : X_i = \eta\}. \tag{6}$$

Hence, $\mathbf{Q}$ is a vector of relative frequencies of the elements of the sample space $\mathcal{X}$. The sample space $\mathcal{Q}$ of $\mathbf{Q}$ consists of vectors $q = (q_0, \ldots, q_M)$, where $q_0, \ldots, q_M \in N^{-1}\{0, 1, \ldots, N\}$ and $q_0 + \cdots + q_M = 1$. The random vector $\mathbf{Q}$ follows the multinomial distribution $\mathrm{Multi}(N, s)$, with probabilities

$$\Pr(\mathbf{Q} = q) = \frac{N!}{\prod_{\eta=0}^{M}(q_\eta N)!} \prod_{\eta=0}^{M} s_\eta^{Nq_\eta}, \quad \text{for all} \quad q \in \mathcal{Q}. \tag{7}$$

Since for each $z \in \mathbb{Z}_2^m$, the observed values $Ux_t \oplus Wy_t$, $t = 1, \ldots, N$ are realisations of i.i.d. random variables following p^z, the empirical p.d. q calculated using (3) is a realisation of a random vector $\mathbf{Q}$ that has multinomial distribution $\mathrm{Multi}(N, p^z)$. Using (2), we have for all $z \in \mathbb{Z}_2^m$,

$$(p * \mathbf{Q})_z = \sum_{\eta \in \mathbb{Z}_2^m} p_{\eta \oplus z} \mathbf{Q}_\eta = \sum_{\eta \in \mathbb{Z}_2^m} p_\eta^z \mathbf{Q}_\eta. \tag{8}$$

Hence, maximising $G(z)$ in (5) is equivalent to finding $z' \in \mathbb{Z}_2^m$ that maximises

$$\sum_{\eta \in \mathbb{Z}_2^m} p_\eta^z q_\eta. \tag{9}$$

By (8) the convolution method has the same statistical behaviour as the method using (9). The next lemma gives the distribution of (8).

Lemma 1. *Let $\lambda_0, \ldots, \lambda_M$ be any real numbers and $\mathbf{Q} = (\mathbf{Q}_0, \ldots, \mathbf{Q}_M)$ be a multinomially distributed random vector with distribution $\mathrm{Multi}(N, s)$. Then the linear combination $N \sum_{\eta=0}^M \lambda_\eta \mathbf{Q}_\eta$ is asymptotically normal with mean and variance given by*

$$\mu = N \sum_{\eta=0}^M \lambda_\eta s_\eta \qquad \sigma^2 = N \sum_{\eta=0}^M \lambda_\eta^2 s_\eta - \mu^2.$$

The proof is given in Appendix 7. Since the lemma does not require the assumption about statistical independence, the assumption is also not needed when using full Biryukov or convolution method.

The concept of capacity was introduced in [5] and it was used in simplifying the formulas of the data complexities:

Definition 1. *The capacity between two p.d.'s $p = (p_0, \ldots, p_M)$ and $q = (q_0, \ldots, q_M)$ is defined by*

$$C(p, q) = \sum_{\eta=0}^M (p_\eta - q_\eta)^2 q_\eta^{-1}.$$

If q is the uniform distribution, we denote $C(p, q) = C(p)$.

5.2   Multiple Hypothesis Testing Problems

Let $\mathbf{X}_1, \ldots, \mathbf{X}_N$, be a sequence of i.i.d. random variables drawn from sample space $\mathcal{X} = \{0, 1, \ldots, M\}$, where M is a positive integer, and let $x_1, \ldots, x_N$, be the corresponding realisations. Assume $d \geq 2$ simple hypotheses, where each hypothesis H_i states that the sample is drawn according to a p.d. $p^i = (p_0^i, \ldots, p_M^i)$, $i = 1, \ldots, d$, and $p^i \neq p^j$, if $i \neq j$. Equivalently, each hypothesis H_i states that the vector $\mathbf{Q}$ defined by (6) is multinomial distributed as $\mathrm{Multi}(N, p^i)$.

The simple d-ary hypothesis testing problem is to determine which hypothesis is correct. Hence, one hypothesis is accepted and the others are rejected. In Bayesian statistics, each hypothesis is given an *a priori* probability $\Pr(H_i)$ for all $i = 1, \ldots, d$. We assume that the *a priori* probabilities are equal.

Let $q = (q_0, \ldots, q_M)$ be the empirical p.d. calculated from the observed values $x_1, \ldots, x_N$, by

$$q_\eta = N^{-1} \#\{t = 1, \ldots, N : x_t = \eta\}, \quad \text{for all} \quad \eta \in \mathcal{X}.$$

A distinguisher is a rule that based on the observed data $x_1, \ldots, x_N$, or, equivalently, q, outputs which hypotheses is accepted:

$$\delta(x_1, \ldots, x_N) = \delta(q) = i, \quad \text{if} \quad H_i \quad \text{is accepted, for} \quad i = 1, \ldots, d$$

The distinguisher is defined using a suitable test statistic $T(q; p^i)$, where p^i (or i) is considered as the parameter and q is the variable.

Let $f(i) = T(q; p^i)$ be a function of the parameter i for given empirical data q. The distinguisher outputs j if it gives the maximum (or minimum) of $f(i)$, for given q. The statistic T should be easy to compute in practice and accurate such that the total error

$$P_e = \sum_{i=1}^{d} \Pr(H_i) \Pr(\delta(\mathbf{Q}) \neq i \mid H_i) \tag{10}$$

is as small as possible. An optimal distinguisher minimising the error probability exists for simple hypotheses testing problems.

Consider first the simple binary hypothesis testing problem with $d = 2$. By Neyman-Pearson lemma in classical statistics and Chernoff's theorem in Bayesian statistics [11], the optimal distinguisher for distinguishing between H_1 and H_2, or p^1 and $p^2 \neq p^1$, equivalently, is given by the log-likelihood ratio (LLR) test statistic

$$\text{LLR}(q; p^1, p^2) = \sum_{\eta \in \mathcal{X}} N q_\eta \log \frac{p_\eta^1}{p_\eta^2}.$$

The distinguisher accepts H_1, that is, outputs p^1 (or accepts H_2 and outputs p^2, respectively) if $\text{LLR}(q; p^1, p^2) \geq \tau$ ($< \tau$) where τ is the threshold that depends on P_e. Obviously, using LLR is the same as finding for given q the maximum of the function

$$l(i) = \sum_{\eta \in \mathcal{X}} q_\eta \log p_\eta^i, \ i = 1, 2.$$

If $p^1, p^2 \neq \theta$ this is equivalent to finding the maximum of

$$L(i) = l(i) + \log(M + 1) = \text{LLR}(q, p^i, \theta), \ i = 1, 2.$$

In Bayesian theory Chernoff's theorem [11] states that $P_e = \mathcal{O}\left(2^{-ND^*(p^1, p^2)}\right)$, where $D^*(p^1, p^2)$ is the Chernoff information between p^1 and p^2 given by

$$D^*(p^1, p^2) = -\min_{0 \leq \lambda \leq 1} \log \left(\sum_{\eta=0}^{M} (p_\eta^1)^\lambda (p_\eta^2)^{1-\lambda} \right). \tag{11}$$

Assume now a d-ary hypothesis testing problem with $d \geq 3$ simple hypotheses. Moreover, assume that $p^i \neq \theta$ for all $i = 1, \ldots, d$. The optimal distinguisher that minimises P_e chooses the hypothesis with the largest conditional probability $\Pr(H_i \mid \mathbf{Q} = q)$, see [12]. Equivalently, by Bayes' theorem, the distinguisher chooses the hypothesis that maximises $\Pr(\mathbf{Q} = q \mid H_i)$.

Consider the likelihood function $\mathcal{L}(p^i) = \Pr(\mathbf{Q} = q \mid H_i)$ that should reach its maximum for the right p.d. p^i, given data q. Using the formula (7) of the p.d. of the multinomial distribution the likelihood function can be written as

$$\mathcal{L}(p^i) = \frac{N!}{\prod_{\eta=0}^{M}(q_\eta N)!} \prod_{\eta=0}^{M} (p_\eta^i)^{N q_\eta}.$$

Taking logarithm and omitting the terms not depending on p^i gives an equivalent test statistics

$$L(i) = \sum_{\eta \in \mathcal{X}} q_\eta \log p^i_\eta + \log(M+1) = N^{-1} \operatorname{LLR}(q, p^i, \theta). \qquad (12)$$

Hence, LLR-statistics gives the optimal distinguisher for a multiple hypothesis testing problem for $d \geq 3$, also. The LLR measures whether the data is drawn from p^i or the uniform distribution. High values imply that the data q is closer to p^i than θ. Hence, we have a theoretical justification for the heuristic LLR-method presented in [6].

Both convolution method and the LLR have the form of a general linear method [7] using the statistic

$$T(\mathbf{Q}; z) = N \sum_{\eta \in \mathcal{X}} \lambda^z_\eta \mathbf{Q}_\eta,$$

where the coefficients $\lambda^z_0, \ldots, \lambda_M$, depend on the parameter z. Comparing the coefficients in the formulas (9) and (12) shows that the LLR-method and convolution method are not equivalent. Hence, the convolution method is not optimal in theory.

Consider the definition (10) of the error probability when distinguishing $d \geq 3$ hypothesis. Each term $\Pr(\delta(\mathbf{Q}) \neq i \mid H_i)$ in the sum is equal to

$$\Pr(\delta(\mathbf{Q}) \neq i \mid H_i) = \sum_{j \neq i, j=1,\ldots,d} \Pr(\delta(\mathbf{Q}) = j \mid H_i).$$

But each probability $\Pr(\delta(\mathbf{Q}) = j \mid H_i)$ corresponds to the binary hypothesis testing problem of distinguishing parameter i from $j \neq i$. Hence, if for given P_e two distinguishers have same data complexity for the binary hypothesis testing problem, then they are also equally efficient in the multiple hypothesis testing setting.

It remains to show that for a given error probability, if the p.d. p is nearly uniform (but not uniform), then the data complexity of the convolution method is of the same order of magnitude as the data complexity of the LLR-method. We study the complexities in the next section.

6 Complexity Analysis

6.1 Data Complexity

To compare the LLR and convolution methods, we have to calculate the data complexity N for given error probability P_e. We know by Sect. 5.2 that the LLR-method is optimal, i.e., for given P_e it has the smallest data complexity. However, based on the tests made in [5] and [13], we suspect that the data complexities of the convolution method and the LLR-method are practically the same as long as the p.d.'s do not variate much from the uniform distribution. More accurately, we assume that there exists ϵ, $0 < \epsilon < 0.5$ such that each p.d. p^z, $z \in \mathbb{Z}_2^m$, satisfies the following conditions:

$$|p^z_\eta - 2^{-m}| \leq \epsilon 2^{-m} \quad \text{for all} \quad z, \eta \in \mathbb{Z}_2^m \quad \text{and}$$
$$|p^{z_1}_\eta - p^{z_2}_\eta| \leq \epsilon p^{z_2}_\eta \quad \text{for all} \quad z_1 \neq z_2 \quad \text{and} \quad z_1, z_2, \eta \in \mathbb{Z}_2^m. \qquad (13)$$

Then for all $z, z_1, z_2 \in \mathbb{Z}_2^m$ the capacities $C(p^z) = C(p) = \epsilon^2 < 1$ and $C(p^{z_1}, p^{z_2}) = \epsilon^2 < 1$, if $z_1 \neq z_2$. The condition (13) holds for all practical ciphers. For example the experiments with reduced round Serpent in [8] showed that for $m \leq 12$, the condition held with the parameter value $\epsilon \approx 1/150$. In general, the value ϵ should be so small that it is possible to approximate the Chernoff information $D^*(p^{z_1}, p^{z_1})$ between two distinct distributions p^{z_1} and p^{z_2} using their capacity: $D^*(p^{z_1}, p^{z_1}) \approx (8 \ln 2)^{-1} C(p^{z_1}, p^{z_2})$, see Theorem 7 in [14].

As noted in the previous section, we only have to consider the distinguishing between two keys z_1 and $z_2 \neq z_1$. Denote for simplicity $p = p^{z_1}$ and $s = p^{z_2}$. If the p.d.'s satisfy condition (13), then by definition (11), the data complexity of the LLR-method is proportional to

$$N = C(p, s)^{-1}. \tag{14}$$

See also [15] for another proof. We now show that (14) holds also for the convolution method, provided that the distributions p and s satisfy condition (13).

The cumulative distribution function of the normed, normal distribution is

$$\Phi(x) = \int_{-\infty}^{x} \frac{1}{\sqrt{2\pi}} e^{-t^2/2} \, dt \, .$$

By Lemma 1 we obtain that the probability of choosing $z_2 \neq z_1$ when H_{z_1} is true is

$$\Pr(\delta(\mathbf{Q}) = y \mid H_{z_1}) = \Pr(T(\mathbf{Q}; y) > T(\mathbf{Q}; z) \mid H_{z_1}) = \Phi\left(\sqrt{N}\frac{\mu}{\sigma}\right),$$

where the expected value μ and variance σ^2 are given by

$$\mu = \sum_{\eta \in \mathbb{Z}_2^m} (p_\eta - s_\eta)p_\eta \qquad \sigma^2 = \sum_{\eta \in \mathbb{Z}_2^m} (p_\eta - s_\eta)^2 p_\eta - \mu^2.$$

The mean μ can be approximated by

$$\mu \approx 2^{-m} \sum_{\eta \in \mathbb{Z}_2^m} (p_\eta - s_\eta)\frac{p_\eta}{s_\eta} = 2^{-m} \sum_{\eta \in \mathbb{Z}_2^m} \left((p_\eta - s_\eta)\frac{p_\eta}{s_\eta} - (p_\eta - s_\eta)\right) = 2^{-m} C(p, s).$$

Moreover,

$$\sum_{\eta \in \mathbb{Z}_2^m} (p_\eta - s_\eta)^2 p_\eta = \sum_{\eta \in \mathbb{Z}_2^m} \frac{(p_\eta - s_\eta)^2}{s_\eta} p_\eta s_\eta \approx 2^{-2m} C(p, s). \tag{15}$$

As $C(p, s) < 1$, the dominating term of σ^2 is given by (15). Hence, $\sigma^2 \approx 2^{-2m} C(p, s)$ and the data complexity is proportional to

$$N = \frac{2^{-2m} C(p, s)}{2^{-2m} C(p, s)^2} = C(p, s)^{-1}.$$

As the number of hypotheses grows, the data complexity N is increased in both cases [5]. For $d = 2^m$ it is proportional to $m/C_{\min}(p)$, where $C_{\min}(p) = \min_{z_1 \neq z_2} C(p^{z_1}, p^{z_2})$.

In [3] efficiency of key ranking was also discussed and the measure *gain* to quantify success in key ranking as a function of data complexity was introduced. Later, in [6] it was proposed to use the measure *advantage*. While Biryukov, et al., need the assumption about statistical independence of the linear approximations in all their theoretical derivations, Hermelin, et al., can do without it, but instead, must make another unrealistic assumption that the ranking statistics for each key candidate are statistically independent. This assumption can be fulfilled if, for each key candidate value, new fresh data is generated to compute the ranking statistic, which will result in overestimating the data complexity. Hence, it is not known exactly in the general case, what the success probabilities of key ranking are for Algorithm 1. Nevertheless, the above analysis applies to key ranking also, and we can conclude that the LLR method and the convolution method have practically the same advantage.

6.2   Time and Memory Complexities

In [8] the Alg. 2 was divided to two phases: the on-line phase and the off-line phase. We follow the division in this paper. The on-line phase is independent of the statistics used in the attack and its sole purpose is to obtain the empirical p.d. q from the N plaintext-ciphertext pairs. The time complexity is Nm and memory complexity is 2^m. We now assume that given data N, we have obtained the empirical p.d. q.

Figures 1, 2 and 3 depict the off-line phase for the full Biryukov method, LLR-method and convolution method, respectively.

Input: empirical correlation vector $\rho = (\rho(0), \ldots, \rho(2^m - 1))$ and theoretical
       correlations $c(0), \ldots, c(2^m - 1)$, of the linear approximation (1) ;
Output: the best key candidate;
for $z = 0, \ldots, 2^m - 1$ **do**
    compute $B(z) = \sum_{a \in \mathbb{Z}_2^m} ((-1)^{a \cdot z} c(a) - \rho(a))^2$;
end
find z' that maximises $B(z)$;
output z';

Fig. 1. Off-line phase of Alg. 1 using full Biryukov method

Input: empirical p.d. q and theoretical p.d. p of the linear approximation (1) ;
Output: the best key candidate;
for $z = 0, \ldots, 2^m - 1$ **do**
    compute p^z, a permutation of p;
    compute $L(z) = \mathrm{LLR}(q, p^z, \theta)$;
end
find z' that maximises $L(z)$;
output z';

Fig. 2. Off-line phase of Alg. 1 using LLR-method

Input: empirical p.d. q and theoretical p.d. p of the linear approximation (1) ;
Output: the best key candidate;
compute $p * q$ using FFT;
find mode z' of $p * q$;
output z';

Fig. 3. Off-line phase of Alg. 1 using convolution method

For each $z \in \mathbb{Z}_2^m$, both full Biryukov method and LLR-method take time 2^m to evaluate. Hence, the time complexity of both the full Biryukov and the LLR-method is 2^{2m}.

In the convolution method the computation of the convolution $p*q$ is done only once. Using FFT, that is, left hand side of (4), it takes time $m2^m$. Hence, the convolution method is much faster than the LLR or the full Biryukov, while all three methods have the same data complexities.

If all the correlations $c(a)$, $a \in \mathbb{Z}_2^m$, are non-negligible, then all three methods have the same memory complexity 2^m. In practice the full linear span of the linear approximations contains many approximations with zero or negligible correlations. Such approximations do not contribute to the capacity and hence are discarded. This has certain effect to the complexities of the algorithms.

Let l be the number of linear approximations used, $m \leq l \leq 2^m$. Then the memory requirement of the off-line phase of the Biryukov method will be reduced from 2^m to l and the time complexity becomes $2^m l$. Since the convolution is computed using the correlations by (4), the same reduction of memory is possible also for the convolution method if we use the correlations instead of the distribution in evaluating the statistic.

We run some experiments on the four-round Serpent, see [16] for an accurate description for the cipher. The test settings were the same as in [6]. To compare the LLR and convolution method in practice, we measured the advantage by Selçuk [17]. In Figure 4 we have plotted the empirical advantage as a function of the data complexity, for $m = 7$. The curves are indistinguishable.

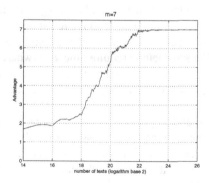

Fig. 4. The empirical advantage as a function of data complexity using LLR and convolution method with $m = 7$ for 4-round Serpent. The curves are equal.

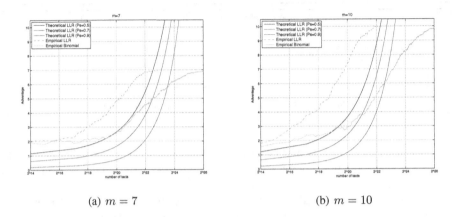

(a) $m = 7$ (b) $m = 10$

Fig. 5. The theoretical and empirical advantage as a function of data complexity using LLR-method for the 4-round Serpent

Figure 5 shows the empirical and theoretical advantage of the LLR-method for $m = 7$ and $m = 10$. The convolution method gives exactly the same results. The theoretical prediction is slightly more pessimistic than the empirical results. However, they are still consistent.

7   Conclusions

We proposed a new method, which we call the convolution method, to perform multi-dimensional linear attacks. The convolution method is expected to give the same result with the same data complexity as the Biryukov method in case the set of linear approximations is completed to contain all approximations with significant correlations within the linear span of the set.

In the convolution method we form the convolution between the empirical and the theoretical p.d related to the multidimensional linear approximation of a block cipher. The right key class is determined as the mode of the resulting p.d. The data complexities of both LLR and the convolution method are of the same magnitude. Moreover, the LLR-method and full Biryukov method require time 2^{2m}, where m is the dimension of the approximation, whereas the convolution method only needs time $m2^m$. Hence, the convolution method is the most efficient in practice. Also, there is no need to assume statistical independence.

In [3] the measure gain and in [6] the measure advantage was used in studying the success of key ranking. The gain requires the assumption of statistical independence of base approximations whereas the advantage requires that the ranking statistics corresponding to different keys should be statistically independent. The latter condition can be satisfied for Alg. 2 [8] but seems to result in an unrealistic and unnecessary increase of the data complexity for Alg. 1. However, the efficiency of the convolution or LLR-methods is not affected by the assumption that is needed in calculating the advantage. If needed, the advantage can be determined approximately also for the Alg.1. The calculations in that case are the same as in [6] and the convolution method remains the most efficient method in practice.

Acknowledgements

We thank Joo Yeon Cho for producing the experiments.

References

1. Matsui, M.: Linear Cryptanalysis Method for DES Cipher. In: Helleseth, T. (ed.) EURO-CRYPT 1993. LNCS, vol. 765, pp. 386–397. Springer, Heidelberg (1994)
2. Burton, S., Kaliski, J., Robshaw, M.J.B.: Linear Cryptanalysis Using Multiple Approximations. In: Desmedt, Y.G. (ed.) CRYPTO 1994. LNCS, vol. 839, pp. 26–39. Springer, Heidelberg (1994)
3. Biryukov, A., Cannière, C.D., Quisquater, M.: On Multiple Linear Approximations. In: Franklin, M. (ed.) CRYPTO 2004. LNCS, vol. 3152, pp. 1–22. Springer, Heidelberg (2004)
4. Murphy, S.: The Independence of Linear Approximations in Symmetric Cryptology. IEEE Transactions on Information Theory 52(12), 5510–5518 (2006)
5. Hermelin, M., Nyberg, K., Cho, J.Y.: Multidimensional Linear Cryptanalysis of Reduced Round Serpent. In: Mu, Y., Susilo, W., Seberry, J. (eds.) ACISP 2008. LNCS, vol. 5107, pp. 203–215. Springer, Heidelberg (2008)
6. Hermelin, M., Cho, J.Y., Nyberg, K.: Statistical Tests for Key Recovery Using Multidimensional Extension of Matsui's Algorithm 1. In: Joux, A. (ed.) EUROCRYPT 2009 - POSTER SESSION. LNCS, vol. 5479. Springer, Heidelberg (2009)
7. Vaudenay, S.: An experiment on DES statistical cryptanalysis. In: CCS 1996: Proceedings of the 3rd ACM conference on Computer and communications security, pp. 139–147. ACM, New York (1996)
8. Hermelin, M., Cho, J.Y., Nyberg, K.: Multidimensional Extension of Matsui's Algorithm 2. In: Dunkelman, O. (ed.) Fast Software Encryption. LNCS, vol. 5665, pp. 209–227. Springer, Heidelberg (2009)
9. Gérard, B., Tillich, J.: On linear cryptanalysis with many linear approximations (2009)
10. Hermelin, M., Nyberg, K.: Multidimensional Linear Distinguishing Attacks and Boolean Functions. In: Fourth International Workshop on Boolean Functions: Cryptography and Applications (2008)
11. Cover, T.M., Thomas, J.A.: 11. Wiley Series in Telecommunications and Signal Processing. In: Elements of Information Theory, 2nd edn. Wiley Interscience, Hoboken (2006)
12. McDonough, R.N., Whalen, A.D.: 5. In: Detection of Signals in Noise, 2nd edn. Academic Press, London (1995)
13. Collard, B., Standaert, F.X., Quisquater, J.J.: Experiments on the Multiple Linear Cryptanalysis of Reduced Round Serpent. In: Nyberg, K. (ed.) FSE 2008. LNCS, vol. 5086, pp. 382–397. Springer, Heidelberg (2008)
14. Baignères, T., Vaudenay, S.: The Complexity of Distinguishing Distributions (Invited Talk). In: Safavi-Naini, R. (ed.) ICITS 2008. LNCS, vol. 5155, pp. 210–222. Springer, Heidelberg (2008)
15. Baignères, T., Junod, P., Vaudenay, S.: How Far Can We Go Beyond Linear Cryptanalysis? In: Lee, P.J. (ed.) ASIACRYPT 2004. LNCS, vol. 3329, pp. 432–450. Springer, Heidelberg (2004)

16. Biham, E., Anderson, R., Knudsen, L.: Serpent: A New Block Cipher Proposal. In: Vaudenay, S. (ed.) FSE 1998. LNCS, vol. 1372, pp. 222–238. Springer, Heidelberg (1998)
17. Selçuk, A.A.: On probability of success in linear and differential cryptanalysis. Journal of Cryptology 21(1), 131–147 (2008)
18. Xiao, G.Z., Massey, J.L.: A Spectral Characterization of Correlation-Immune Combining Functions. IEEE Transactions on Information Theory 34(3), 569–571 (1988)
19. Rohatgi, V.K.: 6.7. Wiley Series in Probability and Mathematical Statistics. In: Statistical Inference, 1st edn. John Wiley & Sons, New York (1984)

Appendix

A   Proof of Theorem 2

The Piling Up lemma [1] that has traditionally been used in calculating correlations of linear combinations of statistically independent linear approximations has a converse. This converse of the Piling Up lemma offers a natural criterion for verifying statistical independence of linear approximations. Given a set of linear approximations it is not sufficient to verify that all linear approximations in the set are pairwise statistically independent. We must also verify that the correlations (or imbalances [3] [4])

$$c(a) = c(a \cdot (Ux \oplus Wy \oplus VK)), \ a \in \mathbb{Z}_2^m.$$

of all linear combinations of the linear approximations must be of certain small magnitude as given by the following theorem.

Theorem 2. *Let* $m \geq 2$ *be an integer. The binary random variables* $\mathbf{X}_1, \mathbf{X}_2,$ *$\ldots, \mathbf{X}_m$, with correlations* $c_i = c(\mathbf{X}_i)$, $i = 1, \ldots, m$ *are statistically independent, if and only if for all index sets* $I \subset \{1, 2, \ldots, m\}$,

$$c\left(\bigoplus_{i \in I} \mathbf{X}_i\right) = \prod_{i \in I} c_i. \tag{16}$$

The *only if* part follows from the Piling Up lemma. The proof of the *if* part, that is, the converse of the Piling Up lemma, is given below, using the Xiao-Massey lemma [18]:

Lemma 2 (Xiao-Massey lemma). *The discrete random variable* $\mathbf{Z}$ *is independent of the* m *independent binary random variables* $\mathbf{X}_1, \ldots, \mathbf{X}_m$ *if and only if* $\mathbf{Z}$ *is independent of the sum* $b_1\mathbf{X}_1 \oplus \cdots \oplus b_m\mathbf{X}_m$, *for every choice of* $b_1, \ldots, b_m \in \{0, 1\}$, *and not all coefficient* b_i *is zero.*

Proof (Converse of the Piling Up lemma). We assume that the random variables $\mathbf{X}_1, \ldots, \mathbf{X}_m$ satisfy condition (16). We do the proof with induction on m. Let $m = 2$. We assume $c(\mathbf{X}_1 \oplus \mathbf{X}_2) = c_1 c_2$ and we have to prove that for all pairs $t = (t_1, t_2) \in \{0, 1\} \times \{0, 1\}$, the probability $\Pr(\mathbf{X}_1 = t_1, \mathbf{X}_2 = t_2) = \Pr(\mathbf{X}_1 = t_1) \Pr(\mathbf{X}_2 = t_2)$.

Denote $\mathbf{X} = (\mathbf{X}_1, \mathbf{X}_2)$. Using the definition of the correlation we have

$\Pr(\mathbf{X}_1 = t_1) \Pr(\mathbf{X}_2 = t_2)$

$= (1/2 + (-1)^{t_1} c_1)(1/2 + (-1)^{t_2} c_2)$

$= 1/4 + (-1)^{(0,1) \cdot t} c_1 + (-1)^{(1,0) \cdot t} c_2 + (-1)^{(1,1) \cdot t} c_1 c_2$

$= c((0,0) \cdot \mathbf{X}) + (-1)^{(0,1) \cdot t} c((1,0) \cdot \mathbf{X}) + (-1)^{(1,0) \cdot t} c((1,0) \cdot \mathbf{X}) + (-1)^{(1,1) \cdot t} c((1,1) \cdot \mathbf{X})$

$= \sum_{a \in \mathbb{Z}_2^2} (-1)^{a \cdot t} c(a \cdot \mathbf{X}).$

But by Lemma 2.1 in [10], the last sum is equal to $\Pr(\mathbf{X} = t) = \Pr(\mathbf{X}_1 = t_1, \mathbf{X}_2 = t_2)$.

Assume now that the claim holds for $2, \ldots, m-1$ binary random variables and let $\mathbf{X}_1, \ldots, \mathbf{X}_m$, satisfy condition (16). By the induction assumption random variables $\mathbf{X}_2, \ldots, \mathbf{X}_m$, are s.i. Hence, it suffices to show that $\mathbf{X}_1$ is s.i. of the $m-1$ random variables $\mathbf{X}_2, \ldots, \mathbf{X}_m$.

Choose any binary coefficients $b_2, \ldots, b_m \in \{0, 1\}$, not all zero, and let $I = \{i = 2, \ldots, m : b_i = 1\}$ be the index set of non-zero coefficients b_i. Denote $\mathbf{Z}_I = b_2 \mathbf{X}_2 \oplus \cdots \oplus b_m \mathbf{X}_m$. By the Xiao-Massey lemma, we must show that the random variable $\mathbf{X}_1$ is s.i. of $\mathbf{Z}_I$ for all index sets $I \subset \{2, 3, \ldots, m\}$. By the induction assumption and Xiao-Massey lemma, the claim holds already for all $I \neq \{2, 3, \ldots, m\}$ and we only have to consider the set $J = \{2, 3, \ldots, m\}$. By the condition (16), the correlation $c(\mathbf{Z}_J) = \prod_{i=2}^m c_i$ and $c(\mathbf{X}_1 \oplus \cdots \oplus \mathbf{X}_m) = \prod_{i=1}^m c_i$. Hence, the random variables $\mathbf{X}_1$ and $\mathbf{Z}_J$ satisfy

$$c(\mathbf{X}_1 \oplus \mathbf{Z}_J) = \prod_{i=1}^m c_i = c_1 c(\mathbf{Z}_J).$$

But since the theorem holds for $m = 2$, the random variables $\mathbf{X}_1$ and $\mathbf{Z}_J$ must be s.i. $\qquad \square$

B   Proof of Lemma 1

Proof. The expected values, variances and covariances of elements of $\mathbf{Q}$ are [19]

$$E(\mathbf{Q}_\eta) = s_\eta \quad \mathrm{Var}(\mathbf{Q}_\eta) = s_\eta(1 - s_\eta) \quad \mathrm{Cov}(\mathbf{Q}_\eta, \mathbf{Q}_\nu) = -s_\eta s_\nu, \qquad (17)$$

for all $\eta, \nu = 0, 1, \ldots, M$ and $\nu \neq \eta$. The normality follows from the law of large numbers. The expected value follows from linearity and (17). The variance is obtained by

$$\sigma^2 = \sum_{\eta=0}^M \mathrm{Var}(\lambda_\eta \mathbf{Q}_\eta) + \sum_{\eta, \nu = 0, \nu \neq \eta}^M \mathrm{Cov}(\lambda_\eta \mathbf{Q}_\eta, \lambda_\nu \mathbf{Q}_\nu)$$

$$= \sum_{\eta=0}^M \lambda_\eta^2 s_\eta(1 - s_\eta) - \sum_{\eta, \nu = 0, \nu \neq \eta} \lambda_\eta \lambda_\nu s_\eta s_\nu = \sum_{\eta=0}^M \lambda_\eta^2 s_\eta - \mu^2.$$

$\qquad \square$

Practical Key Recovery Attack
against Secret-IV EDON-$\mathcal{R}$

Gaëtan Leurent

École Normale Supérieure – Département d'Informatique,
45 rue d'Ulm, 75230 Paris Cedex 05, France
Gaetan.Leurent@ens.fr

Abstract. The SHA-3 competition has been organized by NIST to select a new hashing standard. EDON-$\mathcal{R}$ was one of the fastest candidates
in the first round of the competition. In this paper we study the security
of EDON-$\mathcal{R}$, and we show that using EDON-$\mathcal{R}$ as a MAC with the secret-
IV or secret-prefix construction is unsafe. We present a practical attack
in the case of EDON-$\mathcal{R}$256, which requires 32 queries, 2^{30} computations,
negligible memory, and a precomputation of 2^{52}. The main part of our
attack can also be adapted to the tweaked EDON-$\mathcal{R}$ in the same settings:
it does not yield a key-recovery attack, but it allows a selective forgery
attack.

This does not directly contradict the security claims of EDON-$\mathcal{R}$ or
the NIST requirements for SHA-3, since the recommended mode to build
a MAC is HMAC. However, we believe that it shows a major weakness
in the design.

Keywords: Hash functions, SHA-3, EDON-$\mathcal{R}$, MAC, secret IV, secret
prefix, key recovery.

1   Introduction

In 2005, a team of researchers led by X. Wang produced breakthrough attacks
against many widely used hash functions, including MD5 [12] and SHA-1 [11].
This has led NIST to call for a new hash function design, and to launch the
SHA-3 competition [7]. This competition has focused the attention of many
cryptographers, and NIST received 64 submissions. 51 designs were accepted to
the first round.

EDON-$\mathcal{R}$ was one of the fastest candidates in the first round of the competition.
It has received some attention from the cryptographic community, resulting in
various attacks on the compression function. There is also a preimage attack
on the full hash function, but it requires of huge amount of memory making it
debatable.

In this paper we show a new attack on EDON-$\mathcal{R}$, when used in the secret-IV
or secret-prefix MAC construction. This mode of operation is not claimed to be
secure by the designers, but our attack has no memory requirement, and is even
practical attack, while previous attacks are largely theoretical. Our approach

J. Pieprzyk (Ed.): CT-RSA 2010, LNCS 5985, pp. 334–349, 2010.

is similar to the one followed by Wang *et al.* who studied a similar MAC used with SHA-1 [10]: we use a non-standard MAC to show weaknesses of the hash function. Note that attacks on hash-based MACs are usually harder to build than attacks on the hash function itself because part of the state is unknown.

Our attack was devised during the first round of the competition and it has been made public before the selection of the second round candidates. Since then, NIST selected 14 candidates for the second round of the competition, based on the cryptanalytic results available at that time. EDON-$\mathcal{R}$ was not selected for the second round.

1.1   MAC Constructions

A Message Authentication Code (MAC) is a symmetric signature algorithm. The sender and the receiver share a secret key k, and each message M is sent together with a short tag $\mathrm{MAC}_k(M)$. The receiver recomputes the tag on his end, and checks whether the tag is correct. It should be hard for an adversary to forge a message with a valid tag without knowing the secret key k. In this paper we consider chosen message attacks, where the adversary has access to a MAC oracle and can ask for the MAC of any message of his choice. He must then produce a forge for a new message. We consider the following attacks, from the strongest to the weakest:

Key recovery: After some interactions with the MAC oracle, the adversary outputs the key k.

Universal forgery: After some interactions with the MAC oracle, the adversary obtains enough information to compute the MAC of any message. This is usually achieved by recovering an equivalent key which allow to compute MACs as efficiently as the original key.

Selective forgery: The adversary is given a challenge message M^* (possibly in some prescribed set), and after some interaction with the MAC oracle, he has to produce the MAC of M^*. Of course, the adversary is not allowed to query the MAC of M^*.

Exitential forgery: After some interactions with the MAC oracle, the adversary produce a forge of a new message of his choice.

We expect a good MAC algorithm to be secure against all these attacks, even existential forgery. Complexity of generic attack against iterated MAC algorithm are given in Table 1.

In this paper, we study MAC algorithms based on EDON-$\mathcal{R}$. We consider two MAC constructions, the secret-IV construction and the secret-prefix construction:

$$\mathrm{IV\text{-}MAC}_k(M) = \text{EDON-}\mathcal{R}_k(M)$$
$$\mathrm{SP\text{-}MAC}_k(M) = \text{EDON-}\mathcal{R}(k\|M)$$

The secret-IV method uses the key as the initial value in the iterative construction of EDON-$\mathcal{R}$, while the secret-prefix method prepends the key to the message to be authenticated. Both constructions are quite similar, and the basic idea is

Table 1. Complexity of generic attacks against iterated MAC. The key-length n is assumed to be equal to the tag length, while m is the size of the inner state.

Attack	Complexity
Key recovery	2^n
Universal forgery	2^n
Selective forgery	2^n
Existential forgery	$\min(2^{m/2}, 2^n)$ [9]

to randomize the state of the hash function with the key before mixing the message into the state. The secret-IV construction is easier to analyse, but the secret prefix construction is more practical because it does not require a modification of the hash function; it can be used with any implementation of EDON-$\mathcal{R}$. For efficiency purpose, it is advisable to pad the key to a full block when using SP-MAC.

This kind of construction is used in some old protocols, like RFC2069 [2] (RFC2069 uses SP-MAC without padding the key to a full block). It is well known that those constructions are weak, because length extension attacks can be used for forgeries, but the key is not expected to leak. Moreover, EDON-$\mathcal{R}$ is a wide-pipe design, so the length extension issue does not apply. In fact, if the hash function is wide-pipe and the compression function is modeled as a random oracle, those constructions are provably secure [1]. Therefore, breaking EDON-$\mathcal{R}$ in this setting is expected be as hard a the generic complexities given in Table 1.

1.2   Road Map

Section 2 will describe EDON-$\mathcal{R}$ and discuss previous analysis. In Section 3, we show how to use a pair a related queries to gather information on both the input and the output of the compression function. The idea is similar to the length extension attack against Merkle-Damgård hash functions. This reduces the key-recovery problem to solving a small equation. In Section 4, we show how to solve this equation. We use simple linear algebra techniques to identify truncated differentials in the main operations of EDON-$\mathcal{R}$, and this leads to an attack with complexity $2^{5n/8}$ using only two queries to the MAC oracle. In Section 5 we use more queries to the MAC oracle to build more equations, and solve the equations using a guess-and-determine technique. This gives a very efficient attack, which is even practical in the case of EDON-$\mathcal{R}$224/256. Finally, in Appendix 6, we show how to extend these results to attack against the secret-prefix construction, and attacks against MACs based on the tweaked version of EDON-$\mathcal{R}$.

2   Description of EDON-$\mathcal{R}$

EDON-$\mathcal{R}$ is a wide-pipe iterative design, based on a compression function $\mathcal{R}$, with a final truncation $\mathcal{T}$. The EDON-$\mathcal{R}$ family is based on two main designs:

Edon-$\mathcal{R}256$ uses 32-bits words, while Edon-$\mathcal{R}512$ uses 64-bit words. Let w denote the size of the words, and n denote the output size ($n = 8w$). We give a description of Edon-$\mathcal{R}$ where the variables are elements of $(\mathbb{F}_2^w)^8$, *i.e.*, 8-tuples of w-bit words. The compression function is based on a quasi-group operation $*$, which take two inputs X and Y in $(\mathbb{F}_2^w)^8$ and compute one output in $(\mathbb{F}_2^w)^8$. The quasi-group operation is just the sum of two permutations, and we will use a permutation based description of Edon-$\mathcal{R}$ in this paper:

$$X * Y = \mu(X) + \nu(Y)$$
$$= Q_0(R_0(P_0(X))) + Q_1(R_1(P_1(Y)))$$

where

- $+$ is a component-wise addition modulo 2^w (w is the word size);
- μ and ν are the permutations defining $*$; we rewrite then with Q_i, R_i, P_i;
- P_0 and P_1 are linear over $\mathbb{Z}_{2^w}^8$, each output word is the sum of five inputs;
- R_0 and R_1 are component-wise rotations of w-bit words;
- Q_0 and Q_1 are linear over $(\mathbb{F}_2^w)^8$, each output word is the xor of three inputs;
- We identify $\mathbb{Z}_{2^w}^8$ and $(\mathbb{F}_2^w)^8$ with the natural mapping between them;
- We also define $\bar{\mu}(X^{[0]}, X^{[1]}, ...X^{[7]}) = \mu(X^{[7]}, X^{[6]}, ...X^{[0]})$.

Note that the quasi-group operation is very easy to invert: given X and $X * Y$, we can compute Y as $\nu^{-1}(X * Y - \mu(X))$.

The compression function takes as input 16 message ($M_{i,0}$ and $M_{i,1}$) words and 16 words of chaining value ($H_{i,0}$ and $H_{i,1}$) and produces 16 words of new chaining value ($H_{i+1,0}$ and $H_{i+1,1}$). The full compression function is described in Figure 1. For more details, see [4].

2.1 Previous Analysis of Edon-$\mathcal{R}$

Previous work [5,6] has shown various weaknesses of the compression function:

- given $M_{i,0}$, $M_{i,1}$, $H_{i+1,0}$ and $H_{i+1,1}$, it is easy to compute $H_{i,0}$ and $H_{i,1}$;
- given $H_{i,0}$, $H_{i,1}$, $M_{i,0}$, and $H_{i+1,0}$, it is easy to compute $M_{i,1}$, and $H_{i+1,1}$;
- given $H_{i+1,1}$, $H_{i,0}$ and $M_{i,0}$, we can find a value of $H_{i,1}$, $H_{i+1,0}$, and $M_{i,1}$ with $2^{n/2}$ operations.

These results can be used to mount various attacks on the hash function:

- We can apply generic attacks against narrow-pipe hash functions: multi-collisions, second preimages on long message, fixed points, ...
- There is a preimage attack with complexity $2^{2n/3}$ and $2^{2n/3}$ memory.

The preimage attack requires less computations than a generic attack, but due to the large memory requirements, the machine to carry out this attack might be more expensive than a machine to perform a parallel brute force, so it is unclear whether this should be considered as an attack.

However, these results show that the compression function of Edon-$\mathcal{R}$ is quite weak, and the security of Edon-$\mathcal{R}$ cannot be based on a security proof of the Merkle-Damgård mode.

More recently, a work by Novotney and Ferguson [8] showed detectable biases in the output of the compression function.

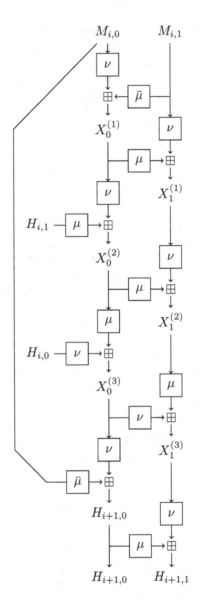

Fig. 1. EDON-$\mathcal{R}$ compression function

2.2 Our Results

We present a new attack on the compression which allow to recover the full chaining value when half of the input chaining value and half of the output chaining value are known:

– given $M_{i,0}$, $M_{i,1}$, $H_{i,1}$ and $H_{i+1,1}$, we can compute $H_{i,0}$ and $H_{i+1,0}$.

In this paper we will describe two attacks: one that requires only two queries and a lot of computations, and a second with more queries and a practical complexity:

	Queries	Time	Memory	Precomputation
EDON-$\mathcal{R}$224/256	2	2^{160}	-	-
EDON-$\mathcal{R}$224/256	32	$\simeq 2^{30}$	-	2^{52}
EDON-$\mathcal{R}$384/512	2	2^{320}	-	-
EDON-$\mathcal{R}$384/512	32	$\simeq 2^{32}$	-	2^{100}

The attack on the compression function can be used to mount the following attacks on MAC constructions, with the same complexities as the attack on the compression function:

- A *key-recovery* attack against secret-IV EDON-$\mathcal{R}$;
- A *universal forgery* attack against secret-prefix EDON-$\mathcal{R}$ when the key is padded to full block;
- A *selective forgery* attack against secret-prefix EDON-$\mathcal{R}$ if the key is not padded to a full block (we can attack any message such that $k\|M$ takes more than one block after the padding);
- A *selective forgery* attack against the secret-prefix and secret-IV constructions when used with the tweaked EDON-$\mathcal{R}$ (we can attack any message that include a valid padding).

Our attacks only needs a few queries and negligible memory. They can easily be parallelized. Those attacks are the first attacks on the full EDON-$\mathcal{R}$ to clearly beat parallel generic attacks.

3   IV Recovery Using Related Queries

The first step of the IV-recovery attack is to gather information about the chaining values. We will make two calls to the MAC oracle, with two related messages, such that after the padding step, the first message is a prefix of the second one. The first message M is chosen arbitrarily such that after the padding it fits in one block pad(M). The second message M' has pad(M) as its first block, and has to fit in two blocks after the padding. This is similar to the length extension attack on narrow-pipe hash function. Applied to a wide-pipe design such as EDON-$\mathcal{R}$, this gives us some information on the input and output of the second compression function (see Fig 2):

- $M_{1,0}$ and $M_{1,1}$ are known;
- $H_{1,1}$ is known;
- $H_{2,1}$ is known.

We will show how to recover $H_{1,0}$. Then $H_{0,0}$ and $H_{0,1}$ can be recovered from $H_{1,0}, H_{1,1}$ and $M_{0,0}, M_{0,1}$ because the compression function of EDON-$\mathcal{R}$ is easy to invert [5]. Since there are 8 unknown words in the input of the compression

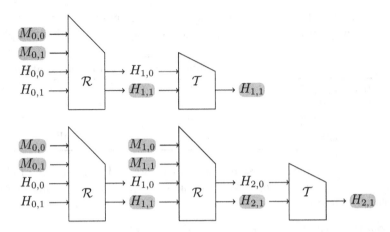

Fig. 2. The first message $\text{pad}(M) = M_{0,0}M_{0,1}$ allow to recover $H_{1,1}$ while the second message $\text{pad}(M') = M_{0,0}M_{0,1}M_{1,0}M_{1,1}$ allows to recover $H_{2,1}$

function ($H_{1,0}$) and we know 8 words of the output of the compression function ($H_{2,1}$), we expect one solution on average. In this setting, a preimage attack will be able to recover *the* value of $H_{1,0}$ and not merely *a* value that gives the same output.

If we look at the description of the compression function [4], we have:

$$
\begin{aligned}
H_{2,1} &= H_{2,0} * X_1^{(3)} \\
&= (\overline{M_{1,0}} * X_0^{(3)}) * (X_1^{(2)} * X_0^{(3)}) \\
&= (\bar{\mu}(M_{1,0}) + \nu(X_0^{(3)})) * (\mu(X_1^{(2)}) + \nu(X_0^{(3)})) \\
&= (U + C_0) * (U + C_1)
\end{aligned}
$$

where $U = \nu(X_0^{(3)})$ is unknown, and $C_0 = \bar{\mu}(M_{1,0})$, $C_1 = \mu(X_1^{(2)})$ are known constants.

If we are able to solve the equation $H = (U + C_0) * (U + C_1)$ where U is the unknown, then we can recover $X_0^{(3)} = \nu^{-1}(U)$, and this will give us $H_{1,0} = \nu^{-1}(X_0^{(3)} - \mu(X_0^{(2)}))$.

4   Solving the Equation $H = (U + C_0) * (U + C_1)$

The main step of the attack is to solve the equation

$$
\begin{aligned}
H &= (U + C_0) * (U + C_1) \\
&= Q_0(R_0(P_0(U + C_0))) + Q_1(R_1(P_1(U + C_1)))
\end{aligned}
$$

All the variables are 8-tuples of w bit words, and U is the unknown. To solve this equation, we will express U over a basis of $\mathbb{Z}_{2^w}^8$ such that some of the basis vectors do not affect some words of $(U + C_0) * (U + C_1)$. Then we can solve the equation more efficiently than by brute force because we do not need to explore the full space.

More precisely, P_0, P_1 are defined by the following matrices over $\mathbb{Z}_{2^w}$ (*i.e.*, the sums are modular additions):

$$P_0 = \begin{bmatrix} 1 1 1 0 1 0 0 1 \\ 1 1 0 1 1 0 0 1 \\ 1 1 0 0 1 0 1 1 \\ 0 0 1 1 0 1 1 1 \\ 0 1 1 1 0 1 1 0 \\ 1 0 1 1 1 1 0 0 \\ 1 1 0 0 0 1 1 1 \\ 0 0 1 1 1 1 1 0 \end{bmatrix} \qquad P_1 = \begin{bmatrix} 1 1 1 0 0 1 0 1 \\ 1 1 0 1 1 0 1 0 \\ 1 1 1 1 0 1 0 0 \\ 0 0 1 1 1 0 1 1 \\ 1 1 0 1 1 1 0 0 \\ 0 0 1 0 1 1 1 1 \\ 0 1 1 0 0 1 1 1 \\ 1 0 0 1 1 0 1 1 \end{bmatrix}$$

We will use three vectors U_0, U_1, U_2 in the kernels of some sub-matrices of P_0 and P_1:

$$U_0 = \begin{bmatrix} 0 & 0 & 0 & 0 & 0 & 0 & 1 & -1 \end{bmatrix}^{\mathrm{T}}$$
$$U_1 = \begin{bmatrix} 2 & 2 & 2 & 2 & 2^{31} - 3 & 2^{31} - 3 & 0 & 2^{31} - 1 \end{bmatrix}^{\mathrm{T}}$$
$$U_2 = \begin{bmatrix} 1 & 0 & 0 & 0 & 2^{31} - 1 & 2^{31} & 0 & 2^{31} \end{bmatrix}^{\mathrm{T}}$$

Then we have (the question marks represent values for which we do not have any useful information):

$$P_0 \cdot U_0 = \begin{bmatrix} ? ? 0 0 ? 0 0 ? \end{bmatrix}^{\mathrm{T}} \qquad P_1 \cdot U_0 = \begin{bmatrix} ? ? 0 0 0 0 0 0 \end{bmatrix}^{\mathrm{T}} \qquad (1)$$
$$P_0 \cdot U_1 = \begin{bmatrix} ? ? 0 0 ? 0 0 ? \end{bmatrix}^{\mathrm{T}} \qquad P_1 \cdot U_1 = \begin{bmatrix} ? ? ? 0 0 ? 0 0 \end{bmatrix}^{\mathrm{T}} \qquad (2)$$
$$P_0 \cdot U_2 = \begin{bmatrix} 0 0 0 0 ? 0 ? ? \end{bmatrix}^{\mathrm{T}} \qquad P_1 \cdot U_2 = \begin{bmatrix} ? ? ? ? 0 ? 0 0 \end{bmatrix}^{\mathrm{T}} \qquad (3)$$

Q_0, Q_1 are defined by the following matrices over $\mathbb{F}_2^w$ (*i.e.*, the sums are exclusive or):

$$Q_0 = \begin{bmatrix} 1 1 0 0 1 0 0 0 \\ 1 0 0 0 1 0 0 1 \\ 0 1 0 0 0 0 1 1 \\ 0 0 1 1 1 0 0 0 \\ 1 1 0 0 0 0 0 1 \\ 0 0 0 1 0 1 1 0 \\ 0 0 1 0 0 1 1 0 \\ 0 0 1 1 0 1 0 0 \end{bmatrix} \qquad Q_1 = \begin{bmatrix} 1 1 0 0 0 1 0 0 \\ 0 0 1 0 0 0 1 1 \\ 1 1 0 1 0 0 0 0 \\ 1 0 0 1 1 0 0 0 \\ 0 1 1 0 0 1 0 0 \\ 0 0 0 1 1 0 1 0 \\ 0 0 1 0 0 1 0 1 \\ 0 0 0 0 1 0 1 1 \end{bmatrix}$$

Due to the positions of the zeros in $P_i \cdot U_j$, we have, for all $\alpha, \beta \in \mathbb{Z}_{2^w}$:

$$Q_0(R_0(P_0(X + \alpha U_0))) \oplus Q_0(R_0(P_0(X))) = \begin{bmatrix} ? ? ? ? ? 0 0 0 \end{bmatrix}^{\mathrm{T}} \qquad (4)$$
$$Q_0(R_0(P_0(X + \alpha U_1))) \oplus Q_0(R_0(P_0(X))) = \begin{bmatrix} ? ? ? ? ? 0 0 0 \end{bmatrix}^{\mathrm{T}} \qquad (5)$$
$$Q_0(R_0(P_0(X + \alpha U_2))) \oplus Q_0(R_0(P_0(X))) = \begin{bmatrix} ? ? ? ? ? ? ? 0 \end{bmatrix}^{\mathrm{T}} \qquad (6)$$

$$Q_1(R_1(P_1(Y + \beta U_0))) \oplus Q_1(R_1(P_1(Y))) = [? ? ? ? ? 0 \, 0 \, 0]^T \tag{7}$$

$$Q_1(R_1(P_1(Y + \beta U_1))) \oplus Q_1(R_1(P_1(Y))) = [? ? ? ? ? 0 \, ? \, 0]^T \tag{8}$$

$$Q_1(R_1(P_1(Y + \beta U_2))) \oplus Q_1(R_1(P_1(Y))) = [? ? ? ? ? ? ? 0]^T \tag{9}$$

This proves that the vectors U_0, U_1, U_2 do not affect some of the output words. This property can be seen as a truncated differential for the $*$ operation:

$$(X + \alpha U_0) * (Y + \beta U_0) \oplus X * Y = [? ? ? ? ? 0 \, 0 \, 0]^T \tag{10}$$

$$(X + \alpha U_1) * (Y + \beta U_1) \oplus X * Y = [? ? ? ? ? 0 \, ? \, 0]^T \tag{11}$$

$$(X + \alpha U_2) * (Y + \beta U_2) \oplus X * Y = [? ? ? ? ? ? ? 0]^T \tag{12}$$

This is a very important part of the attack, so let us explain in more detail what equation (12) means. Using notations similar to the ones from [4], the last output word of $X * Y$ is computed as:

$$(X * Y)^{[7]} = (T_X^{[2]} \oplus T_X^{[3]} \oplus T_X^{[5]}) + (T_Y^{[4]} \oplus T_Y^{[6]} \oplus T_Y^{[7]})$$

where

$$T_X^{[2]} = (X^{[0]} + X^{[1]} + X^{[4]} + X^{[6]} + X^{[7]}) \lll 8$$

$$T_X^{[3]} = (X^{[2]} + X^{[3]} + X^{[5]} + X^{[6]} + X^{[7]}) \lll 13$$

$$T_X^{[5]} = (X^{[0]} + X^{[2]} + X^{[3]} + X^{[4]} + X^{[5]}) \lll 22$$

$$T_Y^{[4]} = (Y^{[0]} + Y^{[1]} + Y^{[3]} + Y^{[4]} + Y^{[5]}) \lll 15$$

$$T_Y^{[6]} = (Y^{[1]} + Y^{[2]} + Y^{[5]} + Y^{[6]} + Y^{[7]}) \lll 25$$

$$T_Y^{[7]} = (Y^{[0]} + Y^{[3]} + Y^{[4]} + Y^{[6]} + Y^{[7]}) \lll 27$$

We now consider $X' = X + \alpha U_2$ and $Y' = Y + \beta U_2$:

$$(X' * Y')^{[7]} = (T'^{[2]}_X \oplus T'^{[3]}_X \oplus T'^{[5]}_X) + (T'^{[4]}_Y \oplus T'^{[6]}_Y \oplus T'^{[7]}_Y)$$

where

$$T'^{[2]}_X = (X^{[0]} + \alpha + X^{[1]} + X^{[4]} + \alpha(2^{31} - 1) + X^{[6]} + X^{[7]} + \alpha 2^{31}) \lll 8$$

$$T'^{[3]}_X = (X^{[2]} + X^{[3]} + X^{[5]} + \alpha 2^{31} + X^{[6]} + X^{[7]} + \alpha 2^{31}) \lll 13$$

$$T'^{[5]}_X = (X^{[0]} + \alpha + X^{[2]} + X^{[3]} + X^{[4]} + \alpha(2^{31} - 1) + X^{[5]} + \alpha 2^{31}) \lll 22$$

$$T'^{[4]}_Y = (Y^{[0]} + \beta + Y^{[1]} + Y^{[3]} + Y^{[4]} + \beta(2^{31} - 1) + Y^{[5]} + \beta 2^{31}) \lll 15$$

$$T'^{[6]}_Y = (Y^{[1]} + Y^{[2]} + Y^{[5]} + \beta 2^{31} + Y^{[6]} + Y^{[7]} + \beta 2^{31}) \lll 25$$

$$T'^{[7]}_Y = (Y^{[0]} + \beta + Y^{[3]} + Y^{[4]} + \beta(2^{31} - 1) + Y^{[6]} + Y^{[7]} + \beta 2^{31}) \lll 27$$

We see that the α and β terms cancels out:

$$T_X^{[2]} = T'^{[2]}_X \qquad\qquad T_X^{[3]} = T'^{[3]}_X \qquad\qquad T_X^{[5]} = T'^{[5]}_X$$
$$T_Y^{[4]} = T'^{[4]}_Y \qquad\qquad T_Y^{[6]} = T'^{[6]}_Y \qquad\qquad T_Y^{[7]} = T'^{[7]}_Y$$

and as a consequence $(X' * Y')^{[7]} = (X * Y)^{[7]}$. This works because U_2 was chosen in the kernel of the linear forms that define $T_X^{[2]}$, $T_X^{[3]}$, $T_X^{[5]}$, $T_Y^{[4]}$, $T_Y^{[6]}$, and $T_Y^{[7]}$. Similarly, U_1 is in the kernel of the linear forms involved in the computation of $(X * Y)^{[5,7]}$ and U_0 is in the kernel of the linear forms involved in the computation of $(X * Y)^{[5,6,7]}$.

Thanks to this property, we can do an exhaustive search with early abort. We extend U_0, U_1, U_2 into a basis $U_0, U_1, ... U_7$[1] of $\mathbb{Z}_{2^w}^8$, and we will represent U in this basis: $U = \sum_{i=0}^7 \alpha_i U_i$. We define $V = (U + C_0) * (U + C_1)$. Due to the properties of U_0, U_1, U_2, we know that:

- α_0 has no effect on $V^{[5]}$, $V^{[6]}$ and $V^{[7]}$;
- α_1 has no effect on $V^{[5]}$ and $V^{[7]}$;
- α_2 has no effect on $V^{[7]}$.

The full algorithm is given by Algorithm 1 and is quite simple. We first iterate over $\alpha_3, \alpha_4, ... \alpha_7$ and we filter the elements such that $V = (U + C_0) * (U + C_1)$ matches H on the last coordinates. If it does not match, we do not need to iterate over $\alpha_0, \alpha_1, \alpha_2$ because this will not modify $V^{[7]}$, so we can abort this branch. For the choices that match, we iterate over α_2 and check $V^{[5]}$. If it matches $H^{[5]}$, we iterate over α_1 and check $V^{[6]}$. If it matches $H^{[6]}$, we can then iterate over α_0.

The time complexity is $2^{5w} = 2^{5n/8}$:

- the first loop is executed 2^{5w} times;
- each matching reduces the number of candidates to 2^{4w};
- each subsequent loop raises the number of candidates to 2^{5w}.

The memory requirements are negligible because we do not need to store a list of candidate. We just perform a breath-first search and we prune the bad branches to reduce the size of the tree.

Once we have recovered $U = \nu(X_0^{(3)})$, it is easy to invert the permutations and recover $X_0^{(3)}$. From that we find $H_{1,0}$ by inverting a quasi-group operation, and we have all the variables of the compression function. We can then recover the key $H_{0,0}, H_{0,1}$ by inverting the first compression function (it is easy when the output and the message are known).

---

[1] For instance, we can use:

$$U_3 = [0,0,1,0,0,0,0,0]^T \quad U_5 = [0,0,0,0,1,0,0,0]^T \quad U_7 = [0,0,0,0,0,0,0,1]^T$$
$$U_4 = [0,0,0,1,0,0,0,0]^T \quad U_6 = [0,0,0,0,0,1,0,0]^T$$

Algorithm 1. Solving $H = (U + C_0) * (U + C_1)$

Input: C_0, C_1, H
Output: U
1: **for all** $\alpha_3, \alpha_4, \ldots \alpha_7 \in \mathbb{Z}_{2^w}$ **do**
2: $\quad U \leftarrow \sum_{i=3}^{7} \alpha_i U_i$
3: $\quad V \leftarrow (U + C_0) * (U + C_1)$
4: $\quad$ **if** $V^{[7]} = H^{[7]}$ **then**
5: $\quad\quad$ **for all** $\alpha_2 \in \mathbb{Z}_{2^w}$ **do**
6: $\quad\quad\quad U \leftarrow \sum_{i=2}^{7} \alpha_i U_i$
7: $\quad\quad\quad V \leftarrow (U + C_0) * (U + C_1)$
8: $\quad\quad\quad$ **if** $V^{[5]} = H^{[5]}$ **then**
9: $\quad\quad\quad\quad$ **for all** $\alpha_1 \in \mathbb{Z}_{2^w}$ **do**
10: $\quad\quad\quad\quad\quad U \leftarrow \sum_{i=1}^{7} \alpha_i U_i$
11: $\quad\quad\quad\quad\quad V \leftarrow (U + C_0) * (U + C_1)$
12: $\quad\quad\quad\quad\quad$ **if** $V^{[6]} = H^{[6]}$ **then**
13: $\quad\quad\quad\quad\quad\quad$ **for all** $\alpha_0 \in \mathbb{Z}_{2^w}$ **do**
14: $\quad\quad\quad\quad\quad\quad\quad U \leftarrow \sum_{i=0}^{7} \alpha_i U_i$
15: $\quad\quad\quad\quad\quad\quad\quad V \leftarrow (U + C_0) * (U + C_1)$
16: $\quad\quad\quad\quad\quad\quad\quad$ **if** $V = H$ **then**
17: $\quad\quad\quad\quad\quad\quad\quad\quad U$ is a solution

---

5   Using More Queries

In this section, we improve this attack using more queries to the MAC oracle. We gather more equations of the form $H = (U + C_0) * (U + C_1)$, and this enables us to mount a very efficient attack. In the case of EDON-$\mathcal{R}256$, it requires about 32 queries and can recover the secret key with about 2^{30} computations after a precomputation of about 2^{52} operations, which makes it a practical attack.

5.1   Building the Queries

To get new equations, we will query the MAC oracle with new messages $M^{(i)}$ so that pad(M) is a prefix of all the $M^{(i)}$'s. Each query will give some equation involving the same $H_{1,0}$, and we will deduce an equation of the form $H^{(i)} = (U + C_0^{(i)}) * (U + C_1^{(i)})$ as in the previous section. Remember that we have $U = \nu(X_0^{(3)}) = \nu(\nu(H_{1,0}) + \mu(X_0^{(2)}))$. We will build our messages so that the value of $X_0^{(2)}$ is the same for all the $M^{(i)}$'s, or equivalently, $X_0^{(1)}$ is the same for all the $M^{(i)}$'s. This means that all the equations will involve the same U, and recovering this U will allow to recover $H_{1,0}$.

Let us assume that we have two such equations, and let us further assume that $C_0^{(i)} = C_0^{(j)}$. Then:

$$H^{(i)} = Q_0(R_0(P_0(U + C_0^{(i)}))) + Q_1(R_1(P_1(U + C_1^{(i)})))$$

$$H^{(j)} = Q_0(R_0(P_0(U + C_0^{(j)}))) + Q_1(R_1(P_1(U + C_1^{(j)})))$$

$$H^{(i)} - H^{(j)} = Q_1(R_1(P_1(U + C_1^{(i)}))) - Q_1(R_1(P_1(U + C_1^{(j)})))$$

since P_1 is linear over $\mathbb{Z}_{2^w}^8$, we can consider $\tilde{U} = P_1 \cdot U$ and $\tilde{C}_1^{(i)} = P_1 \cdot C_1^{(i)}$

$$H^{(i,j)} = H^{(i)} - H^{(j)} = Q_1(R_1(\tilde{U} + \tilde{C}_1^{(i)})) - Q_1(R_1(\tilde{U} + \tilde{C}_1^{(j)})) \tag{13}$$

If we consider $\tilde{U} = P_1 \cdot U$ to be the unknown, this gives a simpler equation than in the previous section, where $H^{(i,j)}$, $\tilde{C}_1^{(i)}$ and $\tilde{C}_1^{(j)}$ are known constants.

However, if we have a pair of messages $M^{(i)}, M^{(j)}$ where $C_0^{(i)} = C_0^{(j)}$ and $X_0^{(1)}$ is constant, then we have $M^{(i)} = M^{(j)}$ and we can only build a trivial equation. Instead, we use messages such that only some words of $C_0^{(i)}$ and $C_0^{(j)}$ are equal. Namely, if we have

$$(P_0 \cdot C_0^{(i)})^{[2,3,5]} = (P_0 \cdot C_0^{(j)})^{[2,3,5]} \tag{14}$$

then

$$
\begin{aligned}
H^{(i,j)[7]} &= \left(\mu(U + C_0^{(i)}) + \nu(U + C_1^{(i)})\right)^{[7]} - \left(\mu(U + C_0^{(j)}) + \nu(U + C_1^{(j)})\right)^{[7]} \\
&= \left(\nu(U + C_1^{(i)}) - \nu(U + C_1^{(j)})\right)^{[7]} + \left(\mu(U + C_0^{(i)}) - \mu(U + C_0^{(j)})\right)^{[7]} \\
&= \left(\nu(U + C_1^{(i)}) - \nu(U + C_1^{(j)})\right)^{[7]} \\
&\quad + \left(P_0(U + C_0^{(i)})^{[2]} \ggg 8 \oplus P_0(U + C_0^{(i)})^{[3]} \ggg 13 \oplus P_0(U + C_0^{(i)})^{[5]} \ggg 22\right) \\
&\quad - \left(P_0(U + C_0^{(j)})^{[2]} \ggg 8 \oplus P_0(U + C_0^{(j)})^{[3]} \ggg 13 \oplus P_0(U + C_0^{(j)})^{[5]} \ggg 22\right) \\
H^{(i,j)[7]} &= Q_1(R_1(\tilde{U} + \tilde{C}_1^{(i)}))^{[7]} - Q_1(R_1(\tilde{U} + \tilde{C}_1^{(j)}))^{[7]} \tag{15}
\end{aligned}
$$

The two gray terms cancel out by linearity of P_0 over $\mathbb{Z}_{2^w}^8$. We can see (15) as a weaker version of (13): we only have an equation on one word, instead of eight. We can build similar equations restricted to any word by choosing appropriate relations between $C_0^{(i)}$ and $C_0^{(j)}$: if we want an equation restricted to word k we just need to have an equality between $P_0 \cdot C_0^{(i)}$ and $P_0 \cdot C_0^{(j)}$ on the three words used in the computation of $Q_0^{[k]}$.

5.2   Dealing with the Padding

Another problem that we face to gather these equations is the padding. Edon-$\mathcal{R}$ uses a padding with Merkle-Damgård strengthening, so there are 65 bits in $M_{1,1}$ that must be kept untouched (129 bits in Edon-$\mathcal{R}$384/512).

To find proper messages, we use a preprocessing step. First, we fix some arbitrary value for $X_0^{(1)}$. Then we take a set of random $M_{1,1}$ satisfying the padding, we compute the corresponding $M_{1,0}$ and we look for a collision in three words of $P_0 \cdot C_0^{(i)}$ according to (14). Each collision costs 2^{48} computations on average (2^{96} for Edon-$\mathcal{R}$384/512), and gives one equation. Note that this is independent of the key we are attacking. It can be done as a preprocessing step, and we only need to store a the message pairs that will be used to extract the equations. Since we need 16 collisions, the time complexity of this preprocessing step will be 16×2^{48} for Edon-$\mathcal{R}$256 and 16×2^{96} for Edon-$\mathcal{R}$512.

5.3 Solving

To recover the value of U, we gather several equation of the type of (15). We can rewrite them as:

$$\left((\tilde{U}^{[4]} + \tilde{C}_1^{(i)[4]}) \ggg 17 \oplus (\tilde{U}^{[6]} + \tilde{C}_1^{(i)[6]}) \ggg 7 \oplus (\tilde{U}^{[7]} + \tilde{C}_1^{(i)[7]}) \ggg 5\right) - \tag{16}$$

$$\left((\tilde{U}^{[4]} + \tilde{C}_1^{(j)[4]}) \ggg 17 \oplus (\tilde{U}^{[6]} + \tilde{C}_1^{(j)[6]}) \ggg 7 \oplus (\tilde{U}^{[7]} + \tilde{C}_1^{(j)[7]}) \ggg 5\right) = H^{(i,j)[7]}$$

We will solve these equations using a guess-and-determine approach. First we guess the 18 lower bits of $\tilde{U}^{[4]}$, the 8 lower bits of $\tilde{U}^{[6]}$, and the 6 lower bits of $\tilde{U}^{[7]}$. This allows us to compute the least significant bit of the left hand side of (16), and we check this bit against the right hand side. If we have enough equations, we can filter out many bad candidates. Then we guess one more bit of $\tilde{U}^{[4]}$, $\tilde{U}^{[6]}$, and $\tilde{U}^{[7]}$. We can now compute one more bit of (16), and again reduce the number of candidates. We repeat this step until all the bits of $\tilde{U}^{[4]}$, $\tilde{U}^{[6]}$, and $\tilde{U}^{[7]}$ have been guessed. Each time we guess some bits, the number of candidates grows, but it will shrink when we check the new bit of (16). The cost of this step is at least 2^{32} because we have to guess 32 bits in the beginning. If we have enough equations and they give an independent filtering, we expect the complexity to be about 2^{32}. We did some experiments with random constants to check our assumptions. Experiments shows that with only 10 equations we can solve (16) for EDON-$\mathcal{R}$256 by exploring slightly more than 2^{32} nodes. This take a few minutes on a desktop PC. For EDON-$\mathcal{R}$512, we have to guess 56 bits, and we expect a complexity of 2^{56}.

Another way to solve this system is to guess the carries instead of guessing the low order bits. In this case, we only use 4 equations, because we have to guess the carries in each equations. We have only 24 carry bits to guess, but the 4 equations have many solutions, so we use extra equations to check each of these solutions until a single solution is left. According to our experiments, this takes about one minute on a desktop PC, and we have about 2^{16} solutions when using 4 equations (the search goes through 2^{30} nodes). Note that the complexity of this technique is independent of the rotation amounts, so it can be applied with any output word, not necessarily the seventh as in (15). More importantly, it is about as efficient on EDON-$\mathcal{R}$512: it take about 20 minutes to explore 2^{33} nodes, and gives about 2^{20} solutions.

This first step gives us $\tilde{U}^{[4]}$, $\tilde{U}^{[6]}$, and $\tilde{U}^{[7]}$. Next, we use an equation similar to (15), but involving the fifth word instead of the seventh:

$$\left((\tilde{U}^{[3]} + \tilde{C}_1^{(i)[3]}) \ggg 21 \oplus (\tilde{U}^{[4]} + \tilde{C}_1^{(i)[4]}) \ggg 17 \oplus (\tilde{U}^{[6]} + \tilde{C}_1^{(i)[6]}) \ggg 7\right) - \tag{17}$$

$$\left((\tilde{U}^{[3]} + \tilde{C}_1^{(j)[3]}) \ggg 21 \oplus (\tilde{U}^{[4]} + \tilde{C}_1^{(j)[4]}) \ggg 17 \oplus (\tilde{U}^{[6]} + \tilde{C}_1^{(j)[6]}) \ggg 7\right) = H^{(i,j)[5]}$$

Since this equation only involves one unknown word $\tilde{U}^{[3]}$, it is quite easy to solve. We use the same technique as previously: we guess the carry bits. We only have 2 carry bits to guess so this step is negligible. We will repeat this using different equations involving different words of $\tilde{U}$, so as to recover the words of $\tilde{U}$ one by one. Then, we can recover $U = P_1^{-1} \cdot \tilde{U}$, and finally $H_{1,0}$.

The number of queries needed for the attack is 30: 2×10 to recover three words in the first step and 2 for each subsequent word.

6   Conclusion

We have shown a practical key-recovery attack against secret-IV EDON-$\mathcal{R}$ and various forgery attacks on secret-prefix EDON-$\mathcal{R}$. Moreover, we show that a selective forgery attack can still be done against the tweaked EDON-$\mathcal{R}$. While those constructions are not required to be secure by NIST, it is a natural construction that is used in some protocols. We believe that a strong cryptographic hash function should not leak the key when used in this setting.

Acknowledgement

Part of this work is supported by the Commission of the European Communities through the IST program under contract IST-2002-507932 ECRYPT, by the French government through the Saphir RNRT project, and by the French DGA.

References

1. Chang, D., Nandi, M.: Improved Indifferentiability Security Analysis of chopMD Hash Function. In: Nyberg, K. (ed.) FSE 2008. LNCS, vol. 5086, pp. 429–443. Springer, Heidelberg (2008)
2. Franks, J., Hallam-Baker, P., Hostetler, J., Leach, P., Luotonen, A., Sink, E., Stewart, L.: RFC2069: An extension to HTTP: Digest access authentication. Internet RFCs (1997)
3. Gligoroski, D., Klima, V.: Official Comment: EDON-$\mathcal{R}$. SHA-3 forum (May 2009)
4. Gligoroski, D., Ødegråd, R.S., Mihova, M., Knapskog, S.J., Kocarev, L., Drápal, A., Klima, V.: Cryptographic Hash Function EDON-R. Submission to NIST (2008)
5. Khovratovich, D., Nikolić, I., Weinmann, R.P.: Cryptanalysis of Edon-R. Available online (2008)
6. Klima, V.: Multicollisions of EDON-R hash function and other observations (2008)
7. National Institute of Standards and Technology: Cryptographic Hash Algorithm Competition, http://csrc.nist.gov/groups/ST/hash/sha-3/index.html
8. Novotney, P., Ferguson, N.: Detectable correlations in edon-r. Cryptology ePrint Archive, Report 2009/378 (2009), http://eprint.iacr.org/
9. Preneel, B., van Oorschot, P.C.: On the Security of Iterated Message Authentication Codes. IEEE Transactions on Information Theory 45(1), 188–199 (1999)
10. Wang, X., Wang, W., Jia, K., Wang, M.: New Distinguishing Attack on MAC using Secret-Prefix Method. In: Dunkelman, O. (ed.) FSE 2009. LNCS, vol. 5665, pp. 363–374. Springer, Heidelberg (2009)
11. Wang, X., Yin, Y.L., Yu, H.: Finding Collisions in the Full SHA-1. In: Shoup, V. (ed.) CRYPTO 2005. LNCS, vol. 3621, pp. 17–36. Springer, Heidelberg (2005)
12. Wang, X., Yu, H.: How to Break MD5 and Other Hash Functions. In: Cramer, R. (ed.) EUROCRYPT 2005. LNCS, vol. 3494, pp. 19–35. Springer, Heidelberg (2005)

Appendix

A   Extension to Other Settings

The IV-recovery attack can be use to break the secret-prefix construction when used with EDON-$\mathcal{R}$, and the secret-IV and secret-prefix construction when used with the tweaked version of EDON-$\mathcal{R}$. If this section we describe the attacks with only two queries, but the attacks with 32 queries can be adapted in the same way.

A.1   Secret-Prefix EDON-$\mathcal{R}$

If the key is padded to a full block, we can recover the chaining value H_1 after processing the key. This chaining value will allow an attacker to compute the MAC of any message:

$\text{pad}(k\|M)$ is a prefix of $k\|M'$.

We apply the attack on the third compression function. We can recover H_2, and compute H_1 by inverting the second compression function.

If the key is not padded to a full block, we have a selective forgery attack. Given a message M^* such that $\text{pad}(k\|M^*)$ has at least two blocks, we use a message M such that the first block of $\text{pad}(k\|M)$ is equal to the first block of $\text{pad}(k\|M^*)$.

The first block is the same.

$\text{pad}(k\|M)$ is a prefix of $k\|M'$.

We apply the attack on the third compression function. Again, we can recover H_2, and compute H_1 by inverting the second compression function. Then, we can forge the MAC of M^*.

A.2   The Tweaked Version of EDON-$\mathcal{R}$

In [3], Gligoroski and Klima proposed a tweak to address the attacks found against EDON-$\mathcal{R}$. The tweak is described as:

Instead of the old compression function $\mathcal{R}(oldPipe, M)$, now the compression function have the following feedback: $\mathcal{R}(oldPipe, M) \oplus oldPipe \oplus M'$, where M is represented in two parts i.e. $M = (M_0, M_1)$, and $M' = (M_1, M_0)$.

It is easy to see that our attack on the compression function can still be applied: if we know the right half of $oldPipe$, the right half of the output of the compression function, and the message block, we can compute the right half of $\mathcal{R}(oldPipe, M)$ from the right half of $\mathcal{R}(oldPipe, M) \oplus oldPipe \oplus M'$. However, we can no longer invert the compression function. Therefore, in the IV-recovery attack from Section 3, we can recover $(H_{1,0}, H_{1,1})$ using the attack on the compression function, but we can not recover the key $(H_{0,0}, H_{0,0})$.

Still, we have a selective forgery attack on the secret-prefix and secret-IV constructions. Let us describe the attack on the secret-prefix construction. The forgery will work for messages M^* such that some prefix of $k \| M^*$ is a valid padded message. For instance, we can fix 65 bits (129 in the case of EDON-$\mathcal{R}512$) at the end of the second block. A random message of 2^ℓ blocks can be attacked with probability $2^{\ell-65}$ ($2^{\ell-129}$ for EDON-$\mathcal{R}512$). Given such a message, we use a message M such that $\mathrm{pad}(k \| M)$ is a prefix of $k \| M^*$:

k	M^*	M^*	M^* pad

H_0 ‖ H_1 ‖ H_2 $\mathrm{pad}(k \| M)$ is a prefix of $k \| M^*$.

k	M	M pad		

H_0 ‖ H_1 ‖ H_2 H_3 $\mathrm{pad}(k \| M)$ is a prefix of $k \| M'$.

k	M'	M'	M' pad

We apply the attack on the following block and we recover the inner state after processing $\mathrm{pad}(k \| M)$. Since $\mathrm{pad}(k \| M)$ is a prefix of M^*, we can use this information to forge the MAC of M^*.

Rebound Attacks on the Reduced Grøstl Hash Function*

Florian Mendel[1], Christian Rechberger[2], Martin Schläffer[1], and Søren S. Thomsen[3]

[1] Institute for Applied Information Processing and Communications (IAIK)
Graz University of Technology, Inffeldgasse 16a, A-8010 Graz, Austria
[2] Dept. of Electrical Engineering ESAT/COSIC, K.U. Leuven,
and Interdisciplinary Institute for BroadBand Technology (IBBT),
Kasteelpark Arenberg 10, B–3001 Heverlee, Belgium
[3] Department of Mathematics, Technical University of Denmark
Matematiktorvet 303S, DK-2800 Kgs. Lyngby, Denmark
martin.schlaeffer@iaik.tugraz.at

Abstract. Grøstl is one of 14 second round candidates of the NIST SHA-3 competition. Cryptanalytic results on the wide-pipe compression function of Grøstl-256 have already been published. However, little is known about the hash function, arguably a much more interesting cryptanalytic setting. Also, Grøstl-512 has not been analyzed yet. In this paper, we show the first cryptanalytic attacks on reduced-round versions of the Grøstl hash functions. These results are obtained by several extensions of the rebound attack. We present a collision attack on 4/10 rounds of the Grøstl-256 hash function and 5/14 rounds of the Grøstl-512 hash functions. Additionally, we give the best collision attack for reduced-round (7/10 and 7/14) versions of the compression function of Grøstl-256 and Grøstl-512.

Keywords: hash function, cryptanalysis, collisions, rebound attack.

1 Introduction

In the last few years the cryptanalysis of hash functions has become an important topic within the cryptographic community. The attacks on the MD4 family of hash functions (e.g., MD5 [12, 15], SHA-1 [2, 14]) have especially weakened the confidence in the security of this design strategy. Many new and interesting hash function designs have been proposed as part of the NIST SHA-3 competition [11]. Most submissions are constructed using specific underlying building

---

* This work was supported in part by the European Commission through the ICT programme under contract ICT-2007-216676 ECRYPT II and the fourth author is supported by a grant from the Villum Kann Rasmussen Foundation. Parts of this work were carried out while the third author was visiting Technical University of Denmark, supported by a grant from DCAMM International Graduate Research School, Danish Center for Applied Mathematics and Mechanics.

J. Pieprzyk (Ed.): CT-RSA 2010, LNCS 5985, pp. 350–365, 2010.

blocks like permutations, explicit compression functions, or block ciphers. Sometimes, proofs are devised to show that some desirable properties of the hash function (like collision resistance) can be reduced to a property of an underlying building block.

In turn, many cryptanalytic results have been published which consider these building blocks. Often, the resulting attacks are not applicable to the hash function itself. While these results are important to analyze the security of a specific design, it is very difficult to compare the results of different hash function proposals. How can we measure and compare the security margin of different designs? In addition to the (reduced) security parameter that is used for the best attack, a number of other issues heavily influence the answer: Is the design wide-pipe, is it based on the sponge model, or does it use an MD-style iteration? Is the hash function based on an ideal block cipher or a random permutation? All these considerations can be bypassed if we compare cryptanalytic results of the complete hash function instead of different underlying building blocks. Thus, a comparison of different designs is made easier.

In this paper we analyze the *hash* function Grøstl [4], which is one of the remaining 2nd-round candidates of the NIST SHA-3 competition. Grøstl has very competitive hardware implementation characteristics (see e.g., Tillich et al. [13] for a comparison), is the fastest among the remaining AES-like designs on most platforms, and naturally deserves cryptanalytic attention.

Grøstl is based on a wide-pipe compression function that is iterated in an MD-style manner. Since the wide-pipe *compression* function of Grøstl is known to be non-random, many distinguishers exist and the hash function has been designed with this fact in mind. With ℓ denoting the output size of the compression function, even collision attacks in $2^{\ell/3}$ time or $2^{\ell/4}$ permutation queries, memoryless preimage attacks in time $2^{\ell/2}$, and very efficient distinguishers (only two calls) are known [4]. Hence a strong output transformation with truncation is an important part of the design.

Shortcut collision attacks on round-reduced versions of the compression function of Grøstl-256 have been presented in a series of papers [5,8,9]. As discussed above, additional distinguishers on the compression function are meaningless. However, showing non-random properties of the underlying permutations or the output transformation can have some significance. See e.g., Mendel et al. [8] for results along those lines, where among others, a distinguisher for 7 rounds of the output transformation with complexity 2^{56} is given.

However, little is known about the hash function, which is arguably a more interesting cryptanalytic setting. Only half of the degrees of freedom are available to an attacker for direct manipulation compared to a compression function attack. Also, Grøstl-512 has not been considered yet. In this paper, we first improve the rebound attack as originally applied to the Grøstl-256 compression function [9]. Using the rebound attack, we give results for the Grøstl-512 compression function and present the first analysis of the reduced Grøstl hash functions. Our results and the best previously known results are summarized in Table 1.

Table 1. Summary of rebound analysis for the round-reduced Grøstl hash and compression functions

Target	Hash Size	Rounds	Time	Memory	Type	Reference
hash	224,256	4/10	2^{64}	2^{64}	collision	Sect. 5.1
function	384,512	5/14	2^{176}	2^{64}	collision	Sect. 5.2
	256	6/10	2^{120}	2^{64}	semi-free-start collision	[9]
compression	224,256	6/10	2^{64}	2^{64}	semi-free-start collision	[8]
function	256	7/10	2^{120}	2^{64}	semi-free-start collision	Sect. 5.3, [5]
	384,512	7/14	2^{152}	2^{64}	semi-free-start collision	Sect. 5.4

We start the paper by recalling the relevant parts of the Grøstl specification in Section 2 and give the basics of the rebound attack in Section 3. The new ideas and improvements which are the basis for our results are presented in Section 4. The results for the hash function and compression function for both Grøstl-256 and Grøstl-512 are given in Section 5. Finally, we conclude in Section 6.

2   Description of Grøstl

The hash function Grøstl was designed by Gauravaram et al. as a candidate for the SHA-3 competition [4]. It is an iterated hash function with a compression function built from two distinct permutations P and Q, which are based on the same principles as the AES round transformation [10]. Grøstl is a wide pipe design with security proofs for the collision and preimage resistance of the compression function [3]. In the following, we describe the Grøstl hash function and the permutations of Grøstl-256 and Grøstl-512 in more detail.

2.1   The Grøstl Hash Function

The input message M is padded and split into blocks $M_1, M_2, \ldots, M_t$ of ℓ bits with $\ell = 512$ for Grøstl-256 and $\ell = 1024$ for Grøstl-512. The initial value H_0, the intermediate hash values H_i, and the permutations P and Q are of size ℓ as well. The message blocks are processed via the compression function f, which accepts two inputs of size ℓ bits and outputs an ℓ-bit value. The compression function f is defined via the permutations P and Q as follows:

$$f(H, M) = P(H \oplus M) \oplus Q(M) \oplus H.$$

The compression function is iterated in the usual way with $H_0 = IV$ and $H_i \leftarrow f(H_{i-1}, M_i)$ for $1 \leq i \leq t$. The output H_t of the last call of the compression function is processed by an output transformation g defined as $g(x) = \mathrm{trunc}_n(P(x) \oplus x)$, where n is the output size of the hash function and $\mathrm{trunc}_n(x)$ discards all but the least significant n bits of x. Hence, the digest of the message M is defined as $h(M) = g(H_t)$.

2.2   The Grøstl-256 Permutations

As mentioned above, two permutations P and Q are defined for Grøstl-256. The permutations differ only in the used constants. Both permutations operate on a 512-bit state, which can be viewed as an 8×8 matrix of bytes. Each permutation of Grøstl-256 consists of 10 rounds, where the following four AES-like [10] round transformations are applied to the state in the given order:

- AddRoundConstant (AC) XORs a constant to one byte of the state. The constant changes in every round and is different for P and Q.
- SubBytes (SB) applies the AES S-box to each byte of the state.
- ShiftBytes (SH) cyclically rotates the bytes of row i to the left by i positions.
- MixBytes (MB) is a linear diffusion layer, which multiplies each column with a constant 8×8 circulant MDS matrix.

For details on the round transformations we refer to the Grøstl specification [4]. Note that AddRoundConstant is the only transformation that distinguishes P from Q. The properties of the round transformations which are used in the following attacks are similar to those of the AES (see Section 3 for more details).

2.3   The Grøstl-512 Permutations

The permutations used in Grøstl-512 are of size $\ell = 1024$ bits and the state is viewed as an 8×16 matrix of bytes. The permutations use the same round transformations as in Grøstl-256 except for ShiftBytes: Since the permutations are larger, row j is cyclically shifted j positions to the left for $0 \leq j \leq 6$ and row 7 is shifted 11 positions to the left. The number of rounds is increased to 14.

3   The Rebound Attack on Grøstl

The rebound attack was published by Mendel et al. in [9] and is a new tool for the cryptanalysis of hash functions. It can be applied to both block cipher based and permutation based constructions. The idea of the rebound attack is to divide an attack into two phases, an inbound and outbound phase. The inbound phase is an efficient meet-in-the-middle phase, which exploits the available degrees of freedom in the middle of a (truncated) differential path to guarantee that the expensive part of a differential path holds. In the (mainly) probabilistic outbound phase the solutions of the inbound phase are computed backwards and forwards to obtain an attack on the hash or compression function. In the following, we explain the rebound attack using the 6 round semi-free-start collision attack on Grøstl-256. For a more detailed description, we refer to the original paper [9].

3.1   The Truncated Differential Path

The rebound attack on 6 rounds of the Grøstl-256 compression function uses a truncated differential path with a high number of active bytes in the middle

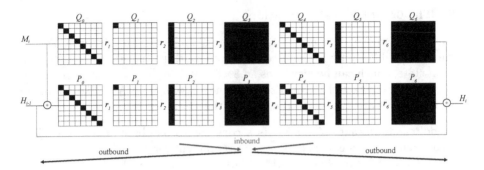

Fig. 1. Overview of the rebound attack on 6 rounds of the Grøstl-256 compression function. Black bytes are active

and a low number of active bytes at the input and output of each permutation. Due to the wide-trail design strategy, such a path can easily be constructed for Grøstl-256. For the attack on 6 rounds, a full active state is placed in the middle of each permutation. The detailed path is given in Fig. 1 and the sequence of active bytes between each round r_i is as follows:

$$8 \xrightarrow{r_1} 1 \xrightarrow{r_2} 8 \xrightarrow{r_3} 64 \xrightarrow{r_4} 8 \xrightarrow{r_5} 8 \xrightarrow{r_6} 64$$

By using the same truncated differential path in both permutations P and Q, we can construct a semi-free-start collision for the compression function of Grøstl-256 reduced to 6 rounds. In the following, we will show how to find input pairs for P and Q that follow the 6-round differential trail given above by applying a rebound attack.

3.2 The Inbound Phase

We start the rebound attack with the inbound phase in round r_3 and r_4 and deterministically propagate to the full active SubBytes layer in the middle. Hence, we search for differences and values conforming to the truncated differential path shown in Fig. 2. We first choose random differences for the 8 active bytes in P_4. These differences are linearly propagated backward to 64 active bytes at the output of the previous SubBytes layer (P_4^{SB}). Then, we choose random differences for each active byte prior to the MixBytes transformation in P_3^{SH} and linearly propagate forward to the full active input of SubBytes (P_4^{SB}). Note that we can compute each column independently. Next, we need to check whether the input/output differential of all 64 active S-boxes are possible.

For a single S-box, the probability that a random S-box differential exists is about one half, which can be verified by computing the difference distribution table (DDT) of the AES S-box (see [9] for more details). For each valid S-box differential, we get at least two (in some cases 4) possible byte values such that the differential holds. For each column, we try all 2^8 non-zero differences of the according byte in P_3^{SH} and thus, expect one valid differential for all 8 S-boxes

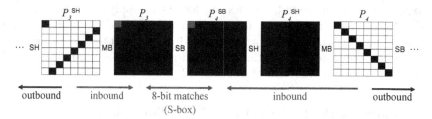

Fig. 2. The inbound phase of the attack on the Grøstl-256 compression function using 8-bit S-box matches. The input and output of one S-box is highlighted.

of that column. With two independent solutions for each S-box, we get at least 2^8 pairs for one column. Hence, the average complexity to find a valid pair is 1. We repeat this for all 8 active bytes of P_3^{SH} and get about 2^{64} solutions for the inbound phase.

Note that we can choose from about 2^{64} differences for the active bytes in P_4. Hence, we can construct up to 2^{128} pairs that follow the truncated differential path of the inbound phase between state P_3^{SH} and P_4.

3.3 The Outbound Phase

In the outbound phase, we probabilistically propagate the pairs of the inbound phase outwards, to match the differences at the input and output of the permutations. The probability for the propagation from 8 to 1 active byte through the MixBytes transformation in round r_2 is 2^{-56}. Hence, we can construct one pair conforming to the truncated differential path for each of P and Q with a complexity of 2^{56}.

To get a semi-free-start collision, the differences at the input and output of P and Q need to be equal. Note that we can construct pairs for P and Q independently. Hence, we can do a standard birthday attack to match the 8-byte difference at the input, and the 8-byte difference at the output before MixBytes with a complexity of 2^{64}. Since MixBytes is the same linear transformation in both P and Q, the 64 active bytes at the output will match if the differences at the input of MixBytes are equal. Hence, the total complexity for the semi-free-start collision on 6 rounds of the compression function of Grøstl-256 is 2^{120} compression function evaluations and 2^{64} memory due to the birthday attack.

4 Extending the Rebound Attack

In this section, we describe three improvements for the rebound attack on Grøstl. The first improvements uses 64-bit SuperBoxes [1] instead of 8-bit S-boxes to match the differences in the inbound phase. This idea has already been applied in the improved attack on the Whirlpool hash function in Lamberger et al. [7, Appendix A] and was independently observed in Gilbert and Peyrin [5]. This allows us to extend the inbound phase of the compression function attack on 6 rounds of Grøstl-256 by one round. The second idea is to apply the rebound attack to the Grøstl hash function by using a common inbound phase at the input of both P and Q. The third contribution addresses Grøstl-512. We have constructed new truncated differential paths and apply the rebound attack to the hash and compression function of Grøstl-512.

4.1 Improving the Inbound Phase Using SuperBoxes

In the standard inbound phase, the differences are computed inwards through MixBytes to the input and output of the intermediate SubBytes layer. Then, each S-box is checked for a valid differential (see Fig. 2). If we consider SuperBoxes instead of S-boxes we can extend the inbound phase by one full active state and get the following sequence of active bytes (instead of $8 \rightarrow 64 \rightarrow 8$):

$$8 \rightarrow 64 \rightarrow 64 \rightarrow 8$$

A SuperBox of Grøstl is defined similar to the SuperBox of the AES [1]. For Grøstl, the SuperBox consists of 8 parallel S-boxes, followed by one MixBytes transformation and another 8 parallel S-boxes: SB - MB - SB . Note that the SubBytes and ShiftBytes transformations can be interchanged. Hence, a Super-Box behaves like a non-linear 64-bit S-box. Unfortunately, the differential distribution table (DDT) of the SuperBox has 2^{128} entries which is too much for a collision attack on Grøstl-256. However, if the input and output differences of the SuperBox are fixed, we can iterate through all 2^{64} input values to check if a given differential holds.

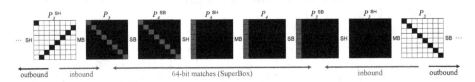

Fig. 3. The inbound phase on the Grøstl-256 compression function using 64-bit matches with one SuperBox being highlighted

In the following, we show how we can still find one solution (pair) for the extended inbound phase with an average complexity of one. We start the inbound phase at state P_3^{SH} and P_5 (see Fig. 3) and proceed as follows:

1. Start with all 2^{64} differences in state P_3^{SH}, compute forwards through MixBytes to state P_3, and store the resulting differences in list L_1.
2. Choose a random difference for state P_5 and compute backward through MixBytes and ShiftBytes to state P_5^{SB}.
3. Connect the output differences of the 8 parallel SuperBoxes (state P_5^{SB}) with the corresponding input differences of the SuperBoxes (state P_3):

(a) For each SuperBox (column) at state P_5^{SB}, take all 2^{64} possible values and compute both values and differences backward to state P_3.

(b) We get 2^{64} input differences for each SuperBox in state P_3 and store the resulting differences and values in list L_2.

(c) To find a solution for the inbound phase, we need to match the 8-byte differences in list L_2 with the corresponding differences of list L_1. Since both lists have 2^{64} entries and we have a condition on 64 bits, we get $2^{64} \times 2^{64} \times 2^{-64} = 2^{64}$ solutions (differences and values) and update L_1 accordingly.

(d) Repeat this for every SuperBox (column) of state P_5^{SB} and in each case we get 2^{64} solutions again.

4. For the whole inbound phase, we expect 2^{64} solutions with a complexity of 2^{64} in time and memory.

All in all, we can find one solution for the inbound phase with an average complexity of one. Note that we can still choose from 2^{64} differences for state P_5. Hence, we can find up to 2^{128} pairs according to the truncated differential path of the extended inbound phase. In other words, in the inbound phase we can construct up to 2^{128} starting points for the probabilistic outbound phase of the attack.

4.2 Rebound Attack on the Grøstl Hash Function

The main idea of the rebound attack on the Grøstl hash function is to do one half of the inbound phase in each P and Q. We then need to match the differences over the input of the two permutations in the inbound phase (see Fig. 4). The truncated differential path used is similar to the one of the previous section, but "wraps around" the input of P and Q. In this case, the chaining input or IV can be a predefined constant and only the message input (values and differences) is defined by the attack. Note that we use two full active states in each of P and Q since the first ShiftBytes in P and Q cancel out when going around. Hence, the columns of almost two rounds can be solved independently in the inbound phase.

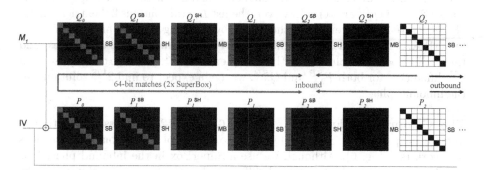

Fig. 4. The inbound phase of the attack on the hash function Grøstl-256 with one 64-bit match (two SuperBoxes) being highlighted

The technique is very similar to the previous section, since we can use independent 64-bit matches again. These two consecutive SuperBoxes (in both P and in Q) are completely independent between state Q_2^{SB} and P_2^{SB}. Again, we can find one solution (pair) for the inbound phase with an average complexity of one. We start the inbound phase with a random difference for state P_2 and compute backward to state P_2^{SB}. Next. we take all 2^{64} nonzero differences in state Q_2, compute backwards to state Q_2^{SB} and store the resulting differences in list L_1. Similar as in Section 4.1, we connect the output difference of the 8 parallel SuperBoxes of P (state P_2^{SB}) with the corresponding output differences of the SuperBoxes of Q (state Q_2^{SB}) by merging lists of size 2^{64}. We get 2^{64} solutions with a complexity of 2^{64} in time and memory. Again, we can repeat the inbound phase about 2^{64} times with other starting differences in P_2. Hence, we can construct up to 2^{128} starting points for the subsequent probabilistic outbound phase of the attack.

4.3   Constructing Truncated Differential Paths for Grøstl-512

The difficult part of the rebound attack on Grøstl-512 is to find a "good" truncated differential path. However, using a match-in-the-middle on the SuperBox, we can construct a path with similar properties as for Grøstl-256. The complexity of the rebound attack is determined by the outbound phase. Hence, we need a truncated differential path with as few active bytes in the outbound phase as possible. Similar to Grøstl-256, a straightforward truncated differential path starts with (a minimum of) 8 active bytes at both ends of the inbound phase. In the following, we show how the inbound phase of such a path works for the hash function, and how to get a valid truncated differential path for the inbound phase of the compression function as well.

The Hash Function. For the rebound attack on the Grøstl-512 hash function, the truncated differential path of the inbound phase is given in Fig. 5. Due to the symmetry of the ShiftBytes transformations in P and Q, we can again do the 64-bit matches over each two SuperBoxes independently (see Section 4.2). Contrary to the Grøstl-256 case, some output differences of the SuperBoxes in state P_2^{SB} and Q_2^{SB} are zero. However, the list L_1 still contains 2^{64} entries and we also generate 2^{64} differences for the list L_2 by iterating through all values of each SuperBox. Again, we have a condition on 64 bits (including zero differences) and thus, still expect 2^{64} solutions with a complexity of 2^{64}. Since we can choose from 2^{64} differences for both P_2^{SB} and Q_2^{SB}, we again expect to find 2^{128} solutions for the inbound phase.

The Compression Function. For the Grøstl-512 compression function, a differential path with 8 active bytes at each end of the inbound phase does not work (see Fig. 6). Although we use a SuperBox in the inbound phase this results in an impossible truncated differential path. For most columns of the MixBytes transition in the middle, the sum of active bytes at input and output is below 9, which is not possible according to the MDS property of MixBytes.

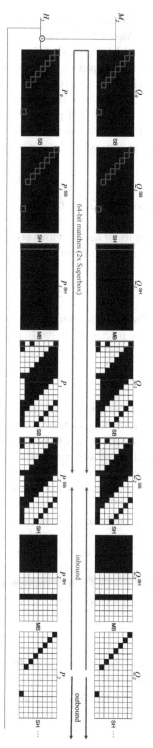

Fig. 5. Inbound phase of the attack on the Grøstl-512 hash function with one 64-bit match (two SuperBoxes) being highlighted.

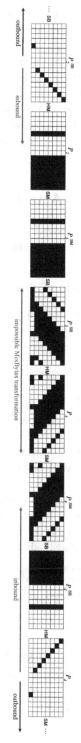

Fig. 6. Impossible inbound phase of the attack on the Grøstl-512 compression function.

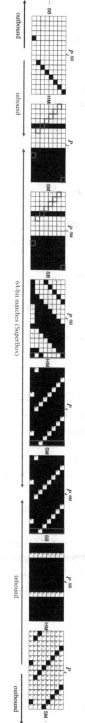

Fig. 7. Inbound phase of the attack on the Grøstl-512 compression function with one 64-bit match (SuperBox) being highlighted.

With only 8 active bytes in state P_3^{SH} and P_5, we do not get enough active bytes for a valid MixBytes transformation in round r_4. Also rotating the position of active bytes in state P_3^{SH} or P_5 does not give a valid truncated differential path. However, we can add a second active column at the output of the inbound phase (see Fig. 7). This results in an almost full active state in round r_4 and the truncated differential path is valid. Again, we can apply the same technique as in the previous section and expect 2^{64} solutions of the inbound phase with a complexity of 2^{64} by merging lists of size 2^{64}. Note that with 24 active bytes in P_3^{SH} and P_5, we can get up to 2^{192} solutions (starting points in the outbound phase) in the inbound phase.

5 Results of Rebound Attacks on Reduced Grøstl

In this section, we apply the improved inbound techniques of the previous section to the round-reduced Grøstl hash functions and compression functions.

5.1 Collisions for 4 Rounds of Grøstl-256

The complete truncated differential path for the collision attack on 4 rounds of the Grøstl-256 hash function is given in Fig. 8. The sequence of active bytes in each round for both, P and Q are given as follows:

$$64 \xrightarrow{r_1} 64 \xrightarrow{r_2} 8 \xrightarrow{r_3} 8 \xrightarrow{r_4} 64$$

The details for the inbound phase of the attack are given in Section 4.2. Remember that we get 2^{64} pairs with a complexity of 2^{64} conforming to the truncated differential path up to round r_2. In the outbound phase, each of these pairs propagate to the output of the permutations according to the truncated differential path given in Fig. 8 with a probability of one. To get a zero output difference of the hash function, the 8-byte differences prior to the last MixBytes need to be the same (see Section 3.3). Since we have 2^{64} solutions for the inbound phase, and we have a 64-bit condition in the outbound phase, we expect to get one pair which results in a collision. The complexity of this collision attack on the Grøstl-256 hash function is thus, 2^{64} in both time and memory.

Note that using the previous techniques a collision attack on 5 rounds according to the following truncated differential path for both, P and Q is not possible:

$$64 \xrightarrow{r_1} 64 \xrightarrow{r_2} 8 \xrightarrow{r_3} 1 \xrightarrow{r_4} 8 \xrightarrow{r_5} 64$$

Each of the two $8 \rightarrow 1$ transitions of MixBytes in round r_3 have a probability of 2^{-56}. Together with the probabilistic match on 64 bits at the end of the path, the total complexity is $2^{56+56+64} = 2^{176}$ which exceeds the generic complexity for a collision attack on Grøstl-256.

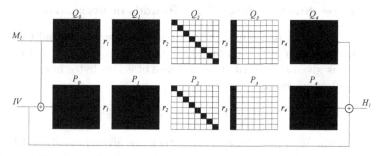

Fig. 8. Truncated differential path for the collision attack on 4 rounds of the Grøstl-256 hash function

5.2 Collisions for 5 Rounds of Grøstl-512

Contrary to the collision attack on Grøstl-256 we can extend the truncated differential path for Grøstl-512 to 5 rounds, with the following number of active bytes in each, P and Q:

$$128 \xrightarrow{r_1} 64 \xrightarrow{r_2} 8 \xrightarrow{r_3} 1 \xrightarrow{r_4} 8 \xrightarrow{r_5} 64$$

The complexity of the outbound phase is given by the two probabilistic $8 \to 1$ transitions of MixBytes in round r_3 of P and Q, and the match of the 64-bit differences prior to the last MixBytes transformation in round r_5. Hence, the total complexity of the attack is $2^{56+56+64} = 2^{176}$ compression function evaluations. Note that we need to construct 2^{176} solutions in the inbound phase for the attack to succeed. However, as shown in Section 4.3, we can only find up to 2^{128} pairs for the inbound phase.

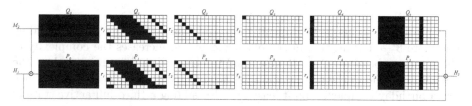

Fig. 9. Truncated differential path for the collision attack on 5 rounds of the Grøstl-512 hash function. An additional first block is used to generate enough freedom for the attack to succeed.

We can get the needed additional freedom for a 5 round collision attack by prepending a first message block. The collision attack works as follows. First we choose an arbitrary first message block. Then, we repeat the inbound phase for all 2^{128} possible starting points to get 2^{128} solutions. Since the probability of the outbound phase is 2^{-176} we need to repeat the inbound phase with 2^{48} different first message blocks to find a collision for 5 rounds. The total complexity of the attack is about $2^{64+56+56} = 2^{176}$ compression function evaluations and 2^{64} memory.

5.3   Semi-Free-Start Collision for 7 Rounds of Grøstl-256

The improved inbound phase using the SuperBox allows to extend the 6-round semi-free-start collision attack on Grøstl-256 by one round. The truncated differential path is given in Fig. 10. The sequence of active bytes in each round for both, P and Q are given as follows:

$$8 \xrightarrow{r_1} 1 \xrightarrow{r_2} 8 \xrightarrow{r_3} 64 \xrightarrow{r_4} 64 \xrightarrow{r_5} 8 \xrightarrow{r_6} 8 \xrightarrow{r_7} 64$$

The details of the inbound phase of the attack are given in Section 4.1 and we can get one pair with an average complexity of one. The solutions of the inbound phase are propagated outwards as in the attack on 6 rounds (see Section 3.3). We have one $8 \rightarrow 1$ MixBytes transition in round r_2 with probability 2^{-56}, and a birthday match on $2 \cdot 64$ bits at the input and output with complexity 2^{64}. Hence, the total complexity of the attack is 2^{120} compression function evaluations and 2^{64} memory.

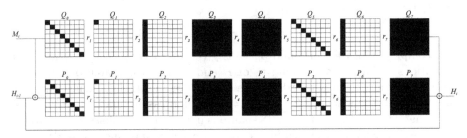

Fig. 10. The truncated differential path for the semi-free-start collision on 7 rounds of the compression function of Grøstl-256

Note that it seems to be difficult to extend this attack to 8 rounds. Adding one more $8 \rightarrow 1$ transition in the outbound phase, increases the complexity of the attack to be above 2^{128}. If we extend the truncated differential path at the beginning or end of the permutation, we need to match a full active state which has a birthday complexity of at least 2^{256}. By adding a third full active state in the middle, the columns in the match-in-the-middle phase are not independent anymore and we would need to match the differences of a full active state.

5.4   Semi-Free-Start Collision for 7 Rounds of Grøstl-512

The truncated differential path for the inbound phase of the rebound attack on the Grøstl-512 compression function has 8 active bytes in round r_3 and 16 active bytes in round r_5. The resulting 7-round truncated differential path is similar to the Grøstl-256 case (see Fig. 11) and the sequence of active bytes is given as follows:

$$8 \xrightarrow{r_1} 1 \xrightarrow{r_2} 8 \xrightarrow{r_3} 64 \xrightarrow{r_4} 110 \xrightarrow{r_5} 16 \xrightarrow{r_6} 16 \xrightarrow{r_7} 110$$

In the inbound phase, we connect the differences between the input of SubBytes of round r_4 and the output of SubBytes of round r_5 by using the SuperBox again. We get one solution with an average complexity of one.

The complexity of the attack is determined by the outbound phase. We have one probabilistic $8 \rightarrow 1$ MixBytes transition in round r_2, and do a birthday match in 8 active bytes at the beginning and 16 active bytes at the end of the path. Hence, the total complexity for the collision attack on 7 rounds is $2^{56+32+64} = 2^{152}$ with memory requirements of 2^{64} due to the inbound phase and birthday match.

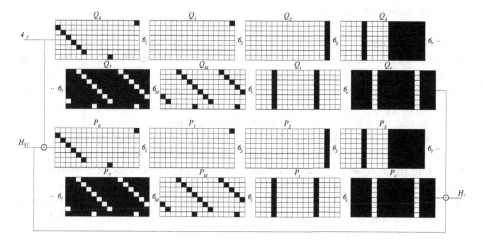

Fig. 11. Truncated differential path for the semi-free-start collision on 7 rounds of Grøstl-512

Although we could construct an 8-round truncated differential path with the following number of active bytes, we cannot find enough pairs for a collision attack on the compression function:

$$8 \xrightarrow{r_1} 1 \xrightarrow{r_2} 8 \xrightarrow{r_3} 64 \xrightarrow{r_4} 110 \xrightarrow{r_5} 16 \xrightarrow{r_6} 2 \xrightarrow{r_7} 8 \xrightarrow{r_8} 64$$

The path is constructed by carefully placing the positions of active bytes in round r_6 such that the two active bytes are shifted into the same column in round r_7. With three $8 \rightarrow 1$ MixBytes transitions and a birthday match on $2 \cdot 64$ bits at the input and output, we would get a total complexity of $2^{3 \cdot 56 + 2 \cdot 32} = 2^{232}$. Note that we get only $2^{3 \cdot 64} = 2^{192}$ solutions for the inbound phase (see Section 4.3). After the three probabilistic MixBytes transitions, we get only $2^{192-3 \cdot 56} = 2^{24}$ valid pairs for each permutation. Contrary to the Grøstl-512 hash function, we cannot use the freedom of a previous message block in the middle of the compression function. Hence, this attack on 8-rounds of Grøstl-512 compression function does not work.

6   Conclusion

In this work, we have presented a variety of new results on the SHA-3 candidate Grøstl. We improve the rebound attack on the compression function of Grøstl-256 by one round and provide the first results for Grøstl-512. Most importantly, we give the first cryptanalytic results for the Grøstl hash function and achieve 4 out of 10 rounds for Grøstl-256, and 5 out of 14 rounds for Grøstl-512. This allows to reason about the security margin of Grøstl and compare it with other hash functions based on different building blocks. However, for many candidates, only results on their underlying compression function, permutation or block cipher are known at this point.

The given results allow for the first time a high-level comparison between permutation based and block-cipher based hashing from a cryptanalytic perspective. The block-cipher based Whirlpool hash function and the permutation based Grøstl hash function share a number of similarities: 8-bit S-boxes arranged in an 8x8 geometry and AES-like round transformations. The S-boxes are different, but their exact specification does not make a difference with respect to the attacks we consider here. Whereas the rebound attack can break up to 8 rounds of the Whirlpool hash function [6,7] with complexity below 2^{128}, it can only break 4 rounds of the Grøstl hash function with complexity below 2^{128}. The main reason is the fact that in most block-cipher designs round keys are added at several places during the computation, also in the block cipher at the core of Whirlpool. Used in an unkeyed setting, this mixing of inputs during the computation gives an attacker easier access for manipulating internal state variables, and in turn allows more efficient attacks.

The ideas presented in this paper are also applicable to other AES-based hash functions like ECHO, SHAvite-3, LANE, and Cheetah. Additionally, future work will include the application of the rebound idea to other hash function constructions. This may require more sophisticated tools to obtain appropriate (truncated) differential paths first, whereas for the so far considered AES-based constructions, good differentials are easily obtainable "by hand".

References

1. Daemen, J., Rijmen, V.: Understanding Two-Round Differentials in AES. In: De Prisco, R., Yung, M. (eds.) SCN 2006. LNCS, vol. 4116, pp. 78–94. Springer, Heidelberg (2006)
2. De Cannière, C., Rechberger, C.: Finding SHA-1 Characteristics: General Results and Applications. In: Lai, X., Chen, K. (eds.) ASIACRYPT 2006. LNCS, vol. 4284, pp. 1–20. Springer, Heidelberg (2006)
3. Fouque, P.A., Stern, J., Zimmer, S.: Cryptanalysis of Tweaked Versions of SMASH and Reparation. In: Avanzi, R.M., Keliher, L., Sica, F. (eds.) Selected Areas in Cryptography. LNCS, vol. 5381, pp. 136–150. Springer, Heidelberg (2009)
4. Gauravaram, P., Knudsen, L.R., Matusiewicz, K., Mendel, F., Rechberger, C., Schläffer, M., Thomsen, S.S.: Grøstl – a SHA-3 candidate. Submission to NIST (2008), http://www.groestl.info

5. Gilbert, H., Peyrin, T.: Super-Sbox Cryptanalysis: Improved Attacks for AES-like permutations. Cryptology ePrint Archive, Report 2009/531 (2009), http://eprint.iacr.org/
6. Lamberger, M., Mendel, F., Rechberger, C., Rijmen, V., Schläffer, M.: Cryptanalysis of the Whirlpool Hash Function (manuscript)
7. Lamberger, M., Mendel, F., Rechberger, C., Rijmen, V., Schläffer, M.: Rebound Distinguishers: Results on the Full Whirlpool Compression Function. In: Matsui, M. (ed.) ASIACRYPT 2009. LNCS, vol. 5912, pp. 126–143. Springer, Heidelberg (2009)
8. Mendel, F., Peyrin, T., Rechberger, C., Schläffer, M.: Improved Cryptanalysis of the Reduced Grøstl Compression Function, ECHO Permutation and AES Block Cipher. In: Rijmen, V. (ed.) SAC 2009. LNCS, vol. 5867, pp. 16–35. Springer, Heidelberg (2009)
9. Mendel, F., Rechberger, C., Schläffer, M., Thomsen, S.S.: The Rebound Attack: Cryptanalysis of Reduced Whirlpool and Grøstl. In: Dunkelman, O. (ed.) FSE 2009. LNCS, vol. 5665, pp. 260–276. Springer, Heidelberg (2009)
10. National Institute of Standards and Technology: FIPS PUB 197, Advanced Encryption Standard (AES). Federal Information Processing Standards Publication 197, U.S. Department of Commerce (November 2001)
11. National Institute of Standards and Technology: Announcing Request for Candidate Algorithm Nominations for a New Cryptographic Hash Algorithm (SHA-3) Family. Federal Register Notice (November 2007), http://csrc.nist.gov
12. Stevens, M., Lenstra, A.K., de Weger, B.: Chosen-Prefix Collisions for MD5 and Colliding X.509 Certificates for Different Identities. In: Naor, M. (ed.) EUROCRYPT 2007. LNCS, vol. 4515, pp. 1–22. Springer, Heidelberg (2007)
13. Tillich, S., Feldhofer, M., Kirschbaum, M., Plos, T., Schmidt, J.M., Szekely, A.: High-Speed Hardware Implementations of BLAKE, Blue Midnight Wish, CubeHash, ECHO, Fugue, Grøstl, Hamsi, J.H., Keccak, Luffa, Shabal, SHAvite-3, SIMD, and Skein. Cryptology ePrint Archive, Report 2009/510 (2009), http://eprint.iacr.org/
14. Wang, X., Yin, Y.L., Yu, H.: Finding Collisions in the Full SHA-1. In: Shoup, V. (ed.) CRYPTO 2005. LNCS, vol. 3621, pp. 17–36. Springer, Heidelberg (2005)
15. Wang, X., Yu, H.: How to Break MD5 and Other Hash Functions. In: Cramer, R. (ed.) EUROCRYPT 2005. LNCS, vol. 3494, pp. 19–35. Springer, Heidelberg (2005)

The Sum of CBC MACs Is a Secure PRF

Kan Yasuda

NTT Information Sharing Platform Laboratories, NTT Corporation
9-11 Midoricho-3-chome Musashino-shi, Tokyo 180-8585 Japan
yasuda.kan@lab.ntt.co.jp

Abstract. We present a new message authentication code (MAC) based on block ciphers. Our new MAC algorithm, though twice as slow as an ordinary CBC MAC, can be proven to be a pseudo-random function secure against $O(2^{2n/3})$ queries, under the assumption that the underlying n-bit block cipher is a secure pseudo-random permutation. Our design is quite simple, being similar to Algorithm 5 (and 6) of ISO/IEC 9797-1:1999—we just take the sum (xor) of two encrypted CBC MACs. We remark that no proof of security above the birthday bound $(2^{n/2})$ has been known for the sum of CBC MACs. The sum construction now becomes the first realization of a block-cipher-based, deterministic, stateless MAC algorithm being provably secure beyond the birthday bound of $O(2^{n/2})$ and running with practical efficiency.

Keywords: PRP, PRF, sum construction, ISO/IEC 9797-1:1999, collision, game-playing proof, lazy sampling, 64-bit block cipher.

1 Introduction

Message Authentication Codes, or MAC algorithms, fall into one of the three categories: universal-hash-based, compression-function-based, or block-cipher-based. The first type of MAC—based on a universal hash function—tends to be either software-oriented [7] or hardware-oriented [32], but not both. The high performance crucially depends on the choice of platforms, which is a problem already addressed in [9]. This type of MAC also requires a larger footprint, as it employs a universal hash function plus a finalization algorithm, which is usually a compression function or block cipher. The second type of MAC—based on a compression function—has not come to maturity, because the theory of designing a good compression function has not been established, as evidenced by the initiation of the SHA-3 competition [28].[1] Although we have a sound compression-function-based mode like HMAC [2], which is widely standardized, deployed and studied, the lack of a sound compression function momentarily stops us from choosing this type of MAC algorithm.[2] The last type of MAC—based on a block cipher—has a balanced performance between software and

---

[1] Indeed, many of the SHA-3 candidate algorithms design their compression functions based on (big) block ciphers or permutations.

[2] HMAC is based on the Merkle-Damgård construction, but we do not know if SHA-3 will be natively equipped with such a construction.

J. Pieprzyk (Ed.): CT-RSA 2010, LNCS 5985, pp. 366–381, 2010.

hardware, and the security of block ciphers has been well-studied for a couple of decades. As a result, we have quite a few promising block ciphers like AES-128 [26]. Hence, block-cipher-based MACs are often preferred to other types of MAC algorithms.

However, block-cipher-based MACs also have a problem. The problem is that the block size is too small. This is a serious problem particularly for 64-bit block ciphers. Recall that, due to the generic birthday attack [31] on iterative MACs, the security of ordinary block-cipher-based MACs is limited to the $2^{n/2}$ query complexity (more precisely, $2^{n/2}$-many queries), where n is the block size. When $n = 64$, the figure $2^{n/2}$ corresponds to 32 GByte, which is rather small. It seems increasingly urgent to resolve this problem, as we explain in the following:

1. **Triple DES.** The Triple-DES algorithm continues to be used in legacy applications. For example, in the financial systems, Triple-DES is planned to be used at least for the next decade [1]. In legacy applications, the size of each message might be much smaller than 32 GByte, but through multiple transactions the total query complexity is likely to reach this bound eventually.

2. **New 64-Bit Block Ciphers.** Due to the market needs for lightweight cryptography, such as RFIDs and smart cards, new 64-bit block ciphers including HIGHT [14] and PRESENT [10] were developed recently. These ciphers are expected to be used not only within the low-power devices but also in a wide variety of applications, often exceeding the 32 GByte limit of data to be processed.

3. **New Birthday Attack.** Recently, a new type of birthday attack [19] has been reported on a class of block-cipher-based MACs including CBC MACs. The new attack is much more powerful than the previously-known generic birthday attack [31], because the new attack realizes *universal* forgery rather than existential. The new attack warns us that the birthday bound is not only a theoretical interest but also a practical limit, at least for block-cipher-based MACs.

So the motivation behind the current work is clear. We would like to develop a new, highly secure MAC algorithm which can be used with 64-bit block ciphers. By "highly secure" we mean secure beyond the birthday bound of $2^{n/2}$. We successfully come up with one such construction, following an old idea that dates back to ISO/IEC 9797-1:1999 [21]. Our construction has the following features:

1. **Simple.** Our construction is quite simple. It is just a sum (xor) of two encrypted CBC MACs, using independent keys. So we call our construction SUM-ECBC. The idea of taking the sum of two CBC MACs originates from Algorithm 5 (and 6) of ISO/IEC 9797-1:1999. We note that the SUM-ECBC MAC algorithm remains to be deterministic; SUM-ECBC does not make use of randomization or nonce.

2. **Efficient.** Our SUM-ECBC algorithm is practically efficient. SUM-ECBC is twice as slow as ordinary CBC MACs ("rate-2"), but the overhead can be reduced via parallel implementations. Other rate-2 CBC MACs have been already standardized in ISO/IEC 9797-1:1999 [21], which means that this type of MAC is efficient enough for industrial use.

3. **Secure.** The security of SUM-ECBC is based on the assumption that the underlying block cipher is a good PRP (Pseudo-Random Permutation). Under this assumption, SUM-ECBC becomes a secure PRF (Pseudo-Random Function) and achieves security of $O(2^{2n/3})$.[3] When $n = 64$, the figure $2^{2n/3} \approx 2^{42.7}$ corresponds to about 51 TByte. This is a significant gain over the 32-GByte limit (derived from $2^{n/2} = 2^{32}$ for $n = 64$).

SUM-ECBC is the first block-cipher-based deterministic MAC algorithm that both becomes provably secure beyond the birthday bound and runs at practical speed. Previous beyond-the-birthday-bound constructions tended to be of theoretical interest only; see Sect. 2. We remark that our results almost immediately imply that Algorithm 6 of ISO/IEC 9797-1 is also provably secure up to the $O(2^{2n/3})$ bound. Algorithm 6 runs slightly less efficiently than our SUM-ECBC and requires more keys.

Organization of the Paper. In Sect. 2 we review previous constructions of block-cipher-based MAC algorithms. After the essential preliminaries in Sect. 3, we define our SUM-ECBC MAC algorithm in Sect. 4 and prove its security in Sect. 5. In Sect. 6 we discuss possible directions of future work.

2   Previous Work

We point out that most of the recent work on block-cipher-based MACs has been devoted to either reducing the number of keys or weakening the assumptions about the underlying primitive (i.e., unpredictability rather than pseudorandomness). We emphasize that our goal is to improve security above the birthday bound without making the construction randomized or stateful.

History of CBC MACs. The plain CBC MAC was formally analyzed in [4]. The plain CBC MAC was able to handle only prefix-free queries. This problem was solved by EMAC [30], which was able to handle arbitrary varying-length messages by using two keys. EMAC was not padding-efficient; i.e., it caused an extra invocation to the block cipher when the original message length (before padding) was a multiple of the block length. XCBC [8] also handled arbitrary varying-length messages. XCBC required three keys but was padding-efficient. TMAC [22] improved upon XCBC by reducing the number of keys by one. OMAC [16] (now known as the NIST standard CMAC [27]) further reduced the number of keys, realizing a single-key, padding-efficient CBC MAC. Recently, a new MAC algorithm GCBC [25] has been introduced as an alternative to CMAC. All of these constructions have security only up to the birthday bound of $O(2^{n/2})$.

RMAC [18] was an unusual CBC MAC algorithm in that it achieved a bound close to the full security $O(2^n)$. However, RMAC was a randomized

---

[3] It is intriguing to note that this bound $O(2^{2n/3})$ has appeared in several pieces of previous work on block-cipher-based schemes, such as Lucks' SUM^2 construction [23], Iwata's CENC mode of operation [15] and Minematsu's double-length block cipher [24].

algorithm, and its proof of security was done in the ideal-cipher model (rather than in the standard model).

Enciphered-CBC [11] was twice as slow (rate-2) as ordinary CBC MACs but had some extra security features. Recently, Dodis and Steinberger [12] have proposed a new mode of operation which is three times slower (rate-3) than ordinary CBC MACs but has a better bound than Enciphered-CBC. These two constructions do not achieve security above the birthday limit (but rather achieve security under the weakened assumption of unpredictability).

The idea of taking the sum of two CBC MACs originates from Algorithm 5 (and 6) of ISO/IEC 9797-1:1999. Algorithm 5 is the sum of two plain one-key CBC MACs, which results in a two-key algorithm. Algorithm 6 is the sum of two three-key CBC MACs, which results in a six-key algorithm. The reason behind introducing these sum constructions was to thwart generic birthday attacks (which are applicable to single-chain CBC MACs) and hence to increase security against forgery attacks [21]. Algorithm 5 was shown to be vulnerable to a birthday attack (due to Joux et al. [20]) of $O(2^{n/2})$ complexity. However, the security of other sum constructions (such as Algorithm 6) remained to be studied; we did not know whether or not other parallel instances of two CBC chains increase security above the birthday bound. In this sense our results complement Joux et al.'s attack [20].

Other Constructions. Some block-cipher-based MAC algorithms aimed for parallelizability. These were PMAC [9], XOR MAC [3] and XECB [13]. The first one has security only up to the birthday bound. The latter two utilizes nonce, randomization, or state information.

Generic Approaches. There are some generic approaches to attaining beyond-the-birthday-bound security. For example, one could combine the SUM^2 construction [23] with Feistel network of six rounds [29]. This method was mentioned in [17]. Unfortunately, such a construction would be extremely inefficient. The double-length block cipher [24] based on tweakable block ciphers would be a more efficient approach to this problem, but one would need to treat the key input of the underlying block cipher like the data input, frequently updating the key schedule.

3   Security Definitions and Proof Tools

In this section we arrange necessary preliminaries for the presentation of the work. We give security definitions. We also introduce a system of notation, which will be used throughout our analyses.

Adversaries and Resources. An adversary A is an oracle machine. An adversary A has access to its oracle $Q(\cdot)$ and after interaction with the oracle outputs a bit, 1 or 0. We write

$$A^{Q(\cdot)} = 1$$

to denote the event that A outputs 1 after interacting with $Q(\cdot)$.

We measure the resources of A in terms of time and query complexities. In order to measure the running time of an adversary A, we fix a model of computation. The running time of A includes the time to execute its overlying experiment (game) and also the size of its description (code). In order to measure the code size, we fix a method of encoding. The query complexity is measured in terms of the number of queries made to the oracle and also in terms of the maximum length of each query. The length of a query is measured in blocks (i.e., n bits), rather than in bits, and includes the length of its padding bits.

Random Permutations and Lazy Sampling. Let $\mathrm{Perm}(n)$ denote the set of permutations on $\{0,1\}^n$. We say that $\pi : \{0,1\}^n \to \{0,1\}^n$ is a *random permutation* if it is drawn uniformly at random from the set $\mathrm{Perm}(n)$. We write

$$\pi \xleftarrow{\mathrm{R}} \mathrm{Perm}(n),$$

where $\xleftarrow{\mathrm{R}}$ means uniformly random sampling.

We often perform *lazy sampling* for specifying a random permutation π. That is, the description of π is initially undefined, and when the value $\pi(x)$ becomes necessary at some point in the game, the corresponding range point y is randomly selected. We implicitly maintain two sets, $\mathrm{Dom}(\pi)$ and $\mathrm{Rng}(\pi)$, which keep the record of already-defined domain points and that of range points, respectively. Therefore, if $x \notin \mathrm{Dom}(\pi)$, then we perform $y \xleftarrow{\mathrm{R}} \{0,1\}^n \setminus \mathrm{Rng}(\pi)$, which establishes $y = \pi(x)$, adds x to the set $\mathrm{Dom}(\pi)$, and adds y to the set $\mathrm{Rng}(\pi)$.

Block Ciphers and Pseudo-Random Permutations (PRPs). Our building block is a block cipher E. A block cipher E takes a random key K from its key space, and for each key K the specified function $E_K : \{0,1\}^n \to \{0,1\}^n$ is a permutation.

Informally, we say that a block cipher E is a *(secure) pseudo-random permutation (PRP)* if it is indistinguishable from a random permutation $\pi : \{0,1\}^n \to \{0,1\}^n$. Specifically, we consider the advantage function

$$\mathbf{Adv}_E^{\mathrm{prp}}(A) := \Pr\left[A^{E_K(\cdot)} = 1\right] - \Pr\left[A^{\pi(\cdot)} = 1\right],$$

and if this quantity is "small" for a class of adversaries, then we say that E is a PRP. Here note that the first probability is defined over the random choice of K, whereas the second probability is defined over the random choice of π (and internal coin tosses of A, if any). We further define

$$\mathbf{Adv}_E^{\mathrm{prp}}(t, q, \ell) := \max_A \mathbf{Adv}_E^{\mathrm{prp}}(A),$$

where the max runs over all adversaries A whose running time is at most t, making at most q queries to its oracle, each query being at most ℓ blocks. If there is no computational primitives involved, then we omit the parameter t. If the function accepts only fixed-length inputs, then we omit the parameter ℓ.

MACs and Pseudo-Random Functions (PRFs). Our goal is to construct a *pseudo-random function (PRF)* that accepts varying-length messages. Any PRF can be used as a secure MAC. Let $F_K : \{0,1\}^* \rightarrow \{0,1\}^n$ be a keyed function. Informally, F is a secure PRF if it is indistinguishable from a *random function* $\mathcal{R} : \{0,1\}^* \rightarrow \{0,1\}^n$ (drawn uniformly at random from the set of functions mapping $\{0,1\}^*$ to $\{0,1\}^n$).[4] Precisely, we define

$$\mathbf{Adv}_F^{\mathrm{prf}}(A) := \Pr\big[A^{F_K(\cdot)} = 1\big] - \Pr\big[A^{\mathcal{R}(\cdot)} = 1\big].$$

We also define $\mathbf{Adv}_F^{\mathrm{prf}}(t, q, \ell)$ in a similar way. Note that a random function $\mathcal{R} : \{0,1\}^* \rightarrow \{0,1\}^n$ can be also lazily sampled, as $y \xleftarrow{\mathrm{R}} \{0,1\}^n$ for $y = \mathcal{R}(x)$.

4   Specifications of SUM-ECBC MAC Algorithm

In this section we give the complete description of our SUM-ECBC MAC algorithm. The algorithm SUM-ECBC$[E] : \{0,1\}^* \rightarrow \{0,1\}^n$ employs an n-bit block cipher E with four independent keys K_1, K_2, K_3 and K_4. The SUM-ECBC algorithm takes as its input a message $m \in \{0,1\}^*$ and produces an n-bit tag τ.

See Alg. 1 for the precise definition. In the definition, the SUM-ECBC algorithm calls a subroutine CBC, which is described in Alg. 2. In the specifications

---

Algorithm 1. SUM-ECBC$[E](m)$

1: $v \leftarrow \mathrm{CBC}[E_{K_1}](m)$
2: $w \leftarrow \mathrm{CBC}[E_{K_3}](m)$
3: $\tau \leftarrow E_{K_2}(v) \oplus E_{K_4}(w)$
4: **return** τ

Algorithm 2. CBC$[E_K](m)$

1: $m_1 \cdots m_r \leftarrow m\|10^*;\ v_0 \leftarrow 0^n$
2: **for** $i = 1$ to r **do**
3: $\quad v_i \leftarrow E_K\big(v_{i-1} \oplus m_i\big)$
4: **end for**
5: **return** v_r

---

of the CBC algorithm, the notation $m_1 \cdots m_r \leftarrow m\|10^*$ means the padding and decomposing operations. Namely, we first pad the message m with appending bits $10 \cdots 0$ with the minimum number of zeros so that the length becomes a multiple of n. We then decompose the padded message $m\|10^*$ into n-bit blocks $m_1, \ldots, m_r$ so that we have $m_1\| \cdots \|m_r = m\|10^*$. The initial value v_0 is set to the zero string $0^n = 0 \cdots 0 \in \{0,1\}^n$. The symbol $\oplus$ denotes the xor operation. See also Fig. 1 for an illustration of the SUM-ECBC algorithm.

---

[4] Strictly speaking, we can only define a random function $\mathcal{R} : \{0,1\}^{\ell n} \rightarrow \{0,1\}^n$ rather than $\mathcal{R} : \{0,1\}^* \rightarrow \{0,1\}^n$. This formality issue does not cause a serious problem for the arguments in this paper.

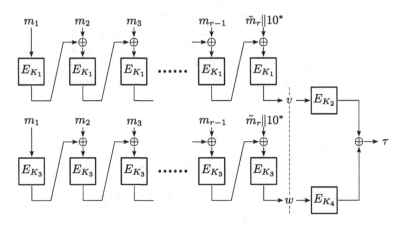

Fig. 1. Our SUM-ECBC MAC algorithm using four block-cipher keys K_1, K_2, K_3 and K_4. A message $m \in \{0,1\}^*$ is decomposed as $m = m_1 \| \cdots \| m_{r-1} \| \tilde{m}_r$, and the scheme utilizes the standard 10^* padding method.

5   Security Proof of SUM-ECBC MAC Algorithm

The security analyses of our construction are interesting in that bounds of the form $O(2^{2n/3})$ arise from multiple places in the proof. Roughly speaking, we divide our proof into four cases. Out of the four, two cases are identical, which leads to essentially three different cases. We give separate analyses to the three cases, and each case yields an $O(2^{2n/3})$ bound.

5.1   From Computational Setting to Information-Theoretic Scenario

As usual, we begin with replacing the block ciphers E_{K_1}, E_{K_2}, E_{K_3} and E_{K_4} with independent random permutations π_1, π_2, π_3 and $\pi_4 : \{0,1\}^n \rightarrow \{0,1\}^n$, respectively. We write SUM-ECBC$[\pi]$ for the resulting scheme. A standard argument shows the following result:

Lemma 1. *We have*

$$\mathbf{Adv}^{\mathrm{prf}}_{\mathrm{SUM\text{-}ECBC}[E]}(t,q,\ell) \leq \mathbf{Adv}^{\mathrm{prf}}_{\mathrm{SUM\text{-}ECBC}[\pi]}(q,\ell) + 4 \cdot \mathbf{Adv}^{\mathrm{prp}}_{E}(t',\ell q),$$

where the time complexity t' is about the original running time t plus the time to compute the block cipher E for ℓq times.

5.2   Main Theorem

Now we prove our main theorem. We have two different bounds. The first one is of the form $\ell^4 q^3/2^{2n}$. Although this bound is only up to the birthday limit in terms of ℓ, it should be noted that $\ell^4/2^{2n}$ gives us much better quantitative evaluation than the usual birthday bound of the form $\ell^2/2^n$. The second bound, which is of the form $\ell^3 q^3/2^{2n}$, is better than the first one and is beyond the birthday limit. Unfortunately, the second bound is valid only for relatively short messages,[5] i.e., $\ell \leq 2^{2n/5}$.

Theorem 1. *We have*

$$\mathbf{Adv}^{\mathrm{prf}}_{\mathrm{SUM\text{-}ECBC}[\pi]}(q, \ell) \leq \frac{12\ell^4 q^3}{2^{2n}}.$$

Moreover, if $\ell \leq 2^{2n/5}$, then we have a better bound

$$\mathbf{Adv}^{\mathrm{prf}}_{\mathrm{SUM\text{-}ECBC}[\pi]}(q, \ell) \leq \frac{40\ell^3 q^3}{2^{2n}}.$$

Proof. We first give an outline of the proof. The proof is divided into four different cases. Detailed analyses for each of the four cases will be given later in Sect. 5.3, 5.4, 5.5 and 5.6, respectively.

Let A be an adversary that makes at most q queries, each query being at most ℓ blocks. The goal of A is to distinguish between the SUM-ECBC$[\pi](\cdot)$ oracle and a truly random function $\mathcal{R} : \{0,1\}^* \to \{0,1\}^n$. Upon a query $m \in \{0,1\}^*$, we consider the code of SUM-ECBC$[\pi](m)$ as described in Alg. 3. In the code,

---

Algorithm 3. Main

1: $v \leftarrow \mathrm{CBC}[\pi_1](m)$
2: $w \leftarrow \mathrm{CBC}[\pi_3](m)$
3: **if** $v \notin \mathrm{Dom}(\pi_2)$ and $w \notin \mathrm{Dom}(\pi_4)$ **then**
4:     go to **Case A** (this computes τ and may set a **bad** flag)
5: **end if**
6: **if** $v \in \mathrm{Dom}(\pi_2)$ and $w \notin \mathrm{Dom}(\pi_4)$ **then**
7:     go to **Case B** (this computes τ and may set a **bad** flag)
8: **end if**
9: **if** $v \notin \mathrm{Dom}(\pi_2)$ and $w \in \mathrm{Dom}(\pi_4)$ **then**
10:    go to **Case C** (this computes τ and may set a **bad** flag)
11: **end if**
12: **if** $v \in \mathrm{Dom}(\pi_2)$ and $w \in \mathrm{Dom}(\pi_4)$ **then**
13:    go to **Case D** (this always sets a **bad** flag and computes τ anyway)
14: **end if**
15: **return** τ

---

the random permutations π_1 through π_4 are lazily sampled. Depending on the behavior right after a **bad** flag gets set, this code gives us either the simulation of the SUM-ECBC$[\pi](\cdot)$ oracle or the random oracle $\mathcal{R}(\cdot)$. In other words, we

---

[5] When $n = 64$ we have $2^{2n/5} = 2^{25.6}$, which corresponds to about 388 MByte.

see that these two oracles behave exactly the same until a bad event occurs. Therefore, by the fundamental lemma of game playing [6], we get

$$\mathbf{Adv}^{\mathrm{prf}}_{\mathrm{SUM\text{-}ECBC}[\pi]}(A) = \Pr\left[A^{\mathrm{SUM\text{-}ECBC}[\pi](\cdot)} = 1\right] - \Pr\left[A^{\mathcal{R}(\cdot)} = 1\right]$$
$$\le \Pr\left[A \text{ sets } \mathbf{bad}\right].$$

Since we are working in an information-theoretic scenario, involving no computational primitives, we can assume that A is deterministic. Moreover, the adversary A learns nothing from the values returned by the oracles, as the values are mere random strings and do not help A set **bad** flags (until one of the flags gets set). Therefore, we may further assume that A is non-adaptive. That is, we only need to consider a fixed sequence of queries $m^{(1)}, \ldots, m^{(q)}$ output by A.

It amounts to computing the probabilities that the sequence $m^{(1)}, \ldots, m^{(q)}$ sets **bad** flags. The probabilities will be given by the detailed analyses later, as follows:

Case A. This case can be handled easily owing to the technique of *fair sets* developed in [23]. The probability is at most $2q^3/2^{2n}$.

Case B. For this case we bound the probability by $2\ell^2 q^3/2^{2n}$.

Case C. This case is identical to Case B.

Case D. We obtain two different bounds for this case. One is $6\ell^4 q^3/2^{2n}$. The other is $34\ell^3 q^3/2^{2n}$ with the restriction $\ell \le 2^{2n/5}$.

Finally we sum up the probabilities as

$$\frac{2q^3}{2^{2n}} + \frac{2\ell^2 q^3}{2^{2n}} + \frac{2\ell^2 q^3}{2^{2n}} + \frac{6\ell^4 q^3}{2^{2n}} \le \frac{12\ell^4 q^3}{2^{2n}}, \text{ and}$$
$$\frac{2q^3}{2^{2n}} + \frac{2\ell^2 q^3}{2^{2n}} + \frac{2\ell^2 q^3}{2^{2n}} + \frac{34\ell^3 q^3}{2^{2n}} \le \frac{40\ell^3 q^3}{2^{2n}},$$

as desired. □

5.3 Detailed Analysis of Case A: $v \notin \mathrm{Dom}(\pi_2)$ and $w \notin \mathrm{Dom}(\pi_4)$

We utilize the technique of *fair sets* [23] developed in the security proof of the SUM^2 construction $\pi_2(x) \oplus \pi_4(x)$. We perform the operation of sampling two range points $\pi_2(v)$ and $\pi_4(w)$, as $y \xleftarrow{\mathrm{R}} \{0,1\}^n \setminus \mathrm{Rng}(\pi_2)$ and $z \xleftarrow{\mathrm{R}} \{0,1\}^n \setminus \mathrm{Rng}(\pi_4)$. The proof amounts to computing the probability that $y \oplus z$ deviates from the uniformly random distribution.

Lemma 2. *After q queries, the probability that $\pi_2(v) \oplus \pi_4(w)$ in Case A can be distinguished from a truly random distribution is at most*

$$\frac{2q^3}{2^{2n}},$$

where $q \le 2^{n-1}$.

Algorithm 4. Case A

1: $Y \leftarrow \{0,1\}^n \setminus \mathrm{Rng}(\pi_2)$; $Z \leftarrow \{0,1\}^n \setminus \mathrm{Rng}(\pi_4)$
2: Choose a fair set $U \subset Y \times Z$
3: $(y, z) \xleftarrow{\mathrm{R}} Y \times Z$
4: **if** $(y, z) \notin U$ **then**
5:     bad $\leftarrow$ **true** $\boxed{(y, z) \xleftarrow{\mathrm{R}} U}$
6: **end if**
7: $\tau \leftarrow y \oplus z$
8: **return** τ

Proof. The proof is almost exactly the same as the one for Lucks' SUM^2 construction $\pi_2(x) \oplus \pi_4(x)$ [23]. The fact that we have $v \neq w$ does not have much effect on the computation of the probability. Specifically, we consider the simulation of $\pi_2(v) \oplus \pi_4(w)$ as depicted in Alg. 4. The code without the boxed statement corresponds with $\pi_2(v) \oplus \pi_4(w)$. The code with the boxed statement corresponds with the random oracle $\mathcal{R}$, because the set U is fair; that is, U is chosen so that the number of pairs $(y, z) \in U$ satisfying $\tau = y \oplus z$ is the same for each value $\tau \in \{0,1\}^n$. In the code, we choose a fair set U as follows. Enumerate $\mathrm{Rng}(\pi_2)$ as $\{y_1, \ldots, y_\alpha\}$ and $\mathrm{Rng}(\pi_4)$ as $\{z_1, \ldots, z_\beta\}$. For each i and j such that $1 \leq i \leq \alpha$ and $1 \leq j \leq \beta$ we choose arbitrarily representatives $(y_i', z_j') \in Y \times Z$ such that $y_i' \oplus z_j' = y_i \oplus z_j$. We then define

$$U \leftarrow Y \times Z \setminus \bigcup_{i,j} \{(y_i', z_j')\}.$$

We see that, for each value $\tau \in \{0,1\}^n$, we have $|\{(y, z) \in U \mid y \oplus z = \tau\}| = 2^n - \alpha - \beta$, so U is indeed a fair set.

After q queries to the SUM-ECBC$[\pi](\cdot)$ oracle, the overall probability that a bad event occurs becomes

$$
\begin{aligned}
\Pr[\mathbf{bad}] &\leq \sum_{i=1}^{q} \frac{|(Y \times Z) \setminus U|}{|Y \times Z|} \\
&= \sum_{i=1}^{q} \frac{\alpha\beta}{(2^n - \alpha)(2^n - \beta)} \\
&\leq \sum_{i=0}^{q-1} \frac{i^2}{(2^n - q)^2} \\
&\leq \frac{1}{(2^n - q)^2} \cdot \sum_{i=0}^{q-1} i^2 \leq \frac{1}{(2^{n-1})^2} \cdot \frac{q(q-1)(2q-1)}{6} \leq \frac{2q^3}{2^{2n}},
\end{aligned}
$$

where we used the condition $q \leq 2^{n-1}$. $\qquad\square$

5.4   Detailed Analysis of Case B: $v \in \mathrm{Dom}(\pi_2)$ and $w \notin \mathrm{Dom}(\pi_4)$

In this case a collision occurs at the input value of π_2. The input value w of π_4 is fresh, which launches the random sampling operation $z \xleftarrow{\mathrm{R}} \{0,1\}^n \setminus \mathrm{Rng}(\pi_4)$, so that $z = \pi_4(w)$. The SUM-ECBC MAC algorithm outputs the value $y \oplus z$, where $y = \pi_2(v)$ is an already-defined range point. The distribution $y \oplus z$ is almost uniformly random, unless a bad event occurs. We aim at bounding the probability that a collision occurs at the input of π_2 *and* subsequently a bad event occurs during the sampling operation of a range point for π_4.

In order to evaluate the collision probability at the input of π_2, we recall the following results from previous work:

Lemma 3 (CBC Collision). *For any two distinct messages $m, m' \in \{0,1\}^*$, each being at most ℓ blocks, the collision probability*

$$\epsilon := \Pr\Big[\mathrm{CBC}[\pi](m) = \mathrm{CBC}[\pi](m') \;\Big|\; \pi \xleftarrow{\mathrm{R}} \mathrm{Perm}(n)\Big]$$

can be bounded as follows:

$$\epsilon \leq \frac{4\ell^2}{2^n}, \ \ or \tag{1}$$

$$\epsilon \leq \frac{2\ell}{2^n} + \frac{8\ell^4}{2^{2n}}. \tag{2}$$

Note that we wrote $\mathrm{CBC}[\pi]$ to denote the CBC algorithm using a random permutation $\pi : \{0,1\}^n \to \{0,1\}^n$. The probability is defined over the choice of π.

Proof. A proof of the first bound (1) can be found in [8]. The second bound (2) was shown by Bellare et al. in [5].   □

We next examine the sampling operation of z. Note that the distribution $y \oplus z$ would be uniformly random if the sampling operation were $z \xleftarrow{\mathrm{R}} \{0,1\}^n$. So we consider the simulation of $\pi_2(v) \oplus \pi_4(w)$ as described in Alg. 5. In the code, the simulation of π_4, which is part of SUM-ECBC$[\pi]$, is with the boxed statement, while the random oracle $\mathcal{R}$ corresponds with the code without the boxed statement. We observe that for each sampling operation the **bad** flag gets set with a probability $|\mathrm{Rng}(\pi_4)|/2^n$.

Algorithm 5. Case B

1: $y \leftarrow \pi_2(v)$
2: $z \xleftarrow{\mathrm{R}} \{0,1\}^n$
3: **if** $z \in \mathrm{Rng}(\pi_4)$ **then**
4:     **bad** $\leftarrow$ **true** $\boxed{z \xleftarrow{\mathrm{R}} \{0,1\}^n \setminus \mathrm{Rng}(\pi_4)}$
5: **end if**
6: $\tau \leftarrow y \oplus z$
7: **return** τ

We now compute the overall probability that the bad event occurs. Let $m^{(1)}, \ldots, m^{(q)}$ be a sequence of messages. Then the probability that this sequence sets the **bad** flag can be bounded as

$$\sum_{i=2}^{q} \sum_{j=1}^{i-1} \Pr\left[v^{(i)} = v^{(j)} \wedge \mathbf{bad} \mid \pi_1, \pi_2, \pi_3, \pi_4 \xleftarrow{\text{R}} \mathrm{Perm}(n) \right]$$

$$= \sum_{i=2}^{q} \sum_{j=1}^{i-1} \Pr\left[v^{(i)} = v^{(j)} \mid \pi_1 \xleftarrow{\text{R}} \mathrm{Perm}(n) \right] \cdot \Pr\left[\mathbf{bad} \mid v^{(i)} = v^{(j)}, \pi_4 \xleftarrow{\text{R}} \mathrm{Perm}(n) \right]$$

$$\leq \sum_{i=2}^{q} \sum_{j=1}^{i-1} \frac{4\ell^2}{2^n} \cdot \frac{\left| \mathrm{Rng}(\pi_4) \right|}{2^n} \leq \sum_{i=2}^{q} \sum_{j=1}^{i-1} \frac{4\ell^2}{2^n} \cdot \frac{q}{2^n} \leq \frac{q^2}{2} \cdot \frac{4\ell^2}{2^n} \cdot \frac{q}{2^n} = \frac{2\ell^2 q^3}{2^{2n}},$$

where we wrote $v^{(i)} := \mathrm{CBC}[\pi_1]\big(m^{(i)}\big)$ and $v^{(j)} := \mathrm{CBC}[\pi_1]\big(m^{(j)}\big)$. Note that we used (1) for bounding the collision probability $\Pr\left[v^{(i)} = v^{(j)} \right]$.

5.5 Detailed Analysis of Case C: $v \notin \mathrm{Dom}(\pi_2)$ and $w \in \mathrm{Dom}(\pi_4)$

This case is identical to Case B. We simply change the roles of (π_1, π_2) for the ones of (π_3, π_4). Similarly to Case B, we obtain

$$\frac{2\ell^2 q^3}{2^{2n}}$$

as an upper bound for the probability of the bad event in Case C.

5.6 Detailed Analysis of Case D: $v \in \mathrm{Dom}(\pi_2)$ and $w \in \mathrm{Dom}(\pi_4)$

This case itself is a bad event, giving us the trivial code (see Alg. 6) that involves no sampling. Let $m^{(1)}, \ldots, m^{(q)}$ be a sequence of messages. We would like to

---

Algorithm 6. Case D

1: $\mathbf{bad} \leftarrow \mathbf{true}$
2: $y \leftarrow \pi_2(v)$
3: $z \leftarrow \pi_4(w)$
4: $\tau \leftarrow y \oplus z$
5: **return** τ

---

compute the probability that at the i-th query $m^{(i)}$ we get $v^{(i)} \in \mathrm{Dom}(\pi_2)$ and $w^{(i)} \in \mathrm{Dom}(\pi_4)$. The event implies that there exists some earlier queries $m^{(j)}$ and $m^{(k)}$ (m and k may be equal) such that $v^{(j)} = v^{(i)}$ and $w^{(k)} = w^{(i)}$. We compute

$$\Pr\left[v^{(j)} = v^{(i)} \wedge w^{(k)} = w^{(i)} \mid \pi_1, \pi_2, \pi_3, \pi_4 \xleftarrow{\text{R}} \mathrm{Perm}(n) \right]$$

$$= \Pr\left[v^{(j)} = v^{(i)} \mid \pi_1 \xleftarrow{\text{R}} \mathrm{Perm}(n) \right] \cdot \left[w^{(k)} = w^{(i)} \mid \pi_3 \xleftarrow{\text{R}} \mathrm{Perm}(n) \right]. \quad (3)$$

On one hand, using (1), we obtain

$$(3) \leq \left(\frac{4\ell^2}{2^n}\right)^2 = \frac{16\ell^4}{2^{2n}}.$$

We evaluate the overall probability by running indices i, j and k. We get

$$\sum_{i=2}^{q}\sum_{j=1}^{i-1}\sum_{k=1}^{i-1}\Pr\left[v^{(j)} = v^{(i)} \wedge w^{(k)} = w^{(i)}\right] \leq \sum_{i=2}^{q}\sum_{j=1}^{i-1}\sum_{k=1}^{i-1}\frac{16\ell^4}{2^{2n}}$$

$$\leq \frac{q^3}{3} \cdot \frac{16\ell^4}{2^{2n}} \leq \frac{6\ell^4 q^3}{2^{2n}}.$$

On the other hand, we also see that

$$(3) \leq \left(\frac{2\ell}{2^n} + \frac{8\ell^4}{2^{2n}}\right)^2 = \frac{4\ell^2}{2^{2n}} + \frac{32\ell^5}{2^{3n}} + \frac{64\ell^8}{2^{4n}} \leq \frac{\ell^3}{2^{2n}} \cdot \left(4 + \frac{32\ell^2}{2^n} + \frac{64\ell^5}{2^{2n}}\right), \quad (4)$$

using the bound (2). Now we make the assumption $\ell \leq 2^{2n/5}$. Then we are able to proceed as

$$(4) \leq \frac{\ell^3}{2^{2n}} \cdot (4 + 32 + 64) = \frac{100\ell^3}{2^{2n}}.$$

For the overall probability in this case we compute

$$\sum_{i=2}^{q}\sum_{j=1}^{i-1}\sum_{k=1}^{i-1}\Pr\left[v^{(j)} = v^{(i)} \wedge w^{(k)} = w^{(i)}\right] \leq \sum_{i=2}^{q}\sum_{j=1}^{i-1}\sum_{k=1}^{i-1}\frac{100\ell^3}{2^{2n}}$$

$$\leq \frac{q^3}{3} \cdot \frac{100\ell^3}{2^{2n}} \leq \frac{34\ell^3 q^3}{2^{2n}},$$

as desired.

6   Future Directions of Research

The research subject of block-cipher-based MACs has been much discussed, but we believe the subject needs to be further studied from various aspects. In this paper we have proven that SUM-ECBC, a rate-2 CBC MAC, is secure up to a bound of $O(2^{2n/3})$. We consider that this work is the beginnings of research on the high security of deterministic CBC MACs. We hope our results facilitate further progress in this field and offer a range of possibilities for future work. So we list some of them below.

Sum of Other Type of CBC MACs. We have shown that the sum of two encrypted CBC MACs does indeed improve security. A question is whether the same holds true for other types of CBC MAC. We notice that our proof techniques can be easily adapted to Algorithm 6 of ISO/IEC 9797-1:1999. So Algorithm 6 is also secure up to the $O(2^{2n/3})$ bound.[6] On the other hand, our results

---

[6] As mentioned in Sect. 2 already, Algorithm 6 requires six keys rather than four and runs slightly slower than our SUM-ECBC algorithm.

do not seem to be directly applicable to the sum of one-key CBC MACs such as CMAC or GCBC. The problem of reducing the number of keys remains open.

Tightness of the Bound $O(2^{2n/3})$. Our SUM-ECBC achieves a bound of $O(2^{2n/3})$, which implies that Joux et al.'s attack [20] is not immediately applicable (Recall that the attack works at the complexity of $O(2^{n/2})$). We know neither if an attack of $O(2^{2n/3})$ exists for SUM-ECBC, nor if our proof can be improved to obtain a bound better than $O(2^{2n/3})$, leaving the tightness problem open.

Better Constructions. Another question is whether we can come up with a block-cipher-based construction whose security bound is better than $O(2^{2n/3})$. It should be noted that increasing the number of parallel instances (triple, quadruple, etc.) would improve the bound (cf. Lucks' SUM^d construction [23]). Unfortunately it would become less and less efficient. The problem of breaking this tradeoff line between security and efficiency remains open. In particular, we do not know whether we can construct a rate-1 block-cipher-based MAC algorithm that achieves security above the birthday bound.

Acknowledgments

The author is grateful to the anonymous reviewers for their insightful comments. Also, the author would like to thank Yu Sasaki for helping understand the contents of Joux et al.'s attack paper [20].

References

1. ANSI. Triple Data Encryption Algorithm modes of operation. X9.52:1998 (1998)
2. Bellare, M., Canetti, R., Krawczyk, H.: Keying hash functions for message authentication. In: Koblitz, N. (ed.) CRYPTO 1996. LNCS, vol. 1109, pp. 1–15. Springer, Heidelberg (1996)
3. Bellare, M., Guérin, R., Rogaway, P.: XOR MACs: New methods for message authentication using finite pseudorandom functions. In: Coppersmith, D. (ed.) CRYPTO 1995. LNCS, vol. 963, pp. 15–28. Springer, Heidelberg (1995)
4. Bellare, M., Kilian, J., Rogaway, P.: The security of cipher block chaining. In: Desmedt, Y.G. (ed.) CRYPTO 1994. LNCS, vol. 839, pp. 341–358. Springer, Heidelberg (1994)
5. Bellare, M., Pietrzak, K., Rogaway, P.: Improved security analyses for CBC MACs. In: Shoup, V. (ed.) CRYPTO 2005. LNCS, vol. 3621, pp. 527–545. Springer, Heidelberg (2005)
6. Bellare, M., Rogaway, P.: The security of triple encryption and a framework for code-based game-playing proofs. In: Vaudenay, S. (ed.) EUROCRYPT 2006. LNCS, vol. 4004, pp. 409–426. Springer, Heidelberg (2006)
7. Black, J., Halevi, S., Krawczyk, H., Krovetz, T., Rogaway, P.: UMAC: Fast and secure message authentication. In: Wiener, M. (ed.) CRYPTO 1999. LNCS, vol. 1666, pp. 216–233. Springer, Heidelberg (1999)

8. Black, J., Rogaway, P.: CBC MACs for arbitrary-length messages: The three-key constructions. In: Bellare, M. (ed.) CRYPTO 2000. LNCS, vol. 1880, pp. 197–215. Springer, Heidelberg (2000)

9. Black, J., Rogaway, P.: A block-cipher mode of operation for parallelizable message authentication. In: Knudsen, L.R. (ed.) EUROCRYPT 2002. LNCS, vol. 2332, pp. 384–397. Springer, Heidelberg (2002)

10. Bogdanov, A., Knudsen, L.R., Leander, G., Paar, C., Poschmann, A., Robshaw, M.J.B., Seurin, Y., Vikkelsoe, C.: PRESENT: An ultra-lightweight block cipher. In: Paillier, P., Verbauwhede, I. (eds.) CHES 2007. LNCS, vol. 4727, pp. 450–466. Springer, Heidelberg (2007)

11. Dodis, Y., Pietrzak, K., Puniya, P.: A new mode of operation for block ciphers and length-preserving MACs. In: Smart, N.P. (ed.) EUROCRYPT 2008. LNCS, vol. 4965, pp. 198–219. Springer, Heidelberg (2008)

12. Dodis, Y., Steinberger, J.P.: Message authentication codes from unpredictable block ciphers. In: Halevi, S. (ed.) Advances in Cryptology - CRYPTO 2009. LNCS, vol. 5677, pp. 267–285. Springer, Heidelberg (2009)

13. Gligor, V.D., Donescu, P.: Fast encryption and authentication: XCBC encryption and XECB authentication modes. In: Matsui, M. (ed.) FSE 2001. LNCS, vol. 2355, pp. 92–108. Springer, Heidelberg (2002)

14. Hong, D., Sung, J., Hong, S., Lim, J., Lee, S., Koo, B., Lee, C., Chang, D., Lee, J., Jeong, K., Kim, H., Kim, J., Chee, S.: HIGHT: A new block cipher suitable for low-resource device. In: Goubin, L., Matsui, M. (eds.) CHES 2006. LNCS, vol. 4249, pp. 46–59. Springer, Heidelberg (2006)

15. Iwata, T.: New blockcipher modes of operation with beyond the birthday bound security. In: Robshaw, M.J.B. (ed.) FSE 2006. LNCS, vol. 4047, pp. 310–327. Springer, Heidelberg (2006)

16. Iwata, T., Kurosawa, K.: OMAC: One-key CBC MAC. In: Johansson, T. (ed.) FSE 2003. LNCS, vol. 2887, pp. 129–153. Springer, Heidelberg (2003)

17. Iwata, T., Yasuda, K.: HBS: A single-key mode of operation for deterministic authenticated encryption. In: Dunkelman, O. (ed.) FSE 2009. LNCS, vol. 5665, pp. 394–415. Springer, Heidelberg (2009)

18. Jaulmes, É., Joux, A., Valette, F.: On the security of randomized CBC-MAC beyond the birthday paradox limit: A new construction. In: Daemen, J., Rijmen, V. (eds.) FSE 2002. LNCS, vol. 2365, pp. 237–251. Springer, Heidelberg (2002)

19. Jia, K., Wang, X., Yuan, Z., Xu, G.: Distinguishing and second-preimage attacks on CBC-like MACs. In: Garay, J.A., Miyaji, A., Otsuka, A. (eds.) CANS 2009. LNCS, vol. 5888, pp. 349–361. Springer, Heidelberg (2009)

20. Joux, A., Poupard, G., Stern, J.: New attacks against standardized MACs. In: Johansson, T. (ed.) FSE 2003. LNCS, vol. 2887, pp. 170–181. Springer, Heidelberg (2003)

21. JTC1. ISO/IEC 9797-1:1999 Information technology—Security techniques—Message Authentication Codes (MACs)—Part 1: Mechanisms using a block cipher (1999)

22. Kurosawa, K., Iwata, T.: TMAC: Two-key CBC MAC. In: Joye, M. (ed.) CT-RSA 2003. LNCS, vol. 2612, pp. 33–49. Springer, Heidelberg (2003)

23. Lucks, S.: The sum of PRPs is a secure PRF. In: Preneel, B. (ed.) EUROCRYPT 2000. LNCS, vol. 1807, pp. 470–484. Springer, Heidelberg (2000)

24. Minematsu, K.: Beyond-birthday-bound security based on tweakable block cipher. In: Dunkelman, O. (ed.) Fast Software Encryption. LNCS, vol. 5665, pp. 308–326. Springer, Heidelberg (2009)

25. Nandi, M.: Fast and secure CBC-type MAC algorithms. In: Dunkelman, O. (ed.) FSE 2009. LNCS, vol. 5665, pp. 375–393. Springer, Heidelberg (2009)
26. NIST. Advanced Encryption Standard (AES). FIPS 197 (2001)
27. NIST. Recommendation for block cipher modes of operation: The CMAC mode for authentication. SP 800-38B (2005)
28. NIST. Request for candidate algorithm nominations for a new cryptographic hash algorithm (SHA-3) family. Federal Register Notice, November 2 (2007)
29. Patarin, J.: Security of random Feistel schemes with 5 or more rounds. In: Franklin, M. (ed.) CRYPTO 2004. LNCS, vol. 3152, pp. 106–122. Springer, Heidelberg (2004)
30. Petrank, E., Rackoff, C.: CBC MAC for real-time data sources. J. Cryptology 13(3), 315–338 (2000)
31. Preneel, B., van Oorschot, P.C.: MDx-MAC and building fast MACs from hash functions. In: Coppersmith, D. (ed.) CRYPTO 1995. LNCS, vol. 963, pp. 1–14. Springer, Heidelberg (1995)
32. Satoh, A., Sugawara, T., Aoki, T.: High-speed pipelined hardware architecture for Galois Counter Mode. In: Garay, J.A., Lenstra, A.K., Mambo, M., Peralta, R. (eds.) ISC 2007. LNCS, vol. 4779, pp. 118–129. Springer, Heidelberg (2007)

On Fast Verification of Hash Chains

Dae Hyun Yum[1], Jin Seok Kim[2], Pil Joong Lee[1], and Sung Je Hong[2]

[1] Information Security Lab, POSTECH, Republic of Korea
{dhyum,pjl}@postech.ac.kr
[2] High Performance Computing Lab, POSTECH, Republic of Korea
{treasure,sjhong}@postech.ac.kr

Abstract. A hash chain H for a hash function $\mathsf{hash}(\cdot)$ is a sequence of hash values $\langle x_n, x_{n-1}, \ldots, x_0 \rangle$, where x_0 is a secret value, x_i is generated by $x_i = \mathsf{hash}(x_{i-1})$ for $1 \leq i \leq n$, and x_n is a public value. Hash values of H are disclosed gradually from x_{n-1} to x_0. The correctness of a disclosed hash value x_i can be verified by checking the equation $x_n \stackrel{?}{=} \mathsf{hash}^{n-i}(x_i)$. To speed up the verification, Fischlin introduced a check-bit scheme at CT-RSA 2004. The basic idea of the check-bit scheme is to output some extra information cb, called a check-bit vector, in addition to the public value x_n, which allows each verifier to perform only a fraction of the original work according to his or her own security level. We revisit the Fischlin's check-bit scheme and show that the length of the check-bit vector cb can be reduced nearly by half. The reduced length of cb is close to the theoretic lower bound.

Keywords: Hash chain, progressive verification, check-bit scheme.

1 Introduction

Hash chains have been used as an important cryptographic tool for various applications including payment systems [1,2], one-time password systems [3], multicast authentication [4,5], secure routing [6], and on-line auctions [7]. Hash chains make use of computation-effective hash functions that can be implemented even in small mobile devices. Despite the computational efficiency of hash functions, the performance improvement of hash chains is a challenging research topic for chains of long length.

Researchers have studied algorithmic aspects of hash chains in two ways: efficient generation and verification. A naïve generation algorithm of a hash chain is to recompute each hash value from the secret value x_0. That is, one can simply calculate $x_{n-i} = \mathsf{hash}^{n-i}(x_0)$ at time period i, which has the storage complexity of $O(1)$ but computation complexity of $O(n)$. Another trivial generation algorithm is to precompute and store all hash values in memory and output x_{n-i} at time period i by executing lookup operations. The storage complexity of this generation algorithm is $O(n)$. More sophisticated generation technique is the so-called "fractal" generation algorithms [8,9,10,11,12] that have the storage complexity of $O(\log n)$ and computation complexity of $O(\log n)$. Fractal generation algorithms store $O(\log n)$ intermediary hash values in advance and change

J. Pieprzyk (Ed.): CT-RSA 2010, LNCS 5985, pp. 382–396, 2010.

their values as time elapses. With amortization techniques, the computational cost of $O(\log n)$ can be achieved.

While efficient generation of hash chains has drawn a lot of attention, there is only a single work on the efficient verification of hash chains. At CT-RSA 2004, Fischlin [13] introduced a check-bit scheme to speed up the verification by outputting a check-bit vector cb in addition to the public value.[1] The check-bit vector cb allows to improve the verification time when the verifier is presented an allegedly correct chain value. For security bound T and ϵ on the adversarial running time and success probability, the check-bit scheme allows to decide correctness after a fraction p of the original workload. Here, the original workload is i hash function evaluations if x_{n-i} is given. An interesting property is that the security parameters T and ϵ can be chosen individually by any verifier, even differently for each verification run. As the security parameters T and ϵ determine the fraction $p = p(T, \epsilon)$ of hash chain computations, the more liberal the verifier chooses the security level the less work he or she has to carry out. In this sense, the property is related to the notion of progressive verification [14].

We revisit the Fischlin's check-bit scheme and show that the length of the check-bit vector cb can be reduced nearly by half (with a minor modification). While the original proof uses bounds and approximations, we try to compute exact values with a combinatoric approach. The reduced length of cb is no more than a factor of two away from the theoretic lower bound. In [13], Fischlin leaves an open problem of providing other check-bit schemes with shorter check-bit vectors with comparable simplicity. We answer affirmatively not by providing a new scheme but by giving a new analysis.

2   Preliminaries

We review the terminology and definitions mainly from [13]. A hash chain H for a cryptographic hash function $\mathsf{hash}(\cdot)$ is a sequence of hash values $\langle x_n, \ldots, x_0 \rangle$, where x_0 is a seed or a randomly chosen value and x_i is generated iteratively by $x_i = \mathsf{hash}(x_{i-1})$ for $1 \leq i \leq n$. A single hash value x_i is referred to as a link. The beginning link x_0 is a secret value and the end link x_n is a public value. As x_n is known publicly, the length of $H = \langle x_n, x_{n-1}, \ldots, x_0 \rangle$ is defined as n (not including x_n) or the number of links that are to be disclosed later. Hash values of H are disclosed gradually from x_{n-1} to x_0; at time period i, the link x_{n-i} is disclosed. Pictorially, we represent the hash chain H as $(n + 1)$ points that are equally spaced in a horizontal line as Fig. 1, where check bits, which will be explained later, are also depicted.

In the most simple form, a hash chain can be described by two algorithms $\mathcal{G}$ and $\mathcal{V}$. The generation algorithm $\mathcal{G}$ simply chooses a random x_0 and computes the chain up to x_n. The verification algorithm $\mathcal{V}$ takes as input x_n and some purported link x for time period i and checks that $\mathsf{hash}^i(x) \overset{?}{=} x_n$.

---

[1] Since construction with absolute work bound is trivial and not of much interest, we focus on check-bit schemes with relative bound.

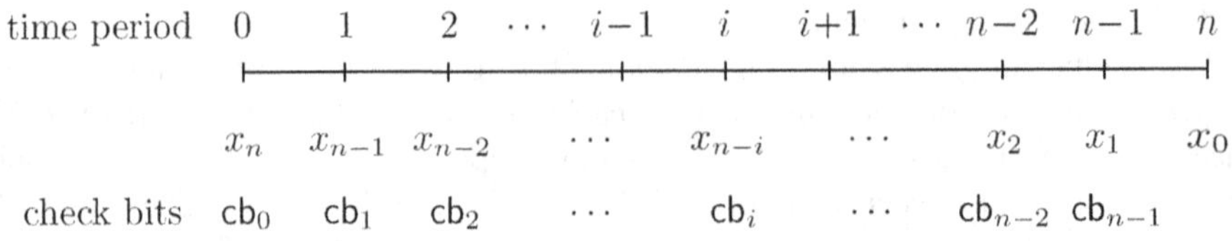

Fig. 1. Hash chain

The basic generation and verification algorithms, for speeding up the verification process, can be augmented to output some extra information. Algorithm $\mathcal{G}$, when generating the chain for a seed x_0, repeatedly runs a deterministic selection algorithm $\mathcal{S}$ as subroutine for each hash function iteration. For each such execution, for $i = n - 1$ down to 0, algorithm $\mathcal{S}$ produces a string cb_i (possibly the empty string $\perp$), which is determined by the intermediate value $x_{n-i} = \mathsf{hash}^{n-i}(x_0)$, the time period number i, and the preceding strings $cb_{i+1}, \ldots, cb_{n-1}$. Fig. 1 will be helpful for understanding the relation between links, time periods, and check bits. The check-bit vector cb is the concatenation of $cb_0, cb_1, \ldots, cb_{n-1}$, i.e., $cb = cb_0 \| cb_1 \| \cdots \| cb_{n-1}$. The notations $cb_{\geq i}$ and $cb_{>i}$ are defined by $cb_{\geq i} = cb_i \| \cdots \| cb_{n-1}$ and $cb_{>i} = cb_{\geq i+1}$ for $i \leq n - 1$ (where $cb_{>n-1} = \perp$).

The augmented verification algorithm $\mathcal{V}$ takes as input the end link x_n, the check-bit vector cb, a purported link x, and a time period i as well as T and ϵ. The verification algorithm $\mathcal{V}$ verifies that x is the correct link for time period i; for each hash function iteration in time period j (from $j = i - 1$ to $j = 0$), the selection algorithm $\mathcal{S}(\mathsf{hash}^{i-j}(x), j, cb_{>j})$ is executed and the result is compared to the given cb_j. Two parameters T and ϵ represent the bounds on the adversarial running time and success probability.

Definition 1. *A check-bit scheme for parameter n is a triple $(\mathcal{G}, \mathcal{V}, \mathcal{S})$ of algorithms (of which $\mathcal{G}$ is probabilistic) such that*

Algorithm $\mathcal{G}$
- *picks a seed x_0 according to some efficiently samplable distribution,*
- *computes $x_i = \mathsf{hash}(x_{i-1})$ for $i = 1, 2, \ldots, n$,*
- *computes $cb_i = \mathcal{S}(x_{n-i}, i, cb_{>i})$ for $i = n - 1, \ldots, 0$,*
- *outputs (x_0, x_n, cb).*

Algorithm $\mathcal{V}$
- *gets inputs x_n, cb, x and a time period i as well as T and ϵ,*
- *repeats the following until $i = 0$ or halt:*
  - *if $cb_i \neq \mathcal{S}(x, i, cb_{>i})$ then reject and stop*
  - *else set $i \leftarrow i - 1$ and $x \leftarrow \mathsf{hash}(x)$*
  - *if $x = x_n$ then accept, else reject.*

Algorithm $\mathcal{S}$
- *takes x, i and $\mathsf{cb}_{>i}$ as input,*
- *computes and returns $\mathsf{cb}_i = \mathcal{S}(x, i, \mathsf{cb}_{>i})$.*

In addition, the scheme is complete, i.e., the verifier never rejects a valid input x_n, cb, x and i, where $x = x_{n-i}$ has been produced by $\mathcal{G}$.

To measure the running time T of an adversary $\mathcal{A}$, only the hash function evaluations are counted. Formally, the adversary is given access to an oracle $\mathsf{hash}(\cdot)$, where generating correct images without querying the oracle is assumed to be infeasible. Let ϵ ($0 \leq \epsilon \leq 1$) denote a bound on the adversary's success probability and p ($0 \leq p \leq 1$) denote a bound on the fraction of the original work the verifier performs, at least if the position of the given link exceeds a certain distance Δ from the end link. The offset Δ allows to overcome the problem of verifying links within the given bound if the purported links are too close to the end link, e.g., checking x_{n-1} with less than 50% of the work. It may depend on the adversarial bounds T and ϵ.

We now define the following experiment for a check-bit scheme $(\mathcal{G}, \mathcal{V}, \mathcal{S})$ with parameter n:

Experiment $\mathrm{Exp}_{\mathcal{A}}(T, \epsilon, p, \Delta)$:

- Algorithm $\mathcal{G}$ generates (x_0, x_n, cb).
- The adversary $\mathcal{A}$ gets as input (x_n, cb). The adversary also gets access to the hash oracle $\mathsf{hash}(\cdot)$ and an oracle $\mathsf{Release}(\cdot)$ which takes an integer j as input and returns x_{n-j}. Let r denote the maximum over all queries to $\mathsf{Release}$ (where $r = 0$ if $\mathcal{A}$ has never queried the oracle).
- In addition to oracle queries, the adversary performs internal computations and finally outputs (x, k), where $1 \leq k \leq n$.
- The verifier $\mathcal{V}$ is invoked on $(x_n, \mathsf{cb}, x, k, T, \epsilon)$ and returns a decision after v hash function evaluations.

Definition 2. *We say that the adversary $\mathcal{A}$ wins experiment $\mathrm{Exp}_{\mathcal{A}}(T, \epsilon, p, \Delta)$ if the following conditions are satisfied,*

- *$\mathcal{A}$ makes at most T hash function evaluations,*
- *$\mathcal{A}$ has queried the hash oracle about $\mathsf{hash}^i(x)$ for all $i = 0, 1, \ldots, \lfloor kp \rfloor - 1$,*
- *$\mathcal{A}$ has queried the oracle $\mathsf{Release}$ only about values smaller than k, i.e., if $k > r$,*
- *if the position k exceeds the offset, i.e., if $k \geq \Delta$, and*
- *if the verifier does not reject within $v \leq kp$ hash function evaluations.*

The mere purpose of letting the adversary query about the output (i.e., the second condition) is to charge the adversary's running time also for the time to verify the output. A check-bit scheme is (T, ϵ, p, Δ)-verifiable if no adversary running in time T can cause the verifier to perform a fraction p or more of the work with probability more than ϵ. Here, the work refers to the number k of hash function evaluations required to verify the correct link x_{n-k} at time period k. As the security bound is chosen by the verifier, p and Δ may be functions of the security parameters T and ϵ.

Definition 3. *A check-bit scheme $(\mathcal{G}, \mathcal{V}, \mathcal{S})$ with parameter n is (T, ϵ, p, Δ)-verifiable if, for any adversary $\mathcal{A}$ running in time at most T, the probability of $\mathcal{A}$ winning experiment $\mathrm{Exp}_{\mathcal{A}}(T, \epsilon, p, \Delta)$ is at most ϵ. The scheme is (p, Δ)-verifiable if, for any adversary $\mathcal{A}$ and any T, ϵ, the probability of $\mathcal{A}$ winning experiment $\mathrm{Exp}_{\mathcal{A}}(T, \epsilon, p, \Delta)$ is at most ϵ.*

3 Fischlin's Check-Bit Scheme

For the construction, the time period number is also input to the hash function. Let $\mathsf{hash}'(\cdot)$ be a hash function and $[\![\cdot]\!]$ be some fixed-length encoding such that the encoding is one-to-one for integers $0, 1, \ldots, n$. Then a hash function $\mathsf{hash}(\cdot)$ for inputs $x_i = [\![i]\!] \| x_i'$ is defined by $\mathsf{hash}(x_i) = [\![i+1]\!] \| \mathsf{hash}'([\![i]\!] \| x_i')$. For random x_0', the beginning link is set as $x_0 = [\![0]\!] \| x_0'$ and the chain is derived by iterating $\mathsf{hash}(\cdot)$ on x_0. The verifier, when presented a link $x_i = [\![i]\!] \| x_i'$, should check that i matches the claimed time period. If $\mathsf{hash}'(\cdot)$ is an appropriate hash function, the least significant bits of the chain links are still well distributed. Let $x|_b$ denote the b least significant bits of a string x. Let $[i, j]$ denote a segment of consecutive points at $i, i+1, \ldots, j$ and $|[i, j]| = j - i + 1$ be the length of $[i, j]$.

The idea of Fischlin's check-bit scheme is to increase the density of check bits towards the end link of the chain as the workload of the verifier is in proportion to the distance to the end link. The hash chain of length n is first partitioned into $I = \log_2 n$ intervals of length $1, 2, 2^2, \ldots, \frac{n}{2}$, where n is assumed to be a power of two. For $\ell = 1, 2, \ldots, I$, each interval $\mathcal{I}_\ell = [2^{\ell-1}, 2^\ell - 1]$ ranges from $2^{\ell-1}$ to $2^\ell - 1$ (and the point n is added to interval $\mathcal{I}_I$). In interval $\mathcal{I}_I$, the selection algorithm $\mathcal{S}_{b,B}$ outputs the b least significant bits of the intermediate links at positions $\frac{jn}{B}$ for $j \in \{\frac{1}{2}B, \frac{1}{2}B+1, \ldots, B-1\}$, where B is an even number determining the density of the check bits. For ease of implementation, B should be a power of two. In interval $\mathcal{I}_{I-1}$, $\mathcal{S}_{b,B}$ outputs the b least significant bits of links at positions $\frac{jn}{2B}$, i.e., the density becomes double. In general, $\mathcal{S}_{b,B}$ outputs the b least significant bits of x_{n-i} for $i \in \mathcal{I}_\ell$ if $i = \frac{jn}{2^{I-\ell}B} = \frac{j2^\ell}{B}$ (with appropriate rounding). Fischlin's check-bit scheme is given as follows.

Construction 4 (Fischlin). The check-bit scheme $(\mathcal{G}_{b,B}, \mathcal{V}_{b,B}, \mathcal{S}_{b,B})$ with parameter $n > B$ is described by the following selection algorithm $\mathcal{S}_{b,B}(x, i)$:

Algorithm $\mathcal{S}_{b,B}(x, i)$

    if $(i \in \mathcal{I}_\ell) \wedge (i = \lfloor \frac{j2^\ell}{B} \rfloor)$ for $j \in \{\frac{1}{2}B, \frac{1}{2}B+1, \ldots, B-1\}$ **then**

        output $\mathsf{cb}_i = x|_b$

    else

        output $\mathsf{cb}_i = \perp$

    end if

For each interval $\mathcal{I}_\ell$ such that $|\mathcal{I}_\ell| = 2^{\ell-1} \geq \frac{B}{2}$, the variable j runs through $\frac{B}{2}$ values and $x|_b$ is output for such j. The number of check bits in these intervals is $\frac{1}{2}bB \cdot (I - \lceil \log_2 B \rceil + 1)$. For the remaining intervals with $\ell < \log_2 B$, the selection algorithm always outputs $x|_b$ and thus (at most) bB bits are output. The total length of a check-bit vector is bound by:

$$\frac{1}{2}bB \cdot (I - \lceil \log_2 B \rceil + 1) + bB = \frac{1}{2}bB \cdot (\log_2 n - \lceil \log_2 B \rceil + 3).$$

If B is a power of two, then $\frac{1}{2}bB$ bits are needed for intervals with $\ell < \log_2 B$ and the total length of a check-bit vector is:

$$\frac{1}{2}bB \cdot (I - \log_2 B + 1) + \frac{1}{2}bB = \frac{1}{2}bB \cdot (\log_2 n - \log_2 B + 2).$$

The Fischlin's check-bit scheme in Construction 4 can also be rewritten as follows.

Construction 5. The check-bit scheme $(\mathcal{G}_{b,B}, \mathcal{V}_{b,B}, \mathcal{S}_{b,B})$ with parameter $n > B$ is described by the following selection algorithm $\mathcal{S}_{b,B}(x,i)$, where $\ell_{min} = \lceil \log_2 B \rceil$:

Algorithm $\mathcal{S}_{b,B}(x,i)$
Find ℓ such that $i \in \mathcal{I}_\ell$
if $\ell \geq \ell_{min}$ then
    if $i = \lfloor \frac{j2^\ell}{B} \rfloor$ for $j \in \{\frac{1}{2}B, \frac{1}{2}B + 1, \ldots, B - 1\}$ then
        output $\mathsf{cb}_i = x|_b$
    else
        output $\mathsf{cb}_i = \bot$
    end if
else
    output $\mathsf{cb}_i = x|_b$
end if

The security of the Fischlin's check-bit scheme is based on two assumptions. The first assumption basically says that finding pre-images for the given chain is impossible at least within a given bound like $T_0 \approx 2^{60}$ and $\epsilon_0 \approx 2^{-40}$. In particular, giving away some bits of pre-images of chain values must not facilitate the search. The values T_0, ϵ_0 are large bounds for adequate security against inverters and collision-finders of the underlying hash function $\mathsf{hash}(\cdot)$. They are determined by the choice of $\mathsf{hash}(\cdot)$ and are usually fixed.

Assumption 6. *Let $\mathcal{A}$ be an adversary running in time at most T_0 against the scheme $(\mathcal{G}_{b,B}, \mathcal{V}_{b,B}, \mathcal{S}_{b,B})$ in Construction 4. Then, the probability of $\mathcal{A}$ winning experiment $\mathrm{Exp}_\mathcal{A}(T_0, \epsilon_0, 1, 1)$, i.e., the experiment with bounds $p = 1$ and $\Delta = 1$, is at most ϵ_0.*

Note that the bound $p = 1$ implies that the verifier can check the complete chain and, in particular, also compares the final value to the original end link. The assumption therefore says that the adversary is not able to fine (x, k) within the success bound (T_0, ϵ_0) such that $x_n = \mathsf{hash}^k(x)$ and x has not been released before. Although the adversary may not be able to find a pre-image of the chain's end link, it might still be possible to find a related pre-image such that large parts of the check-bit vector coincide. The following assumption rules this out.

Assumption 7. *For any two seeds x_0, y_0 such that $\mathsf{hash}^i(x_0) \neq \mathsf{hash}^i(y_0)$ for all $i = 0, 1, \ldots, n$, we assume that the check-bit vectors $\mathsf{cb}(x_0)$ and $\mathsf{cb}(y_0)$ generated by $\mathcal{S}_{b,B}$ are uniformly and independently distributed strings of the corresponding length (where the probability is over the choice of the hash function).*

Both assumptions are satisfied if $\mathsf{hash}(\cdot)$ is for example modeled as a random oracle [15]. Fischlin showed that Construction 4 is a secure check-bit scheme under the stated assumptions. Note that in Theorem 8, we describe p as a function of w instead of a function of (T, ϵ) merely for notational convenience.

Theorem 8 (Fischlin). *Under Assumption 6 (for parameters T_0, ϵ_0) and Assumption 7, the check-bit scheme $(\mathcal{G}_{b,B}, \mathcal{V}_{b,B}, \mathcal{S}_{b,B})$ in Construction 4 is a (p, Δ)-verifiable check-bit scheme for*

$$p(w) = \begin{cases} \frac{2w+2b}{bB} & if \ 0 \leq w \leq \frac{bB-2b}{2} \\ 1 & otherwise \end{cases} \quad where \ \ w = \log_2 T + \log_2 \frac{1}{\epsilon - \epsilon_0} ,$$

$$\Delta = 2^{\log_2 B - 1}$$

if $T \leq T_0$ and $\epsilon_0 \leq \epsilon$ (and $p = 1$ otherwise). For chains of length n, the check-bit vector cb has at most

$$\frac{1}{2} bB \cdot (\log_2 n - \lceil \log_2 B \rceil + 3)$$

bits. If B is a power of two, the check-bit vector cb has

$$\frac{1}{2} bB \cdot (\log_2 n - \log_2 B + 2)$$

bits.

For example, Theorem 8 says that Construction 4 with parameters $n = 1024$, $B = 128 \ (= 2^7)$, and $b = 2$ requires $p \approx 48.44\%$, $\Delta = 64$ and $|\mathsf{cb}| = 640$ for $T = 2^{40}$, $\epsilon = 2^{-20}$ and $\epsilon_0 \approx 0$.

4   Check-Bit Scheme with Reduced Check-Bit Vector

If we set $b = 1$ (for simplifying the explanation), Fischlin's scheme (i.e., Construction 5) inserts $\frac{B}{2}$ check bits equidistantly into each interval $\mathcal{I}_\ell = [2^{\ell-1}, 2^\ell - 1]$ for $\ell \geq \ell_{min}$. The check bits of $\mathcal{I}_\ell$ is twice as dense as those of $\mathcal{I}_{\ell+1}$ because $|\mathcal{I}_{\ell+1}| = 2^\ell = 2 \cdot 2^{\ell-1} = 2|\mathcal{I}_\ell|$. In contrast, $|\mathcal{I}_\ell| = 2^{\ell-1}$ check bits are inserted into interval $\mathcal{I}_\ell$ for $\ell < \ell_{min}$ and the density is uniform in $[1, \Delta]$.

As we intend to reduce the required number of the check bits, we have to insert fewer check bits into each interval. With a tight analysis of Theorem 8, we find that the check bits of $\mathcal{I}_\ell$ for $\ell \geq \ell_{min}$ can actually be reduced by half without degrading the security level. In other words, the original proof of Theorem 8 underestimates the performance of the check-bit scheme. However, if we reduce the check bits of $\mathcal{I}_\ell$ for $\ell \geq \ell_{min}$ by half, positions near Δ becomes problematic because the density is uniform in $[1, \Delta]$. A possible solution is to make the density in $\mathcal{I}_\ell$ twice of that in $\mathcal{I}_{\ell+1}$ for $[1, \Delta]$ just as for $[\Delta, n]$. Our choice is to increase the density of $[\frac{\Delta}{4}, \Delta]$ only. Specifically, when $b = 1$, we insert $\frac{B}{2}$ check bits into $\mathcal{I}_{\ell_{min}-1}$ and another $\frac{B}{2}$ check bits into $\mathcal{I}_{\ell_{min}-2}$. The interval lengths of $\mathcal{I}_{\ell_{min}-1}$ and $\mathcal{I}_{\ell_{min}-2}$ are less than $\frac{B}{2}$ and therefore $|\mathsf{cb}_i|$ for $i \in \mathcal{I}_{\ell_{min}-1}$ and $i \in \mathcal{I}_{\ell_{min}-2}$ should be increased (according to the interval length). Similarly, one may also increase the density of $[1, \frac{\Delta}{4}]$ but this implies that $|\mathsf{cb}_i|$ in $[1, \frac{\Delta}{4}]$ becomes large and a considerable amount of information on x_i could be leaked. We put no check bits near the end link, i.e., $\mathsf{cb}_i = \perp$ for $i < 2^{\ell_{min}-2}$ because benefits of inserting check bits in $[1, \frac{\Delta}{4}]$ are not very significant. The following Construction 9 is a variant of Construction 5 with above-mentioned modifications.

Construction 9. The check-bit scheme $(\mathcal{G}_{b,B}, \mathcal{V}_{b,B}, \mathcal{S}_{b,B})$ with parameter $n > B$ is described by the following selection algorithm $\mathcal{S}_{b,B}(x, i)$, where $\ell_{min} = \lceil \log_2 B \rceil$:

Algorithm $\mathcal{S}_{b,B}(x, i)$
  Find ℓ such that $i \in \mathcal{I}_\ell$
  if $\ell \geq \ell_{min}$ **then**
    if $i = \lfloor \frac{j2^\ell}{B} \rfloor$ for $j \in \{\frac{1}{2}B, \frac{1}{2}B + 1, \ldots, B - 1\}$ **then**
      output $\mathsf{cb}_i = x|_b$
    else
      output $\mathsf{cb}_i = \perp$
    end if
  else if $\ell = \ell_{min} - 1$ **then**
    output $\mathsf{cb}_i = x|_{2b}$
  else if $\ell = \ell_{min} - 2$ **then**
    output $\mathsf{cb}_i = x|_{4b}$
  else
    output $\mathsf{cb}_i = \perp$
  end if

For each interval $\mathcal{I}_\ell$ for $\ell \geq \ell_{min}$, the selection algorithm of Construction 9 is exactly the same as that of Construction 5, which requires $\frac{1}{2}bB \cdot (I - \lceil \log_2 B \rceil + 1)$ check bits. For $\mathcal{I}_{\ell_{min}-1}$ and $\mathcal{I}_{\ell_{min}-2}$, the selection algorithm outputs (at most) $2bB$ bits. Therefore, the total length of a check-bit vector is bound by:

$$\frac{1}{2}bB \cdot (I - \lceil \log_2 B \rceil + 1) + 2bB = \frac{1}{2}bB \cdot (\log_2 n - \lceil \log_2 B \rceil + 5).$$

If B is a power of two, we have:

$$\frac{1}{2}bB \cdot (I - \log_2 B + 1) + bB = \frac{1}{2}bB \cdot (\log_2 n - \log_2 B + 3).$$

In Theorem 10, we show that Construction 9 requires a reduced check-bit vector for given security levels. Theorem 10 says that for example, Construction 9 with parameters $n = 1024$, $B = 128$ $(= 2^7)$ and $b = 1$ requires $p \approx 48.39\%$, $\Delta = 64$ and $|cb| = 384$ for $T = 2^{40}$, $\epsilon = 2^{-20}$ and $\epsilon_0 \approx 0$. Recall that Theorem 8 requires $p \approx 48.44\%$ and $|cb| = 640$. In this example, while maintaining the security level and verification efficiency, $b = 1$ instead of $b = 2$ can be used and thus only half number of check bits are inserted into $\mathcal{I}_\ell$ for $\ell \geq \ell_{min}$. Although Construction 9 uses more check bits for the segment $[1, \Delta]$, this entails relatively small overhead because Δ is usually small. Note also that the lower bound of Theorem 13 in Appendix A requires $|cb| \geq 210$. Thus, we are (practically speaking) no more than a factor of two away from the optimal solution in this case, which is sometimes called "almost optimal" [8].

Theorem 10. *Under Assumption 6 and Assumption 7, the check-bit scheme $(\mathcal{G}_{b,B}, \mathcal{V}_{b,B}, \mathcal{S}_{b,B})$ in Construction 9 is a (p, Δ)-verifiable check-bit scheme for*

$$p(w) = \begin{cases} \frac{2w}{2w+bB} & if \ 0 \leq w \leq \frac{bB}{2} \\ 1 - \frac{bB}{4w} & if \ \frac{bB}{2} \leq w \leq bB \\ 1 & otherwise \end{cases} \quad where \ \ w = \log_2 T + \log_2 \frac{1}{\epsilon - \epsilon_0},$$

$$\Delta = 2^{\lceil \log_2 B \rceil - 1}$$

if $T \leq T_0$ and $\epsilon_0 \leq \epsilon$ (and $p = 1$ otherwise). For chains of length n, the check-bit vector cb has at most

$$\frac{1}{2}bB \cdot (\log_2 n - \lceil \log_2 B \rceil + 5)$$

bits. If B is a power of two, the check-bit vector cb has

$$\frac{1}{2}bB \cdot (\log_2 n - \log_2 B + 3)$$

bits.

Before giving the proof, we draw a brief comparison between Theorem 8 and Theorem 10. Let $p_1(\cdot)$ denote $p(\cdot)$ of Theorem 8 and $p_2(\cdot)$ denote $p(\cdot)$ of Theorem 10. We use the approximation of $p_1(w) = \frac{2w+2b}{bB} \approx \frac{2w}{bB}$.

$$p_1(w) \approx \begin{cases} 0 & if \ w = 0 \\ 0.25 & if \ w = \frac{1}{8}bB \\ 0.5 & if \ w = \frac{1}{4}bB \\ 1 & if \ w = \frac{1}{2}bB \\ 1 & if \ w > \frac{1}{2}bB \end{cases}, \quad p_2(w) = \begin{cases} 0 & if \ w = 0 \\ 0.25 & if \ w = \frac{1}{6}bB \\ 0.5 & if \ w = \frac{1}{2}bB \\ 0.75 & if \ w = bB \\ 1 & if \ w > bB \end{cases} \quad (1)$$

From Eq. (1), one can see that $p_1(w)$ takes $0 \sim 1$ for $0 \leq w \leq \frac{1}{2}bB$ and $p_2(w)$ takes $0 \sim 1$ for $0 \leq w \leq bB$. In this sense, bB of $p_2(w)$ can be reduced by half for given security levels, which results in the reduction of check bits in intervals $\mathcal{I}_\ell$

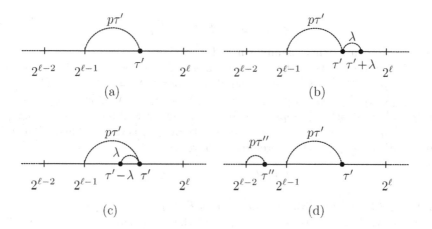

Fig. 2. Case I

for $\ell \geq \ell_{min}$. An interesting fact on $p_2(w)$ is that it does not take values between 0.75 and 1, i.e., $p_2(w)$ takes $0 \sim 1 - \frac{1}{2^2}$ as well as 1. This is because we do not put check bits in $\mathcal{I}_\ell$ for $\ell < \ell_{min} - 2$. If additional bB check bits are inserted in each interval $\mathcal{I}_{\ell_{min}-i}$ for $2 < i \leq j$, then $p_2(w)$ can take $0 \sim 1 - \frac{1}{2^j}$ as well as 1.

Proof. Let $\mathcal{A}$ be an adversary that runs at most T steps and tries to come up with (x, τ) for $\Delta \leq \tau \leq n$ which can pass the verification test with a non-negligible probability ϵ. We show how to set the parameter p to bound the adversary's success probability below the given ϵ.

To check the validity of $\mathcal{A}$'s output (x, τ), i.e., a link x at time period τ, verifier computes and compares cb_i for $\tau - p\tau \leq i < \tau$. We define a function $\mathcal{F}(\tau, p) = [\tau - p\tau, \tau]$ for $\tau \in \mathbb{N}$ and $0 \leq p \leq 1$ and let $\#[i, j]$ be the number of check bits in the segment $[i, j]$. If $\#\mathcal{F}(\tau, p) \geq w = \log_2 T + \log_2 \frac{1}{\epsilon - \epsilon_0}$, the success probability of the adversary $\mathcal{A}$ is below ϵ because $\mathcal{A}$ can compute at most T hash function evaluations. For given w, our goal is to find the smallest p such that $\#\mathcal{F}(\tau, p) \geq w$ where $\Delta \leq \tau \leq n$. We assume that B is a power of two for simplicity's sake, which gives $\ell_{min} = \log_2 B$. If we set $\Delta = 2^{\log_2 B - 1} = \frac{B}{2}$, we have $\Delta = 2^{\ell_{min}-1}$ and $\mathcal{I}_{\ell_{min}} = [2^{\ell_{min}-1}, 2^{\ell_{min}} - 1] = [\Delta, 2\Delta - 1]$.

The proof consists of two cases of $0 \leq w \leq \frac{bB}{2}$ and $\frac{bB}{2} \leq w \leq bB$ with $T \leq T_0$ and $\epsilon_0 \leq \epsilon$; other cases are trivial. Let q be $q = 1 - p$. As each interval $\mathcal{I}_\ell = [2^{\ell-1}, 2^\ell - 1]$ for $\ell \geq \ell_{min} - 2$ contains $\frac{bB}{2}$ check bits, $0 \leq w \leq \frac{bB}{2}$ implies $0 \leq p \leq \frac{1}{2}$ and $\frac{bB}{2} \leq w \leq bB$ implies $\frac{1}{2} \leq p \leq \frac{3}{4}$.

CASE I: $0 \leq w \leq \frac{bB}{2}$ ($0 \leq p \leq \frac{1}{2}$ and $\frac{1}{2} \leq q \leq 1$)

First, we consider $\#\mathcal{F}(\tau, p)$ for $\tau \in \mathcal{I}_\ell$ for $\ell_{min} + 1 \leq \ell \leq I$, where $I = \log_2 n$. Let τ' be such that $\mathcal{F}(\tau', p) = [2^{\ell-1}, \tau']$ or $\tau' - p\tau' = 2^{\ell-1}$ in Fig. 2-(a). If τ' moves λ points to the right (but still $\tau' + \lambda \in \mathcal{I}_\ell$) as in Fig. 2-(b), we have

$$\#\mathcal{F}(\tau',p) \leq \#\mathcal{F}(\tau' + \lambda, p) \tag{2}$$

because the density in $\mathcal{I}_\ell$ is uniform and $\tau' < \tau' + \lambda$.

If τ' moves λ points to the left (but still $\tau' - \lambda \in \mathcal{I}_\ell$) as in Fig. 2–(c), we have $\mathcal{F}(\tau' - \lambda, p) = [2^{\ell-1} - q\lambda, \tau' - \lambda]$ from $\tau' - \lambda - p(\tau' - \lambda) = (\tau' - p\tau') - (1-p)\lambda = 2^{\ell-1} - q\lambda$. When compared with $\#\mathcal{F}(\tau',p)$, $\#\mathcal{F}(\tau' - \lambda, p)$ loses $\#[\tau' - \lambda, \tau']$ in $\mathcal{I}_\ell$ and gains $\#[2^{\ell-1} - q\lambda, 2^{\ell-1}]$ in $\mathcal{I}_{\ell-1}$. Here, the relation $\#[\tau' - \lambda, \tau'] \leq \#[2^{\ell-1} - q\lambda, 2^{\ell-1}]$ holds, because the density in $\mathcal{I}_{\ell-1}$ is the double of that in $\mathcal{I}_\ell$ and q satisfies $\frac{1}{2} \leq q \leq 1$. Therefore, we have

$$\#\mathcal{F}(\tau',p) \leq \#\mathcal{F}(\tau' - \lambda, p). \tag{3}$$

From Eq. (2) and Eq. (3), we know that $\#\mathcal{F}(\tau',p) \leq \#\mathcal{F}(\tau,p)$ for $\tau \in \mathcal{I}_\ell$.

Similarly, if we choose $\tau'' \in \mathcal{I}_{\ell-1}$ such that $\mathcal{F}(\tau'',p) = [2^{\ell-2}, \tau'']$ or $\tau'' - p\tau'' = 2^{\ell-2}$ in Fig. 2–(d), we have $\#\mathcal{F}(\tau'',p) \leq \#\mathcal{F}(\tau,p)$ for $\tau \in \mathcal{I}_{\ell-1}$. Let's compare $\#\mathcal{F}(\tau',p)$ with $\#\mathcal{F}(\tau'',p)$. From $\tau' - p\tau' = 2^{\ell-1}$ and $\tau'' - p\tau'' = 2^{\ell-2}$, we get $\tau' = \frac{1}{q}2^{\ell-1}$ and $\tau'' = \frac{1}{q}2^{\ell-2}$ that give $|\mathcal{F}(\tau',p)| = p\tau' = \frac{p}{q}2^{\ell-1}$ and $|\mathcal{F}(\tau'',p)| = p\tau'' = \frac{p}{q}2^{\ell-2}$. As the the density in $\mathcal{I}_{\ell-1}$ is the double of that in $\mathcal{I}_\ell$, we have

$$\#\mathcal{F}(\tau',p) = \#\mathcal{F}(\tau'',p). \tag{4}$$

From Eq. (2), Eq. (3) and Eq. (4), we can see that $\#\mathcal{F}(\tau',p) \leq \#\mathcal{F}(\tau,p)$ for $\Delta \leq \tau \leq n$.

Now, we only have to find p such that $\#\mathcal{F}(\tau',p) \geq w$ for $\tau' = \frac{1}{q}2^{\ell-1}$. As $\frac{bB}{2}$ check bits are inserted into $\mathcal{I}_\ell$ and w satisfies $0 \leq w \leq \frac{bB}{2}$, we choose p from Fig. 2–(a) as follows.

$$p = \frac{\frac{w}{\frac{bB}{2}}|\mathcal{I}_\ell|}{2^{\ell-1} + \frac{w}{\frac{bB}{2}}|\mathcal{I}_\ell|} = \frac{\frac{2w}{bB}2^{\ell-1}}{2^{\ell-1} + \frac{2w}{bB}2^{\ell-1}} = \frac{2w}{bB + 2w} \tag{5}$$

Theorem 10 for $0 \leq w \leq \frac{bB}{2}$ follows from Eq. (5).

CASE II: $\frac{bB}{2} \leq w \leq bB$ ($\frac{1}{2} \leq p \leq \frac{3}{4}$ and $\frac{1}{4} \leq q \leq \frac{1}{2}$)

As before, we first consider $\#\mathcal{F}(\tau,p)$ for $\tau \in \mathcal{I}_\ell$ for $\ell_{min} + 1 \leq \ell \leq I$. Let τ' be such that $\mathcal{F}(\tau',p) = [2^{\ell-2}, \tau']$ or $\tau' - p\tau' = 2^{\ell-2}$ in Fig. 3–(a). If τ' moves λ points to the right (but still $\tau' + \lambda \in \mathcal{I}_\ell$) as in Fig. 3–(b), we have $\mathcal{F}(\tau' + \lambda, p) = [2^{\ell-2} + q\lambda, \tau' + \lambda]$ from $\tau' + \lambda - p(\tau' + \lambda) = (\tau' - p\tau') + (1-p)\lambda = 2^{\ell-2} + q\lambda$. When compared with $\#\mathcal{F}(\tau',p)$, $\#\mathcal{F}(\tau' + \lambda, p)$ loses $\#[2^{\ell-2}, 2^{\ell-2} + q\lambda]$ in $\mathcal{I}_{\ell-1}$ and gains $\#[\tau', \tau' + \lambda]$ in $\mathcal{I}_\ell$. Here, the relation $\#[2^{\ell-2}, 2^{\ell-2} + q\lambda] \leq \#[\tau', \tau' + \lambda]$ holds, because the density in $\mathcal{I}_{\ell-1}$ is the double of that in $\mathcal{I}_\ell$ and q satisfies $\frac{1}{4} \leq q \leq \frac{1}{2}$. Therefore, we have

$$\#\mathcal{F}(\tau',p) \leq \#\mathcal{F}(\tau' + \lambda, p). \tag{6}$$

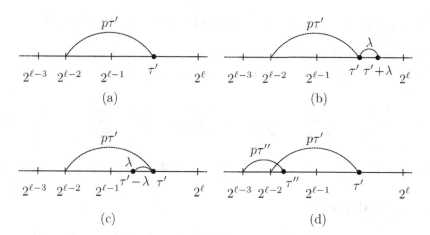

Fig. 3. Case II

If τ' moves λ points to the left (but still $\tau' - \lambda \in \mathcal{I}_\ell$) as in Fig. 3–(c), we have $\mathcal{F}(\tau' - \lambda, p) = [2^{\ell-2} - q\lambda, \tau' - \lambda]$ from $\tau' - \lambda - p(\tau' - \lambda) = (\tau' - p\tau') - (1 - p)\lambda = 2^{\ell-2} - q\lambda$. When compared with $\#\mathcal{F}(\tau', p)$, $\#\mathcal{F}(\tau' - \lambda, p)$ gains $\#[2^{\ell-2} - q\lambda, 2^{\ell-2}]$ in $\mathcal{I}_{\ell-2}$ and loses $\#[\tau' - \lambda, \tau']$ in $\mathcal{I}_\ell$. Here, the relation $\#[2^{\ell-2} - q\lambda, 2^{\ell-2}] \geq \#[\tau' - \lambda, \tau']$ holds, because the density in $\mathcal{I}_{\ell-2}$ is the quadruple of that in $\mathcal{I}_\ell$ and q satisfies $\frac{1}{4} \leq q \leq \frac{1}{2}$. Therefore, we have

$$\#\mathcal{F}(\tau', p) \leq \#\mathcal{F}(\tau' + \lambda, p). \tag{7}$$

From Eq. (6) and Eq. (7), we know that $\#\mathcal{F}(\tau', p) \leq \#\mathcal{F}(\tau, p)$ for $\tau \in \mathcal{I}_\ell$.

Similarly, if we choose $\tau'' \in \mathcal{I}_{\ell-1}$ such that $\mathcal{F}(\tau'', p) = [2^{\ell-3}, \tau'']$ or $\tau'' - p\tau'' = 2^{\ell-3}$ in Fig. 3–(d), we have $\#\mathcal{F}(\tau'', p) \leq \#\mathcal{F}(\tau, p)$ for $\tau \in \mathcal{I}_{\ell-1}$. Let's compare $\#\mathcal{F}(\tau', p)$ with $\#\mathcal{F}(\tau'', p)$. From $\tau' - p\tau' = 2^{\ell-2}$ and $\tau'' - p\tau'' = 2^{\ell-3}$, we get $\tau' = \frac{1}{q}2^{\ell-2}$ and $\tau'' = \frac{1}{q}2^{\ell-3}$. Then, it follows that

$$|\mathcal{F}(\tau', p)| = p\tau' = \frac{p}{q}2^{\ell-2} = 2^{\ell-2} + \frac{p-q}{q}2^{\ell-2} = |\mathcal{I}_{\ell-1}| + \frac{p-q}{q}2^{\ell-2} \text{ and}$$

$$|\mathcal{F}(\tau'', p)| = p\tau'' = \frac{p}{q}2^{\ell-3} = 2^{\ell-3} + \frac{p-q}{q}2^{\ell-3} = |\mathcal{I}_{\ell-2}| + \frac{p-q}{q}2^{\ell-3}.$$

As $\#\mathcal{I}_\ell = \#\mathcal{I}_{\ell-1} = \#\mathcal{I}_{\ell-2} = \frac{bB}{2}$, we have

$$\#\mathcal{F}(\tau', p) = \#\mathcal{I}_{\ell-1} + \frac{\frac{p-q}{q}2^{\ell-2}}{2^{\ell-1}} \cdot \#\mathcal{I}_\ell = \frac{bB}{2}\left(1 + \frac{p-q}{2q}\right) \text{ and}$$

$$\#\mathcal{F}(\tau'', p) = \#\mathcal{I}_{\ell-2} + \frac{\frac{p-q}{q}2^{\ell-3}}{2^{\ell-2}} \cdot \#\mathcal{I}_{\ell-1} = \frac{bB}{2}\left(1 + \frac{p-q}{2q}\right),$$

which gives

$$\#\mathcal{F}(\tau', p) = \#\mathcal{F}(\tau'', p). \tag{8}$$

From Eq. (6), Eq. (7) and Eq. (8), we can see that $\#\mathcal{F}(\tau',p) \leq \#\mathcal{F}(\tau,p)$ for $\Delta \leq \tau \leq n$.

Now, we find p such that $\#\mathcal{F}(\tau',p) \geq w$ for $\tau' = \frac{1}{q}2^{\ell-2}$. From $\#\mathcal{I}_{\ell-1} = \#\mathcal{I}_\ell = \frac{bB}{2}$ and $\frac{bB}{2} \leq w \leq bB$, we can compute p of Fig. 3–(a) as follows.

$$p = \frac{|\mathcal{I}_{\ell-1}| + \frac{w-\frac{bB}{2}}{\frac{bB}{2}}|\mathcal{I}_\ell|}{2^{\ell-1} + \frac{w-\frac{bB}{2}}{\frac{bB}{2}}|\mathcal{I}_\ell|} = \frac{2^{\ell-2} + \frac{2w-bB}{bB}2^{\ell-1}}{2^{\ell-1} + \frac{2w-bB}{bB}2^{\ell-1}} = \frac{1 + \frac{4w-2bB}{bB}}{2 + \frac{4w-2bB}{bB}} = 1 - \frac{bB}{4w} \qquad (9)$$

Theorem 10 for $\frac{bB}{2} \leq w \leq bB$ follows from Eq. (9).   Q.E.D. ∎

Acknowledgements

Dae Hyun Yum and Pil Joong Lee were supported by Basic Science Research Program through the National Research Foundation of Korea (NRF) funded by the Ministry of Education, Science and Technology (2009-0075147) and the Brain Korea 21 Project. Jin Seok Kim and Sung Je Hong were supported by the MKE (The Ministry of Knowledge Economy), Korea, under the HNRC (Home Network Research Center) – ITRC (Information Technology Research Center) support program supervised by the NIPA (National IT Industry Promotion Agency) NIPA-2009-C1090-0902-0035.

References

1. Anderson, R.J., Manifavas, C., Sutherland, C.: Netcard - a practical electronic-cash system. In: Lomas, M. (ed.) Security Protocols 1996. LNCS, vol. 1189, pp. 49–57. Springer, Heidelberg (1997)
2. Rivest, R.L., Shamir, A.: Payword and micromint: Two simple micropayment schemes. In: Lomas, M. (ed.) Security Protocols 1996. LNCS, vol. 1189, pp. 69–87. Springer, Heidelberg (1997)
3. Haller, N.: The s/key one-time password system. RFC 1760, Internet Engineering Task Force (1995)
4. Perrig, A., Canetti, R., Song, D.X., Tygar, J.D.: Efficient and secure source authentication for multicast. In: NDSS 2001, The Internet Society (2001)
5. Perrig, A., Canetti, R., Tygar, J.D., Song, D.X.: Efficient authentication and signing of multicast streams over lossy channels. In: IEEE Symposium on Security and Privacy, pp. 56–73. IEEE Computer Society, Los Alamitos (2000)
6. Hu, Y.C., Perrig, A., Johnson, D.B.: Ariadne: A secure on-demand routing protocol for ad hoc networks. Wireless Networks 11(1-2), 21–38 (2005)
7. Stubblebine, S.G., Syverson, P.F.: Fair on-line auctions without special trusted parties. In: Franklin, M.K. (ed.) FC 1999. LNCS, vol. 1648, pp. 230–240. Springer, Heidelberg (1999)
8. Coppersmith, D., Jakobsson, M.: Almost optimal hash sequence traversal. In: Blaze, M. (ed.) FC 2002. LNCS, vol. 2357, pp. 102–119. Springer, Heidelberg (2003)
9. Jakobsson, M.: Fractal hash sequence representation and traversal. In: IEEE International Symposium on Information Theory, pp. 437–444. IEEE, Los Alamitos (2002)

10. Kim, S.R.: Improved scalable hash chain traversal. In: Zhou, J., Yung, M., Han, Y. (eds.) ACNS 2003. LNCS, vol. 2846, pp. 86–95. Springer, Heidelberg (2003)
11. Sella, Y.: On the computation-storage trade-offs of hash chain traversal. In: Wright, R.N. (ed.) FC 2003. LNCS, vol. 2742, pp. 270–285. Springer, Heidelberg (2003)
12. Yum, D.H., Seo, J.W., Eom, S., Lee, P.J.: Single-layer fractal hash chain traversal with almost optimal complexity. In: Fischlin, M. (ed.) Topics in Cryptology – CT-RSA 2009. LNCS, vol. 5473, pp. 325–339. Springer, Heidelberg (2009)
13. Fischlin, M.: Fast verification of hash chains. In: Okamoto, T. (ed.) CT-RSA 2004. LNCS, vol. 2964, pp. 339–352. Springer, Heidelberg (2004)
14. Fischlin, M.: Progressive verification: The case of message authentication. In: Johansson, T., Maitra, S. (eds.) INDOCRYPT 2003. LNCS, vol. 2904, pp. 416–429. Springer, Heidelberg (2003)
15. Bellare, M., Rogaway, P.: Random oracles are practical: A paradigm for designing efficient protocols. In: ACM Conference on Computer and Communications Security, pp. 62–73 (1993)

Appendix

A   On Lower Bound

We briefly review Fischlin's theoretic lower bound for position-driven check-bit schemes, to which all known check-bit schemes belong, and give a simple remark.

Definition 11. *Let $(\mathcal{G}, \mathcal{V}, \mathcal{S})$ be a check-bit scheme (for parameter n). Algorithm $\mathcal{S}$ is position-driven if for any two seeds x_0, x_0', we have $|\mathsf{cb}_i(x_0)| = |\mathsf{cb}_i(x_0')|$.*

Assumption 12. *Let $(\mathcal{G}, \mathcal{V}, \mathcal{S})$ be a check-bit scheme for parameter n. Then, for any i, we assume that for random x_0 and x_0', the probability that $\mathsf{cb}_i(x_0) = \mathsf{cb}_i(x_0')$ is at least $2^{\min(|\mathsf{cb}_i(x_0)|, |\mathsf{cb}_i(x_0')|)}$, where the probability is over the choices of x_0 and x_0'.*

Let n be a power of $1/q$ where $q = p - 1$ and α_ℓ be defined by $\alpha_0 = 0$ and $\alpha_\ell = q^{I-\ell}$ for $I = \log_{1/q} n$ and $\ell \geq 1$. For $\ell = 1, \ldots, I$, define the ℓ-th interval $\hat{\mathcal{I}}_\ell$ to be a segment $[\alpha_{\ell-1}n+1, \alpha_\ell n]$. For a seed x_0 chosen by $\mathcal{G}$, let c_ℓ be the number of check-bit positions in the interval $\hat{\mathcal{I}}_\ell$. Fischlin [13] proved the following lower bound for position-driven check-bit schemes.

Theorem 13 (Fischlin). *Let $(\mathcal{G}, \mathcal{V}, \mathcal{S})$ be a check-bit scheme with parameter n, where $\mathcal{S}$ is a position-driven selection algorithm. Assume that Assumption 12 holds and the computation of a chain of length n requires at most hn hash function evaluations. If $(\mathcal{G}, \mathcal{V}, \mathcal{S})$ is a (T, ϵ, p, Δ)-verifiable check-bit scheme, the length of the check-bit vector is given by*

$$|\mathsf{cb}| = \sum_{\ell=\lceil \log_{1/(1-p)} \Delta \rceil}^{I} c_\ell, \qquad where \;\; c_\ell \geq \log_2 T - \log_2 hn - \log_2 \ln \frac{1}{1-\epsilon}. \quad (10)$$

We observe that Fischlin's proof of Theorem 13 actually allows $\log_2 hn$ to be replaced by $\log_2 h|\hat{\mathcal{I}}_\ell|$ in Eq. (10). (We do not go into details because this is trivial if one reads Fischlin's proof in [13].) Therefore, we can obtain a little tighter bound as

$$|\mathsf{cb}| = \sum_{\ell=\lceil \log_{1/(1-p)} \Delta \rceil}^{I} c_\ell, \qquad where \;\; c_\ell \geq \log_2 T - \log_2 h|\hat{\mathcal{I}}_\ell| - \log_2 \ln \frac{1}{1-\epsilon}. \quad (11)$$

With respect to our example (i.e., $T = 2^{40}$, $\epsilon = 2^{-20}$, $p = 50\%$, $\Delta = 64$ and $h = 1$), a check-bit vector of approximately 210 bits is required from the bound of Eq. (11). Recall that Construction 9 requires $|\mathsf{cb}| = 384$ bits. However, if only the case of $p = 50\%$ is considered, check bits of $\mathcal{I}_{\ell_{min}-2}$ in Construction 9 can be removed in this example[2] and the length of the check-bit vector can be reduced to $|\mathsf{cb}| = 384 - 64 = 320$ bits.

---

[2] In this example, we have $\mathcal{F}(\Delta, \frac{1}{2}) \cap \mathcal{I}_{\ell_{min}-2} = \phi$ where $\mathcal{F}(\cdot, \cdot)$ is defined in the proof of Theorem 10.

Author Index